MICROBIAL INSECTICIDES: PRINCIPLES AND APPLICATIONS

INSECTS AND OTHER TERRESTRIAL ARTHROPODS: BIOLOGY, CHEMISTRY AND BEHAVIOR

Additional books in this series can be found on Nova's website
under the Series tab.

Additional E-books in this series can be found on Nova's website
under the E-books tab.

AGRICULTURE ISSUES AND POLICIES

Additional books in this series can be found on Nova's website
under the Series tab.

Additional E-books in this series can be found on Nova's website
under the E-books tab.

INSECTS AND OTHER TERRESTRIAL ARTHROPODS: BIOLOGY, CHEMISTRY AND BEHAVIOR

MICROBIAL INSECTICIDES: PRINCIPLES AND APPLICATIONS

J. FRANCIS BORGIO,
K. SAHAYARAJ
AND
I. ALPER SUSURLUK
EDITORS

Nova Science Publishers, Inc.
New York

For permission to use material from this book please contact us:
Telephone 631-231-7269; Fax 631-231-8175
Web Site: http://www.novapublishers.com

NOTICE TO THE READER

The Publisher has taken reasonable care in the preparation of this book, but makes no expressed or implied warranty of any kind and assumes no responsibility for any errors or omissions. No liability is assumed for incidental or consequential damages in connection with or arising out of information contained in this book. The Publisher shall not be liable for any special, consequential, or exemplary damages resulting, in whole or in part, from the readers' use of, or reliance upon, this material. Any parts of this book based on government reports are so indicated and copyright is claimed for those parts to the extent applicable to compilations of such works.

Independent verification should be sought for any data, advice or recommendations contained in this book. In addition, no responsibility is assumed by the Publisher for any injury and/or damage to persons or property arising from any methods, products, instructions, ideas or otherwise contained in this publication.

This publication is designed to provide accurate and authoritative information with regard to the subject matter covered herein. It is sold with the clear understanding that the Publisher is not engaged in rendering legal or any other professional services. If legal or any other expert assistance is required, the services of a competent person should be sought. FROM A DECLARATION OF PARTICIPANTS JOINTLY ADOPTED BY A COMMITTEE OF THE AMERICAN BAR ASSOCIATION AND A COMMITTEE OF PUBLISHERS.

Additional color graphics may be available in the e-book version of this book.

LIBRARY OF CONGRESS CATALOGING-IN-PUBLICATION DATA

Microbial insecticides : principles and applications / editors: J. Francis Borgio, K. Sahayaraj, I. Alper Susurluk.
      p. cm.
 Includes index.
 ISBN 978-1-61209-223-2 (hardcover)
 1. Microbial insecticides. 2. Insect pests--Biological control. I.
Borgio, J. Francis. II. Sahayaraj, K. III. Susurluk, I. Alper (Ismail Alper)

 SB933.34M53 2011
 632'.9517--dc22
                         2010047060

Published by Nova Science Publishers, Inc. † New York

CONTENTS

PREFACE

The field of Microbial Insecticides encompasses highly diverse life forms—bacteria, fungi, nematodes, and viruses. They have a profound influence on insect pests on crops: they play an essential role in the management of pests in cultivated crops and play a crucial role in the life of farmers and agricultural industry. The literature associated with Microbial Insecticides, of necessity, tends to be specialized and focused. It is difficult to find sources that provide broad perspectives on a wide range of Microbial pesticides. The contributors of Microbial Insecticides: Principles and Applications have aimed to fulfill it.

The concept behind this venture is to provide a single reference volume with appeal to microbial insecticidist on all levels and fields, including those working in research, teaching, government, industries etc. This is not a textbook of entomopathogenic fungi, nematodes, bacteria and viruses or of interactions, but a bit of each of these!

This handbook is divided into five major parts. Part I characterizes the entomopathogenic fungi. A chapter on the methods used in isolation, identification and preservation of entomopathogenic fungi comes early in this part. Besides the conventional biological control potential described in this handbook, modern fields like toxicity, enzymology, proteomics and phylogenetic analysis have also been included.

Parts II deals with the entomopathogenic nematodes. After a chapter on ecology of entomopathogenic nematodes, there follows three chapters covering various aspects on isolation, identification, characterization and preservation, including mode of action and field efficacy. The final chapter is on the genetics of entomopathogenic nematodes. Part III is on entomopathogenic viruses, which calls on various mass production technologies and their biological control potential. Part IV describes entomopathogenic bacteria, the bioassay procedures and mass production technology and approach to providing safe biological control potential, and the role of restriction modification systems for the production of entomopathogenic preparations. Part V delineates interactions between entomopathogens. There is a chapter devoted entirely to microbial biocontrol agents and their interactions with nematodes and an extended chapter on the control of arthropod pests of tropical tree fruits, reflecting the renewed interest in their basics and applications. Each chapter starts with an 'Abstract", and ends with a brief conclusion with the dual aim of giving a flavor of what is coming up and providing a revision aid. Academic, government, and industry professionals who are actively pursuing research on the microbial insecticides have been chosen as the authors of these chapters.

The authors wish to thank the contributors who have made the production of this book possible. The authors thank all those who supplied reprints; they are especially grateful to Dr.

Hyun-Woo Park (Public Health Entomology Research and Education Center, Florida A and M University, USA), Dr. R.A. Humber (Biological IPM Research Unit, RW Holley Center for Agriculture and Health, NY, USA), Dr. Fernando E. Vega (Sustainable Perennial Crops Laboratory, USDA, Agricultural Research Service, Building 011A, BARC-W, Beltsville, MD 20705, USA). The authors are grateful to Dr. S. Ramesh and Prof M, Amuthan (Department of Microbiology, SPKC, Alwarkurichi, TN, India) for their constant encouragement. Grateful thanks also to those who have reviewed the materials of the book and provided valuable feedback. Special thanks to Mrs. B. Bency Borgio for her timely help throughout the preparation period. The authors are sorry that they were unable to include all the topics suggested, but if they had done so the book would have run to several volumes!

Many thanks to all at Nova Science Publishers who have helped the book to come to fruition. Finally, thanks to our families for their support and for their patience during those many hours we spent ensconced in the study.

The authors hope that this book would be useful to you and they are interested to hear your valuable comments. They have tried to ensure that there are no errors, but it is probable that some have slipped through; if you come across any errors please inform us.

Dr. J. Francis Borgio
PG and Research Department of Microbiology,
St. Joseph's College (Autonomous),
Bangalore - 560 027, INDIA.
E-mail: jfborgio@gmail.com

Dr. K. Sahayaraj
Crop Protection Research Centre,
Department of Advanced Zoology and Biotechnology,
St. Xavier's College (Autonomous),
Palayamkottai - 627 002, Tamil Nadu, INDIA.
E-mail: ksraj42@gmail.com

Dr. I. Alper Susurluk
Uludag University, Agriculture Faculty,
Plant Protection Department, 16059 Nilufer-Bursa,
TURKEY.
E-mail: susurluk@uludag.edu.tr

LIST OF CONTRIBUTORS

Dr. I. Alper Susurluk
Uludag University, Agriculture Faculty, Plant Protection Department, 16059 Nilufer-Bursa, TURKEY. E-mail: susurluk@uludag.edu.tr

Dr. Adão Valmir dos Santos
Departmentof Biotechnology, State University of North Fluminense-UENF, Avenida Alberto Lamego 2000, Campos dos Goytacazes, RJ 28013-602, BRAZIL. E-mail: adao@uenf.br

Dr. Ali Mehrvar
Department of Plant Protection, Faculty of Agriculture, University of Maragheh – 55181-83111, Golshahr, Maragheh, East-Azerbaijan, IRAN. E-mail: mehrvar@mhec.ac.ir

Dr. Armando Pérez-Torres
Depto. de Biología Celular y Tisular, Facultad de Medicina, Universidad Nacional Autónoma de México, México D.F. 04510, MEXICO. E-mail: armandop@servidor.unam.mx

Dr. Atwa, A. Atwa
King Abdul Aziz University (KAU), Jeddah, KING OF SAUDI ARABIA (KSA). E-mail: atwaradwan@hotmail.com

Dr. Claudia Dolinski
Universidade Estadual do Norte Fluminense Darcy Ribeiro/CCTA/LEF, Av. Alberto Lamego, 2000, Pq. Califórnia Campos dos Goytacazes, RJ, BRAZIL, 28015-620. E-mail: Claudia.dolinski@censanet.com.br

Dr. Carlos Peres Silva
Department of Biochemistry, Centro de Ciências Biológicas, Universidade Federal de Santa Catarina, Florianópolis, SC, 88040-900, BRAZIL. E-mail: capsilva@ccb.ufsc.br

Dr. Conchita Toriello
Depto. de Microbiología y Parasitología, Facultad de Medicina, Universidad Nacional Autónoma de México, México D.F. 04510, MÉXICO. E-mail: toriello@servidor.unam.mx

Dr. Cafer Eken
Department of Plant Protection, Faculty of Agriculture, Atatürk University, 2 5240 Erzurum, TURKEY. E-mail: ceken@atauni.edu.tr

Dr. Dong Woon Lee
Department of Applied Biology, Kyungpook National University, Sangju, 742-711, Kyungpook, Republic of KOREA.

Dr. Ebrahim Karimi
Microbial Biotechnology and Biosafety Department, Agricultural Biotechnology Research Institute of Iran (ABRII), Mahdasht Road, 31535-1897, Karaj, IRAN. E-mail: ekarimi20@yahoo.com

Dr. Hyeong Hwan Kim
Horticultural Environmental Division, National Horticultural Research Institute, RDA, Suwon, 441 – 440, Gyeonggi, REPUBLIC OF KOREA. E-mail: hychoo@gnu.ac.kr

Dr. Gholamreza Salehi Jouzani
Microbial Biotechnology and Biosafety Department, Agricultural Biotechnology Research Institute of Iran (ABRII), Mahdasht Road, 31535-1897, Karaj, IRAN, E-mail: gsalehi@abrii.ac.ir

Dr. Jichao Fang
Institute of Plant Protection, Jiangsu Academy of Agricultural Sciences, 50 Zhongling Street, Nanjing, Jiangsu 210014, CHINA. E-mail: fangjc@jaas.ac.cn

Dr. Huifang Wang
Jiangsu Provincial Commission of Agriculture, 1316 Agro-Forestry Tower, 8 Moonlight Square, Caochangmen Street, Nanjing, Jiangsu 210036, CHINA. E-mail: im_whf@yahoo.com.cn

Dr. Hortensia Navarro-Barranco
Depto. de Microbiología y Parasitología, Facultad de Medicina, Universidad Nacional Autónoma de México, México D.F. 04510, MÉXICO. E-mail: horte56@yahoo.com.mx

Dra. Ninfa María Rosas-García
Laboratorio de Biotecnología Ambiental. Centro de Biotecnología Genómica- IPN. Blvd. del Maestro s/n Reynosa, Tamp. MÉXICO CP 88710. Email: nrosas@ipn.mx, ninfarosasg@yahoo.com.mx

Dr. Toledo Andrea Vanesa
Centro de Investigaciones de Fitopatología (CIDEFI), Facultad de Ciencias Agrarias y Forestales, Universidad Nacional de La Plata, Avda. 60 y 119 s/n, 1900, La Plata, Buenos Aires, ARGENTINA. E-mail: atoledo@cepave.edu.ar

Dr. Richard Ian Samuels
Department of Entomology and Plant Pathology, State University of North Fluminense-UENF, Avenida Alberto Lamego 2000, Campos dos Goytacazes, RJ 28013-602, BRAZIL. E. mail: richard@uenf.br

Dr. Victor Hernández-Velázquez
Centro de Investigación en Biotecnología, Universidad Autónoma del estado de Morelos, Cuernavaca 62209, MÉXICO. E-mail: victorhv61@yahoo.es

Dr. Simoni C. Dias
Centro de Análises Proteômicas e Bioquímicas, Pós-Graduação em Ciências Genômicas e Biotecnologia UCB, Brasília-DF, BRAZIL,

Dr. Octávio L. Franco
Universidade Federal de Juiz de Fora, Juiz de Fora-MG, Brazil. Universidade Católica de Brasília, Pós-Graduação em Ciências Gênomicas e Biotecnologia, SGAN 916 – Av W5 – Módulo C, Brasília, DF, BRAZIL. E-mail: ocfranco@pos.ucb.br

Dr. Milton A. Typas
Department of Genetics and Biotechnology, Faculty of Biology, University of Athens, Panepistimiopolis, Kouponia, Athens 157 01, GREECE, E-mail: matypas@biol.uoa.gr

Dr. Lawrence A. Lacey
Yakima Agricultural Research Laboratory, USDA-ARS, 5230 Konnowac Pass Rd., Wapato, WA, 98951, USA

Dr. Neiva Knaak
Microbiology Laboratory, PPG-Biology, Universidade do Vale do Rio dos Sinos; Av. Unisinos, 950. 93001-970, São Leopoldo, RS, BRASIL. E-mail: neivaknaak@gmail.com

Dr. Lidia Mariana Fiuza
Microbiology Laboratory, PPG-Biology, Universidade do Vale do Rio dos Sinos; Av. Unisinos, 950. 93001-970, São Leopoldo, RS, BRASIL. E-mail: fiuza@unisinos.br

Dr. Ho Yul Choo
Department of Applied Biology and Environmental Sciences, College of Agriculture and Life Sciences, Gyeongsang National University, Jinju, Gyeongnam, 660-701, Republic of KOREA.

Dr. Sang Myeong Lee
Korea Forest Research Institute, Southern Forest Research Center, Jinju, 660-300, Gyeongnam, Republic of KOREA.

Dr. Yadollah Dalvand
Microbial Biotechnology and Biosafety Department, Agricultural Biotechnology Research Institute of Iran (ABRII), Mahdasht Road, 31535-1897, Karaj, IRAN. E-mail: dalvandyadola@yahoo.com

Dr. Vladimir E. Repin
Institute of chemical biology and fundamental medicine, Novosibirsk, 630090 RUSSIA. E-mail: ver@niboch.nsc.ru

Dr. J. Francis Borgio
PG and Research Department of Microbiology, St. Joseph's College (Autonomous), Bangalore - 560 027, INDIA. E-mail: jfborgio@gmail.com; borgiomicro@gmail.com

Dr. K. Sahayaraj
Crop Protection Research Centre, St. Xavier's College (Autonomous), Palayamkottai - 627 002, Tamil Nadu, INDIA. E-mail: ksraj42@gmail.com

Dr. Mitzi Flores-Ponce
Laboratorio Nacional de Genómica para la Biodiversidad, CINVESTAV-IPN, 36821, Irapuato, MEXICO.

Dr. Rafael Montiel
Laboratorio Nacional de Genómica para la Biodiversidad, CINVESTAV-IPN, 36821, Irapuato, MEXICO, E-mail: montiel@ira.cinvestav.mx

Dr. You-Jin Hao
CIRN, Departamento de Biologia, Universidade dos Acores, Ponta Delgada, 9501-80, Azores, PORTUGAL. Department of Physics, the University of Chicago, Chicago, Illinois, 60637, USA

Dr. Xiaoyi Wu
Institute for the Control of Agrochemicals, Jiangsu Provincial Commission of Agriculture, 1909 Agro-Forestry Tower, 8 Moonlight Square, Caochangmen Street, Nanjing, Jiangsu 210036, CHINA. E-mail: xiaoyiwu.cn1@gmail.com

PART I:
ENTOMOPATHOGENIC FUNGI

In: Microbial Insecticides: Principles and Applications   ISBN: 978-1-61209-223-2
Editors: J. Francis Borgio, K. Sahayaraj, et al.   © 2011 Nova Science Publishers, Inc

Chapter 1

ISOLATION, IDENTIFICATION AND PRESERVATION OF ENTOMOPATHOGENIC FUNGI

Cafer Eken[*]

Department of Plant Protection, Faculty of Agriculture, Atatürk University,
25240 Erzurum, Turkey

ABSTRACT

This chapter describes the various techniques and procedures about the isolation, identification and preservation of entomopathogenic fungi. Isolation is divided into two categories: from soil and isolation from cadavers. Keys to the most important entomopathogenic fungal genera are included here. Short descriptions of the important entomopathogenic species have been also given. There are numerous methodologies available to preservation of entomopathogenic fungi. No single method can be applied to all fungi. The main requirement of a preservation technique is to maintain the fungus in a viable and stable state without morphological, physiological, or genetic change during storage.

1.1. INTRODUCTION

Until now, at least, 90 genera and more than 700 species of entomopathogenic fungi have been identified as closely associated with invertebrates, many of which are important agricultural pests, including whiteflies (Chandler *et al.,* 1993; Gindin *et al.,* 2000; Faria and Wraight, 2001), aphids (Hsiao *et al.,* 1992; Hatting *et al.,* 1999; Shah *et al.,* 2004; Barta and Cagáň, 2006a), acari (Chandler *et al.,* 2000; Bugeme *et al.,* 2009; Eken and Hayat, 2009) and thrips (Gindin *et al.,* 1996; Al-mazra'awi *et al.,* 2009). Fungal entomopathogens almost always attack their hosts by direct penetrating the exoskeleton and do not require ingestion with subsequent infection through the gut wall. Interest in using the insect pathogens as

[*] E-mail: ceken@atauni.edu.tr; Phone: +90 442 231 1467; Fax: +90 442 231 1469

control agents within IPM programs has generated a research in the development of microbial insecticides (Faria and Wraight, 2007).

1.2. ISOLATION

1.2.1. Isolation of Entomopathogenic Fungi from Soil

For most soils, fungi (either saprotrophs, mycorrhizal or pathogens) are the main component of its microbiota (Gams, 1992). Soil factors (temperature, pH or organic content, relative moisture or mineral, organic or biotic components) can affect fungal persistence and activity (Charnley, 1997).

The soil habitat is considered as excellent habitat for insect pathogenic fungi and other microorganisms since it is protected from UV radiation and buffered against extreme biotic and abiotic influences (Keller and Zimmerman, 1989). Most of these fungi, along with a range of bacteria, can grow on artificial media *in vitro*. These abilities have long been exploited to isolate microorganisms from soil samples and specific media have been developed to select for certain groups of microorganisms.

Most bacteria are inhibited by low pH and, in general, fungal growth will be favoured on media where the pH is less than 5. In most instances, however, inhibition of bacteria is achieved by amending media with antibacterial agents (e.g. chloramphenicol, tetracycline or streptomycin) (Goettel and Inglis, 1997). Some media for the selective isolation of entomopathogenic fungi have also been developed. Generally, the genera *Metarhizium*, *Beauveria* and *Paecilomyces* have been investigated the most.

1.2.1.1. Media for Isolation of Beauveria Spp.

For isolation of *Beauveria brongniartii* a selective medium was developed by Strasser *et al.* (1996). This medium has also been successfully used to isolate *B. bassiana* from phylloplanes of different plant species (Meyling and Eilenberg, 2006a).

The medium for isolation of *Beauveria* spp is presented in Box I.

Box I
Media for isolation of *Beauveria* spp. (Meyling, 2007)
Dissolve 5 g Peptone, 10 g Glucose and 6 g Agar in 500 ml distilled water
Adjust the pH to 6.3 with 1 M HCl
Autoclave for 20 min. at 120 °C
Cool the medium to 50-60 °C, add:
0.5 ml a' 0.6 g/ml Streptomycin
0.5 ml a' 0.05 g/ml Tetracycline
0.5 ml a' 0.1 g/ml Dodine
2.5 ml a' 0.05 g/5 ml Cyclohexamide
Invert gently the bottle without making air bobbles and pour the plates

1.2.1.2. Media for Isolation of Metarhizium Spp

Meyling (2007) provide a list of suitable selective media for *Metarhizium*. The medium contains the antibiotics chloramphenicol and streptomycin as well as the fungicide dodine. The procedure to make this medium is described in Box II.

Box II
Media for isolation of Metarhizium spp. (Meyling, 2007)
Suspend 32.5 gram Sabouraud dextrose agar in 500 ml distilled water in a blue cap bottle. Add 1 ml Dodine (a fungicide to inhibit fungal growth) [5 g in 45 ml distilled water] Mix the medium and mark the blue cap bottle with autoclave tape. Autoclave the medium for 20 min at 120 °C 20 bar . Cool the medium after autoclaving to approx. 60°C and add: 500 µl Chloramphenicol (antibiotic, inhibits bacteria) [1g in 10 ml 96% ethanol] 500 µl Streptomycin sulphate (antibiotic, inhibits bacteria) [0,5 g in 10 ml sterile dis. H_2O] Invert the bottle gently and pour the plates.

1.2.2. Soil Dilution Plating

The isolation of entomopathogenic fungi from soil can employ soil dilution plate method. A commonly used procedure is to place 10 g of soil into 90 ml of sterile water in a screw-capped bottle. The contents are mixed thoroughly for 20 min, and then a 1 ml aliquot is withdrawn and added to 9 ml of sterile water and shaken for 4 min. A 1 ml aliquot is removed and added to 49 ml of sterile water and shaken for 2 min. using a wide-mouth pipette, transfer 1 ml of the final dilution to each of five sterile petri dishes and add 15 ml of the desired culture medium which has been cooled to 45 °C. This description is offered only as a general illustration of the technique, and the actual amount of dilution of the soil sample will depend upon the number and type of fungal propagules present.

1.2.3. Soil Plating

A commonly used variant of the soil soil dilution plate technique is the Warcup plate method. For this technique a small soil sample (0.005 – 0.15 g) is transferred to a sterile petri dish and crushed in a drop of water. To the petri dish, add approximately 15 ml of the desired growth medium. Gently swirl the plate to distribute the soil particles in the medium and incubate as appropriate for the particular fungus. Unless a selective medium is used, this usually provides unsatisfactory isolation of entomopathogenic fungi due to over-growth of contaminant fungi (Goettel and Inglis, 1997).

1.2.4. Insect Bait Method

Insect baits may be used to indirectly isolate fungi from soil. The use of selective media exploits the saprotrophic abilities of hypocrealean entomopathogenic fungi. For the method to be feasible insects, which are easily reared and are susceptible to the fungi, must be used.

Box III
Insect bait method (Zimmermann, 1986; Kessler *et al.*, 2003)
Remove the roots and gravels from soil samples. The soil sample passed through 5-mm pore sieve. If the soil samples were too dry, they moistened with sterile distilled water. Plastic boxes (6 cm high, diameter 4.5 cm) filled with 60 g of soil. Five *G. mellonella* late instar larvae are added to the soil. During the first five days, the boxes turned once daily to keep the of the bait larvae moving in the soil. The larvae incubated for approximately 14 days at 22 °C in dark conditions. Dead larvae surface sterilized with 1 % sodium hypochlorite for 3 min and then rinsed twice with sterile distilled water. After removing free water of the larvae surface, placed onto selective medium agar plates.

Although larvae of the wax moth (*Galleria mellonella*, Lepidoptera: Pyralidae) are most commonly used, larvae of other insects such as the mealworm (*Tenebrio molitor*, Coleoptera: Tenebrionidae), the large flour beetle (*Tribolium destructor*, Coleoptera: Tenebrionidae), the turnip root fly (*Delia floralis*, Diptera: Anthomyiidae) and the pine bark beetle (*Acanthocinus aedilis*, Coleoptera: Cerambycidae) may also be used (Zimmermann, 1986; Klingen *et al.*, 2002; Meyling, 2007; Pilz *et al.*, 2008). The insect bait method for the selective isolation of entomopathogenic fungi, numerous studies have been carried out using insect baits, especially *G. mellonella* (Vänninen *et al.*, 1989; Vänninen, 1996; Chandler *et al.*, 1997; Bidochka *et al.*, 1998; Klingen *et al.*, 2002; Keller *et al.*, 2003; Kessler *et al.*, 2003; Meyling and Eilenberg, 2006b; Pilz *et al.*, 2008; Sun and Liu, 2008). The method is presented in Box III.

1.2.5. Isolation of Entomopathogenic Fungi from Cadavers

1.2.5.1. Collection of Dead Insects
Dead insects in the field may indicate the presence of disease.

- Look for insects hanging from plants, under trees and bushes.
- Collect dead insects in sterile glass or plastic containers with screw tops, paper bags or envelopes.
- Leave the containers open for three to four (3-4) days so that the cadavers dry out.
- Don't dry the cadavers artificially.
- Don't leave cadavers in the sun.
- These air dried specimens can be stored at room temperature for a few weeks.
- For longer periods, store specimens in a refrigerator (4°C).
- Never store infected specimens in alcohol.

1.2.5.2. Collection of Live Insects
Collect live insects in the field.
- Keep the insects in cages and feed them.

- Observe the insects.
- You may see insects behaving abnormally; these are signs that disease may be present: not feeding, poor coordination, jerky movements and excessive grooming loss of orientation.
- Insects may also show taxic responses, climbing up high on plants, expose themselves to the sun, or hiding.
- Collect any dead insects (cadavers) daily. If the body is soft and black you may have a bacterial or viral infection. If the body is harder, this may indicate a fungal infection. Incubate the cadaver in a humid chamber (such as a petri dish contains moistered tissue paper) to see if you get external sporulation

1.2.5.3. Isolation of Fungi

Entomopathogenic fungi sporulate on the outside of the host insect under moist conditions and on the inside of the host when the environment is dry. The following steps are suggested for isolating fungi from diseased insects.

Insects with fresh external sporulation;

- Take spores with a fine, sterile needle.
- Streak spores onto several different agar media in Petri dishes with antibiotics: tap water agar, potato carrot agar, malt extract agar.
- Incubate at 20-28°C.
- Examine all cultures daily with a stereoscopic microscope.
- Newly dead insects with no external growth;
- Incubate for several days at high humidity.
- Observe for sporulation.
- Mount spore structures on a slide in water, or use a specific fungal stain e.g. cotton blue in lactophenol

If the insect was not properly dried or has been dead for too long, other contaminating pathogens may hide the growth of the pathogen which killed the insect.

Insects which have been dead for a long time;

- Surface sterilize the insect in sodium hypochlorite for several minutes.
- Rinse in three changes of sterile, distilled water.
- Dissect internal tissues (usually replaced by fungal hyphae).
- Streak spores onto several different agar media with antibiotics: tap water agar, potato carrot agar, malt extract agar.
- Incubate at 20-28°C.
- Examine all cultures daily using a stereoscopic microscope.

Isolation of entomopathogenic Entomophthorales can be undertaken using two basically similar methods (i) The 'descending conidia'showering method (Papierok and Hajek, 1997); Cadavers that are filled with hyphal bodies but from which conidiophores have not yet emerged should be surface sterilized by dipping them successively in 2% sodium hyplochlorite solution (2-3 min).

Insects bearing conidiophores are placed on a moistened piece of tissue, paper towel or filter paper that is attached to the inside of a small sterile Petri dish lid.

After having prepared the infected specimen, the Petri dish lid is inverted over the base of a sterile Petri dish, which can contain culture medium. Conidial production begins within a few hours to a day, depending on whether conidiophores were already formed. Conidia are collected for periods of varying lengths (from a few minutes to a few hours), depending on the intensity of sporulation.

In the case of a Petri dish containing culture medium, the lid with the specimen attached is replaced by a sterile lid. If a Petri dish with an empty base in used, conidia which land on the bottom are transferred to media by rubbing a piece of solid medium across the spores and subsequently using this spore-bearing medium to inoculate a culture tube. (ii) The 'ascending conidia' showering method; Either insects bearing conidiophores already, or insects with no emerging conidiophores that have been surface sterilized with 2% sodium hyplochlorite solution (2-3 min). Insects are placed in the base of small Petri dishes on water or on a moistened piece of tissue or filter paper. A sterile slide or cover glass is placed above the dish in order to collect discharged conidia (Papierok and Hajek, 1997).

Media for isolation; As the most common solid media for growing routinely arthropod-pathogenic Entomophthorales, the most common media for isolating these fungi contain egg yolk and milk.

It contains respectively:

- 60% egg yolk,
- 40% whole milk (which makes, for instance, 3 egg yolks and 30 mL milk, approximatively).

The procedure is as follows (Papierok and Hajek, 1997):

- Sterilize whole milk at 120°C for 30 min, and keep at room temperature,
- Surface sterilize fresh eggs in a mixture of ethanol 90° (200 mL) and 2% sodium hypochlorite (800 mL) for 10 min,
- Break egg shells with forceps. Remove the egg yolk, and pour it in a sterile graduated cylinder,
- Add milk to egg yolk in the graduated cylinder,
- Stir the content with a sterile glass rod, to get an homogeneous mixture,
- Distribute the medium in tubes plugged with sterile cotton,
- Once a tube is filled and plugged, tilt immediately to get the expected regular culture tube,
- Sterilize in the oven at 100°C for 30 min,
- Keep the tubes at room temperature for 24 h,
- Remove cotton plugs and plug securely tubes with sterile rubber ones,
- These tubes can be stored for several months or even a few years, as long as they are kept in the dark.

1.3. OUTLINE CLASSIFICATION OF MAJOR GROUPS OF ENTOMOPATHOGENIC FUNGI

Traditionally, fungal taxonomy has been based on morphological, developmental, and physiological characteristics. The most significant fungus characteristics used for identification are spores and spore-bearing structures (sporophores), and to some extent the characteristics of the fungus body (mycelium). The shape, size, color, and manner of arrangement of spores on the sporophores or in the fruiting bodies, as well as the shape and color of the sporophores or fruiting bodies, are sufficient characteristics to suggest, to one somewhat experienced in the taxonomy of fungi, the class, order, family, and genus to which the particular fungus belongs (Agrios, 2005). The classification of the main groups of fungi have been revised in recent years as more molecular data has become available; the scheme given here is the arrangement adopted in the 10th edition of the Dictionary of the Fungi (Kirk et al., 2008) in which the fungi, in the traditional sense, are distributed through three Kingdoms, *Protozoa*, *Chromista* and *Fungi*. Some fungal-like organisms, often referred to as lower fungi, are now considered to belong to the kingdom Protozoa (e.g., Myxomycetes and Plasmodiophoromycetes) or to the kingdom Chromista (e.g., Oomycetes). True fungi, however (i.e. Chytridiomycota, Zygomycota, Ascomycota, Basidiomycota and Glomeromycota) belong to the kingdom Fungi. The fungi and fungal-like organisms that cause diseases on arthropods are a diverse group (Table I). The following systematic is based on morphological and cytological characteristics. The main classes that contain entomopathogenic species are reviewed in the following sections (Agrios, 2005; James *et al.,* 2006a).

Table 1. Major orders and genera of entomopathogenic fungi (Samson *et al.*, 1988; Tanada and Kaya, 1993; Boucias and Pendland, 1998; Kirk *et al.*, 2008)*

Kingdom	Phylum	Class	Order	Genera
Chromista	Oomycota	Oomycetes	Pythiales	*Lagenidium*
			Leptomitales	*Aphanomycopsis*
			Saprolegniales	*Couchia*
				Leptolegnia
Fungi	Ascomycota	Laboulbeniomycetes	Laboulbeniales	*Filariomyces*
				Hesperomyces
				Trenomyces
		Saccharomycetes	Saccharomycetales	*Blastodendrion*
				Candida
				Metschnikowia
				Mycoderma
				Saccharomyces
		Dothideomycetes	Myriangiales	*Myriangium*
			Pleosporales	*Podonectria*
				Tetracrium
		Eurotiomycetes	Ascosphaerales	*Ascosphaera*
			Eurotiales	*Aspergillus*
				Paecilomyces
		Sordariomycetes	Hypocreales	*Calonectria*

Table 1. (Continued)

Kingdom	Phylum	Class	Order	Genera
				Cordyceps
				Cordycepioideus
				Hypocrella
				Nectria
				Torrubiella
				Aschersonia
				Acremonium
				Akanthomyces
				Beauveria
				Engyodontium
				Fusarium
				Gibellula
				Hirsutella
				Hymenostilbe
				Lecanicillium
				Metarhizium
				Nomuraea
				Sorosporella
				Stilbella
				Syngliocladium
				Tolypocladium
			Ophiostomatales	*Sporothrix*
	Basidiomycota	Agaricomycetes	Polyporales	*Aegerita*
		Pucciniomycetes	Septobasidiales	*Septobasidium*
				Uredinella
	Blastocladiomycota	Blastocladiomycetes	Blastocladiales	*Catenaria*
				Coelomomyces
				Coelomycidium
	Chytridiomycota	Chytridiomycetes	Chytridiales	*Myiophagus*
	Zygomycota	Incertae sedis	Basidiobolales	*Basidiobolus*
			Entomophthorales	*Batkoa*
				Conidiobolus
				Entomophaga
				Entomophthora
				Erynia
				Furia
				Massospora
				Meristacrum
				Neozygites
				Strongwellsea
				Zoophthora
			Mucorales	*Sporodiniella*

* Does not include all entomopathogenic genera

Phylum: Oomycota - Have biflagellate zoospores, with longer tinsel flagellum directed forward and a shorter whiplash flagellum directed backward. Diploid thallus, with meiosis occurring in the developing gametangia. Sexual reproduction can occur between gametangia (antheridia and oogonia) on the same or different hyphae. Gametangial contact produces thick-walled sexual oospore. Cell walls composed of glucans and small amounts of hydroxyproline and cellulose.

Phylum: Ascomycota - Have septate and haploid mycelia and the sexual spores, ascospores (meiospores), are produced in an ascus on a fruiting body, the ascomata. Produce sexual spores, called ascospores, generally in groups of eight within an ascus. Produce asexual spores (conidia) on free hyphae or in asexual fruiting structures (pycnidia, acervuli, etc.).

Phylum: Basidiomycota - Sexual spores, called basidiospores (meiospores), are produced externally on a club-like, one- or four-celled sporereproducing structure called a basidium. There are very few Basidiomycetes that have been implicated in insect pathogenesis.

Phylum: Blastocladiomycota - Zoospore with a single flagellum, side-body complex, nuclear cap of membrane-bounded ribosomes, cone-shaped nucleus that terminates near the kinetosome, microtubules radiate anteriorly from the proximal end of the kinetosome around the nucleus, zoospore flagellum lacks electron-opaque plug in transition zone. Asexual reproduction with zoospores, sexual reproduction through fusion of planogametes, life cycle with sporic meiosis.

Phylum: Chytridiomycota - Chytridiomycota has been defined traditionally on the basis of the presence of a single posteriorly inserted smooth flagellum. Produce zoospores that have a single posterior flagellum. Chrytridiomycetes are characterized by cell walls containing glucans and chitin and no cellulose.

Phylum: Zygomycota - Produce nonmotile asexual spores in sporangia. No zoospores. The Zygomycota are separated on the basis of often nonseptate, multinuclear hyphae, and production of zygospores by copulation between gametangia.

In recent times, classification systems have attempted to take a more phylogenetic approach to systematics (James *et al.,* 2006b; Hibbett *et al.,* 2007), but a true phylogenetic classification had not emerged for most organisms until the advent of molecular biology. Molecular techniques that target RNA and DNA have been applied at every level of the taxonomic hierarchy. Diversity within the entomopathogenic fungi has been analyzed using random amplified polymorphic DNA (RAPD) markers (Bidochka *et al.,* 1994; Urtz and Rice, 1997; Berretta *et al.,* 1998; Castrillo *et al.,* 1999; Kao *et al.,* 2002; Dalzoto *et al.,* 2003; Carneiro *et al.,* 2008), internal transcribed spacers of the ribosomal DNA (rDNA-ITS) sequencing (Gaitan *et al.,* 2002; Aquino de Muro *et al.,* 2003, 2005; Wada *et al.,* 2003; Carneiro *et al.,* 2008) and PCR-RFLP (Coates *et al.,* 2002), mitochondrial DNA (Hegedus and Khachatourians, 1993; Sosa-Gómez *et al.,* 2009), pulsed field gel electrophoresis (Pfeifer and Khachatourians, 1993). There are a number of reviews on the use of molecular characterization for entomogenous fungi (Clarkson, 1992; Driver and Milner, 1998).

1.3.1. A Key to Major Genera of Entomopathogenic Fungi

This key is primarily a synthesis of the most influential taxonomic monographs on the entomopathogenic fungi published recently (Humber, 1998).

1a. Spores and hyphae or other fungal structures visible on exterior of host or host body is obscured by fungus; few or no spores form inside host cadaver   2

1b. Fungal growth and sporulation wholly (or nearly wholly) confined to *interior* of host body   30

2a. Elongated macroscopic structures (synnemata or club-like to columnar stromata) project from host   3

2b. Fungal growth may cover all or part of the host and may spread onto the substrate but large, projecting structures are absent   10

3a. Conidia form on synnemata and/or on mycelium on the host body   4

3b. Flask-like to laterally flattened fruiting structures (perithecia) present whether on or submersed in an erect, dense to fleshy, club-like to columnar stroma or on body of host; if mature, containing elongated asci with thickened apical caps   9

4a. Conidia formed in short to long chains   5

4b. Conidia produced singly on many separate denticles on each conidiogenous cell or, if in some sort of slime, singly (slime sometimes not evident) or in small groups in a slime droplet   7

5a. Conidiogenous cells flask-like, with swollen base and a distinct neck, borne singly or in loose clusters; chains of conidia often long and divergent (when borne on clusters of conidiogenous cells)   *Paecilomyces*

5b. Conidiogenous cells short, with rounded to broadly conical apices (*not* having a distinctly narrowed and extended neck)   6

6a. Conidiogenous cells clustered on more or less swollen vesicle on short to long, conidiophores projecting laterally from synnemata and/or the hyphal mat covering the host; conidia pale to yellow or violet in mass; affecting spiders   *Gibellula*

6b. Conidiogenous cells borne at apices of broadly branched, densely intertwined conidiophores forming a compact hymenium; conidia borne in parallel chains and usually green in mass   *Metarhizium*

7a. Conidiogenous cell with swollen base and elongated, narrow to spine-like neck; conidia formed singly (usually with a distinct slime coating) or small groups in a slime droplet   *Hirsutella*

7b. Conidiogenous cells producing several to many conidia, each formed singly on separate denticles   8

8a. Conidiogenous cell with an extended, denticulate apex (growing apex repeatedly forms a conidium and regrows [rebranches] just below the new conidium)
                    Beauveria

8b. Conidiogenous cell short and compact, cylindrical to broadly clavate, with apex studded by many denticles, each of which bears a single conidium   *Hymenostilbe*

9a. Erect stroma bears perithecia superficial to partially or fully immersed (with only small, circular opening raised above stromatic surface); perithecia scattered or grouped in a more or less differentiated, apical or lateral fertile part; asci (if present) with thickened apical cap perforated by narrow canal and filiform ascospores (that often dissociate into one-celled part-spores); conidia, if also present, are formed on host body, lower portion of stroma, or on synnemata   *Cordyceps*

9b. Perithecia occur *only* on or emerging from a cottony to woolly layer covering host
                    Torrubiella

10a. Fungus covering host is a stroma (fleshy to hard mass of intertwined hyphae); sporulation occurs in cavities below the stromatic surface   11

10b. Host partially to completely covered by wispy, cottony, woolly, or felt-like growth or by a dark-colored, extensive patch having columns and chambers below its surface but *not* forming a stroma   12

11a. Spores are fusoid, one-celled conidia discharged in a slime mass from fertile chambers immersed in the stroma but not set off by a differentiated wall

Aschersonia

11b. Globose to flask-like perithecia delimited by a distinct wall are immersed in stroma and contain elongated asci with thickened apices or, at maturity, a (non-slimy) mass of globose, ovoid or rod-like spores formed by dissociation of multiseptate ascospores; *Aschersonia* conidial state often present on same stroma   *Hypocrella*

12a. Fungus a dark brown to black, sometimes extensive patch on woody plant parts; upper surface dense to felt-like, with elongated or clavate thick-walled cells (teleutospores) remaining attached; open chambers and vertical fungal columns underlie the more or less solid upper surface and shelter living scale insects, some of which contain prominently coiled haustorial hyphae   *Septobasidium*

12b. Fungal hyphae emerging from or covering host are colorless to light colored, wispy to cottony, woolly, felt-like or waxy-looking mat   13

13a. Flask-like to laterally compressed perithecia present, superficial to partially immersed in fungus covering the host; asci elongate, with thickened apex; when mature, filiform multiseptate ascospores tend to dissociate into 1-celled part-spores; conidial state(s) may occur simultaneously on host body or synnemata; especially on spiders or homopterans

Torrubiella

13b. Spores form on external surfaces of the fungus; no sexual structures (perithecia) are present   14

14a. Conidia form on cells with elongated denticulate necks bearing multiple conidia or on awl- to flask-shaped or short blocky conidiogenous cells; conidia form singly or successively in dry chains or slime drops (Hyphomycetes)   15

14b. Conidia forcibly discharged and may rapidly form forcibly or passively dispersed secondary conidia (Entomophthorales)   22

15a. Conidiogenous cell with an extended, denticulate apex (growing apex repeatedly forms an conidium and regrows [rebranches] just below the new conidium)   *Beauveria*

15b. Conidiogenous cells are awl- to flask-shaped, with or without an obvious neck; conidia borne singly, in chains, or in slime drops   16

16a. Conidia single or in chains on apices of conidiogenous cells   17

16b. Conidia aggregate in slime drops at apices of conidiogenous cells   20

17a. Conidia borne singly on conidiogenous cell with swollen base and one or more narrow, elongated necks; conidia globose or, if not, usually having an obvious slime coat; especially on mites   *Hirsutella*

17b. Conidia borne in chains, not covered by any obvious slime   18

18a. Conidiophores much branched in a candelabrum-like manner but very densely intertwined, and forming nearly wax-like fertile areas; conidiogenous cells short, blocky, without apical necks; conidial chains long and, usually, laterally adherent in columns or continuous plates   *Metarhizium*

18b. Conidiophores individually distinct and unbranched or with a main axis and short side branches bearing single or clustered conidiogenous cells   19

19a. Conidiogenous cells flask-like, with swollen base and a distinct neck, borne singly or in loose clusters; chains of conidia often long and divergent (when borne on clusters of conidiogenous cells)   *Paecilomyces*

19b. Conidiogenous cells short and blocky with little obvious neck, borne in small clusters on short branches grouped in dense whorls on (otherwise unbranched) conidiophores; conidial chains short; especially on Noctuidae (Lepidoptera)   *Nomuraea*

20a. Conidia aggregating in slime droplets with morphology either (1) macroconidia, elongated, gently to strongly curved with somewhat pointed ends, one or more transverse septa and usually a short (basal) bulge or bend ('foot') and/or (2) microconidia aseptate, with variable morphology; conidiogenous cells often distinctly thicker than vegetative hyphae; hyphae often with terminal or intercalary chlamydospores (thick-walled spore-like swellings of vegetative cells; surface smooth or decorated)   *Fusarium*

20b. Conidiogenous cells little thicker than hyphae, occurring singly or grouped into regular clusters and/or whorls; conidia one-celled; mycelium highly uniform in diameter
21

21a. Conidiogenous cells usually tapering uniformly from base to truncate apex, usually without a swollen base or distinct neck; occurring singly, in pairs or whorled along hyphae or in terminal clusters   *Lecanicillium*

21b. Conidiogenous cells with a swollen to flask-like base and a (usually short) neck often bent out of axis of the conidiogenous cell; conidiogenous cells borne singly, clustered, or in whorls aggregating in loose 'heads' on erect apically branching conidiophores poorly differentiated from vegetative hyphae   *Tolypocladium*

22a. In aceto-orcein, primary conidia obviously uninucleate and sometimes seen to be bitunicate (with outer wall layer lifting partially off of spores in liquid mounts)   23

22b. In aceto-orcein, primary conidia obviously multinucleate or nuclei not readily seen
26

23a. Conidia long clavate to obviously elongated (length/width ratio usually $\geq$ 2.5), papilla broadly conical, often with a slight flaring or ridge at junction with basal papilla
24

23b. Conidia ovoid to clavate; papilla rounded and frequently laterally displaced from axis of conidium   25

24a. Conidia readily forming elongate secondary capilliconidia attached laterally to and passively dispersed from capillary conidiophores; rhizoids and cystidia not thicker than hyphae; rhizoids numerous, often fasciculate or in columns   *Zoophthora*

24b. Conidia never forming secondary capilliconidia; conidia often strongly curved and/or markedly elongated; rhizoids and/or cystidia 2-3x thicker than hyphae; especially on dipterans (or other insects) in wet habitats (on wetted rocks, in or near streams, etc.)
Erynia

25a. Conidia never producing secondary capilliconidia; rhizoids 2-3x thicker than hyphae, ending with a discoid holdfast; cystidia at base 2-3x thicker than hyphae, tapering towards apex   *Pandora*

25b. Conidia never producing secondary capilliconidia; rhizoids not thicker than hyphae, numerous, solitary to fasciculate, with weak terminal branching system or sucker-like holdfasts; cystidia as thick as hyphae, often only weakly tapered   *Furia*

26a. In aceto-orcein, nuclei staining readily, with obviously granular contents   27

26b. In aceto-orcein, nuclei not readily visible or not staining   29

27a. Conidia with apical point and broad flat papilla; discharged by cannon-like expulsion of fluid from conidiogenous cell forming halo-like zone around conidia after discharge   *Entomophthora*

27b. Conidia without apical projection and discharged by eversion of a rounded (not flat) papilla   28

28a. Conidia pyriform with papilla merging smoothly into spore outline; formed by direct expansion of tip of conidiogenous cell (with no narrower connection between conidiogenous cell and conidium); rhizoids never formed   *Entomophaga*

28b. Conidia globose with papilla emerging abruptly from spore outline; formed on conidiogenous cells with a narrowed neck below the conidium; if present, rhizoids 2-3x thicker than hyphae, with discoid terminal holdfast   *Batkoa*

29a. Conidia globose to pyriform, papilla rounded, with many (inconspicuous) nuclei; secondary conidia (a) single, forcibly discharged and resembling primaries, (b) single, passively dispersed capilliconidia formed in axis of capillary conidiophore, or (c) numerous on a primary conidium, small, forcibly discharged (microconidia)   *Conidiobolus*

29b. Conidia globose to pyriform, papilla flattened, usually 4-nucleate; secondary conidia (a) forcibly discharged, resembling primary or (b) almond- to drop-shaped, laterally attached to a capillary conidiophore with a sharp subapical bend; especially on aphids or mites

Neozygites

30a. Affecting larval bees (Apidae and Megachilidae), causing chalkbrood; fungus in cadavers is white or black, organized as large spheres (spore cysts) containing smaller walled spherical groups (asci) of (asco)spores   *Ascosphaera*

30b. Affecting insects other than bees; spores formed individually rather than in spherical groups of inside larger spheres   31

31a. Spores formed *inside* a fungal cell, in a more or less loosely fitted outer (sporangial) wall   32

31b. Spores forming directly at apices of hyphae or hyphal bodies by budding or intercalary (thick walled but not confined loosely inside remnant of another cell)   33

32a. Spores (oospores) thick-walled, smooth walled, colorless; formed inside irregularly shaped cell (oogonia); some cells in thick mycelium producing narrow tube through cuticle with evanescent terminal vesicle from which motile, biflagellate zoospores are released; affecting mosquitoes   *Lagenidium*

32b. Spores (resistant sporangia) globose or subglobose, golden-brown with hexagonally reticulated surface; formed inside close fitting thin (but evanescent) outer wall

Myiophagus

33a. Affecting gregarious cicadas (Homoptera: Cicadidae); terminal segments of abdominal exoskeleton drop off to expose loose to compact, colorless to colored fungal mass; spores thin-walled or, if thick-walled, with strongly sculptured surface   *Massospora*

33b. Not on cicadas; spores throughout body, not confined to terminal abdominal segments)   34

34a. Spores (zygospores or azygospores) with outer surfaces smooth or with surface irregularly roughened, warted, or spinose; colorless to pale or deeply colored (various colors possible), brown, gray, or black   35

34b. Spores (thick-walled resistant sporangia) with surface regularly decorated with ridges, pits, punctations, striations, reticulations; yellow-brown to golden-brown   37

35a. Resting spores gray, brown or black (outer wall is colored; inner wall is hyaline), with smooth or rough surface; binucleate but nuclei often not staining strongly in aceto-orcein if spore wall is cracked; infected hosts from which conidia were discharged and then produced almond- to drop-shaped secondary capilliconidia should be evident in the infected population; affecting aphids, scales, or mites   *Neozygites*

35b. Resting spores colorless, colored, or dark, surfaces smooth or rough; infected host population may or may not include cadavers producing conidia but, if present, conidia not as above   36

36a. When spores are gently crushed in aceto-orcein (to crack walls and partially extrude cytoplasm), nuclei are poorly stained (or unstained) and, if seen, do not have obviously granular contents (Ancylistaceae)   *Conidiobolus*

36b. When spores are gently crushed in aceto-orcein (to crack walls and partially extrude cytoplasm), nuclei stain well and have obviously granular contents
                                                              Entomophthoraceae

37a. Sporangia ellipsoid (not globose), with a preformed dehiscence slit (may not be obvious); wall very thick, golden-brown, pitted to elaborately sculptured; affecting larvae/pupae of mosquitoes (or midges)   *Coelomomyces*

37b. Sporangia globose or subglobose, with no visible dehiscence slit; wall relatively thin; surface with low (hexagonally) reticulated ridges; affecting terrestrial insects
                                                                    Myiophagus

1.3.2. Diagnoses of Major Entomopathogenic Fungi

1.3.2.1. Beauveria bassiana (Balsamo) Vuillemin

After 14 days on potato dextrose agar (PDA) or malt agar (MA) colonies look velvet to powdery, rarely forming synnemata. At the beginning these colonies have white mycelial margins which become pale yellow or sometimes reddish. The underneath surface of the colonies is colourless, or yellow to reddish. *B. bassiana* reproduces by production of dry spores conidiophores that grow sympodially. Conidia are round to ovoid (globose), smooth walled, 2-3 x 2-2.5 µm; carried singly on conidiogenous cells with flask-like base extending into denticulate rachis; passively discharged (de Hoog, 1972).

B. bassiana infects a wide range of hosts and survives in the soil as a saprophyte. It differs from *B. brongniartii* (Saccardo) Petch by the more clustered conidiogenous cells and the globose conidia. Entomopathogenic fungal species *B. bassiana* and *B. brongniartii* were described for the first time about 170 and 110 years ago, respectively (Zimmermann, 2007a). Agostino Bassi was the first to demonstrate experimentally the infectious nature of insect disease in his 1835 study of the white muscardine disease of silkworms (*Bombyx mori* L.), caused by *B. bassiana*.

1.3.2.2. Conidiobolus obscurus (Hall and Dunn) Remaudie`Re and Keller

Primary conidia (30.0)-35.79-(42.0) x (27.0)-30.99-(36.0) µm (Hatting *et al.* 1999); basal hemispherical papilla emerging abruptly from the spore; forcibly discharged. Secondary conidia: morphologically similar to primary conidia; produced on a short germ tube arising from primary conidia, (28.5)-30.2-(32.1) x (19.1)-20.5-(23.2) μm (Scorsetti *et al.,* 2007). Conidiophores: simple, unbranched. Resting spores spherical to subspherical, (30.6)-33.5-(39.3).

The fungus has been widely studied as a potential aphid control agent (Latgé, 1980; Latgé *et al.,* 1983), although a part of older data also concerns the species. The culture of this pathogen can be established on Sabouraud's dextrose agar or media based on egg yolk and milk (Keller, 1987).

1.3.2.3. Entomophaga aulicae (Reichardt IN Bail) Batko

Primary conidia (22.8)-28.0-(35.0) x (17.5)-19.5-(23.3) μm (Kalkar and Carner, 2005); generally pyriform to obovate with a uniformly oval apex, and broad, rounded papilla at the base; forcibly discharged by papillary eversion. Conidia were formed directly on the apex of simple conidiophores without a neck or constriction. Secondary conidia were similar in shape, and smaller than, primary conidia. Secondary conidia (17.5)-18.5-(22.7) x (10.0)-14.1-(17.5) μm; forcibly discharged and adhered to the walls of the diet cup. Resting spores (19.0)-34.5-(40.0) μm in diameter (Kalkar and Carner, 2005). The fungus has been reported as an important natural enemy of *Lambdina fiscellaria* (Guenée) (Otvos *et al.,* 1973; McDonald and Nolan, 1995), *Helicoverpa zea* (Boddie) and *Celama sorghiella* (Riley) (Hamm, 1980), *Pseudaletia separata* (Walker) (Ohbayashi and Iwabuchi, 1991), *Chionarctia nivea* (Ménétriès) (Yamazaki *et al.,* 2004) and *Plathypena scabra* (F.) (Kalkar and Carner, 2005).

1.3.2.4. Entomophthora planchoniana Cornu

Primary conidia (16.0)-18.48-(22.0) x (13.9)-15.54-(19.0) μm; bell-shaped with broad flat papilla and pointed apex; forcibly discharged. Secondary conidia: budding from the primary conidia; slightly smaller, (12.0)-13.66-(15.0) x (10.0)-11.03-(12.0) μm; nonapiculate with more rounded papillae. Conidiophores simple and unbranched. Rhizoids: mostly fasciculate, spreading out onto the substrate in different random directions; almost the same diameter as conidiophores. *E. planchoniana* causes widespread epizootics in the aphid populations throughout the world (Humber and Feng, 1991). The fungus has been reported as an important natural enemy of aphid populations (Hatting *et al.,* 1999; Freimoser *et al.,* 2001; Barta and Cagáň, 2006a; Scorsetti *et al.,* 2007).

1.3.2.5. Lecanicillium lecanii (Zimmermann) Zare and W. Gams

Colony diameters reached 29 mm in 10 days on malt extract agar (MEA) at 24 °C in darkness, rather compact, white, with clear yellow on reverse. Phialides relatively short, (13.9)-21.8-(30.7) x (0.9)-1.15-(1.9) μm, conical, aculeate, produced singly or in whorls of up to six. Conidia hyaline, short ellipsoidal, (1.9)-3.5-(5.4) x (0.8)-1-(1.5) μm, formed in heads at the apex of the phialides (Scorsetti *et al.,* 2008).

L. lecanii (as *Verticillium lecanii*) is a common fungal pathogen of aphids, scales, whiteflies and several other insects (Hall, 1981).

1.3.2.6. Metarhizium anisopliae (Metschnikoff) Sokorin

Colonies on PDA have a white mycelial margin with clumps of conidiophors, which become coloured with the development of spores. The colour of the spores ranges from yellow green, dark herbage green, to sometimes pink. Spores are produced in columnar stands. Conidiophors are abundant, and arise from vegetative hyphae branching irregularly, usually with 2-3 branches at each node. Conidia are formed in chains, narrow cylindrical smooth, aseptae. *M. anisopliae* occurs in two different forms based on conidial size: *M. anisopliae* var. *anisopliae* (3.5-9.0 x 1.5-3.5 μm) and *M. anisopliae* var. *major* (9.0- 18.0 x 1.8-4.5 μm) (Tulloch, 1976). The green muscardine fungus, *M. anisopliae*, has a wide host range of insects and is common in nature. For about 130 years, entomopathogenic fungi and especially *M. anisopliae*, have been used for biocontrol of pest insects (Zimmermann, 2007b).

1.3.2.7. Neozygites Fresenii (Nowakowski) Remaudie`Re and Keller

Primary conidia (15.5)-17.71-(20.5) x (12.5)-14.83-(16.5) μm; nearly spherical to ovoid with a flattened basal papilla; forcibly discharged. Secondary conidia: capilliconidia (19.0)-23.54-(29.0) x (10.0)-13.4-(16.0) μm; carried on capillary conidiophores arising from primary conidia; passively discharged from capillary conidiophores; almond-shaped with a mucoid drop at the tip. Conidiophores simple, unbranched. Resting spores: black to smoky-grey in color; ovoid; arising from conjugation between two spherical gametangia; (23.5)-27.3-(30.0) x (17.5)-20.77-(23.0) μm. Isolation: not cultured *in vitro* (Hatting *et al.,* 1999).

N. fresenii is a widely occurring fungal pathogen of aphids and has been recorded in various countries. Aphids of the genus *Aphis*, especially the species *Aphis fabae* Scopoli (Barta and Cagáň, 2002, 2006b) and *Aphis gossypii* Glover (Steinkraus *et al.,* 1995; Steinkraus and Boys, 2005; Scorsetti *et al.,* 2007; Abney *et al.,* 2008) are considered the major host organisms for the pathogen.

1.3.2.8. Pandora neoaphidis (Remaudi`Ere and Hennebert) Humber

Primary conidia (17.0)-20.92-(27.0) x (9.5)-10.97-(13.0) μm (Hatting *et al.,* 1999); generally ovoid to clavate; uninucleate with basal papilla displaced laterally from the spore axis; forcibly discharged by papillary eversion, often creating a white halo around the cadaver. Secondary conidia (13.0)-15.1-(19.0) x (7.8)-11.1-(13.4) μm (Scorsetti *et al.,* 2007): produced singly on primary conidia; similar to or more nearly globose than the latter. Conidiophores: digitately branched at their apices. Cystidia: distally tapering, 2- to 33 thicker than conidiophores; generally produced before the formation of the hymenium. Rhizoids: 2- to 33 diameter of hyphae; each ending in a discoid-like expansion. The species probably does not produce resting spores (Hatting *et al.,* 1999). The fungus can be easily isolated and cultivated on solid media based on Sabouraud's dextrose agar enriched with cow's milk and egg yolk (Keller, 1991) as well as in liquid media (Latgé *et al.,* 1983; Li *et al.,* 1993).

1.4. PRESERVATION OF CULTURES

The aim of preserving a fungus is to maintain it in a viable state without change to its genetic, physiological, or anatomical characters. There is no universal method for storing entomopathogenic fungi. Selection of a method must be based on the nature of the pathogen and the advantages and disadvantages of each method. If the pathogen is not well understood, preservation by more than one method should be done. Before storing, culture purity must be verified and its morphology, growth characteristics, and pathogenic behavior known in order to detect any changes while in storage (Humber, 1997). Fungi can lose their virulence if repeatedly grown on artificial media. Avoid the loss of virulence by using cultures from a stock of inoculum, or by reisolating from infected insects You must store the original infected insect. Isolate the fungus by growing it on agar in a Petri dish, until it is free from contaminants. All of the following methods are not suitable for all species of fungi that grow in artificial culture. Trial and error is often required to establish the optimum method.

1.4.1. Agar Slopes

The methods are simple and inexpensive because specialized equipment is not required. Inoculum is transferred from an actively growing fungus culture to test tubes (screw cap or plugged with cotton or foam) containing an agar medium of choice. Alternating nutrient-rich with nutrient-poor media at each transfer helps to maintain healthy cultures. Slopes (test tubes) are inoculated with the fungus and after incubation can be stored in a cool dust-free environment. A refrigerator or a cold room at 5-8°C is suitable. Some strains are cold sensitive; 15°C is usually suitable for cold sensitive strains. Tube cultures will need transferring to fresh media at least every six months, maybe more frequently. The main disadvantage is frequent transfer may lead to such deleterious changes as losses of pathogenicity, virulence, or sporulation. Cultures must be checked periodically for contamination and desiccation. Cultures of invertebrate pathogens must be inspected carefully for such morphological changes and bioassayed periodically for changes of essential pathological characteristics (Humber, 1997).

1.4.2. Mineral Oil

Fungi actively growing on agar media in tubes remain viable for long periods when covered with mineral oil. This method is not require expensive apparatus or chemicals and is easy to use. The method is applicable to many fungi. Cultures kept under mineral oil may remain viable for decades (Cavalcanti, 1991; Mendes da Silva *et al.*, 1994). Pathogenicity of entomopathogenic fungi may be undiminished after several months of storage (Balardin and Loch, 1988), but whether pathogenicity or virulence decline after many years of storage under mineral oil remains uncertain. The disadvantage is that subculturing from oil stocks is messy.

Set up for storage (Humber, 1997);

- Use vigorously growing culture slants in glass culture tubes.
- Autoclave a supply of heavy mineral oil and reautoclave 24-48 h later to kill any bacterial spores activated by the first autoclaving.
- Aseptically cover the culture slant with sterile oil to the depth of 1 cm.
- Under aseptic conditions, cover the cultures with sterile mineral oil to 1 cm above the edge of the agar.
- Cover tubes with tight caps or plugs and apply a couple of layers of paraffin film as a further vapour barrier.
- Store either in a refrigerator or at room temperature.

Culture recovery;

- With a sterile scalpel, loop, or needle, recover an explant from the submerged culture.
- Drain excess oil on a sterile paper towel from the explant and place on fresh medium. The first subculture is slow growing due to the presence of the oil; two to three transfers are needed to restore the original growth rate.
- Reseal and return the tube to long-term storage.
- Monitor the culture for viability and/or contamination.

1.4.3. Water Storage

Many diverse fungi can be stored using this method. Apparently, the water suppresses morphological changes in most fungi. Cultures have been stored this way for up to 20 years (Hartung de Capriles *et al.*, 1989) although many fungi lose viability much sooner. This technique requires having only sterile screwcap tubes or vials and sterilized water.

Set up for storage (Humber, 1997);

- Use vigorously growing, relatively young fungal cultures for inoculum.
- Autoclave a supply of distilled water and screwcap vials or tubes.
- Dispense sterile water into sterile storage tubes.
- Inoculate tubes with small (ca. 1 mm^3) blocks of a culture and/or an aqueous suspension of spores. The volume of water must be ≥ 40 times the total volume of culture inoculum preserved in it.
- Incubate at room temperature for 7 days.
- Cover tubes with tight fitting caps. For more security, also seal with paraffin film or by dipping tube tops in melted paraffin.
- Tubes can be stored at room temperature or at about 15-18°C in the dark.

Culture recovery;

- Recover a block from the tube with a sterile scalpel, loop, or needle and place (fungus side down) onto fresh medium.
- Monitor the culture for viability and/or contamination.
- Reseal and return the tube to long-term storage.

1.4.4. Freeze-Drying (Lyophilization)

One of the favoured methods of preserving fungal cultures is by freeze-drying. Freeze-drying may be the most widely used of the 'technologically sophisticated' approaches to preserving fungal germplasm and is the primary technique used at most general service culture collections. The apparatus is very expensive to buy and maintain. Experienced staff are required to operate the system and in preparation of the material for preservation. Freeze-drying is effective for many diverse fungi. The technique is not suitable for preserving nonsporulating fungi, although some workers have reported limited success when lyophilizing mycelium (Tan, 1997). In this stable form viability can be for 10 years or more.

Set up for storage (Humber, 1997);

- Use sporulating cultures for inoculums.
- Sterilize and dry cotton-plugged ampoules.
- Cover sporulating cultures with sterilized skim milk solution or another nutritionally complex carrier, and suspend spores and hyphae. Transfer small quantities of this suspension to sterile ampoules and 'cure' the contents for several hours in a refrigerator.
- Rapidly freeze the preparation in a mixture of dry ice and either ethanol or propylene glycol or by placing ampoules in an ultracold freezer or in the vapour phase in a liquid nitrogen dewar.
- Attach frozen ampoules to a strong vacuum on the lyophilizer. Preparations must remain frozen during the initial stages of vacuum desiccation. After desiccation, ampoules are sealed under high vacuum and stored at 4 °C.
- They may also kept at ambient temperature if necessary but viability will probably decline sooner than if refrigerated.

Culture recovery;

- Most culture collections that send out lyophilized cultures provide detailed directions on how to open and reconstitute such preparations.
- If an ampoule is not pre-scored, score the neck with a file or diamond pencil.
- Surface sterilize in 70% ethanol or sodium hypochlorite solution (e.g. a 1:1 dilution of commercial bleach), wrap the scored ampoule in a sterile paper wipe moistened (but not soaking!) with ethanol and break at the scoring.
- Add sterile water or liquid medium (the type and quantity of liquid will usually be indicated by the culture's sender) to reconstitute the culture.
- Rest the material in a sterile hood for 1-30 min to soften and to rehydrate the dried pellet. Resuspend and pipette the reconstituted mixture onto fresh culture medium.
- Monitor the culture for viability and/or contamination.

1.4.5. Silica Gel

Silica gel is the preferred method of storage in some laboratories because it does not require expensive apparatus and it is easy to use. Cultures can be repeatedly taken from a single storage tube, although contamination must be avoided. The major disadvantage is a gradual decline in viability but this can be overcome by replacement with fresh material (Windels *et al.,* 1993). The advantage of silica gel is that it prevents all fungal growth and metabolism.

Storage of spores on sterile anhydrous silica gel crystals is the only storage technique for microorganisms that recommends use of -20 °C freezers. It is possible to store fungus-inoculated silica gel at room temperature if freezer space is unavailable but fungal viability will usually decrease sooner. Fungal spores stored on silica gel may remain viable for more than ten years (Windels *et al.*, 1993), and this method is inexpensive, simple and reliable for many fungi (Smith, 1993) including entomopathogens (Bell and Hamalle, 1974).

Set up for storage (Humber, 1997);

- Use sporulating culture for inoculums.
- Fill 25x200mm screw-cap glass tubes one-third full with 6-12 mesh, medium grade, and uncoloured anhydrous silica gel crystals.
- Sterilize the tubes and silica gel in an oven at 180 °C for 1.5 h to assure that the silica gel is both sterile and fully anhydrous. Do not autoclave the silica, it will become wet!
- A cold 5-7% skimmed milk spore suspension is poured onto sterilised cooled silica gel crystals. Tilt tubes during inoculation to expose the greatest possible surface area and rotate or agitate tubes while adding the inoculum suspension.
- The gel is allowed to dry at room temperature until the crystals separate –about 14 days. Both conidia and milk solids are absorbed onto the silica gel surface but the fungus does not colonise the substrate and so the chance of degeneration is restricted (Windels *et al.*, 1988, 1993).
- Store at -20 °C although tubes may also be kept at room temperature.
- Check viability and sterility of the stored preparation after ca. 1-2 weeks.

Culture recovery;

- Sprinkle a few granules of inoculated silica gel from a tube onto fresh culture medium. The cultures may take a bit longer than normal to start growing.
- Tightly reseal tube and return to long-term storage.

1.4.6. Cryopreservation

Cultures, mycelia or spore suspensions are treated with a cryoprotectant such as 10% glycerol or dimethylsulphoxide (DMSO), before aseptic transfer into sterile ampoules and frozen to ultra low temperatures usually in the vapour phase of liquid nitrogen.

Liquid nitrogen is -196 °C and is assumed that the metabolic activity is almost at a stand stil at this temperature. Fungi that cannot withstand lyophilization often can be stored in liquid nitrogen. Storage in liquid nitrogen is not mutagenic, and does not influence morphological and pathogenicity characters (Sinclair and Dhingra, 1995). The cryopreservation of microfungi at the ultra-low temperature of −196°C in liquid nitrogen or the vapor above is currently regarded as the best method of preservation (Ryan and Smith, 2004). It can be widely applied to sporulating and nonsporulating cultures. This technique requires large and expensive equipment and a reliable source of liquid nitrogen.

Set up for storage (Humber, 1997);

- For cultures grown on solid media: Dispense sterile cryoprotectant into storage units (vials, straws, etc.) and then add 2-4 cubes (1-3 mm on a side); make sure that there is ≥40:1 ratio of cryoprotectant to inoculum.
- For cultures grown in liquid media: Add an appropriate volume of undiluted, sterile cryoprotectant directly to cultures and disperse quickly by gentle agitation to minimize osmotic damage. Aliquot directly into sterile cryovials.
- 'Cure' fungus in cryoprotectant at 4 $^{\circ}$C for 2-48 h to allow uptake of the cryoprotectant.
- Place the cryovials in the neck of the nitrogen refrigerator at –35 $^{\circ}$C for 45 min.
- Transfer the cryovials into storage racks held in the vapor phase of the liquid nitrogen, this cools them to below –150 $^{\circ}$C.
- After at least 1 d retrieve an ampoule from the refrigerator to test viability and purity of the fungus.

Culture recovery;

- Warm the cryovial rapidly by immersion in a water bath at 37 $^{\circ}$C. Remove immediately on completion of thawing and do not allow it to warm up to the temperature of the bath. Alternatively, thaw the vials in a controlled-rate cooler on a warming cycle.
- Opening of the cryovial and the transfer to media should be carried out in an appropriate level microbiological safety cabinet. Surface sterilize the cryovials by immersion or wiping with 70% (v/v) alcohol. Aseptically transfer the contents using a Pasteur pipet and transfer on to suitable growth medium.

CONCLUSION

The fungal diseases in insect populations are common and widespread. They can often destroy insect populations in spectacular epizootics and thus attract man's attention. Many of these fungi are considered an important factor regulating pest insect populations. Traditionally, insect pathogenic fungi have been studied by both entomologists and mycologists. Generally speaking, entomologists have been focused on the practical aspects of using a fungal entomopathogen in the field, while mycologists are more interested in fungal taxonomy, mode of action, and phylogenetics. Molecular techniques is important in the analytical tool box to study and investigate the biology and ecology of fungi. Today, fungal molecular systematics is a mature discipline in which multi-locus datasets, extensive taxon sampling, and rigorous analytical approaches are standard.

REFERENCES

Abney, M.R., Ruberson, J.R., Herzog, G.A., Kring, T.J., Steinkraus, D.C. and Roberts, P.M. 2008. Rise and fall of cotton aphid (Hemiptera: Aphididae) populations in southeastern cotton production systems. *Journal of Economic Entomology*, 101: 23-35.

Agrios, G.N., 2005. Plant Pathology. Elsevier Academic Press, San Diego, CA, USA.

Al-mazra'awi, M.S., Al-Abbadi, A., Shatnawi, M.A. and Ateyyat, M. 2009. Effect of application method on the interaction between *Beauveria bassiana* and neem tree extract when combined for *Thrips tabaci* (Thysanoptera: Thripidae) control. *Journal of Food Agriculture and Environment*, 7: 869-873.

Aquino de Muro, M., Elliott, S., Moore, D., Parker, B.L., Skinner, M., Reid, W. and El Bouhssini, M. 2005. Molecular characterisation of *Beauveria bassiana* isolates obtained from overwintering sites of Sunn Pests (*Eurygaster* and *Aelia* species). *Mycological Research*, 109: 294-306.

Aquino de Muro, M., Mehta, S. and Moore, D. 2003. The use of amplified fragment length polymorphism for molecular analysis of *Beauveria bassiana* isolates from Kenya and other countries, and their correlation with host and geographical origin. *FEMS Microbiology Letters*, 229: 249-257.

Balardin, R.S. and Loch, L.C. 1988. Methods for inoculum production and preservation of *Nomuraea rileyi* (Farlow) Samson. *Summa Phytopathologica*, 14: 144-151.

Barta, M. and Cagáň L., 2002. Prevalence of natural fungal mortality of black bean aphid, *Aphis fabae* Scopoli, on primary host and two secondary hosts. *Acta Fytotechnica et Zootechnica*, 5: 57–64.

Barta, M. and Cagáň L., 2006a. Aphid-pathogenic Entomophthorales (their taxonomy, biology and ecology). *Biologia*, 61: 543-616.

Barta, M. and Cagáň L., 2006b. Observations on the occurrence of Entomophthorales infecting aphids (Aphidoidea) in Slovakia. *BioControl*, 51: 795-808.

Bell, J.V. and Hamalle, R.J. 1974. Viability and pathogenicity of entomogenous fungi after prolonged storage on silica gel at -20 degrees C. *Canadian Journal of Microbiology*, 20: 639-642.

Berretta, M.F., Lecuona, R.E., Zandomeni, R.O. and Grau, O. 1998. Genotyping isolates of the entomopathogenic fungus *Beauveria bassiana* by RAPD with fluorescent labels. *Journal of Invertebrate Pathology*, 71: 145-150.

Bidochka, M.J., Kasperski, J.E. and Wild, G.A.M. 1998. Occurrence of the entomopathogenic fungi *Metarhizium anisopliae* and *Beauveria bassiana* in soils from temperate and near-northern habitats. *Canadian Journal of Botany*, 76: 1198-1204.

Bidochka, M.J., McDonald, M.A., St Leger, R.J. and Roberts, D.W. 1994. Differentiation of species and strains of entomopathogenic fungi by random amplification of polymorphic DNA (RAPD). *Current Genetics*, 25: 107-113.

Boucias, D.G. and Pendland, J.C. 1998. *Principles of Insect Pathology*. Kluwer Academic Publishers, Boston, Dordrecht, London.

Bugeme, D.M., Knapp, M., Boga, H.I., Wanjoya, A.K. and Maniania, N.K. 2009. Influence of temperature on virulence of fungal isolates of *Metarhizium anisopliae* and *Beauveria bassiana* to the two-spotted spider mite *Tetranychus urticae*. *Mycopathologia*, 167: 221-227.

Carneiro, A.A., Gomes, E.A., Guimarães, C.T., Fernandes, F.T., Carneiro, N.P. and Cruz, I. 2008. Molecular characterization and pathogenicity of isolates of *Beauveria* spp. to fall armyworm. *Pesquisa Agropecuária Brasileira,* 43: 513-520.

Castrillo, L.A., Wiegmann, B.M. and Brooks, W.M. 1999. Genetic variation in *Beauveria bassiana* populations associated with the darkling beetle *Alphitobius diaperinus*. *Journal of Invertebrate Pathology*, 73: 269-275.

Cavalcanti, M.A.D.Q., 1991. Viability of Basidiomycotina cultures preserved in mineral oil. *Revista Latinoamericana de Microbiologia,* 32: 265-268.

Chandler, D., Davidson, G., Pell, J.K., Ball, B.V., Shaw, K. and Sunderland, K.D. 2000. Fungal biocontrol of Acari. *Biocontrol Science and Technology,* 10: 357-384.

Chandler, D., Hay, D. and Reid, A.P. 1997. Sampling and occurrence of entomopathogenic fungi and nematodes in UK soils. *Applied Soil Ecology*, 5: 133-141.

Chandler, D., Heale, J.B. and Gillespie, A.T. 1993. Germination of the entomopathogenic fungus *Verticillium lecanii* on scales of the glasshouse whitefly *Trialeurodes vaporariorum*. *Biocontrol Science and Technology*, 3: 161-164.

Charnley, A.K. 1997. Entomopathogenic fungi and their role in pest control. In: Wicklow, D.T. and Söderström, B. (Eds.), *The mycota IV. Environmental and microbial relationships*. Springer, Berlin Heidelberg, Germany, pp. 185-201.

Clarkson, J.M. 1992. Molecular approaches to the study of entomopathogenic fungi. In: Lomer, C.J. and Prior, C. (Eds.), *Biological control of locusts and grasshoppers*. CABI Press, Wallingford, pp. 191-199.

Coates, B.S., Hellmich, R.L. and Lewis, L.C. 2002. *Beauveria bassiana* haplotype determination based on nuclear rDNA internal transcribed spacer PCR-RFLP. *Mycological Research*, 106: 40-50.

Dalzoto, P.R., Glienke-Blanco, C., Kava-Cordeiro, V., Araujo, W.L. and Azevedo, J.L. 2003. RAPD analyses of recombination processes in the entomopathogenic fungus *Beauveria bassiana*. *Mycological Research*, 107: 1069-1074.

de Hoog, G.S. 1972. The genera *Beauveria, Isaria, Tritirachium* and *Acrodontium* gen. nov. *Studies in Mycology*, 1: 1-41.

Driver, F. and Milner, R.J. 1998. PCR applications to the taxonomy of entomopathogenic fungi. In: Bridge, P.D., Arora, D.K., Reddy, C.A. and Elander, R.P. (Eds.), *Applications of PCR in mycology*. CABI Press, Wallingford, pp. 153-186.

Eken, C. and Hayat, R. 2009. Preliminary evaluation of *Cladosporium cladosporioides* (Fresen.) de Vries in laboratory conditions, as a potential candidate for biocontrol of *Tetranychus urticae* Koch. *World Journal of Microbiology and Biotechnology*, 25: 489-492.

Faria, M. and Wraight, S.P. 2001. Biological control of *Bemisia tabaci* with fungi. *Crop Protection,* 20: 767-778.

Faria, M. and Wraight, S.P. 2007. Mycoinsecticides and Mycoacaricides: A comprehensive list with worldwide coverage and international classification of formulation types. *Biological Control*, 43: 237-256.

Freimoser, F.M., Jensen, A.B., Tuor, U., Aebi, M. and Eilenberg, J. 2001. Isolation and in vitro cultivation of the aphid pathogenic fungus *Entomophthora planchoniana*. *Canadian Journal of Microbiology,* 47: 1082-1087.

Gaitan, A., Valderrama, A.M., Saldarriaga, G., Velez, P. and Bustillo, A. 2002. Genetic variability of *Beauveria bassiana* associated with the coffee berry borer *Hypothenemus hampei* and other insects. *Mycological Research*, 106:1307-1314.

Gams, W. 1992. The analysis of communities of saprophytic microfungi with special reference to soil fungi. In: Winterhoff, W. (Ed.), *Fungi in vegetation science*. Kluwer Academic Publication, Dordrecht, Netherland, pp. 183-223.

Gindin, G., Barash, I., Raccah, B., Singer, S., Ben-Ze'ev, I. and Klein, M. 1996. The potential of some entomopathogenic fungi as biocontrol agents against the onion thrips, *Thrips tabaci* and the western flower thrips, *Frankliniella occidentalis*. *Folia Entomologica Hungarica*, 62: 37-42.

Gindin, G., Geschtovt, N.U., Raccah, B. and Barash, I. 2000. Pathogenicity of *Verticillium lecanii* to different developmental stages of the silverleaf whitefly, *Bemisia argentifolii*. *Phytoparasitica*, 28: 229-239.

Goettel, M.S. and Inglis, G.D. 1997. Fungi: Hyphomycetes. In: Lacey, L.A. (Ed.), *Manual of techniques in insect pathology*. Academic Press, San Diego, USA, pp. 213-249.

Hall, R.A. 1981. The fungus *Verticillium lecani* as a microbial insecticide against aphids and scales. In: Burges, H.D. (Ed.), *Microbial control of pests and plant diseases 1970-1980*. Academic Press, London, pp. 483-498.

Hamm, J.J. 1980. Epizootics of the *Entomophthora aulicae* in lepidopterous pests of sorghum. *Journal of Invertebrate Pathology*, 36: 60-63.

Hartung de Capriles, C., Mata, S. and Middelveen, M. 1989. Preservation of fungi in water (Castellani): 20 years. *Mycopathologia*, 106: 73-79.

Hatting, J.L., Humber, R.A., Poprawski, T.J. and Miller R.M. 1999. A survey of fungal pathogens of aphids from South Africa, with special reference to cereal aphids. *Biological Control*, 16: 1-12.

Hegedus, D.D. and Khachatourians, G.G. 1993. Identification of molecular variants in mithocondrial DNAs of members of the genera *Beauveria*, *Verticillium*, *Paecilomyces*, *Tolypocladium* and *Metarhizium*. *Applied and Environmental Microbiology*, 59: 4283-4288.

Hibbett, D.S., Binder, M., Bischoff, J.F., Blackwell, M., Cannon, P.F., Eriksson, O.E., Huhndorf, S., James, T., Kirk, P.M., Lucking, R., Lumbsch, H.T., Lutzoni, F., Matheny, P.B., Mclaughlin, D.J., Powell, M.J., Redhead, S., Schoch, C.L., Spatafora, J.W., Stalpers, J.A., Vilgalys, R., Aime, M.C., Aptroot, A., Bauer, R., Begerow, D., Benny, G.L., Castlebury, L.A., Crous, P.W., Dai, Y.C., Gams, W., Geiser, D.M., Griffith, G.W., Gueidan, C., Hawksworth, D.L., Hestmark, G., Hosaka, K., Humber, R.A., Hyde, K.D., Ironside, J.E., Koljalg, U., Kurtzman, C.P., Larsson, K.H., Lichtwardt, R., Longcore, J., Miadlikowska, J., Miller, A., Moncalvo, J.M., Mozley-Standridge, S., Oberwinkler, F., Parmasto, E., Reeb, V., Rogers, J.D., Roux, C., Ryvarden, L., Sampaio, J.P., Schussler, A., Sugiyama, J., Thorn, R.G., Tibell, L., Untereiner, W.A., Walker, C., Wang, Z., Weir, A., Weiss. M., White, M.M., Winka, K., Yao, Y.J. and Zhang, N. 2007. A higher-level phylogenetic classification of the Fungi. *Mycological Research,* 111: 509-547.

Hsiao, W.F., Bidochka, M.J. and Khachatourians, G.G. 1992. Effect of temperature and relative humidity on the virulence of the entomopathogenic fungus, *Verticillium lecanii*, towards the oat-bird berry aphid, *Rhopalosiphum padi* (Homoptera : Aphididae). *Journal of Applied Entomology*, 114: 484-490.

Humber, R.A. 1997. Fungi: preservation of cultures. In: Lacey, L.A. (Ed.), *Manual of techniques in insect pathology*. Academic Press, San Diego, USA, pp. 269-279.

Humber, R.A. 1998. Entomopathogenic fungal identification. In: Humber, R.A. and Steinkraus, D. (Eds.), *APS/ESA Joint Annual Meeting*, November 8-12, 1998, Las Vegas, NV.

Humber, R.A. and Feng, M.G. 1991. *Entomophthora chromaphidis* (Entomophthorales): the correct identification of an aphid pathogen in the Pacific Northwest and elsewhere. *Mycotaxon*, 41: 497-504.

James, T.Y., Kauff, F., Schoch, C., Matheny, P.B., Hofstetter, V., Cox, C.J., Celio, G., Geuidan, C., Fraker, E., Miadlikowska, J., Lumbsch, H.T., Rauhut, A., Reeb, V., Arnold, A.E., Amtoft, A., Stajich, J.E., Hosaka, K., Sung, G.H., Johnson, D., O'Rourke, B., Crockett, M., Binder, M., Curtis, J.M., Slot, J.C., Wang, Z., Wilson, A.W., Schüßler, A., Longcore, J.E., O'Donnell, K., Mozley-Standridge, S., Porter, D., Letcher, P.M., Powell, M.J., Taylor, J.W., White, M.M., Griffith, G.W., Davies, D.R., Humber, R.A., Morton, J.B., Sugiyama, J., Rossman, A., Rogers, J.D., Pfister, D.H., Hewitt, D., Hansen, K., Hambleton, S., Shoemaker, R.A., Kohlmeyer, J., Volkmann-Kohlmeyer, B., Spotts, R.A., Serdani, M., Crous, P.W., Hughes, K.W., Matsuura, K., Langer, E., Langer, G., Untereiner, W.A., Lücking, R., Büdel, B., Geiser, D.M., Aptroot, A., Diederich, P., Schmitt, I., Schultz, M., Yahr, R., Hibbett, D.S., Lutzoni, F., McLaughlin, D.J., Spatafora, J.W. and Vilgalys, R. 2006b. Reconstructing the early evolution of fungi using a six-gene phylogeny. *Nature (London)*, 443: 818-822.

James, T.Y., Letcher, P.M., Longcore, J.E., Mozley-Standridge, S.E., Porter, D., Powell, M.J., Griffith, G.W. and Vilgalys, R. 2006a. A molecular phylogeny of the flagellated fungi (Chytridiomycota) and description of a new phylum (Blastocladiomycota). *Mycologia*, 98: 860-871.

Kalkar, Ö. and Carner, G.R. 2005. Characterization of the fungal pathogen, *Entomophaga aulicae* (Zygomycetes: Entomophthorales) in larval populations of the green cloverworm, *Plathypena scabra* (Lepidoptera: Noctuidae). *Turkish Journal of Biology*, 29: 243-248.

Kao, S.S., Tsai, Y.S., Yang, P.S., Hung, T.H. and Ko, J.L. 2002. Use of random amplified polymorphic DNA to characterize entomopathogenic fungi, *Nomuraea* spp., *Beauveria* spp., and *Metarhizium anisopliae* var. *anisopliae*, from Taiwan and China. *Formosan Entomologist,* 22: 125-134.

Keller, S. 1987. Arthropod-pathogenic Entomophthorales of Switzerland. I. *Conidiobolus*, *Entomophaga* and *Entomophthora*. *Sydowia*, 40: 122-167.

Keller, S. 1991. Arthropod-pathogenic Entomophthorales of Switzerland. II. *Erynia*, *Eryniopsis*, *Neozygites*, *Zoophthora* and *Tarichium*. *Sydowia*, 43: 39-122.

Keller, S., and Zimmerman, G., 1989. Mycopathogens of soil insects. In: Wilding, N., Collins, N.M., Hammond, P.M. and Webber, J.F. (Eds.), *Insect-fungus interactions*. Academic Press, London, pp. 240–270.

Keller, S., Kessler, P. and Schweizer, C. 2003. Distribution of insect pathogenic soil fungi in Switzerland with special reference to *Beauveria brongniartii* and *Metharhizium anisopliae*. *BioControl*, 48: 307-319.

Kessler, P., Matzke, H. and Keller, S. 2003. The effect of application time and soil factors on the occurrence of *Beauveria brongniartii* applied as a biological control agent in soil. *Journal of Invertebrate Pathology*, 84: 15–23.

Kirk, P.M., Cannon, P.F., Minter, D.W. and Stalpers, J.A. 2008. *Dictionary of the Fungi* 10th edition. CAB International, Wallingford, UK.

Klingen, I., Eilenberg, J. and Meadow, R. 2002. Effects of farming system, field margins and bait insect on the occurrence of insect pathogenic fungi in soils. *Agriculture Ecosystems and Environment*, 91: 191-198.

Latgé, J.P. 1980. Sporulation of *Entomophthora obscura* Hall and Dunn in liquid culture. *Canadian Journal of Microbiology*, 26: 1038-1048.

Latgé, J.P., Silvie, P., Papierok, B., Remaudi`ere, G., Dedryver, C.A. and Rabasse, J.M. 1983. Advantages and disadvantages of *Conidiobolus obscurus* and of *Erynia neoaphidis* in the biological control of aphids. In: Cavalloro, R. (Ed.), *Aphid antagonists*. A.A. Balkema, Rotterdam, the Netherlands, pp. 20-32.

Li, Z., Butt, M., Beckett, A. and Wilding, N. 1993. The structure of dry mycelia of the entomophthoralean fungi *Zoophthora radicans* and *Erynia neoaphidis* following different preparatory treatments. *Mycological Research*, 97: 1315-1323.

McDonald, D.M. and Nolan, R.A. 1995. Effects of relative humidity and temperature on *Entomophaga aulicae* conidium discharge from infected eastern hemlock looper larvae and subsequent conidium development. *Journal of Invertebrate Pathology,* 65: 83-90.

Mendes da Silva, A.M., Borba, C.M. and Oliveira, P.C. de. 1994. Viability and morphological alterations of *Paracoccidioides brasiliensis* strains preserved under mineral oil for long periods of time. *Mycoses*, 37: 165-169.

Meyling, N.V. 2007. Methods for isolation of entomopathogenic fungi from the soil environment. Available: http://orgprints.org/11200.

Meyling, N.V. and Eilenberg, J. 2006a. Isolation and characterisation of *Beauveria bassiana* isolates from phylloplanes of hedgerow vegetation. *Mycological Research*, 110: 188-195.

Meyling, N.V. and Eilenberg, J. 2006b. Occurrence and distribution of soil borne entomopathogenic fungi within a single organic agroecosystem. *Agriculture Ecosystems and Environment*, 113: 336-341.

Ohbayashi, T. and Iwabuchi, K. 1991. Abnormal behavior of the common armyworm *Pseudaletia separata* (Walker) (Lepidoptera: Noctuidae) larvae infected with an entomogenous fungus, *Entomophaga aulicae*, and a nuclear polyhedrosis virus. *Applied Entomology and Zoology,* 26: 579-585.

Otvos, I.S., Macleod, D.M. and Tyrrell, D. 1973. Two species of Entomophthora pathogenic to the eastern hemlock looper (Lepidoptera: Geometridae) in Newfoundland. *Canadian Entomologist*, 105: 1435-1441.

Papierok, B. and Hajek, A.E. 1997. Fungi: Entomophthorales. In: Lacey, L.A. (Ed.), *Manual of techniques in insect pathology*. Academic Press, San Diego, USA, pp. 187-212.

Pfeifer, T.A. and Khachatourians, G.G. 1993. Electrophoretic karyotype of entomopathogenic deuteromycete *Beauveria bassiana*. *Journal of Invertebrate Pathology*, 61: 231-235.

Pilz, C., Wegensteiner, R. and Keller, S. 2008. Natural occurrence of insect pathogenic fungi and insect parasitic nematodes in *Diabrotica virgifera virgifera* populations. *BioControl*, 53: 353-359.

Ryan, M.J. and Smith, D. 2004. Fungal genetic resource centres and the genomic challenge. *Mycological Research,* 108: 1351-1362.

Samson, R.A., Evans, H.C. and Latgé, J.P. 1988. *Atlas of Entomopathogenic Fungi.* Springer-Verlag, Berlin.

Scorsetti, A.C., Humber, R.A., De Gregorio, C. and López Lastra, C.C. 2008. New records of entomopathogenic fungi infecting *Bemisia tabaci* and *Trialeurodes vaporariorum*, pests of horticultural crops, in Argentina. *BioControl*, 53: 787-796.

Scorsetti, A.C., Humber, R.A., García, J.J. and López Lastra, C.C. 2007. Natural occurrence of entomopathogenic fungi (Zygomycetes : Entomophthorales) of aphid (Hemiptera : Aphididae) pests of horticultural crops in Argentina. *BioControl*, 52: 641-655.

Shah, P.A., Clark, S.J. and Pell, J.K. 2004. Assessment of aphid host range and isolate variability in *Pandora neoaphidis* (Zygomycetes: Entomophthorales). *Biological Control*, 29: 90-99.

Sinclair, J.B. and Dhingra, O.D. 1995. *Basic Plant Pathology Methods*. CRC Press, Boca Raton, FL, USA.

Smith, C. 1993. Long-term preservation of test strains (fungus). *International Biodeterioration and Biodegradation*, 31: 227-230.

Sosa-Gómez, D.R., Humber, R.A., Hodge, K.T., Binneck, E. and da Silva-Brandão, K.L. 2009. Variability of the mitochondrial SSU rDNA of *Nomuraea* species and other entomopathogenic fungi from Hypocreales. *Mycopathologia*, 167: 145-154.

Steinkraus, D.C. and Boys, G.O. 2005. Mass harvesting of entomopathogenic fungus, *Neozygites fresenii*, from natural field epizootics in the cotton aphid, *Aphis gossypii*. *Journal of Invertebrate Pathology*, 88: 212-217.

Steinkraus, D.C., Hollingsworth, R.G. and Slaymaker, P.H. 1995. Prevalence of *Neozygites fresenii* (Entomophthorales: Neozygitaceae) on cotton aphids (Homoptera: Aphididae) in Arkansas cotton. *Environmental Entomology*, 24: 465–474.

Strasser, H., Forer, A. And Schinner, F. 1996. Development of media for the selective isolation and maintenance of virulence of *Beauveria brongniartii*. In: Jackson, T.A. and Glare T.R. (Eds.), *Microbial control of soil dwelling pests*. AgResearch, Lincoln, New Zealand, pp. 125-130.

Sun, B.D. and Liu, X.Z. 2008. Occurrence and diversity of insect-associated fungi in natural soils in China. *Applied Soil Ecology*, 39: 100-108.

Tan, C.S. 1997. Preservation of fungi. *Cryptogamic Mycology*, 18: 157-163.

Tanada, Y. and Kaya, H. K. 1993. *Insect Pathology*. Academic Press, Inc. San Diego, CA, USA.

Tulloch, M. 1976. The genus *Metarhizium*. *Transactions of the British Mycological Society*, 66: 407-411.

Urtz, B.E. and Rice, W.C. 1997. RAPD-PCR characterization of *Beauveria bassiana* isolates from the rice water weevil *Lissorhoptrus oryzophilus*. *Letters in Applied Microbiology*, 25: 405-409.

Vänninen, I. 1996. Distribution and occurrence of four entomopathogenic fungi in Finland: Effect of geographical location, habitat type and soil type. *Mycological Research*, 100: 93-101.

Vänninen, I., Husberg, G.B. and Hokkanen, M.T. 1989. Occurrence of entomopathogenic fungi and entomoparasitic nematodes in cultivated soils in Finland. *Acta Entomologica Fennica*, 53: 65-71.

Wada, S., Horita, M., Hirayae, K. and Shimazu, M. 2003. Discrimination of Japanase isolates of *Beauveria brongniartii* (Deuteromycotina: Hyphomycetes) by RFLP or the rDNA-ITS regions. *Applied Entomology and Zoology*, 38: 551-557.

Windels, C.E., Burnes, P.M. and Kommedahl, T. 1988. Five-year preservation of *Fusarium* species on silica gel and soil. *Phytopathology*, 78: 107-109.

Windels, C.E., Burnes, P.M. and Kommedahl, T. 1993. *Fusarium* species stored for silica gel and soil for ten years. *Mycologia*, 85: 21-23.

Yamazaki, K., Sugiura, S. and Fukasawa, Y. 2004. Epizootics and behavioral alteration in the arctiid caterpillar *Chionarctia nivea* (Lepidoptera: Arctiidae) caused by an entomopathogenic fungus, *Entomophaga aulicae* (Zygomycetes: Entomophthorales). *Entomological Science,* 7: 219-223.

Zimmermann, G. 1986. The *Galleria* bait method for detection of entomopathogenic fungi in soil. *Journal of Applied Entomology*, 102: 213-215.

Zimmermann, G. 2007a. Review on safety of the entomopathogenic fungi *Beauveria bassiana* and *Beauveria brongniartii*. *Biocontrol Science and Technology*, 17: 553-596.

Zimmermann, G. 2007b. Review on safety of the entomopathogenic fungus *Metarhizium anisopliae*. *Biocontrol Science and Technology*, 17: 879-920.

ISBN: 978-1-61209-223-2

Chapter 2

BIOLOGICAL CONTROL POTENTIAL OF ENTOMOPATHOGENIC FUNGI

Toledo Andrea Vanesa[*]

Centro de Investigaciones de Fitopatología (CIDEFI), Facultad de Ciencias Agrarias y
Forestales, Universidad Nacional de La Plata, Avda. 60 y 119 s/n, 1900, La Plata, Buenos
Aires, Argentina

ABSTRACT

Fungi have one of the widest host ranges among arthropod pathogens . Theoretical
and practical researches have focused on several aspects of fungi biology, physiology,
ecology, and epidemiology, but predominantly from the viewpoint of their potential in
host population regulation. In an Integrated Pest Management program, fungi are
compatible with some fungicides and many other types of pesticides, although their
activity is strongly influenced by the environment. This chapter shows how some of the
abiotic and biotic factors of the environment, as well as the interactions between fungal
pathogens and their hosts, influence the use of entomopathogenic fungi as biological
control agents.

2.1. INTRODUCTION

Entomopathogenic fungi are usual natural enemies of arthropods worldwide, occupying
virtually every niche in which arthropods are also found. There are more than 700 species of
entomopathogens from within the fungal kingdom, being most of the species from the fungal
divisions Zygomycota and Ascomycota. Most entomopathogens within the Zygomycota
occur within the order Entomophthorales, characterized by the formation of a sexual spore
(zygospore). Many species have a worldwide distribution and often cause epizootics. Most
species are obligate parasites with a restricted host range; however, some species have
relatively wide host ranges, such as *Zoophthora radicans* (Brefeld) Batko. The most common

[*] E-mail: atoledo@cepave.edu.ar, Fax: + 54 221 4252346; Phone: + 54 221 4236758

genera include *Conidiobolus*, *Entomophaga*, *Entomophthora*, *Erynia*, *Pandora*, *Neozygites* and *Zoophthora*. A few genera of entomopathogenic fungi produce an ascomycetous sexual state (*i.e.*, ascospores produced within an ascus), characteristic of the Ascomycota division. The most important of these include *Cordyceps*, *Torrubiella* and *Ascophaera*. Many important species of entomopathogenic fungi appear to have lost most or all of their capacity to produce a sexual state. These include prominent pathogens in the genera *Aspergillus*, *Aschersonia*, *Beauveria*, *Culicomyces*, *Fusarium*, *Gibellula*, *Hirsutella*, *Hymenostilbe*, *Lecanicillium*, *Metarhizium*, *Nomuraea*, *Isaria*, *Sorosporella*, and *Tolypocladium*. All of the asexual (anamorphic) forms of these fungi produce spores, termed conidia, and some produce chlamydospores. Entomopathogenic fungi produce infective spores that attach to, germinate, penetrate, and invade their hosts mostly though the cuticle, although some are capable of breaching the alimentary canal. Once within the host, they proliferate by exploiting the nutritional resources of their hosts, ultimately killing them and producing more infective conidia for spread or resting structures for persistent. In most entomopathogenic fungi, spore dispersal is passive, relying mainly on wind and water. However, in the Entomophthorales, spores are forcibly discharged and can land many centimeters away from the host or disperse long distances on air currents. Fungi have one of the widest host ranges among the pathogens of the arthropods. However, host range vary widely on different fungal species. For instance, most species belonging to the order Hypocreales (Ascomycota), such as *Beauveria bassiana* (Balsamo-Crivelli) Vuillemin and *Metarhizium anisopliae* (Metschnikoff) Sorokin have characteristically broad host ranges in the soil-inhabiting arthropods in temperate regions.

Fundamental researches have focused on many theoretical and practical aspects of fungi biology, physiology, ecology, and epidemiology, but predominantly from the viewpoint of their potential in host population regulation. The use of the entomopathogenic fungi as possible microbial biocontrol agents was considered for the first time at the end of the 19th century. In this period where the synthetic chemical insecticides were still unknown, this possibility of control had become common among the researchers; although the greater quantity of studies were carried out during the 20th century (Cantwell, 1974; Burges, 1981; Davidson, 1981; Ferron, 1985). Since then, fungi have demonstrated considerable potential for insects control, mainly within Integrated Pest Management (IPM) programs (Lacey and Goettel, 1995) due to their restricted host range, with limited harm to non-target organisms such as predators, parasites, and other pathogens (Goettel *et al.*, 1990; 2001; Goettel and Hajek, 2001; Vestergaard *et al.*, 2003). Fungi are also compatible with some fungicides and many other types of pesticides, although their activity is strongly influenced by the biotic and abiotic environmental conditions (Wraight *et al.*, 2007). Their ecology, physiology, and life cycles are highly variable, reflecting adaptation to overcome environmental limitations and the host's defences. This chapter covers how the use of entomopathogenic fungi as biological control agents can be influenced by abiotic and biotic factors, as well as the interactions between the fungal pathogens and these factors.

2.2. COMPATIBILITY WITH FUNGICIDES AND PESTICIDES

Biological control, in particular when accomplished by entomopathogens, should be considered as an important component in IPM programs reducing pest population density.

Therefore, the conservation of such entomopathogens, whether they occur naturally or when they are applied or introduced to control pests, is an interesting practice.

However, the use of incompatible insecticides may inhibit the development and reproduction of these pathogens, thereby reducing their benefic effects in IPM programs (Malo, 1993; Duarte *et al.,* 1992; Anderson and Roberts, 1983; Alves and Lecuona, 1998). On the other hand, the use of selective pesticides in association with entomopathogens can increase the efficiency of insect control, minimizing environmental contamination hazards and the appearance of pest resistance (Moino and Alves, 1998; Quintela and McCoy, 1998). There are numerous studies on the effects of agrochemicals upon entomopathogenic fungi.

For example, Mier *et al.* (2004) studied the effect of the mineral oil citroline (used as agrochemical, insecticide and conidial dispersant agent) upon viability and in proteases and chitinases production of *M. anisopliae* var. *acridum.* Their studies showed a significant difference in fungal viability when compared to the control without citroline (81.4% reduction of colony forming units).

Conversely, numerous studies indicated that entomopathogenic fungi are compatible with many insecticides and fungicides when used in recomended doses (Khalil *et al.,* 1985; Neves *et al.,* 2001; Oliveira *et al.,* 2003; Cuthbertson *et al.,* 2005; Samson *et al.,* 2005; Bahiense *et al.,* 2006; Alizadeh *et al.,* 2007a; b).

2.3. ENHANCED FUNGAL TRANSMISSION

Besides dispersal of fungal spores by wind currents (Shimazu *et al.,* 2002) and rain splashes from soil surfaces (Bruck and Lewis, 2002b), insects could be potentially a means to disperse fungal inoculum. Aphid migratory alates, together with their predators and parasitoids, are known to disperse conidia of *Pandora neoaphidis* (Remaudière and Hennebert) Humber (Zygomycota: Entomophthorales), thereby increasing infection rates in aphid populations (Pell *et al.,* 1997; Roy *et al.,* 1998, 2001; Feng *et al.,* 2004; Feng, *et al.,* 2007; Baverstock *et al.,* 2008a). Regarding hypocrealean entomopathogenic fungi, *Isaria fumosorosea* Wize was dispersed by the ladybird *Hippodamia convergens* Guerin (Coleoptera: Coccinellidae) to individuals of the Russian wheat aphid *Diuraphis noxia* Kurdjumov (Hemiptera: Aphididae) in laboratory experiments (Pell and Vandenberg, 2002). Likewise, *B. bassiana* infections were initiated in the European corn borer *Ostrinia nubialis* Hübner (Lepidoptera: Crambidae) through dispersal by the fungivorous beetle *Carpophilus freemani* Dobson (Coleoptera: Nitidulidae) (Bruck and Lewis, 2002a).

In addition, Meyling *et al.* (2006) investigated the potential of nettle aphids *Microlophium carnosum* (Buckton) (Hemiptera: Aphididae) and their predator *Anthocoris nemorum* L. (Hemiptera: Cimicidae) to disperse conidia of *B. bassiana* from soil to nettles and from sporulating cadavers in the nettle canopy. Within the soil environment, which is a well-known reservoir of *B. bassiana* inoculum (Keller and Zimmerman, 1989), the fungus can be spready by collembolans (Dromph, 2001, 2003). Lastly, dispersal of *B. bassiana* by insect activity has been developed and used for pest management via the auto-dissemination strategy (Meadow *et al.,* 2000; Dowd and Vega, 2003; Vickers *et al.,* 2004).

2.4. Environmental Constraints

The effectiveness of the propagules to spread and infect insects of the infective propagules depends on their ability to remain viable for a period of time appropriate to their function. Viability strictly refers to the potential of a propagule to give rise to one or more individuals, either directly or by production of additional propagules, regardless of the conditions that may be necessary for it to do so. The viability is governed by factors intrinsic and extrinsic to the propagule. The first ones include spore size, shape, internal composition and organization, and wall structure. Extrinsic factors, include temperature, irradiance, availability of water and the influence of microorganisms, which are dynamic and can fluctuate quickly and at a large range. To evaluate the ability of a fungus to control an insect pest population, both its virulence to the targeted insect and its fitness to environmental conditions occurring in the insect habitat must be considered in order to enhance their use as an agent for insect control.

2.4.1. Temperature

Hot and cold weather lessen the use of entomopathogenic fungi as agents for biological control of insects. Most entomopathogenic fungi have optimal growth temperatures between 25 and 35° C (Roberts and Campbell, 1977), although some species grow at a wide temperature range (*e.g.*, *B. bassiana* from 8 to 35° C, with a maximum temperature threshold for growth at 46° C for 6 h) (Fargues *et al.,* 1997, Fernandes *et al.,* 2008). On the other hand, propagules of most of the species survive well at below-zero temperatures and can be stored for long periods at − 20 to − 80° C or in liquid nitrogen (− 196° C). Spores of some species can tolerate very high temperatures for very short periods (*e.g.*, 150° C for 30 sec); however, the maximum threshold for long periods is usually close to 40° C (Wraight *et al.,* 2007).

Effective use of insect pathogens within IPM programs requieres the selection of fungal pathogens tolerant to the temperature range of the ecosystem involved. The influence of temperature on *in vitro* germination, vegetative growth, and sporulation is generally studied in controlled-environment chambers using a variety of culture media (Goettel and Inglis, 1997). Baverstock *et al.* (2008b) reported that under controlled conditions, the aphid-pathogenic fungus *P. neoaphidis* remained active on soil and was able to infect aphids for up to 80 days. However, the percentage of aphids that became infected decreased from 76% on day 1 to 11% on day 80. Whereas there was little difference in the activity of conidia that had been maintained at 4° C and 10° C, activity at 18° C was considerably reduced. On the other hand, under field conditions inoculum activity was strongly influenced by season. On day 49 there was little or no activity during spring, summer or winter. However, during autumn an average proportion of 0.08 aphids still became infected with *P. neoaphidis*. These results suggest that *P. neoaphidis* can remain active on soil under the temperature conditions that occur seasonally in the UK, and that this fungus could persist year round without a resting stage. Fernandes *et al.* (2008) evaluated the thermotolerance and cold activity of 60 entomopathogenic fungal isolates, including species of *Beauveria* spp. and *Metarhizium* spp. High variability in conidial thermotolerance was found among the *Beauveria* spp. isolates after exposure to 45° C for 2 hours, as evidenced by low (0–20%), medium (20–60%), or high germination (60–80%). The thermal death point (0% germination) for three somewhat thermotolerant *B. bassiana* isolates was 46° C for 6 hours. At low temperatures (5° C), with few exceptions, most of the *B. bassiana* isolates germinated well (ca. 100%). On the other hand, only one isolate of *Metarhizium* sp. was cold-active. The dormant conidia of fungi tolerate higher temperatures than mycelium due to higher amounts of saturated fatty acids, which decreases cellular membrane permeability (Crisan, 1973; Pupin *et al.*, 2000; Guerzoni *et al.*, 2001), and higher amounts of trehalose and mannitol, which provides protection against denaturation of proteins and membranes (Thevelein, 1984; Rangel *et al.*, 2006). In designing a bioassay, it is imperative that consideration be given to the temperatures that should match those expected in a field environment.

2.4.2. Solar Radiation

In nature, many fungi are exposed to solar radiation in the visible and non-visible regions (280-400 nm) of the electromagnetic spectrum. This radiation influences many aspects of their behavior such as growth, development, and reproduction. Fungal propagules are highly susceptible to damage by solar radiation. The persistence of these on substrates exposed to direct solar radiation is substantially reduced relative to that of propagules protected by plant canopies or on the undrside of leaves (Wraight *et al.*, 2007). Despite this high inherent susceptibility, entomopathogenic fungal species and strains within species differ significantly in their susceptibility to solar radiation (Morley-Davies *et al.*, 1995; Fargues *et al.*, 1996; Braga *et al.*, 2001). Several observations suggest that heavily pigmented spores are more resistant to damaging radiation than hyaline spores are. Recently Singaravelan *et al.* (2008) investigated the cause-effect relationship between median exposure to UV radiation and the melanin concentration inconidia of the filamentous soil fungus *Aspergillus niger* van Tieghem. The results indicated that mean conidial melanin concentration of African (AS) strains were threefold higher than European (ES) strains and hence the former resisted UVA irradiation better than the latter. Comparisons of melanin in the conidia of *A. niger* strains from sunny and shady microniches on the predominantly sunny AS and predominantly shady ES indicated that shady conditions on the AS have no influence on the selection on melanin; in contrast, the sunny strains from the ES displayed higher melanin concentrations. They conclude that melanin in *A. niger* is an adaptive trait against UVR generated by natural selection. Melanin pigments absorb various kinds of radiation and dissipate energy primarily by undergoing reversible increase in free radicals (Bell and Wheeler, 1986). A number of researchers have measured the persistence of fungal propagules exposed to direct and indirect solar radiation in field (Inglis *et al.*, 1993; 1995; Smits *et al.*, 1996). However, a number of potentially confounding variables (e.g. temperature, precipitation, relative humidity, time of year, cloud cover, etc.) greatly complicate assessments of these limitations (Wraight *et al.*, 2007).

2.4.3. Moisture

The use of many entomopathogenic fungi for biological control may be limited by their requirements for specific environmental conditions, particularly high humidity (Yendol,

1968; Knipling, 1979; Hall and Papierok, 1982). Many studies have attempted to demonstrate the importance of moisture in the initiation of natural fungal epizootics (Carruthers and Soper, 1987), and many past failures in obtaining insect control using fungi have been attributed to dry weather conditions (Wraight *et al.,* 2007). However, fungi such as the entomophthoralean *Neozygites fresenii* (Nowakowski) Remaudière and Keller appear to be a somewhat atypical because of their ability to cause epizootics during relatively dry periods (Gustafsson, 1969; Thoizon, 1970; Dedryver, 1978; Steinkraus *et al.,* 1991; Steinkraus and Slaymaker, 1994). Among the hypocrealean fungi, *M. anisopliae* var. *acridum* is capable to infecting the desert locust *Schistocerca gregaria* (Forskal) (Orthoptera: Acrididae), at a relative humidity as low as 13% (Fargues *et al.,* 1997). The documented capacity of some fungi to operate under dry conditions generally has been attributed to the presence of humid microhabitats in which they are active, like abaxial leaf surfaces or membranous folds of insect cuticle (Wraight *et al.,* 2007).

2.4.4. Rainfall

An important feature in the use of entomopathogenic fungi is their ability to persist in the environment via horizontal transmission between infective and healthy hosts (Anderson, 1982; Harper, 1987; Roberts and Hajek, 1992). Factors that affect transmission of pathogens from infected to healthy insects therefore regulate a critical step in the dynamics of insect diseases. In fact, studies of Arthurs *et al.* (2001) showed that following grasshoppers death, *M. anisopliae* var. *acridum* can be persistent in the environment, sporulate on host cadavers and re-infect new hosts at a realistically low field density, although at least in arid or semi-arid areas, rainfall may be critical to the horizontal transmission of this pathogen.

2.4.5. Influence of Microorganisms

The epicuticle of the host integument, a composite structure containing a wax layer and several lipoproteins layers, is the site of the initial fungus-host interaction. Attachment of conidia to the cuticle depends upon a molecular interaction between the conidial surface and wax layer of the epicuticle (Boucias and Pendland, 1991). For some fungus-insect interactions, the failure of the fungi to invade insect cuticle has been attributed to the presence of inhibitory compounds on the cuticle surface such as phenols, quinones, and lipids (Smith and Grula, 1981; Szafranek *et al.,* 2001; Howard and Lord, 2003; James *et al.,* 2003; Lord and Howard, 2004). Alternatively, Hubner (1958), Walstad *et al.* (1970) and Schabel (1978) also suggest that the failure to infect the host is because of antibiosis caused by the microbiota (*e.g.* other fungi and bacteria) living on the cuticular surface of the host. Toledo *et al.* (2009) have observed, under scanning electron microscopy, the presence of bacillus-like bacteria on the cuticular surface of *Peregrinus maidis* (Ashmead) (Hemiptera: Delphacidae), an importat pest of maize, treated with *B. bassiana* and *M. anisopliae*. Fungal germination on the cuticular surface was observed at 24 hours post-inoculation, but the germination percentages were low confronted with 95.5% and 100% *in vitro* for *B. bassiana* and *M. anisopliae* respectively. Recently Toledo *et al.* (unpublished results) have identified three species of bacteria (*Bacillus licheniformis, B. pumilus* and *B. subtilis*) isolated from cuticular surface of two important maize pests, *Dalbulus maidis* (De Long and Wolcott) (Hemiptera: Cicadellidae) and *Delphacodes kuscheli* Fennah (Hemiptera: Delphacidae), which also revealed that several strains of these bacteria were *in vitro* antagonistic against the growth of *B. bassiana*, with percentages of growth inhibition ranging between 40% to 83%. On the other hand, the observations of Dillon and Charnley (1986) confirmed the existence of antifungal toxins produced by the gut bacterial flora of the desert locust *S. gregaria*. Likewise, bioassays carried out by Ansari *et al.* (2005) revealed that *Photorabdus luminescens* (a symbiotic bacteria of entomopathogenic nematodes), was antagonistic to *M. anisopliae*, *B. bassiana*, *B. brongniartii* (Saccardo) Petch and *I. fumosorosea* by inhibiting their growth and conidial production.

2.5. HOST DEFENSES

Host defenses are often ignored, even though they affect parameters essential to pathogen and host evolution such as transmission and longevity (Moore *et al.,* 1992).

2.5.1. Behavioral Fever

Behavioral fever is the increase of body temperature in infected insects above that normally occurring in uninfected ones. Infected insects achieve this by looking for places in the environment with a higher temperature, aiming for death or suppression of the pathogen or a delay in the time until death. Fever is a common host response to many pathogens. An example of true behavioral fever is that of house flies, *Musca domestica* L. (Diptera: Muscidae), infected with *Entomophthora schizophorae* Keller and Wilding or *E. muscae* (Cohn) Fresen (Zygomycota: Entomophthorales). Behavioral fever in fungal-infected house flies was first documented by Olesen (1984). Further studies demonstrated that the survival time of infected house flies increased if they were exposed to high temperatures shortly after exposure to fungus inoculum (Watson *et al.,* 1993). This phenomenon is associated not only with entomophthoralean fungi but also with species of hypocrealean attacking grasshoppers and locusts. Insect species that can regulate body temperatures restrict pathogen growth without changing their normal behavior (Blanford and Thomas, 1999 a and b; 2000; Inglis *et al.,* 1996; 1997). This effect is enhanced significantly if they also exhibit behavioral fever whereby they change their behavior in direct response to disease challenge by elevating even more their body temperature (Blanford *et al.,* 1998; Inglis *et al.,* 1996; Ouedraogo *et al.,* 2003; Ouedraogo *et al.,* 2004). In recent studies, Springate and Thomas (2005) characterized and determined the thermal biology of the grasshopper *Chorthippus parallelus* (Zetterstedt) (Orthoptera: Acrididae) and the influence of thermoregulatory behavior for resistance against the temperate *B. bassiana* and the tropical *M. anisopliae* var. *acridum* fungal pathogens, respectively. Their results show that grasshopper is an active behavioral thermoregulator, with a preferred temperature range of 32–35° C. Normal thermoregulation was found to reduce virulence and spore production of *B. bassiana* but did not appear to affect *M. anisopliae* var. *acridum*. Their results suggest that the effects of temperature on host resistance depend on the thermal sensitivity of the pathogen and, in this case, derive from direct effects of temperature on pathogen growth rather than indirect effects mediated by host immune response.

2.5.2. Grooming Behavior

In general, host defensive reactions to pathogens are considered from an immunological perspective (Bidochka and Hajek, 1998; Gillespie *et al.,* 2000; Kedra and Bogús, 2006; Bogús *et al.,* 2007). However, there have been many reports of host-mediated behavior reducing pathogen transmission. Host behaviors related to increased resistance to insect pathogens include increased grooming and nest cleaning, secretion of antibiotics, pathogen avoidance, disposal of infected individuals, and relocation of the entire colony (Roy *et al.,* 2006).

Increased grooming in response to fungal pathogens has been documented widely in both solitary and eusocial insects (Oi and Pereira, 1993; Siebeneicher *et al.,* 1992). Removal of fungal conidia by allo-grooming and mutual-grooming can be highly effective. The number of *B. bassiana* conidia on the integument of both larval and adult red imported fire ants, *Solenopsis invicta* (Buren) (Hymenoptera: Formicidae) wass significantly reduced by grooming (Oi and Pereira, 1993). Similarly, *Cordyceps* spores were removed by another species of ant *Cephalotes atratus* L. (Hymenoptera: Formicidae). In recent studies Toledo *et al.* (2007), also observed grooming behavior in individuals of *D. maidis*. Analyzing comparative pathogenicity tests against adults of *P. maidis*, *D. kuscheli* (Hemiptera: Delphacidae), and *D. maidis*, these authors detected that both delphacid species were significantly more susceptible to *B. bassiana* than the cicadellid one. These authors observed that in contrast to the other two species of insects, *D. maidis* individuals tried to clean their body by repeatedly rubbing the third pair of legs on its tegmine. On the other hand, other studies indicate that termites detect conidia of the fungal pathogen *M. anisopliae* and exhibit a striking vibratory display (Rosengaus *et al.,* 1999) that might directly reduce the number of propagules attached to the insect cuticle, thus decreasing the probability of fungal infection. Additionally, Yanagawa and Shimizu (2005; 2007) suggested that the mutual grooming behavior by termite workers is very effective to protect themselves against *M. anisopliae* infection. Recently Yanagawa *et al.* (2008) reported that the termite *Coptotermes formosanus* Shiraki (Isoptera: Rhinotermitidae) were highly susceptible to entomopathogenic fungi, *I. fumosorosea*, *B. brongniartii* and *M. anisopliae* when reared individually, while termites reared in groups were highly resistant to those fungi. Besides, their quantitative assays revealed a significant difference in the number of conidia attached among three entomopathogenic fungi. The conidia of *B. brongniartii* and *I. fumosorosea* bound to termite cuticles more effectively than *M. anisopliae* conidia. Their results suggested that mutual grooming behavior is more effective than self-grooming in the removal of conidia from cuticles of their nestmates. Consideration of hygienic behaviors can be extended beyond grooming to encompass nest cleaning. For example, social insects use a number of hygienic behaviors to reduce fungal transmission within the colony. Necrophoresis, or cadaver removal, is a common behavior of social insects in response to most pathogens and, while this could contribute to reduce infection of insects inside the colony, could increase intercolony transmission (Roy *et al.,* 2006). Imported red fire ants, *S. invicta*, have also been observed to bury nestmates infected with *B. bassiana*, thereby reducing probability of transmission

(Pereira and Stimac, 1992). Lastly, social insects avoid in some circumstances areas with high inoculum density within the nest or establish new nest sites (Oi and Pereira, 1993).

2.5.3. Protection by Endosymbiont

A number of invertebrates have intimate associations with microorganism symbionts such as bacteria or yeast-like organisms (Sasaki *et al.,* 1996; Marzorati *et al.,* 2006; Gruwell *et al.,* 2007; Chanbusarakum and Ullman, 2008). Over the past few years, several microorganisms have been shown to influence their hosts' s ability to defend against natural enemies (Gil-Turnes *et al.,* 1989; Oliver *et al.,* 2003; Bensadia *et al.,* 2006). For example, aphids possess a range of facultative bacterial endosymbionts that may help in defense against parasitoids (Oliver *et al.,* 2003) and influence the aphids' ability to use different plant species as hosts (Tsuchida *et al.,* 2004). Related to this, Scarborough *et al.* (2005) show that the bacterium *Regiella insecticola*, one of the most common facultative symbionts of pea aphid *Acyrthosiphon pisum* (Harris) (Hemiptera: Aphididae), contributes to aphid resistance to the fungus *P. neoaphidis* and lowers the transmission rate of the fungus, increasing the aphids' ability to survive exposure to the fungal pathogen and reduce fungal sporulation on infected aphids. Recently, Yoder *et al.* (2008) suggest that the endosymbiotic fungus *Scopulariopsis brevicaulis* Bainier (Ascomycota : Microascales) provides protection to the American dog tick *Dermacentor variabilis* (Say) (Arachnida: Ixodidae) against *M. anisopliae*. Thus, the *S. brevicaulis*/tick association appears to be a mutualistic symbiosis.

CONCLUSIONS

In this chapter the compatibility of pathogenic fungi with fungicides and pesticides for their use as biocontrol agents against insect pests was reviewed. Additionally, the factors that promote and reduce fungal transmission were also reviewed. The following points summarize the state of our knowledge about some aspects that need to be taken into account when planning an IPM to control insect pests.

- The use of selective insecticides and fungicides in recommended concentrations, in association with entomopathogens can increase the efficiency of control, minimizing environmental contamination hazards and the the risk of insect resistance.
- Beside dispersal by wind currents and rain splash from soil surfaces, insects such as predators, parasitoids, and migratory alates could potentially contribute to spread fungus inoculum.
- Heat, cold, and low humidity restrict the use of entomopathogenic fungi as agents for biological control of insects. Tolerant pathogens to adverse climatic conditions found in the ecosystem involved are needed for their effective use in IPM strategies.
- Rainfall may be critical to the horizontal transmission of fungal pathogens due to in the transmission of fungal pathogens the production of conidia from host cadavers and the ability of conidia to initiate new infections are two crucial factors.

- For some insect-microbe systems, the failure of fungi to invade the insect cuticle has been attributed to the presence of inhibitory compounds present in the cuticular surface of the host, such as phenols, quinones and lipids, additionally to the antibiotic effect of the cuticle-inhabiting microbiota.
- Insects can defend themselves actively from pathogen attack, using mechanisms such as fever and grooming.
- Many insects have symbiotic associations with bacteria or yeast-like microorganisms, that may help in defense against parasitoids or pathogens.

Over the time there have been many attempts to determine the potential of entomopathogenic fungi to reduce the damage caused by insect pests. However, most these efforts have ended up in failure andvery few of them became products found on the market at the present time. The most probable causes for this could have been insufficient knowledge of the complex host-pathogen interaction, and hence to overcome the constraints for successful implementation in field as opposed to laboratory conditions. Fortunately, the amount of research about host-pathogen and pathogen-environment interactions has been increasing, providing information about how to increase the use of the entomopathogenic fungi as alternative to the conventional chemical insecticides.

ACKNOWLEDGMENTS

The author wish to thank Dr. Pablo Carpane (Monsanto Argentina, SAIC) for critical review of the manuscript.

REFERENCES

Alizadeh, A., Samih, M.A and Izadi, H. 2007a. Compatibility of *Verticillium lecani* (Zimm.) with several pesticides. *Communication in Agricultural and Applied Biological Science*, 72(4): 1011-5.

Alizadeh, A., Samih, M.A., Khezri, M. and Riseh, R.S. 2007b. Compatibility of *Beauveria bassiana* (Bals.) Vuill. with Several Pesticides. *International Journal of Agriculture and Biology*, 9 (1): 31-34.

Alves, S.B. and Lecuona, R.E. 1998. Epizootiologia aplicada ao controle microbiano de insetos. In: Alves, S.B. (Ed.), *Controle microbiano de insetos*. Piracicaba: FEALQ, pp. 97-170.

Anderson, R.M. 1982. Theoretical basis for the use of pathogens as biological agents of pest species. *Parasitology*, 84 (4): 3-33.

Anderson, T.E. and Roberts, D.W. 1983. Compatibility of *Beauveria bassiana* strain with formulations used in Colorado Potato Beetle (Coleoptera: Chrysomelidae) control. *Journal of Economic Entomology*, 76 (6): 1437-1441.

Ansari, M.A., Tirry, L. and Moens, M. 2005. Antagonism between entomopathogenic fungi and bacterial symbionts of entomopathogenic nematodes. *BioControl*, 50 (3): 465-475.

Arthurs, S.P., Thomas, M.B. and Lawton, J.L. 2001. Seasonal patterns of persistence and infectivity of *Metarhizium anisopliae* var. *acridum* in grasshopper cadavers in the Sahel. *Entomologia Experimentalis et Applicata*, 100 (1): 69-76.

Bahiense, T.C., Fernandes, E.K., and Bittencourt, V.R. 2006. Compatibility of the fungus *Metarhizium anisopliae* and deltamethrin to control a resistant strain of *Boophilus microplus* tick. *Veterinary Parasitology*, 141(3-4): 319-24.

Baverstock, J., Baverstock, K.E., Clark, S.J. and Pell, J.K. 2008a. Transmission of *Pandora neoaphidis* in the presence of co-occurring arthropods. *Journal of Invertebrate Pathology*, 98 (3): 356-359.

Baverstock, J., Clark, S.J. and Pell, J.K. 2008b. Effect of seasonal abiotic conditions and field margin habitat on the activity of *Pandora neoaphidis* inoculum on soil. *Journal of Invertebrate Pathology,* 97 (3): 282-290.

Bell, A.A. and Wheeler, M.H.. 1986. Biosynthesis and functions of fungal melanins. *Annual Review of Phytopathology,* 24: 411-451.

Bensadia, F., Boudreault, S., Guay, J.F., Michaud, D. and Cloutier, C. 2006. Aphid clonal resistance to a parasitoid fails under heat stress. *Journal of Insect Physiology*, 52 (2): 146-157.

Bidochka, M.J. and Hajek, A.E. 1998. A nonpermissive Entomophthoralean fungal infection increases activation of insect prophenoloxidase. *Journal of Invertebrate Pathology*, 72 (3): 231-238.

Blanford, S. and Thomas, M.B. 1999a. Host thermal biology: the key to understanding host-pathogen interactions and microbial pest control?. *Agricultural and Forest Entomology*, 1 (3): 195-202.

Blanford, S., Thomas, M.B. and Langewald, J. 1998. Behavioural fever in the Senegalese grasshopper *Oedaleus senegalensis*, and its implications for biological control using pathogens. *Ecological. Entomology*, 23 (1): 9-14.

Blanford, S. and Thomas, M.B. 1999b. Role of thermal biology in disease dynamics. *Aspects of Applied Biology*, 53: 73-82.

Blanford, S. and Thomas, M.B. 2000. Thermal behaviour of two acridid species: effects of habitat and season on body temperature and the potential impact on biocontrol with pathogens. *Environmental Entomology*, 29 (5): 1060–1069.

Bogús, M.I., Kedra, E., Bania, J., Szczepanik, M., Czygier, M., Jablonski, P., Pasztaleniec, A., Samborski, J., Mazgajska, J. and Polanowski, A. 2007. Different defense strategies of *Dendrolimus pini*, *Galleria mellonella*, and *Calliphora vicina* against fungal infection. *Journal of Insect Physiology*, 53 (9): 909-922.

Boucias, D.G. and Pendland, J.C. 1991. Attachment of mycopathogens to cuticle. The initial event of mycoses in arthropod hosts. In: Cole, G.T. and Hoch, H.C. (Eds.). *The fungal spore and disease initiation in plants and animals*. Plenum Publishing Corporation, New York, pp. 101-127.

Braga, G.U.L., Flint, S.D., Miller, C.D., Anderson, A.J. and Roberts, D.W. 2001. Variability in response to UV-B among species and strains of *Metarhizium* isolated from sites at latitudes from 61° N to 54° S. *Journal of Invertebrate Pathology*, 78 (2): 98-108.

Bruck, D.J. and Lewis, L.C. 2002a. *Carpophilus freemani* (Coleoptera: Nitidulidae) as a vector of *Beauveria bassiana*. *Journal of Invertebrate Pathology*, 80 (3): 188-190.

Bruck, D.J. and Lewis, L.C. 2002b. Rainfall and crop residue effects on soil dispersion and *Beauveria bassiana* spread to corn. *Applied Soil Ecology*, 20 (3): 183-190.

Burges, H.D. 1981. Microbial control of pests and plant diseases 1970-1980. Academic Press, New York.

Cantwell, G.E. 1974. Insect diseases. Marcel Decker, New York.

Carruthers, R.I. and Soper, R.S. 1987. Fungal diseases. In: Fuxa, J. R. and Tanada, Y. (Eds.), *Epizootiology of Insect Diseases*. John Wiley and Sons, New York, pp. 357-416.

Chanbusarakum, L. and Ullman, D. 2008. Characterization of bacterial symbionts in *Frankliniella occidentalis* (Pergande), Western flower thrips. *Journal of Invertebrate Pathology*, 99 (3): 318-325.

Crisan, E.V. 1973. Current concepts of thermophilism and the thermophilic fungi. Mycologia, 65 (5): 1171-1198.

Cuthbertson, A.G., Walters, K.F., and Deppe, C. 2005. Compatibility of the entomopathogenic fungus *Lecanicillium muscarium* and insecticides for eradication of sweetpotato whitefly, *Bemisia tabaci*. *Mycopathologia*, 160 (1): 35-41.

Davidson, E.W. 1981. Pathogenesis of invertebrate microbial diseases. Allanheld Osmun Publications.

Dedryver, C.A. 1978. Facteurs de limitation des populations d' *Aphis fabae* dans l' ouest de la France. III. Répartition et incidence des différentes espèces d' *Entomophthora* dans les populations. *Entomophaga*, 23 (2): 137-151.

Dillon, R.J. and Charnley, A.K. 1986. Inhibition of *Matarhizium anisopliae* by the gut bacterial flora of the desert locust, *Schistocerca gregaria*: Evidence for an antifungal toxin. *Journal of Invertebrate Pathology*, 47 (3): 350-360.

Dowd, P.F. and Vega, F.E. 2003. Autodissemination of *Beauveria bassiana* by sap beetles (Coleoptera: Nitidulidae) to overwintering sites. *Biocontrol Science and Technology*, 13 (1): 65-75.

Dromph, K.M. 2001. Dispersal of entomopathogenic fungi by collembolans. *Soil Biology and Biochemistry*, 33 (15): 2047-2051.

Dromph, K.M. 2003. Collembolans as vectors of entomopathogenic fungi. *Pedobiologia*, 47 (3): 245-256.

Duarte, A., Menendez, J.M. and Triguero, N. 1992. Estudio preliminar sobre la compatibilidad de *Metarhizium anisopliae* com algunos plaguicidas químicos. *Revista Baracoa,* 22 (2): 31-39.

Fargues, J., Goettel, M.S., Smits, N., Ouedraogo, A., Vidal, C., Lacey, L.A., Lomer, C.J. and Rougier, M. 1996. Variability in susceptibility to simulated sunlight of conidia among isolates of entomopathogenic Hyphomycetes. *Mycopathologia*, 135 (3): 171-181.

Fargues, J., Goettel, M.S., Smits, N., Ouedraogo, A. and Rougier, M. 1997. Effect of temperature on vegetative growth of *Beauveria bassiana* isolates from different origins. *Mycologia*, 89 (3): 383-392.

Feng, M.G., Chen, C. A and Chen, B. 2004. Wide dispersal of aphid-pathogenic Entomophthorales among aphids relies upon migratory alates. *Environmental Microbiology*, 6 (5): 510-516.

Feng, M.G., Chen, C., Shang, S.W., Ying, S.H., Shen, Z.C. and Chen, X.X. 2007. Aphid dispersal flight disseminates fungal pathogens and parasitoids as natural control agents of aphids. *Ecological Entomology*, 32 (1): 97-104.

Fernandes, E.K.K., Range, D.E.N., Moraes, A.M.L., Bittencourt, V.R.E.P. and Roberts, D.W. 2008. Cold activity of *Beauveria* and *Metarhizium*, and thermotolerance of *Beauveria*. *Journal of Invertebrate Pathology*, 98 (1): 69-78.

Ferron, P. 1985. Fungal Control. In: Comprehensive insect physiology, biochemistry and pharmacology (Kerkut, G.A. and Gilbert, L.I. (Eds.),. Pergamon Press, Oxford, pp. 313-346.

Gil-Turnes, M.S., Hay, M.E. and Fenical, W. 1989. Symbiotic marine bacteria chemically defend crustacean embryos from a pathogenic fungus. *Science*, 246 (4926): 116-118.

Gillespie, J.P., Burnett, C. and Charnley, A.K. 2000. The immune response of the desert locust *Schistocerca gregaria* during mycosis of the entomopathogenic fungus, *Metarhizium anisopliae* var *acridum. Journal of Insect Physiology*, 46 (4): 429-437.

Goettel, M.S., Poprawski, T.J., Vandenverg, J.D., Li, Z. and Roberts, D.W. 1990. Safety to nonterget invertebrates of fungal biocontrol agents. In: Laird, M., Lacey, L.A. and Davidson, E.W. (Eds.), *Safety of Microbial Insecticides.* CRC Press, Boca Raton, FL, pp. 209-232.

Goettel, M.S. and Inglis, D.G. 1997. Fungi: Hyphomycetes. In: Lacey, L. (Ed.), *Manual of Techniques in Insect Pathology.* Academic Press, London, UK, pp. 213-249.

Goettel, M.S., Hajek, A.E., Siegel, J.P. and Evans, H.C. 2001. Safety of Fungal Biocontrol Agents. In: Butt, T., Jackson, C. and Magan, N. (Eds.), *Fungal Biocontrol Agents – Progress, Problems and Potential.* CABBI Press, Wallingford, UK, pp. 347-375.

Goettel, M.S. and Hajek, A.E. 2001. Evaluation of nontarget effects of pathogens used for management of arthropods. In: Wajnberg, E., Scott, J.K. and Quimy, P.C. (Eds.), *Evaluating Indirect Ecological Effects of Biological Control.* CABBI Press, Wallingford, UK, pp. 81-97.

Gruwell, M.E., Morse, G.E. and Normark, B.B. 2007. Phylogenetic congruence of armored scale insects (Hemiptera: Diaspididae) and their primary endosymbionts from the phylum Bacteroidetes. *Molecular Phylogenetics and Evolution*, 44 (1): 267-280.

Guerzoni, M.E., Lanciotti, R. and Cocconcelli, P.S. 2001. Alteration in cellular fatty acid composition as a response to salt, acid, oxidative and thermal stresses in *Lactobacillus helveticus. Microbiology*, 147 (8): 2255-2264.

Gustafsson, M. 1969. On species of the genus *Entomophthora* Fres. In Sweden III. Possibility of usage in biological control. Lantbrukshogskolans Annaler, 35: 235-274.

Hall, R.A. and Papierok, B. 1982. Fungi as biological control agents of arthropods of agricultural and medical importance. *Parasitology*, 84 (4): 205-240.

Harper, J.D. 1987. Applied epizootiology: microbial control of insects. In: Fuxa, J.R. and Tanada, Y. (Eds.), *Epizootiology of Insect Diseases*. Wiley, New York, pp. 473–496.

Howard, R.W. and Lord, J.C. 2003. Cuticular lipids of the booklouse, *Liposcelis bostrychophila*: Hydrocarbons, aldehydes, fatty acids, and fatty acid amides. *Journal of Chemical Ecology*, 29 (3): 615-627.

Hubner, J. 1958. Untersuchungen zur Physiologie insektentötender Pilze. Arch. *Mikrobiol.* 29: 257-276.

Inglis, G.D., Goettel, M.S. and Johnson, D.L. 1993. Persistence of the entomopathogenic fungus, *Beauveria bassiana*, on phylloplanes of crested wheatgrass and alfalfa. *Biological Control*, 3 (4): 258-270.

Inglis, G.D., Goettel, M.S. and Johnson, D.L. 1995. Influence of ultraviolet light proctectants on persistence of the entomopathogenic fungus, *Beauveria bassiana. Biological Control*, 5 (4): 581-590.

Inglis, G.D., Johnson, D.L. and Goettel, M.S. 1996. Effects of temperature and thermoregulation on mycosis by *Beauveria bassiana* in grasshoppers. *Biological Control*, 7 (2) :131-139.

Inglis, G.D., Johnson, D.L., Cheng, K-J. and Goettel, M.S. 1997. Use of pathogen combinations to overcome the constraints of temperature on entomopathogenic hyphomycetes against grasshoppers. *Biological Control*, 8 (2): 143-152.

James, R.R., Buckner, J.S. and Freeman, T.P. 2003. Cuticular lipids and silverleaf whitefly stage affect conidial germination of *Beauveria bassiana* and *Paecilomyces fumosoroseus*. *Journal of Invertebrate Pathology*, 84 (2): 67-74.

Kedra, E. and Bogús, M.I. 2006. The influence of *Conidiobolus coronatus* on phagocytic activity of insect hemocytes. *Journal of Invertebrate Pathology*, 91 (1): 50-52.

Keller, S. and Zimmerman, G. 1989. Mycopathogens of soil insects. In: Wilding, N., Collins, N.M., Hammond, P.M. and Webber, J.F. (Eds.), *Insect-Fungus Interactions*. Academic Press, London, UK, pp. 239-270.

Khalil, S.K., Shah, M.A. and Naeem, M. 1985. Laboratory studies on the compatibility of the entomopathogenic fungus *Verticillium lecanii* with certain pesticides. *Agriculture, Ecosystems and Environment*, 13 (3-4): 329-334.

Knipling, E.F. 1979. The basic principles of insect population supression and management. USDA Agriculture Handbooks No. 512.

Lacey, L.A. and Goettel, M.S. 1995. Current developments in microbial control of insect pests and prospects for the early 21st century. *Entomophaga*, 40 (1): 3-27.

Lord, J.C. and Howard, R.W. 2004. A Proposed Role for the Cuticular Fatty Amides of *Liposcelis bostrychophila* (Psocoptera : Liposcelidae) in Preventing Adhesion of Entomopathogenic Fungi with Dry-conidia. *Mycopathologia*, 158 (2): 211-217.

Malo, A.R. 1993. Estudio sobre la compatibilidad del hongo *Beauveria bassiana* (Bals.) Vuill. con formulaciones comerciales de fungicidas e inseticidas. *Revista Colombiana de Entomologia,* 19: 151-158.

Marzorati, M., Alma, A., Sacchi, L., Pajoro, M., Palermo, S., Brusetti, L., Raddadi, N., Balloi, A., Tedeschi, R., Clementi, E., Corona, S., Quaglino, F., Bianco, P.A., Beninati, T., Bandi, C. and Daffonchio, D. 2006. A novel bacteroidetes symbiont is localized in *Scaphoideus titanus*, the insect vector of Flavescence Dorée in *Vitis vinifera*. *Applied and Environmental Microbiology*, 72 (2): 1467-1475.

Meadow, R., Vandenberg, J.D. and Shelton, A.M. 2000. Exchange of inoculum of *Beauveria bassiana* (Bals.) Vuill. (Hyphomycetes) between adult flies of the cabbage maggot *Delia radicum* L. (Diptera: Anthomyiidae). *Biocontrol Science and Technology*, 10 (4): 479-485.

Meyling, N.V. Pell, J.K. and Eilenberg, J. 2006. Dispersal of *Beauveria bassiana* by the activity of nettle insects. *Journal of Invertebrate Pathology*, 93 (2): 121-126.

Mier, T., Rosas-López, B., Castellanos-Moguel, J., García-Gutiérrez, K. and Toriello, C. 2004. Efecto de la citrolina sobre la viabilidad y la producción de proteasas y quitinasas del hongo entomopatógeno *Metarhizium anisopliae* var. *acridum*. Revista Mexicana de *Microbiología*, 19: 113-115.

Moino Jr., A.R. and Alves, S.B. 1998. Efeito de Imidacloprid e Fipronil sobre *Beauveria bassiana* (Bals.) Vuill. e *Metarhizium anisopliae* (Metsch.) Sorok. e no comportamento de limpeza de *Heterotermes tenuis* (Hagem). *Anais da Sociedade Entomológica do Brasil*, 27 (4): 611-619.

Moore D., Reed M., Le Patourel G., Abraham Y.J. and Prior C. 1992. Reduction of feeding by the desert locust, *Schistocerca gregaria*, after infection with *Metarhizium flavoviride*. *Journal of Invertebrate Pathology*, 60: 304-307.

Morley-Davies, J. Moor, D. and Prior, C. 1995. Screening of *Metarhizium* and *Beauveria* spp. Conidia with exposure to simulated sunlight and range of temperatures. *Mycological Research*, 100 (1): 31-38.

Neves, P.M.O.J., Hirose, E., Tchujo, P. and Moino Jr, A. 2001. Compatibility of Entomopathogenic Fungi with Neonicotinoid Insecticides. *Neotropical Entomology*, 30 (2): 263-268.

Oi, D.H. and Pereira, R.M. 1993. Ant behaviour and microbial pathogens (Hymenoptera: Formicidae). *Florida Entomologist*, 76 (1): 63-74.

Olesen, U.S. 1984. Effect of humidity and temperature on *Entomophthora muscae* infecting the house fly, *Musca domestica*, and the increase of survival of the fly by behavioral fever. MSc thesis, University of Copenhagen, Denmark.

Oliveira, C.N., Neves, P.M.O.J. and Kawazoe, L.S. 2003. Compatibility between the entomopathogenic fungus *Beauveria bassiana* and insecticides used in coffee plantations. *Scientia Agricola*, 60 (4): 663-667.

Oliver, K.M., Russell, J.A., Moran, N.A. and Hunter, M.S. 2003. Facultative bacterial symbionts in aphids confer resistance to parasitic wasps. *Proceedings of the National Academy of Sciences of the United States of America*, 100 (4): 1803-1807.

Ouedraogo, R.M., Cusson, M., Goettel, M.S. and Brodeura, J. 2003. Inhibition of fungal growth in thermoregulating locusts, *Locusta migratoria*, infected by the fungus *Metarhizium anisopliae* var *acridum*. *Journal of Invertebrate Pathology*, 82 (2): 103-109.

Ouedraogo, R.M., Goettel, M.S. and Brodeur, J. 2004. Behavioral thermoregulation in the migratory locust: a therapy to overcome fungal infection. *Oecologia*, 138 (2): 312-319.

Pell, J.K., Pluke, R., Clark, S.J., Kenward, M.G., and Alderson, P.G. 1997. Interactions between two aphid natural enemies, the entomopathogenic fungus *Erynia neoaphidis* Remaudière and Hennebert (Zygomycetes : Entomophthorales) and the predatory beetle *Coccinella septempunctata* L. (Coleoptera: Coccinellidae). *Journal of Invertebrate Pathololgy*, 69 (3): 261–268.

Pell, J.K. and Vandenberg, J.D. 2002. Interactions among the aphid *Diuraphis noxia*, the entomopathogenic fungus *Paecilomyces fumosoroseus* and the coccinellid *Hippodamia convergens*. *Biocontrol Science and Technology*, 12 (2): 217-224.

Pereira, R.M. and Stimac, J.L. 1992. Transmission of *Beauveria bassiana* within nests of *Solenopsis invicta* (Hymenoptera : Formicidae) in the laboratory. *Environmental Entomology*, 21 (6): 1427-1432.

Pupin, A.M., Messias, C.L., Piedrabuena, A.E. and Roberts, D.W. 2000. Total lipids and fatty acids of strains of *Metarhizium anisopliae*. *Brazilian Journal of Microbiology*, 31 (2): 121-128.

Quintela, E.D. and McCoy, C.W. 1998. Synergistic effect of imidacloprid and two entomopathogenic fungi on the behavior and survival of larvae of *Diaprepes abbreviatus* (Coleoptera : Curculionidae) in soil. *Journal of Economical Entomology*, 91 (1): 110-122.

Rangel, D.E.N., Anderson, A.J. and Roberts, D.W. 2006. Growth of *Metarhizium anisopliae* on non-preferred carbon sources yields conidia with increased UV-B tolerance. *Journal of Invertebrate Pathology*, 93 (2): 127–134.

Roberts, D.W. and Campbell, A.S. 1977. Stability of entomopathogenic fungi. *Miscellaneous Publications of the Entomological Society of America*, 10 (3): 19-76.

Roberts, D.W. and Hajek, A.E. 1992. Entomopathogenic fungi as bioinsecticides. In: Leatham, G.F. (Ed.), *Frontiers in Industrial Mycology*. Chapman and Hall, New York, pp. 144–159.

Rosengaus, R.B., Jordan, C., Lefebvre, M.L. and Traniello, J.F.A. 1999. Pathogen alarm behavior in a termite: A new form of communication in social insects. *Naturwissenschaften*, 86 (11): 544-548.

Roy, H.E., Pell, J.K., Clark, S.J. and Alderson, P.G. 1998. Implications of predator foraging on aphid pathogen dynamics. *Journal of Invertebrate Pathology*, 71 (3): 236-247.

Roy, H.E., Pell, J.K. and Alderson, P.G. 2001. Targeted dispersal of the aphid pathogenic fungus *Erynia neoaphidis* by the aphid predator *Coccinella septempunctata*. *Biocontrol Science and Technology,* 11 (1): 99-110.

Roy, H.E., Steinkraus, D.C., Eilenberg, J., Hajek, A.E. and Pell, J.K. 2006. Bizarre interactions and endgames: Entomopathogenic fungi and their arthropod hosts. *Annual Review of Entomology*, 51: 331-357.

Samson, P.R., Milner, R.J., Sanderi, E.D. and Bullard, G.K. 2005. Effect of fungicides and insecticides applied during planting of sugarcane on viability of *Metarhizium anisopliae* and its efficacy against white grubs. *BioControl*, 50 (1): 151-163.

Sasaki, T., Kawamura, M. and Ishikawa, H. 1996. Nitrogen recycling in the brown planthopper, *Nilaparvata lugens*: Involvement of Yeast-like endosymbionts in uric acid metabolism. *Journal of Insect Physiology*, 42 (2): 125-129.

Scarborough, C.L., Ferrari, J. and Godfray, H.C.J. 2005. Aphid Protected from Pathogen by Endosymbiont. *Science*, 310 (5755): 1781.

Schabel, H.G. 1978. Percutaneus infection of *Hylobius pales* by *Metarhizium anisopliae*. *Journal of Invertebrate Pathology*, 31 (2): 180-187.

Siebeneicher, S.R., Vinson, S.B. and Kenerley, C.M. 1992. Infection of the red imported fire ant by *Beauveria bassiana* through various routes of exposure. *Journal of Invertebrate Pathology*, 59 (3): 280-285.

Singaravelan, N., Grishkan, I., Beharav, A., Wakamatsu, K., Ito, S. and Nevo, E. 2008. Adaptive Melanin Response of the Soil Fungus *Aspergillus niger* to UV Radiation Stress at "Evolution Canyon", Mount Carmel, Israel. *PLoS ONE*, 3(8): 1-5.

Shimazu, M., Sato, H. and Maehara, N. 2002. Density of the entomopathogenic fungus, *Beauveria bassiana* Vuillemin (Deuteromycotina: Hyphomycetes) in forest air and soil. *Applied Entomology and Zoology*, 37 (1): 19-26.

Smith, R.J. and Grula, E.A. 1981. Nutritional requirements for conidial germination and hyphal growth of *Beauveria bassiana*. *Journal of Invertebrate Pathology*, 37 (2): 222-230.

Smits, N., Rougier, M., Fargues, J., Goujet, R. and Bonhomme, R. 1996. Inactivation of *Paecilomyces fumosoroseus* conidia by diffuse and total solar radiation. *FEMS Microbiology Ecology*, 21 (3): 167-173.

Springate, S. and Thomas, M.B. 2005. Thermal biology of the meadow grasshopper, *Chorthippus parallelus*, and the implications for resistance to diseas. *Ecological Entomology,* 30 (6): 724–732.

Steinkraus, D.C:, Kring, T.J. and Tugwell, N.P. 1991. *Neozygites fresenii* in *Aphis gossypii* on cotton. *Southwestern Entomologist*, 16: 118-122.

Steinkraus, D.C. and Slaymaker, P.H. 1994. Effect of temperature and humidity on formation, germination, and infectivity of conidia of *Neozygites fresenii* (Zygomycetes : Neozygitaceae) from *Aphis gossypii* (Homoptera : Aphididae). *Journal of Invertebrate Pathology*, 64 (2): 130-137.

Szafranek, B., Maliñski, E., Nawrot, J., Sosnowska, D., Ruszkowska, M., Pihlaja, K., Trumpakaj, Z. and Szafranek, J. 2001. In Vitro effects of cuticular lipids of the aphids *Sitobion avenae, Hyalopterus pruni* and *Brevicoryne brassicae* on growth and sporulation of the *Paecilomyces fumosoroseus* and *Beauveria bassiana*. *ARKIVOC*, 3: 81-94

Thevelein, J.M. 1984. Regulation of trehalose mobilization in fungi. *Microbiological Reviews,* 48 (1): 42–59.

Thoizon, G. 1970. Spécificité de parasitisme des aphides par les Entomophthorales. *Annales de la Société Entomologique de France (N.S.),* 6: 517-562.

Toledo, A.V., Remes Lenicov, A.M.M. de. and López Lastra, C.C. 2007. Pathogenicity of fungal isolates (Ascomycota : Hypocreales) against *Peregrinus maidis, Delphacodes kuscheli* (Hemiptera Histopathology caused by the entomopathogenic fungi, *Beauveria bassiana* and *Metarhizium anisopliae*, in the adult planthopper, *Peregrinus maidis*, a maize virus vector : Delphacidae), and *Dalbulus maidis* (Hemiptera : Cicadellidae), vectors of corn diseases. *Mycopathologia*, 163 (4): 225-232.

Toledo, A.V., Remes Lenicov, A.M.M. de, López and Lastra, C.C. 2009. Histopathology caused by the entomopathogenic fungi, *Beauveria bassiana* and *Metarhizium anisopliae*, in the adult planthopper, *Peregrinus maidis*, a maize virus vector. *Journal of Insect Science,* 10 (35): 1-10.

Tsuchida, T., Koga, R and Fukatsu, T. 2004 Host plant specialization governed by facultative symbiont. *Science*, 303: 1989-1989.

Vestergaard, S., Cherry, A., Keller, S. and Goettel, M. 2003. Hyphomycete fungi as microbial control agents. In: Hokkanen, H.M.T. and Hajek, A.E. (Eds.), *Environmental Impacts of Microbial Insecticides*. Kluwer Academic Publishers, Dordrecht, The Netherland, pp. 35-62.

Vickers, R.A., Furlong, M.J., White, A. and Pell, J.K. 2004. Initiation of fungal epizootics in diamondback moth populations within a large Weld cage: proof of concept for auto-dissemination. *Entomologia Experimentalis et Applicata*, 111 (1): 7-17.

Walstad, J.D., Anderson, R.F. and Stambaugh, W.J. 1970. Effects of environmental conditions on two species of muscardine fungi (*Beauveria bassiana* and *Metarhizium anisopliae*). *Journal of Invertebrate Pathology*, 16 (2): 221-226.

Watson, D.W., Mullens, B.A. and Petersen, J.J. 1993. Behavioral fever response of *Musca domestica* (Diptera: Muscidae) to infection by *Entomophthora muscae* (Zygomycetes: Entomophthorales). *Journal of Invertebrate Pathology*, 61 (1): 10-16.

Wraight, S.P., Inglis, D.G. and Goettel, M.S. 2007. Fungi. In: Lacey, L.A. and Kaya, H.K. (Eds.), *Field Manual of Techniques in Invertebrate Pathology*. Application and Evaluation of Pathogens for Control of Insects and other Invertebrate Pests. Spreinger, Dordrecht, The Netherlands, pp. 223-248.

Yanagawa, A. and Shimizu, S. 2005. Defense strategy of the termite, *Coptotermes foumosanus* Shiraki to entomopathogenic fungi. *Japanese Journal of Environment Entomology and Zoology*, 16: 17-22.

Yanagawa, A. and Shimizu, S. 2007. Resistance of the termite, *Coptotermes formosanus* Shiraki to *Metarhizium anisopliae* due to grooming. *BioControl*, 52 (1): 75-85.

Yanagawa, A., Yokohari, F. and Shimizu, S. 2008. Defense mechanism of the termite, *Coptotermes formosanus* Shiraki, to entomopathogenic fungi. *Journal of Invertebrate Pathology*, 97 (2): 165-170.

Yendol, W.G. 1968. Factors affecting germination of *Entomophthora* conidia. *Journal of Invertebrate Pathology*, 10 (1): 116-121.

Yoder, J.A., Benoit, J.B., Denlinger, D.L., Tank, J.L. and Zettler, L.W. 2008. An endosymbiotic conidial fungus, *Scopulariopsis brevicaulis*, protects the American dog tick, *Dermacentor variabilis*, from desiccation imposed by an entomopathogenic fungus. *Journal of Invertebrate Pathology*, 97 (2): 119-127.

In: Microbial Insecticides: Principles and Applications
Editors: J. Francis Borgio, K. Sahayaraj, et al.

ISBN: 978-1-61209-223-2
© 2011 Nova Science Publishers, Inc

Chapter 3

MASS PRODUCTION OF ENTOMOPATHOENIC FUNGI

*J. Francis Borgio[1] * and K. Sahayaraj[2]*

1 PG and Research Department of Microbiology, St. Joseph's College (Autonomous),
Bangalore - 560 027, India
2 Crop Protection Research Centre, Department of Advanced Zoology and
Biotechnology, St. Xavier's College (Autonomous), Palayamkottai - 627 002,
Tamil Nadu, India

ABSTRACT

Advances in the application of microbe-based technology in insect pest management assist us to counter problems created by the application of chemical pesticides. Among the microbial pesticides, fungal pesticides are now preferred as they are target specific, ecofriendly, lacking in toxic residue and are economical. There is a rising commercial demand for bio-products based on entomopathogenic fungal pathogens since they could be mass-produced cheaply with considerable shelf life and are easy to store and apply.

The method of culture will largely depend on the fungal species and the type of propagules. This chapter describes about the various types of culture media and substrates used for the production of entomopathogenic fungi, small scale and commercial-scale production of several economically important entomopathogenic fungi, several methods of extraction and drying of the conidial powder, post harvest storage and tests to check the quality of the products.

The chapter also deals with the current status of commercial fungal biocontrol agents with particular emphasis on the products available.

3.1. INTRODUCTION

For over 60 years, chemical pesticides have been the prevalent tool for insect, weed, and plant disease control. Interest in the use of biological based pest control measures has been brought about by the development of pest resistance to many chemical pesticides coupled

---

* E-mail: jfborgio@gmail.com

with public concerns about the adverse impact of widespread chemical use on human health, food safety, and the environment. Entomopathogenic fungi are being used worldwide for the control of several agricultural importance pests (Ferron, 1985; Mehrotra, 1989; Nasr *et al.,* 1992; Waterhouse, 1998; Wraight *et al.,* 2001; Rao *et al.,* 2003; Mohan *et al.,* 2005; Sahayaraj and Namasivayam, 2008; van Lenteren *et al.,* 2008; Zhang *et al.,* 2010), because they are easy to store and apply, highly virulent (Toledo *et al.,* 2007;), could be used in an inundate manner (Alves and Pereira, 1998), mass-produced cheaply with considerable shelf-life (Khetan, 2001), safer to birds, fish, or mammals (Zimmermann, 1993) and other non-target organisms (Hokkanen and Lynch, 1995; Rosell *et al.,* 2008), widely distributed (Scholte *et al.,* 2003), could be mass-cultured *in vitro* (Jackson *et al.,* 2000; Shah *et al.,* 2007; Mascarin *et al.,* 2010) etc.

Efficient mass production techniques are a prerequisite for the successful field applications of entomopathogenic fungi (Purwar and Sachan, 2006; Shah *et al.,* 2007). Its mass production does not require high-input technology (Prior, 1988). For an efficient and economic large-scale production, the formulation of fungal propagules into a product with an adequate shelf life, and suitable application strategies are fundamentals for the successful development of mycoinsecticides (Abebe, 2002; Shah *et al.,* 2005; Sun and Liu, 2006; Sallam *et al.,* 2007).

The selection of strains with rapid growth and sporulation capacity on nutritionally poor media is important for mass production in an industrial setup (Feng *et al.,* 1994; Luz *et al.,* 2007).

3.2. TYPES OF CULTURE MEDIA FOR THE PRODUCTION OF FUNGAL PROPAGULES

3.2.1. Surface Culture on Solid Media

Most facultative entomogenous fungi will grow on one or more defined or semi-defined agar-based medium (e.g. Czapek-Dox, Sabouraud) or on natural substrates (e.g. wheat, bran, rice, egg yolk, potato pulp). Specialist fungi are usually fastidious on artificial media and are usually best maintained on their respective hosts. A few can be cultured *in vitro,* but require a complex medium.

For example, *Lagenidium giganteum* can be cultivated using a simple medium but requires sterols to induce oosporogenesis (Kerwin *et al.,* 1991). Entomophthoralean fungi grow well on Sabouraud dextrose or maltose agar fortified with coagulated egg yolk and milk (Papierok, 1978; Wilding, 1981). Petri dishes and autoclavable plastic bags are recommended for small and larger-scale production, respectively.

However, other containers such as pans, glass bottles and inflated plastic tubing have been used (Goettel, 1984). Agar-based media are usually used for routine culture. Alternatively, cheaper substrates such as rice or shelled barley can be used in autoclavable bags or other containers, especially when larger amounts of inoculum are required (Aregger, 1992; Jenkins and Thomas, 1996).

Once the fungus has sporulated, conidia are harvested either by washing off using water or a buffer, direct scraping from the substrate surface (e.g. agar), or by sieving (e.g. rice). For some entomophthoralean fungi, the forcibly discharged conidia are allowed to shower directly on the host (Papierok and Hajek, 1997). To obtain conidia virtually free of nutritive substrate contamination, non-cellulolytic fungi can be grown on a semi-permeable membrane such as cellophane (Goettel, 1984).

Pans containing a nutritive substance such as bran are lined with the cellophane, placed in sterile bags, autoclaved, inoculated and incubated. After sporulation has taken place, the membrane with the adhering sporulating fungus is lifted from the nutritive substrate. Conidia can then be scraped from the cellophane surface.

3.2.2. Fermentation on Semi-Solid Media

Production of fungi on semi-solid media involves impregnation of small particles with nutrients. Typically wheat bran is mixed with an inorganic substance such as vermiculite, although other substances can be used to provide a large surface area for growth. The mixture is then steam sterilized and the moisture content adjusted to 50–70%.

The fermentation process takes place either in a bin or a rotating drum through which sterile, moist air is passed. Primary inoculum is usually grown in liquid medium. Toward the end of the fermentation cycle, the moist air is replaced by dry air to reduce the moisture content of the bran and to encourage sporulation. The temperature is controlled by regulating the circulating air temperature.

More recently, nutrient-impregnated membranes have been shown to reduce production costs of *M. anisopliae* conidia (Bailey and Rath, 1994). A range of membranes impregnated with skimmed milk were screened including blotting paper, fly screen, hessian, and gauze-type fabrics.

Sporulation was profuse on Superwipe (an absorbent fibrous material) soaked in skimmed milk (20 g 121) supplemented with sucrose (2 g 121) or dextrose plus potassium nitrate. Spores could be washed off in a similar way to removal of conidia from grain.

3.2.3. Submerged Fermentation

Submerged fermentation can be used for production of blastospores and submerged conidia of selected isolates of entomogenous fungi. Dimorphic filamentous fungi like *M. anisopliae, B. bassiana, Beauveria brongniartii, V. lecanii, Paecilomyces farinosus* and *Nomuraea rileyi* produce relatively thinwalled blastospores in submerged culture that are infectious but difficult to preserve (Ignoffo, 1981).

Blastospores are produced in relatively large quantities during the log phase of growth. Most often they are spherical, oval or rod-shaped single cells which usually germinate within 2–6 h. Although several species of entomogenous fungi produce blastospores, there is considerable intraspecific variation. Some isolates produce blastospores more readily than others.

The culture medium has a profound influence on blastospore production. Blastospores sometimes are indistinguishable from submerged conidia. For example, some isolates of *M. flavoviride, M. anisopliae,* and *Hirsutella thompsonii* will produce conidia-shaped cells in submerged culture occasionally from phialide-like structures (Jenkins and Prior, 1993). Van

Winkelhof and McCoy (1984) noted that of 14 isolates of *H. thompsonii* only one produced true conidia. The others produced conidia-like cells.

3.2.4. Diphasic Fermentation

Diphasic fermentation entails growth of fungi in liquid culture to the end of log phase followed by surface conidiation on a nutrient or inert carrier. This method has been developed for mass production of *B. bassiana* (Bradley *et al.,* 1992) and *M. flavoviride* (Jenkins and Thomas, 1996). A similar approach was used in the production of dry marcescent entomphthoralean mycelium.

3.2.5. Dry Marcescent Process

The development of the dry marcescent process provides a convenient method for production of fungi, especially fastidious species like *Zoophthora radicans*. This process entails the production of the mycelium by submerged fermentation, harvesting by filtration, coating the harvested mycelium with a protective layer of sugar solution and then drying under controlled conditions.

When hydrated, the mycelium quickly sporulates to produce infectious conidia. The dry marcescent process has been used successfully as a source of inoculum for *M. anisopliae, C. clavisporus, B. bassiana, Z. radicans* and *Erynia neoaphidis* (Rombach *et al.,* 1988; Pereira and Roberts, 1990; Krueger *et al.,* 1992)

3.2.6. Parameters for Selection of Isolates

The development of a good mycopesticide relies on the biological properties of the isolate. The following parameters need to be considered in selecting an isolate as a potential microbial control agent: laboratory virulence, field performance, genetic stability, productivity, and stability of conidia in storage, stability in formulation, field persistence (*i. e.* tolerance to environmental factors such as UV, temperature extremes and desiccation), mammalian safety, low environmental impact.

3.2.7. Maintenance of Isolates

A system of maintaining back-up sources of inoculum representing a master culture is vital. Usually maintenance of the isolate is by regular passaging through the host insect and re-isolation, but it is recommend that no more than three sub-cultures on agar should be permitted prior to use in mass production.

The practice of re-isolation through the host maintains the virulence/pathogenicity of the isolate, but does introduce the risk of evolution of fungal strains that differ from the field-collected population of microbial control agents (MCAs). Without the back-up of a master culture, this could lead to inadvertent loss of desirable properties found in the original strain. There are a large number of long-term culture maintenance techniques that can be employed for the preservation of valuable fungal material. These include relatively High-Technology methods such as cryopreservation (liquid nitrogen), lyophilisation (freeze-drying) and storage at -80^0C.

All these methods involve the use of relatively expensive equipment. As a back up to the above, a large stock of 'master' material should be stored. One simple and reliable method is to dry a large number of infected insect cadavers prior to re-isolation of the fungus.

These sporulated cadavers should be dried over silica gel to a constant weight and stored in a freezer in a sealed plastic bag containing silica gel. This material is likely to remain viable for many years and single insects can be removed when a fresh source of inoculum is required.

3.3. COMMERCIAL-SCALE PRODUCTION OF ENTOMOPATHOGENIC FUNGI

3.3.1. Preparation of Liquid Starter Culture

Liquid starter culture is used as an inoculum for the solid substrate. Liquid medium is composed of 2% yeast extract and 4% glucose (but sucrose is often cheaper) and is inoculated with a small square cut from a 10 day old culture of fungi grown on Sabouraud Dextrose Agar (SDA).

Flasks are placed on a rotary shaker (approx. 150 rpm) for 2 – 4 days. The resulting inoculum is diluted 1:40 in sterile water prior to use. Each 600 g bag of rice is inoculated with 25 ml of diluted inoculum.

In the early stages of growth distinct clumps of mycelium may be formed within an otherwise clear broth. This form of growth is less productive than a homogenous colonisation of the culture broth by evenly dispersed mycelium and hyphal fragments.

Furthermore, the thick mycelial 'soup' that is produced by the latter growth form provides a uniform inoculum for coating the rice substrate.

3.3.2. Substrates Used

Many attempts have been made to screen low-cost ingredients to massproduce industrialized biological pesticides. A systematic investigation of fungal nutrition utilization is very much needed to improve mass production and accelerate commercialization. One of the most commonly used media by insect pathologist is sabouraud dextrose agar (SDA) supplemented with yeast extract (SDAY).

Hyphomycetes including *B. bassiana* and *V. lecanii* grow and sporulate well on this medium (Gopalakrishnan, 2001). However, many other media such as consumed, czapeck - dox, malt extract, potato dextrose agar and sabouraud mactose agar (SMA) are also used often for large scale production of conidia. SMYB (Devi *et al.,* 2001) supports maximum spore production in *N. rielyi*

Cheaper nutritive substrates such as cereals grains or bran could also be used for large scale cultivation of entomopathogenic fungi. Sorghum and ragi are suitable cereals for the mass production of *P. fumosoroseus* (Lakshmi *et al.,* 2001). Gopalakrishnan *et al.* (1999) reported that sorghum was the ideal cereal for the mass multiplication of *Paecilomyces farinosus*. Pearl millet and maize also favors fungal growth.

Moistened substrates are autoclaved in mouthed jar, auto cleavable plastic bags or tin trays. Molasses yeast broth supports good growth and highest spore production in *M. anisopliae*, *B. bassiana* and *B. brongniartii*. The use of rice in poly prepaying bags in currently, the widely used method for production of *M. anisopliae*, *M. flavoviridae* and *B. bassiana*. Parboiled autoclaved rice in plastic bags could also be inoculated with blastospore suspension (Jenkins *et al.,* 1998; Mc Coy, 1990). Tincilley *et al.* (2000) reported that carrot was found to be the cheapest and best suitable media for the large-scale production of deuteromycete fungi.

Jack seeds also act as an ideal medium for the mass production of *B. bassiana* and *V. lecanii*. Rich husk also supports the growth and sporulation of *P. fumosoroseus*. Puzari *et al.* (1997) reported that rice husk supplemented with 2% dextrose solution recorded more sporulation of *M. anisopliae*.

Abundance of glucose and minerals in the coconut water, rice wash water, rice boiled water, fishery waste supplemented with dextrose, also enhances the growth and spore production of fungi (Borgio, 2008).

3.3.3. Preparation of Solid Medium

Autoclaved rice, for example provides a nutritive solid support on which conidiation occurs. Dry, unpolished rice is placed in the autoclavable bags (*e.g.* 200 mm x 500 mm - 600 g rice per bag). 500 ml of water is added to the rice in the bags, which are then heat sealed to ¾ of the way across the top.

The water level within the bags comes well above the level of the rice and the bags are loaded upright in to the autoclaves for sterilization for 1 hour. On completion of the autoclave run, the bags are massaged to break up the rice and the open corner of the bag is folded over twice and fixed with a paper clip.

3.3.4. Inoculation of the Medium

The prepared bags are then moved into an air-conditioned room to cool. Once cool, the bags are inoculated with a fixed amount (*e.g.* 25 ml) of diluted inoculum: an automatic dispenser speeds-up the process and reduces contamination.

After inoculation clips are usually removed and the open corner of the bags swabbed with alcohol. The final moisture content of the rice following inoculation is 50%. The bags are sealed with masking tape and massaged to distribute the inoculum evenly and to break up any lumps. The bags are then laid flat and a slit is cut along half the length of the bag, this is sealed with micro-porous tape. It is then incubated and maintained at a fixed temperature (*e.g.* 25-30^{0}C is often good for *Metarhizium*). Conidiation typically takes place after approximately 10 days and bags are incubated for 14days before being transferred to the drying room.

3.3.5. Extraction of the Conidial Powder

After approximately ten days open drying, the conidia are separated from the substrate using either a hand made sieve or a mechanical agitator fitted to four cyclones.

Extraction by Sieving

A two person sieve, consisting of a 300 μm mesh fixed around a wooden frame is loaded with the conidiated substrate. Plastic sheeting is taped around both the top and bottom edges of the sieve and sealed at the top.

A collecting vessel, such as a bucket is fitted to the plastic sheeting at the bottom of the sieve so as to create a funnel into the collecting vessel. The sieve is shaken until all the loose conidial powder has been removed from the rice and has collected in the vessel below.

The conidial powder is then further sieved using a 106 μm sieve to separate the larger rice dust particles from the conidial powder. Even following this second sieving, some rice dust particles remain in the product. Each lot is therefore carefully checked to ensure that it meets the particle size specifications described in the quality control section below.

Mechanical and Cyclone Extraction

The conidiated rice is loaded into the drum of a modified domestic tumble dryer. A hopper system is fitted to the door of the drier to permit constant loading of the substrate as the drier operates. The dryer is run without heat to agitate the substrate and remove the conidia from the rice grains. From the tumble drier, a powerful fan at the final outlet provides a negative pressure to draw conidia and other fine particles down a manifold and into a parallel array of four dual-cyclone separators, which remove the conidia and rice dust from the air-flow.

Larger particles such as rice dust and mycelium are collected in the outer cyclone chamber, whilst the finely divided pure conidial powder is collected in the inner cyclone chamber and falls into a collecting vessel.

After passage through the cyclones clean air is exhausted to the outside of the equipment and is expelled outside. Conidial powder collected from the inner cyclone easily meets the particle size specifications demanded for quality control.

3.3.6. Drying of the Conidial Powder

The conidial powder resulting from either method of extraction is further dried to a final moisture content of 5% in a locally made spore dryer consisting of a dehumidifier which feeds dry air through a layer of silica gel into a large wooden box containing the conidial powder. Care is taken to ensure that the temperature within the drying box does not exceed 40^0C.

The drying process takes between three and five days. Once dried, the conidial powder is weighed into plastic-lined foil sachets (100-300 g/sachet) and small packets of non-indicating silica gel are added at a rate of 20% (w/w) before the bags are sealed. Once packaged, conidia remain viable for >3 years if stored at 10^0C or lower, or >1 year if stored at 30^0C.

3.3.7. Post Harvest Storage

The post harvest storage conditions greatly affect fungal viability and efficacy. Conidial moisture content is an important factor with respect to temperature tolerance and viability. The tolerance of *M. anisopliae* for high temperatures increases with increasing desiccation. Daoust and Roberts (1983) showed that at 37°C, two isolates of *M. anisopliae* retained most viability after long-term storage at either 0 or 96% RH. Drying conidia in the presence of desiccating agents like silica gel and $CaCl_2$ appears to improve their viability but direct contact with the desiccant can be detrimental (Daoust and Roberts, 1983). Moore *et al.* (1995) found that dried conidia stored in oil formulations remained viable longer than those stored as a dried powder, especially if stored at relatively low temperatures (10–14°C compared with 28–32°C). Addition of silica gel to oil-formulated conidia appears to prolong their shelf life. Undried conidia of *M. flavoviride* lose viability rapidly, with germination dropping below 40% after 9 and 32 weeks at 17°C and 8°C, respectively.

After 127 weeks in storage, germination remained at over 60 and 80% for the dried formulations at 17°C and 8°C, respectively (Moore *et al.,* 1996). These conidia were found to have retained virulence similar to that of freshly prepared formulations. Furthermore, conidia dried to 4–5% moisture content showed greater temperature tolerance than conidia with a higher moisture content (McClatchie *et al.,* 1994; Hedgecock *et al.,* 1995).

3.3.8. Quality Control of the Product

Moisture Content Determination

Box 1
Moisture content determination
Label all universal bottles or glass Petri dishes with marker pen before drying.
Place bottles and lids (separate) in pre-heated oven
Allow to dry for at least one hour at 80°C.
Remove bottles from oven
Quickly replace lids before allowing to cool for 1.5 to 2 hours in a desiccator.
Ensure silica gel is fresh by regeneration for at least 3 hours at 80°C.
Weigh all bottles, complete with lids and record weights.
Add sample of substrate or conidial powder to each bottle, replace the lid and reweigh.
For solids, a sample quantity of 3g is recommended while a 2g sample is adequate for conidial spore powder.
Place bottles containing the samples in an oven-proof vessel.
Colonised substrates require drying for at least 4 hours at 130°C.
Spores dried at 103°C for 17 hours.
After the drying period, remove bottles from oven, and allow to cool to room temperature in a bench top dessicator for 1.5 to 2 hours.
Final dry weights can then be recorded.
The percentage moisture content of the samples can then be recorded as follows :
% Moisture = Wet wt (g) – Dry wt (g) / Sample size (g) x 100
Where,
Wet wt = Wt. of bottle, lid and sample before drying
Dry wt = Wt. of bottle, lid and sample after drying
Sample size = Wet wt – Initial wt. of bottle and lid

Germination Test

Box 2
Germination test
Prepare fresh Sabouraud dextrose agar in small 50 mm Petri dishes (3 plates). Wearing facemask, spectacles and gloves, transfer a small quantity of spores onto the upturned lid or base of a clean, dry, plastic Petri dish and place in a humid chamber for 30 min to allow the spores to re-hydrate. Label each dish. After 30 min use a clean microspatula to add a small quantity of hydrated spores to a labelled, capped, plastic tube containing 9 ml Shellsol T The resultant spore concentration of the suspension should be 10^5 spores /ml. Shake vigorously. Clean the microspatula between samples using 70% alcohol and dry. Dip the microspatula into suspension and spread evenly over the surface of the plates Incubate plates for 24 h at 25°C. After 24 h transfer plates to the fridge, to reduce the growth rate. Remove a maximum of 9 plates from fridge, open lids to allow the agar to dry. Examine plates microscopically under x300 Use separate tally counter for germinated and non-germinated spores Count a total of ≥ 300 spores A germinated spore is defined as a spore having a germ tube. Record results on appropriate record sheet and sign and date. Calculate the percentage germination as follows: % Germination= [a /(a+b)]*100 Where, a= Number of germinating spores b= Number of non germinating spores Calculate the average percentage germination of the three plates.

3.4. Factors Influencing Mass Production of Entomopathogenic Fungi

In the recent years, studies on mass production of entomopathogenic fungi have demonstrated that the quantity and quality of propagules produced are dependent on several factors, such as the isolates, nutrients, inoculum density and environmental conditions (Leite *et al.,* 2003). Here we highlighted the importance of aeration, moisture content, temperature, light and culture media in the mass production of entomopathogenic fungi.

3.4.1. Aeration

All mitosporic fungi are aerobic and require the provision of oxygen for growth and conidiation. However, different fungi vary greatly in their requirements for aeration (Churchill, 1982).

In a simple, labour intensive production system, where electronic monitoring and control of the fermentation parameters are not carried out, it is difficult to determine precisely the requirement for oxygen of a given fungus. Observations can be made on the degree of sporulation of a given fungus in small laboratory scale experiments where air supplies can be increased or decreased to varying degrees, in Erlenmeyer flasks for example. This kind of investigation can give some indication of the likely oxygen requirements of the fungus such as low, medium or high, but will not give precise data on oxygen levels or utilization rates as is possible in standard deep tank liquid fermenters and modern proprietary solid substrate fermenters.

Biological Purity and Colony Count

Box 3
Biological purity and colony count
For each sample, 8 universal bottles containing exactly 9 ml of 0.05% Tween 80 and 1 bottle containing 10 ml of the Tween solution should to be prepared and sterilized. Loosely screw on plastic lids and cover each with aluminium foil before autoclaving for 20 min at 121°C, 15 - 20 psi. Allow bottles to cool after sterilization and add approximately 0.1g of conidial powder or 0.5 - 1g of colonized substrate to the bottle containing 10 ml of 0.05% Tween 80. The exact weight, however, must be recorded to at least 3 decimal places. Agitate vigorously to ensure a homogeneous suspension. Under sterile conditions, carry out serial dilution of stock suspension from 10-1 to 10-8. Transfer 200 µl of the homogenous suspension aseptically onto each of the two Sabouraud Dextrose Agar plates. Incubate plates at 25 - 27°C for 3 - 4 days. Check plates after 1 - 2 days count any bacterial or yeast contaminating colonies which may be present. After 3 - 4 days count the number of fungal colonies of interest and the number of contaminating fungal colonies, if present. Record results on the record sheets. The percentage contamination (all types) of the product sample can be calculated using the following equation: % Contamination = [No. contaminant colonies x dilution / No. product colonies x dilution] x 100 A colony count of the sample under investigation can be calculated as follows to give the number of viable colonies per g of initial product: Colonies/g = [No. product colonies x 5 x dilution x 10 ml] / weight (g) dry sample A total count of conidia/g of spore powder or colonized substrate and be made from the 10-1 or 10-2 dilution using a haemocytometer in combination with the calculation below: Conidia/g = [conidia/ml from haemocytometer count x dilution x 10 ml] / Sample Weight (g)

The addition of forced aeration to a fungal production system generally improves yield (Bradley *et al.,* 1992), has been shown to speed up the process of sporulation (Guillon, 1997) and can help to remove excess heat from the system. However, the introduction of forced aeration into an appropriate technology production system is likely to greatly increase the capital cost of the system and, if not designed and implemented carefully, could introduce complications such as contamination and premature desiccation of the substrate. The air to be introduced into such a system must therefore be sterilized and humidified to minimize these risks. Most simple production systems, such as those in Latin America (Aquino *et al.,* 1977; Alves and Pereira, 1989; Antía- Londoño *et al.,* 1992; Mendonça, 1992) and China (Feng *et al.,* 1994) and the LUBILOSA Programme, rely on the passive exchange of air between the growth chamber and the external environment.

3.4.2. Moisture Content

The optimum moisture content for fungal growth and conidiation needs to be identified for each substrate/isolate combination. Moisture content plays a significant role in the final yield of conidia, although the optimization of this parameter can be complex. Most mitosporic fungi prefer humid situations; however, different substrates vary in their moisture sorption curves and will reach maximum adsorption at different moisture content levels. In general, moisture contents within the range 35% to 60% (calculated as percentage water present in the wet substrate on a weight for weight basis) are most commonly used. The determination of optimal moisture content for a given fungus/substrate combination can be carried out initially on a small scale in conical flasks. However, there is a close relationship between moisture content and oxygen availability. Increases in the moisture content of the substrate tend to decrease oxygen availability as the inter-particular spaces become filled with water and air is forced out (Moo-Young *et al.,* 1983). This problem may be compounded during scale-up due to compaction resulting from larger masses of substrate. It is therefore necessary to identify an optimal balance between these two factors at the production scale. Recently *Nuñez-Gaona et al.* (2010) have demonstrated the impact of moisture and and inoculum on the growth and conidia production by *Beauveria bassiana* on wheat bran.

3.4.3. Incubation Temperature

Incubation temperature of the liquid and solid production stages should be matched to the optimal temperature for the type of growth required. Temperature optima vary greatly not only between fungal species but between isolates of the same species (Thomas and Jenkins, 1997). Additionally, the temperature optima for mycelial growth and conidiation are often different (Alasoadura, 1963). Simple laboratory tests can be used to determine temperature optima for a given isolate, thus allowing the provision of optimal or near optimal incubation conditions. Care should be taken during scale-up as heat build-up during metabolism can cause areas of localized heating way above the temperature optimum.

3.4.4. Light

Some fungi require light for sporulation to occur (Alasoadura, 1963), other fungi appear to be relatively unaffected by normal intensities of daylight, while in others some degree of inhibition of the sporulation process is observed at certain light intensities (Vouk and Klas, 1931). The effect of light on sporulation of a given fungus can be assessed easily in the laboratory using aluminium foil to prevent light reaching the growing culture. Some early work has suggested that conidia of *Beauveria* produced in the dark were more virulent than those produced in the light (Masera, 1936). Williams (1959) reported that conidia of many species of fungi produced larger conidia when grown in the dark in comparison to those grown under continuous illumination. These factors should be taken into account when selecting a suitable container in which to carry out the solid substrate fermentation.

3.5. SMALL-SCALE PRODUCTION OF SELECTED ENTOMOPATHOGENIC FUNGI

3.5.1. *Metarhizium anisopliae*

In the late 19th century, Metchnikoff was the first to describe *Metarhizium anisopliae* "green muscardine" infections on the cereal cockchafer and to suggest the use of the microorganism as a biological control agent for insects It is possible to produce *Metarhizium anisopliae* in plastic fermenters. The isolate *Metarhizium anisopliae* must be passed through target host and then be cultivated on Sabouraud`s liquid medium until the concentration of conidia becomes very high and reach 3.19×10^{10} conidia/1 g after 8 days. Light and darkness do not have any decisive effect on mass production of conidia. Temperature is the most important factor for the yield of conidia. For the first 6 days the best temperature is 24-25°C and then it should be decreased to 22-20°C. Submerged culture produce blastospores after 3-4 days, which are suitable seed like cultures in medium for plastic fermenters.

Harvesting, Formulation and Standardization

Process harvesting must be done with Tween 80 or Triton 100, and then is mixed with Siloxyd 125 - 150 g per liter mud and dried for 24 hours. Shelf life is good during storage at 4 - 10°C. Hundred g of conidial powder (2.5×10^{10} conidia/1 g) must be mixed with Siloxyd and diluted so that the total weight will be 500 g. The recommended dose per ha is in this case is 500 g of formulated product (e.g. 2.5×10^{12} of conidia).

3.5.2. *Verticillium lecanii*

Kybal and Vlèek (1976) used polyethylene cushions made of large thin walled, polyethylene tubing sealed into sections which were partially filled (1 cm high layer absolutely horizontal) with submerged culture of *V. lecanii* after 2 + 1 days cultivation and inflated with sterile air. Product was harvested (ca. after 14-16 days) by discarding the medium and retaining the mad. Yield from a 0.8 % peptone, 1 % sorbitol medium was $1 \times 10^{12\text{-}13}$ conidia/1 m².

Formulated Product

Harvesting of mud is done by mixing it with Siloxyd (ca. 100-125 g/ 1 liter of mad and wetting agent (Tween 80) and, after predrying at 30°C one day, the wet cake is crushed using a meat-mincer, and the product if dry to perfection at 30°C can be stored or milled (Condux Universal Mühle Typ 150/S-D) and packaged like water-dispersable powders which is prepared for dilution with water into a final spray. The application dose per hectare or meter square is determined depending of the conidia content.

Standardization

Despite theoretical strain stability, a commercial product must be shown to have constant potency. This should be measured by viable spore count and bioassay. The viability of *V. lecannii* conidia and blastospores can easily be assessed by an agar-slide technique (Hall, 1976).

Registration

The absence of records of *V. lecanii* in man and other vertebrates is an impressive evidence for its innocuity. All *V. lecanii* strains so far examined by Hall (1981) cannot grow at 37°C and so the likelihood of infecting warm-blooded vertebrates internally is exceedingly remote. As reaction to the dose of 10^6 conidia injected intravenously, no adverse symptoms were observed and, 28 days later, no gross pathological changes were apparent in the internal organs and no signs of the fungus could be found either in sectioned organs or in agar cultures from these. Safety tests are being carried out preparatory to commercial exploitation.

3.5.3. *Beauveria bassiana*

Production of *B. bassiana* spores can be achieved using different methodologies, which can be classified into low input and industrial technologies. However, most production of fungal spores worldwide is carried out using simple technologies that demand low inputs. On agar media conidiogenesis starts after six days, while in liquid culture this takes only 3-4 days (Sam1iòáková, 1966). In stirred liquid cultures employed in mass production of this fungus, so-called "blastospores" developed which are thin-walled, larger (3-5 x 2-3 mm) and less resistant than conidia (Mueller-Koegler and Sam1iòáková, 1970); blastospores germinated in 6-10 hours at 18-24 0C, while conidia require 15-20 hours. The germination of conidia requires a saturated atmosphere and the optimal temperature for growth is in the range 25-30°C, minimum 10°C, and maximum 32°C apparently depending on the geographic origin of the isolate; no germination occurs either below 10°C or above 35°C; the thermal death point of conidia has been determined as 50°C for 10 min in water. The optimal pH for growth is 5.7-5.9, and for conidia formation 7-8 (Goral, 1972).

Culture Medium

Good growth occurs on maltose and sucrose and, among others, on the N sources glutamic and aspartic acids, and ammonium oxalate, citrate or tartrate. A medium recommended for optimal growth contains 2% corn steep liquor, 2.5% glucose, 2.5 starch, 0.5% NaCl, and 0.2% CaCO3. The fungus produces lipase, protease, urease, amylase, chitinase, cellulase and 1,2-á- glucanase. Chitinase can be realized into the medium during autolysis, but was also found to act jointly with other enzymes in decomposing insect integuments. The production of a toxic substance is proteins complex consisting of two fractions with different molecular weights (Kuèera and Sam1iòáková, 1968). This toxic metabolite is produced most abundantly on complex media (e.g. cornmeal, yeast or beef extract); inorganic N sources are ineffective (Kuèera, 1971). A red bibenzoquinone pigment, oosporein, with antifungal properties and the yellow pigments tenellin and bassianin have been found. For small scale production, harvesting and formulation see in *Verticillium lecanii.* Generally, diphasic liquid solid fermentation technologies are incapable of yielding mycoinsecticide concentrations higher than $1x10^{10}$ (Jenkins, 1995). Hence, it is advised to use liquid medium for the mass production of entomopathogenic fungi. They could be grown on cooked rice using diphasic liquid-solid fermentation in plastic bags to produce and harvest spore powder. The germination at 24 hours would be over 75% and hence this methodology would be suitable for laboratory and field studies, but not for industrial production when a high concentration of spores are required for formulation and field applications (Posada-Flórez, 2008). Several nutritional studies have been undertaken to improve the growth and sporulation of *Beauveria bassiana* (Cruz et al., 1993; Torre and Cardenas-Cota, 1996). Although the production of conidia is mostly based on solid state fermentation, it requires weeks to complete all process, thereby increasing the production costs. Aerial conidia tend to be more tolerant to desiccation and more stable as a dry preparation, compared to submerged conidia. Some of the liquid media used include sugarcane molasses + rice broth, rice broth + yeast and sugarcane molasses + yeast + rice broth, which could produce highest viable propagule concentrations (Jackson, 1997).

3.6. COMMERCIAL PRODUCTS

The number of fungal products on the market used to control plant diseases is increasing, and nearly 40 have been reported (Cook   et al., 1996) US Department of Agriculture/Agricultural Research Service/BPDL (USDA/ARS/BPDL) Biocontrol of Plant Diseases Laboratory website http://www.barc.usda.gov/psi/bpdl/bioprod.htm). However, many of these materials are not registered as BCAs (also termed biopesticides); rather, they are sold as some form of 'plant growth promoter' or 'stimulant', 'soil conditioner', 'plant strengthener' or 'wound protectant'. By not claiming fungicidal activity, producers of these materials avoid the need for registration and costs for obtaining efficacy, toxicology and environmental fate data. Although this speeds up entry of the product to the market-place, it also introduces an element of potential environmental and health risks in those cases where extensive experimental background information has not been accumulated. Making pesticidal claims for a product without formal registration and permission can lead to a ban on sales and the imposition of penalties [(Federal Insecticide, Fungicide and Rodenticide Act (FIFRA)]. Brief summaries of some of some of these products are given below. Where the information was provided by the manufacturers directly, they are not referenced further. Significantly, more than half of the products are *Trichoderma*- or *Gliocladium*-based preparations reflecting the widespread occurrence of these fungi, the relative ease of their production, their low toxicity and the huge volume of experimental data on these genera. Products are available for control of pathogens in soil and root, aerial and postharvest microbiomes. These may be considered reasonably well-defined habitats that have distinct physicochemical properties containing characteristic microbial communities. Moreover, the inundative release is the most commonly used strategy for these pathogens. This strategy requires the production of large quantities of infective propagules.

3.6.1. *Metarhizium* spp.

Several commercial products of *M. anisopliae* are available for insect control in different agricultural operations such as Bio-Green and Bio-Cane granules for control of soil grubs of pasture and sugar cane in Australia, Green Muscle for control of locusts in Africa, Ago Biocontrol for control of various pests of ornamental crops in South America, and BioPath for control of cockroaches in United States (Shah and Goettel, 1999; Wraight *et al.,* 2001; Shah *et al.,* 2005; Sun and Liu, 2006; Sallam *et al.,* 2007) and EPA was the registered isolate in US (Cook *et al.,* 1996).

Following an extensive development phase an isolate of *M. anisopliae* var. *acridum*, IMI 330189, has been mass-produced and commercialized under the product name 'Green Muscle'.

3.6.2. *Verticillium lecanii*

Verticillium lecanii strains were first introduced commercially in the U.K. for the control of aphids "Vertalex" and whitefly "Mycotal" on protected ornamental and vegetable crops. The products were discontinued in 1986.

3.6.3. *Coniothyrium minitans*

The fungus *Coniothyrium minitans* is a mycoparasite of sclerotia of *Sclerotinia sclerotiorum* and *Sclerotinia minor* which affecting more than 360 plant species. Two products containing this BCA are available: Contans WG, in Germany and Switzerland, and KONI in Hungary.

Both are granular formulations, but Contends WG is sprayed and incorporated into soil after dispersal in water whereas KONI is incorporated into soil directly. Application must be made several weeks prior to planting crops to allow time for the sclerotia to be destroyed.

Currently, although use is restricted to glasshouses and polyethylene tunnels for a range of high-value crops use on field crops and amenity areas is planned. *C. minitans* strain CON/M/91-08 in Contans WG is undergoing consideration for full European registration under the European Union (EU) Council Directive 91/414 and in spring 2001 received approval from the US-EPA and Austria.

3.6.4. *Trichoderma* spp.

Trichoderma-based biocontrol agents (BCAs) possess better ability to promote plant growth and soil remediation activity compared to their counterparts (virus, bacteria, nematodes, and protozoa). Their capability to synthesize antagonistic compounds (proteins, enzymes, and antibiotics) and micro-nutrients (vitamins, hormones, and minerals) enhance their biocontrol activity. Like other fungal BCAs, conidial mass of Trichoderma is the most proficient propagule, which tolerates downstream processing. (Lewis and Papavizas, 1983; Gupta *et al.,* 1997; Verma *et al.,* 2005).

However, the cost of these raw materials for commercial production of BCAs is one of the major limitations behind the restricted use. To overcome the cost limitation, many researchers have successfully used substrates like corn fiber dry mass (Vlaev *et al.,* 1997), sewage sludge compost (Cotxarrera *et al.* 2002), and cranberry pomace (Zheng and Shetty, 1998). Despite the use of alternate sources, the cost of production was still high, as these raw materials need to be supplemented by other nutrients (Verma *et al.,* 2007).

3.6.5. *Trichoderma virens*

The BCA *Trichoderma virens* has appeared on the market in two formulations, GlioGard™, an alginate prill formulation, and SoilGard™, a granular fluid-bed formulation (Lumsden *et al.,* 1996). These products target damping-off diseases of vegetable and ornamental plant seedlings caused by *Rhizoctonia solani* and *Pythium* spp. Application was confined to greenhouse or interior container use (Lumsden *et al.,* 1996). Only the product SoilGard is now produced and is marketed by Thermo Trilogy Corp., Columbia, Maryland, USA.

3.6.6. *Trichoderma harzianum*

A commercial formulation of *Trichoderma harzianum* strain 1295-22 (T-22) is manufactured by BioWorks, Inc., Geneva, New York, USA, and sold through several distributors as T-22 Planter Box™. This conidial formulation is designed for application to large-seeded crops such as maize, beans, cotton and soybeans, and in most cases can be applied to seeds already treated with fungicides (Harman and Björkman, 1998). The seed treatment delivers the *T. harzianum* inoculant to the growing seedling where it colonizes the spermosphere and also the developing root system, protecting crop plants from damping-off diseases.

Similar products using the same strain 1295-22 (T-22) include a granular formulation used as a greenhouse soil amendment, which is called RootShield™ and con- tains the entire thallus of *T. harzianum* colonized on clay particles. Another product, RootShield drench, consists of conidia and inert ingredients for use as a water-suspensible drench. In either case, the product is thought to colonize the root system of the crop to be protected (Harman and Björkman, 1998). This product is claimed to control root diseases caused by *Fusarium, Rhizoctonia* and *Pythium* spp., but not *Phytophthora* spp. Another *T. harzianum* product available in the Czech Republic and Denmark for glasshouse use is Supresivit. This dispersible powder containing conidia of strain PV5736-89 is applied to soil or potting mixes to control disease complexes causing damping-off or root rots of ornamentals and forest-tree seedlings, and as a pea seed treatment to control damping-off.

3.6.7. *Trichoderma viride*

Trichoderma viride is available as a BCA in India from Hoechst Schering AgrEvo Ltd in a product named Ecofit. It is a talc-based powder sold for the control of root rot, seedling rot, damping-off, collar rot and *Fusarium* wilt in cotton, chick-pea, pigeonpea, Bengal gram, groundnut, sunflower, soybean, tobacco and vegetables.

Depending on the plant and disease of interest, Ecofit can be applied before sowing as a dry powder or slurry seed treatment, before planting as a rhizome, tuber or set dip, or as a soil drench for soil incorporation following a preliminary scale-up procedure involving prior inoculation on to farmyard manure. *Ampelomyces quisqualis*, formulation AQ10, is the first biocontrol fungus developed specifically for controlling powdery mildew. AQ10 is water-dispersable and acts as a mycoparasite on powdery mildews affecting leaves, stems or fruits of plants. The range of plants protected includes strawberry, tomato, grape, tree fruit and ornamentals (Dik *et al.,* 1998; Cavalcante *et al.,* 2008).

As with many other plant diseases, powdery mildews (PMD) have developed resistance to commonly used chemicals such as sulphur and demethylationinhibiting fungicides. AQ10 is useful in powdery mildew management programmes to ward off resistance problems and can extend the usefulness of these chemical treatments for a reduced time and amount of application.

3.6.8. *Phlebiopsis (Peniophora) gigantean*

Phlebiopsis gigantea is a common wood-rotting saprotroph that is applied to freshly cut stumps of pine to prevent their colonization by the root-rotting fungus *Heterobasidion annosum.*

It is not a biocide that kills the target organism but rather it competes for the food base that the pathogen would otherwise use. Commercial products containing oidia are available in the UK from Omex Environmental Ltd and in Finland from Kemira Agro Oy as PG Suspension and Rotstop, respectively. *P.* gigantea is also available in other Scandinavian countries and Poland.

Significantly, after 30 years of field use, PG Suspension has become the first fungal disease BCA approved in the UK under the Control of Pesticides Regulation (COPR) 1986 (Pratt *et al.,* 1998). *Trichoderma harzianum* Strain T39 of *T. harzianum* has been used for greenhouse control of *Botrytis cinerea.*

It is produced by Makhteshim Chemical Works and is marketed as Trichodex™ in Europe and Israel. The strategy for best control involves alternating chemical and biological control treatments. This approach, as with *A. quisqualis*, reduced the use of chemicals and may also reduce the incidence of chemical resistance. developed by *B. cinerea.*

3.6.9. *Trichoderma harzianum* + *Trichoderma polysporum*

A combination of strains IMI 206040 and IMI 206039 of *Trichoderma harzianum* and *Trichoderma polysporum*, respectively, sold as BINAB-T, is one of the oldest commercial biopesticides preparations still available.

Produced by Bio-Innovation Eftr AB in Sweden it has been used for over 20 years. Because of the long period of safe use it will continue to receive exemptions from the new pesticide regulations in Sweden (Kemikalieinspektionen) until a decision is made concerning a current application for registration made under the new regulations. BINAB-T has been registered and used in the past for control of numerous diseases, but currently, in Sweden and Denmark, is used largely for the control of grey mould (*B. cinerea*) on strawberries, with some minor use in glasshouse crops to control soil-borne pathogens. The other main market is Chile, where it is used for the suppression of silver-leaf disease (*Chondrostereum purpureum*) and chlorotic leaf curl (*Eutypa*) in stone fruit and grapes, respectively.

3.6.10. *Cryptococcus albidus*

This yeast was developed for use on pome fruits, especially apples and pears, against grey and blue mould caused by *B. cinerea* and *Penicillium expansum*, respectively. A product, Yield*Plus*, is produced commercially by Anchor Yeast, Capetown, South Africa.

CONCLUSION

It is now clear that a stage has been set for fungal biocontrol agents to play a greater role in agriculture. To make this a reality, a long-term commitment between researchers and scientists, government, producers, retailers and agrochemical companies is required. Cost-benefit analyses, toxicology tests and registration of the products should also be carried out.

ACKNOWLEDGMENT

Catalyzed and Supported by Vision Group on Science and Technology, Department of Science and Technology, Govt. of Karnataka (Ref No. No. VGST/Seed Money 2009-10/LS-N1/10-11). The author (JFB) wishes to thank Mrs. B. Bency Borgio for her constant help. K. Sahayaraj thanks the management of St. Xavier's College for the support and encouragements.

REFERENCES

Abebe, H. 2002. Potential of entomopathogenic fungi for the control of *Macrotermes subhyalinus* (Isoptera: Termitidae). Ph. D. Thesis, Dem Fadhbereich Gartenbau der Universität Hannover, Ethiopia.

Alasoadura, S.O. 1963. Fruiting in *Sphaerobolus* with special reference to light. *Annals of Botany*, 27: 123-145.

Alves, S.B. and Pereira, R.M. 1989. Production of *Metarhizium anisopliae* (Metsch.) Sorok and *Beauveria bassiana* (Bals.) Vuill. in plastic trays. *Ecossistema*, 14: 188-192.

Alves, S.B. and Pereira, R.M. 1998. Produc,ao de fungos entomopatogenicos. In: Alves, S.B. (Eds.), *Controle Microbiano de Insetos*, 2nd ed. Biblioteca de Cieˆncias Agra´rias Luiz de Queiroz, Piracicaba, pp. 845–869.

Antía-Londoño, O.P., Posada-Florez, F., Bustillo-Pardey, A.E. and González-García, M.T. 1992. Produccion en finc a del hongo *Beauveria bassiana* para el control de la broca del cafe. *Cenicafé Avances Téchnicos*, 182: 1-12.

Aquino, M.L.N. Vital, A.F., Cavalcanti, V.L.B. and Nascimento, M.G. 1977. Cultura de *Metarhizium anisopliae* (Metsch) Sorokin en sacos de polipropileno. *Boletim Técnico da CODECAP*, 5: 7-11.

Aregger, E. 1992. Conidia production of the fungus *Beauveria brongniartii* on barley and quality evaluation during storage at 2^0C. *Journal of Invertebrate Pathology*, 59: 2-10.

Bailey, L.A. and Rath, A.C. 1994. Production of *Metarhizium anisopliae* spores using nutrient impregnated membranes and its economic analysis. *Biocontrol Sciences and Technology*, 4: 297-307.

Borgio, J.F. 2008. RAPD analyses of entomopathogenic fungus *Metarhizium anisopliae* (Metsch.) Sorokin (Deuteromycotina: Hyphomycetes) and its bioefficacy on *Dysdercus cingulatus* (Fab.) (Hemiptera: Pyrrhocoridae). Manonmanium Sundaranar University, Tirunelveli, Tamil Nadu, India.

Bradley, C.A., Black, W.E., Kearns, R. and Wood, P. 1992. Role of production technology in mycoinsecticide development. In: Leatham, G.F. (Eds.), *Frontiers in industrial mycology*, Chapman and Hall, London, pp. 160-173.

Churchill, B.W. 1982. Mass production of microorganisms for biological control. In: Charudattan, R. and Walker, H.L. (Eds.), *Biological control of weeds with plant pathogens*. John Wiley, New York, pp. 139-156.

Cook, R.J., Bruckart, W.L., Coulson, J.R., Goettel, M.S., Humber, R.A., Lumsden, R.D., Maddox, J.V., McManus, M.L., Moore, L., Meyer, S.F., Quimby, P.C., Stack, J.P. and Vaughn, J.L. 1996. Safety of microorganisms intended for pest control and plant disease control: A framework for scientific evaluation. *Biological Control*, 7: 333–351.

Cruz, B.P.B.; Abreu, O.C.; Oliveira, A.D. and Chiba, S. (1993), Crescimento de *Metarhizium anisopliae* (Metsch.) Sorokin em meios de cultura naturais líquidos. *O Biológico*, 49, 111-116.

Daoust, R.A. and Roberts, D.W. 1983. Studies on the prolonged storage of *Metarhizium anisopliae* conidia: effect of growth substrate on conidial survival and virulence against mosquitoes. *Journal of Invertebrate Pathology*, 41: 161-170.

Devi, P.S.V., Chowdary, A. and Prasad, Y.G. 2001. Cost effective multiplication of entomopathogenic fungus *Nomuraea rileyi* (Farlow) Samson. *Mycopathology*, 151: 35-39.

Dik, A.J., Verhaar, M.A. and Belangaer, R.R. 1998. Comparison of three biological control agents against cucumber powdery mildew (*Sphaerotheca fuliginea*) in semi-commercial-scale glasshouse trials. *European Journal of Plant Pathology*, 104: 413-423.

Feng, M.G., Poprawski, T.J. and Khachatourians, G.G. 1994. Production, formulation and application of the entomopathogenic fungus *Beauveria bassiana* for insect control: current status. *Biocontrol Science and Technology*, 4: 3–34.

Ferron, P. 1985. Fungal control, In: Kerkut, G.A. and Gilbert L.I. Kerkut, G.A. and Gilbert L.I. *Comprehensive insect physiology, biochemistry and pharmacology* 12 *Insect Control*. Pergamon Press, Oxford, pp. 313-346.

Goettel, M.S. 1984. A simple method for mass culturing entomopathogenic hyphomycete fungi. *Journal of Microbiological Methods,* 3: 15-20.

Gopalakrishnan, C. 2001. Mass production and utilization of microbial agents with special reference to insect pathogens. In: Ignachimuthu, S. and Sen, A. (Eds.), *Micorbials in insect pest management.* Oxford and IBH Publishing Co. Pvt. Ltd., New Delhi., pp. 174.

Gopalakrishnan, C., Anusuya, D. and Narayanan, K. 1999. Occurrence of entomopathogenic fungi *Paecilomyces farinosus* (Holmskiold) Brown and smith and *Zoophthore ragicams* (Brefela) Batko in the field population of *Plutella xylostella* Lyon cabridge. *Entomon,* 24: 363-369.

Goral, W.M. and Lappa, N.V. 1972. The effect of medium pH on growth and virulence of *Beauveria bassiana* (Bals.). *Vuill. Mikrobiol. Zh.,* 34(4): 454-457.

Guillon, M. 1997. Production of biopesticides: scale up and quality assurance. In: *BCPC Symposium Proceedings* No. 68. *Microbial insecticides: novelty or necessity.* The British Crop Protection Council, Farnham, UK, pp. 151-162.

Hajek, A. E. and Ruth C. Plymale. 2010. Variability in azygospore production among *Entomophaga maimaiga* isolates. Jouenal of invertebrate pathology, 104 (2): 157-159.

Hall, R.A. 1976. A bioassay of the pathogenicity of *Verticillium lecanii* conidiospores on the aphid *Macrosiphoniella sanborni. Journal of Invertebrate Pathology,* 27: 41-48.

Hall, R.A., 1981. The fungus *Verticillium lecanii* as a Microbial Insecticide against Aphids and Scales. In: Burges H.D. (Eds.), *Microbial Control of Pests and Plant Deseases* 1970-1980. *Acadamic Press* 1981, pp. 482-498.

Harman, G.E. and Björkman, T. 1998. Potential and existing uses of *Trichoderma* and *Gliocladium* for plant disease control and plant growth enhancement. In: Harman, G. E. and Kubicek, C.P. (Eds.), *Trichoderma* and *Gliocladium: Enzymes, Biological Control and Commercial Applications,* Vol. 2. Taylor and Francis, London, pp. 229-265.

Hedgecock, S., Moore, D., Higgins, P.M. and Prior, C. 1995. Influence of moisture content on temperature tolerance and storage of *Metarhizium flavoviride* conidia in an oil formulation. *Biocontrol Science and Technology,* 5: 371–377.

Hokkanen, H. and Lynch, J.M., 1995. Biological Control: Benefits and Risks. Cambridge University Press, Cambridge, UK.

Ignoffo, C.M. 1981. The fungus Nomuraea rileyi as a microbial insecticide. In: Burges, H.D. (Ed.), *Microbial Control of Pests and Plant Diseases,* Academic Press, London, New York , pp. 513–538.

Jackson, T.A. Alves, S.B. and Pereira, R.M. 2000. Success in biological control of soil-dwelling insects by pathogens and nematodes. In: Gurr, G. and Wratten, S. (Eds.), *Biological Control: Measures of Success,* Kluwer Academic Publishers, Boston, pp. 271–296.

Jackson, M.A. and Jaronski, S. T. 2009. Production of microsclerotia of the fungal entomopathogen *Metarhizium anisopliae* and their potential for use as a biocontrol agent for soil-inhabiting insects. *Mycological Research* 1 1 3: 8 4 2 – 8 5 0.

Jackson MA, McGuire MR, Lacey LA, Wraight SP, 1997. Liquid culture production of desiccation tolerant blastospores of the bioinsecticidal fungus *Paecilomyces fumosoroseus. Mycological Research* 101: 35–41.

Jenkins, N.E. and Prior, C. 1993. Growth and formulation of true conidia by *Metarhizium flavoviride* in a simple liquid medium. *Mycological Research,* 7: 1489–1494.

Jenkins NE. 1995. Studies on mass production and field efficacy of *Metarhizium flavoviride* for biological control of locusts and grasshoppers. *PhD thesis*. Cranfield University, UK.

Jenkins, N.E. and Thomas, M.B. 1996. Effect of formulation and application method on the efficacy of aerial and submerged conidia of *Metarhizium flavoviridae* for locust and grasshopper control. *Pesticide Science*, 46: 299-306.

Jenkins, V.E., Heveifa, G., Langewald, J., Cerry, A.J. and Lower, C.J. 1998. Development of mass production technology for aerial conidia for use as mycopesticides. *Biocontrol News and Information*, 9: 21-31.

Kerwin, J.L. Duddles, N.D. and Washino, R.K. 1991. Effects of exogenous phospholipids on lipid composition and sporulation by three strains of *Lagenidium giganteum. Journal of Invertebrate Pathology*, 58: 408-414.

Khetan, S.K. 2001. Microbial Pest Control. Marcel Dekker, New York.

Krueger, S.R., Villani, M.G., Martins, A.S. and Roberts, D.W. 1992. Efficacy of soil applications of *Metarhizium anisopliae* (Metsch.) Sorokin conidia, and standard and lyophilized mycelial particles against scarab grubs. *Journal of Invertebrate Pathology*, 59: 54–60.

Kucera, M. 1971. Toxins of entomophagous fungus *Beauveria bassiana*. 2. effect on nitrogen sources on formation of the toxic protease in submerged culture. *Journal of Invertebrate Pathology*, 17: 211-320.

Kueera, M. and Samsinakova, A. 1968. Toxins of the entomophagous fungus *Beauveria bassiana. Journal of Invertebrate Pathology*, 12: 316-320.

Kybal, J. and Vleek, V. 1976. A simple Device for Stationary Cultivation of Microorganisms. *Biotechnology and Bioengineering*, 18: 1713 - 1718.

Lakshmi, S.M., Alagammai, P.L. and Jeyaraj, S. 2001. Studies on mass culturing of the entomopathogenic white halo fungus *Verticillium lecanii* on three-grain media and its bioefficacy on *Helicoverpa armigera* (Hubner) on pigeonpea. In: Ignachimuthu, S. and Sen, A. (Eds.), *Micorbials in insect pest management*. Oxford and IBH publishing co. pvt. Ltd., New Delhi, pp. 174.

Leite, L.C.; Batista Filho, A.; Almeida, J.E.M. and Alves, S.B. 2003. *Produção de fungos entomopatogênicos*. Ribeirão Preto, Brazil.

Lumsden, R.D., Walter, J.F.and Baker, C.P. 1996. Development of *Gliocladium virens* for damping-off disease control. *Canadian Journal of Plant Pathology*, 18: 463-468.

Luz, C., Netto, M.C.B. and Rocha, L.F.N. 2007. In vitro susceptibility to fungicides by invertebrate-pathogenic and saprobic fungi. *Mycopathologia*, 164 (1): 39-47.

Masera, E. 1936. Contributo allo studio della virulenza epathogenicità di alcuni entomomiceti. *Annuario della R. Stazione Bacologica Sperimentale di Padova*, 48: 477-491.

Mc Coy, C.W. 1990. Entomogenous fungi as microbial pesticides. In: Baker, R.R. and Dunn, P.E. (Eds.), *New directions in biological control: Alternatives for suppressing agricultural pests and diseases*, Alan R Liss. New York, 112: 139-159.

Mehrotra, K.N. 1989. Pesticide resistance in insect pests: Indian Scenario. *Pesticide Research Journal,* 1: 95– 103.

Mendonça, A.F. 1992. Mass production, application and formulation of *Metarhizium anisopliae* for control of sugarcane froghopper, *Mahanarva posticata*, in Brazil. In: Lomer, C.J. and Prior, C. (Eds.), *Biological control of locusts and grasshoppers*. CAB International, Wallingford, U.K., pp. 239-244.

Mohan, K.G., Kumaraswamy, G., Mathavan, S. and Muraleedharan, D. 2005. Characterization and cDNA cloning of Apolipophorin III Gene in the red cotton bug, *Dysdercus cingulatus*. *Entomon*, 29(4): 373-381.

Moore, D., Bateman, R.P., Carey, M., Prior, C. 1995. Long-term storage of *Metarhizium flavoviride* conidia in oil formulations for the control of locusts and grasshoppers. *Biocontrol Science and Technology*, 5: 193–199.

Moore, R.E., Corbett, T.H., Patterson, G.M.L. and Valeriote, F.A. 1996. The search for new antitumor drugs from blue-green algae. *Curr. Pharm. Des*, 2: 317-330.

Moo-Young, M., Moreira, A.R. and Tengerdy, R.P. 1983. Principles of solid substrate fermentation. In: Smith, J.E., Berry, D.R. and Kristiansen, B. (Eds.), *Fungal technology*. London; Edward Arnold, pp. 117-144.

Muller-Kogler, E. and Samsinakova, A. 1970. Zur Massenkultur des Insektenpathogenen Pilzes *Beauveria bassiana*. *Experientia*, 26: 1400.

Nasr, F.N., Regheb, W.S. and Hanna, M.T. 1992. Microbial control of cottonseed bug *Oxycarenus hyalinipennis* Costa, affecting seed germination in okra *Hibicus esculentus*. *Journal of Pest Control and Environmental Science,* 4(2): 79-90.

Nuñez-Gaona Oscar, Saucedo-Castañeda Gerardo, Alatorre-Rosas Raquel, Loera Octavio. 2010. Effect of moisture content and inoculum on the growth and conidia production by *Beauveria bassiana* on wheat bran. *Braz. arch. biol. technol.* 53(4): 771-777.

Papierok, B. 1978. Obtention *in vivo* des azygospores d' *Entomophthora thaxterianan* Petch, champignon pathogene de pucerons (Homopteres, Aphididae). *Comptes Rendus de l'Academie des Sciences Paris*, 286 (D): 1503-1506.

Papierok, B. and Hajek, A.E. 1997. Fungi: Entomophthorales. In: Lacey, L.A. (Ed.), *Manual of Techniques in Insect Pathology.*Academic Press, London, pp. 187-212.

Pereira, R.M. and Roberts, D.W. 1990. Dry mycelium preparations of the entomopathogenic fungi *Metarhizium anisopliae* and *Beauveria bassiana*. *Journal of Invertebrate Pathology*, 56: 39–46.

Posada-Flórez FJ. 2008. Production of *Beauveria bassiana* fungal spores on rice to control the coffee berry borer, *Hypothenemus hampei*, in Colombia. *Journal of Insect Science* 8:41-53.

Pratt, J.E., Johansson, M. and Hüttermann, A. 1998. Chemical control of *Heterobasidion annosum*. In: Woodward, S., Stenlid, J., Karjalainen, R. and Hüttermann, A. (Eds.), *Heterobasidion annosum: Biology, Ecology, Impact and Control*. CAB International, Wallingford, UK., pp. 259-282.

Prior, C., 1988. Biological pesticides for low external-input agriculture. *Biocontrol News and Information*, 10: 17–22.

Purwar, J.P. and Sachan, G.C. 2006. Insect pest management through entomogenous fungi: a review. *Journal of Applied Bioscience,* 32(1): 1-26.

Puzari, K.C., Sharmah, D.K. and Hazarika, L.K. 1997. Medium for mass production of *Beauveria bassiana* (Balsamo) Vaillemin. *Journal of Biological Control*, 11: 96-100.

Rao, N.V., Maheswari, T.U., Prasad, P.R., Naidu, V.G. and Savithri, P. 2003. Integrated Insect Pest Management, Agrobios India, Jodhpur, India. 248 pp.

Rombach, M.C., Aguda R.M. and Roberts, D.W. 1988. Production of *Beauveria bassiana* (Deuteromycotina: Hyphomycetes) in different liquid media and subsequent conidiation of dry mycelium. *Entomophaga*, 33: 315–324.

Rosell, G., Quero, C., Coll, J. and Guerrero, A. 2008. Biorational insecticides in pest management *Jornal of Pesticide Science*, 33(2): 103–121.

Sahayaraj, K. and Karthick Raja Namasivayam, S. 2008. Mass production of entomopathogenic fungi using agricultural products and byproducts. *African Journal of Biotechnology*, 7(12): 1907 – 1910.

Sallam, M.N., McAvoy, C.A., Samson, P.R. and Bull, J.J. 2007. Soil sampling for *Metarhizium anisopliae* spores in Queensland sugarcane fields. *Bio Control*, 52(4): 491-505.

Samsinakova, A. 1966. Growth and sporulation of submerged cultures of the fungus *Beauveria bassiana* in various media. *Journal of Invertebrate Pathology*, 8: 395-400.

Scholte, E.J., Njiru, B.N., Smallegange, R.C., Takken, W. and Knols, B.G.J. 2003. Infection of adult malaria (*Anopheles gambiae s.s.*) and Wlariasis (*Culex quinquefasciatus*) vectors with the entomopathogenic fungus *Metarhizium anisopliae*. *Malaria Journal*, 2: 29.

Shah, F.A., Prasad, M. and Butt, T.M. 2007. A novel method for the quantitative assessment of the percolation of *Metarhizium anisopliae* conidia through horticultural growing media, *BioControl*, 52: 889–893.

Shah, F.A., Wang, C.S. and Butt, T.M. 2005. Nutrition influences growth and virulence of the insect-pathogenic fungus *Metarhizium anisopliae*. *FEMS Microbiology Letters*, 251: 259–266.

Shah, P.A.and Goettel, M.S. 1999. *Directory of microbial control products and services*, 2nd Edn. Division on Microbial Control, Society for Invertebrate Pathology.

Sun, M. and Liu, X. 2006. Carbon requirements of some nematophagous, entomopathogenic and mycoparasitic Hyphomycetes as fungal biocontrol agents. *Mycopathologia*, 161: 295–305.

Thomas, M.B. and Jenkins, N.E. 1997. Effects of temperature on growth of *Metarhizium flavoviride* and virulence to the variegated grasshopper *Zonocerus variegatus*. *Mycological Research*, 101: 1469- 1474.

Tincilley, A., Easwaramoorthy, S. and Santhalakshmi, G. 2004. Attempts on mass production of *Nomuraea rileyi* on various agricultural products and by products. *Journal of Biological Control*, 18(1): 35-40.

Toledo, A.V., de Remes Lenicov, A.M.M. and Lastra, C.C.L. 2007. Pathogenicity of fungal isolates (Ascomycota: Hypocreales) against *Peregrinus maidis*, *Delphacodes kuscheli* (Hemiptera: Delphacidae), and *Dalbulus maidis* (Hemiptera: Cicadellidae), vectors of corn diseases. *Mycopathologia*, 163: 225–232.

Torre, M. de la and Cardenas-Cota, H.M. 1996, Production of *Paecilomyces fumosoroseus* conidia in submerged culture. *Entomophaga*, 41, 443-453.

van Lenteren, J.C., Antoon, J. M., Loomans, J., Babendreier. D. and Bigler, F. 2008. Harmonia axyridis: an environmental risk assessment for Northwest Europe *BioControl*, 53: 37–54.

Van Winkelhoff, A.J. and McCoy, C.W. 1984 Conidiation of *Hirsutella thompsonii* var. synnematosa in submerged culture. *Journal of Invertebrate Pathology*, 43, 59–68.

Verma, M., Satinder, K. B., Tyagi, R. D., Surampalli, R. Y., and Valeró, J. R. 2005. Wastewater as potential for antagonistic fungus (*Trichoderma* sp): Role of pre-treatment and solids concentration. *Water Reaserch*, 39, 3587–3596.

Vouk, V. and Klas, Z. 1931. Conditions influencing the growth of the insecticidal fungus *Metarhizium anisopliae* (Metsch.) Sor. *International Corn Borer Investigations*, 4: 24-45.

Waterhouse, D.F. 1998. Biological Control of Insect Pests: Southeast Asian Prospects, Australian Centre for International Agricultural Research, Canberra. 523 pp.

Wilding, N. 1981. Pest control by Entomophthorales. In: Burges, H. D. (Ed.), *Microbial Control of Pests and Plant Diseases.* Academic Press, London, pp. 539-554.

Williams, C.N. 1959. Spore size in relation to culture conditions. *Transactions of the British Mycological Society*, 42: 213-222.

Wilson, M.J., Glen, D.M. and George, S.K. 1993. The Rhabditid nematode *Phasmarhabditis hermaphrodita* as a potential biological control agent for slugs. *Biocontrol Science and Technology*, 3: 503-511.

Wraight, S.P., Jackson, M.A. and de Kock, S.L. 2001. Production, stabilization and formulation of fungal biocontrol agents. In: Butt, T.M., Jackson, C.W. and Magan, N. (Eds.), *Fungi as Biocontrol Agents: Progress, Problems and Potential*, CABI Publishing, Wallingford, UK, pp. 253–287.

Zhang, S., Guoxiong Peng. And Yuxian Xia. 2010. Microcycle conidiation and the conidial properties in the entomopathogenic fungus *Metarhizium acridum* on agar medium. *Biocontrol Science and Technology,* 20 (8): 809 – 819.

Zimmermann, G. 1993. The entomopathogenic fungus *Metarhizium anisopliae* and its potential as a biocontrol agent. *Pesticide Science*, 37: 375-379.

In: Microbial Insecticides: Principles and Applications   ISBN: 978-1-61209-223-2
Editors: J. Francis Borgio, K. Sahayaraj, et al.   © 2011 Nova Science Publishers, Inc

Chapter 4

ENZYMOLOGY OF ENTOMOPATHOGENIC FUNGI

*Richard Ian Samuels[1] *, Adão Valmir dos Santos[2] and Carlos Peres Silva[3]*

[1]Department of Entomology and Plant Pathology, State University of North Fluminense-
UENF, Avenida Alberto Lamego 2000, Campos dos Goytacazes, RJ 28013-602
[2]Departmentof Biotechnology, State University of North Fluminense-UENF, Avenida
Alberto Lamego 2000, Campos dos Goytacazes, RJ 28013-602
[3]Department of Biochemistry, Centro de Ciências Biológicas, Universidade
Federal de Santa Catarina, Florianópolis, SC, 88040-900, Brazil

ABSTRACT

Entomopathogenic fungi secrete a range of enzymes intrinsically involved in the
pathogenicity process, mainly during cuticle penetration but these enzymes also have a
role during colonization of the host insect.

The two most important classes of enzymes involved in cuticle penetration are the
peptidases and chitinases, specifically produced by the fungus to breakdown the main
components of the cuticle (protein and chitin). Other enzyme activities detected during
adhesion, germination and penetration, such as that of lipases would appear to be
involved in breakdown of insect defensive barriers. Trehalases may have a fundamental
role in sequestering nutrients for the fungus during the haemolymph colonization phase.
This review describes the role of enzymes in cuticle penetration and subsequently during
the colonization process.

Furthermore, the possible use of molecular modifications of the fungi in relation to
expression of the key enzymes is discussed as method of genetically improving biological
control agents. Finally, the industrial applications of enzymes of fungal origin are
discussed.

* E- mail: richard@uenf.br

4.1. INTRODUCTION

Entomopathogenic fungi infect their insect hosts by penetration of the cuticle, utilizing enzymatic and/or physical mechanisms (Charnley and St. Leger, 1991). Protein and chitin are the major components of insect cuticle representing a significant barrier to the invading fungus, but also representing a valuable source of energy for the developing microorganism. Insect cuticle is composed mainly of crystalline chitin microfibres embedded in a protein matrix (Hepburn, 1985). Therefore it would be predicted that entomopathogenic fungi secrete enzymes that degrade the components of the insect cuticle.

The ability of entomopathogenic fungi to produce extracellular enzymes with activity towards the main chemical constituents of insect cuticle has been known for some years. Gabriel (1968) demonstrated that certain species of Entomphthora were capable of producing lipolytic, proteolytic and chitinolytic enzymes. Similar enzyme activities were subsequently identified in culture filtrates of *Beauveria bassiana* (Leopold and Samsinakova, 1970) and *Metarhizium anisopliae* (Kucera, 1980).

However, a realistic profile of enzyme production was only discovered following ground-breaking research by St. Leger and co-workers (from 1986), utilizing insect cuticle fragments suspended in liquid media to simulate natural conditions. Using this method, enzymes capable of hydrolysing cuticle constituents were detected in the culture filtrates of entomopathogenic fungi (St. Leger *et al.,* 1986a,b). Proteolytic enzymes are thought to attack the cuticle before chitinolytic enzymes (chitinases and β-N-acetylglucosaminidases) as protein masks the chitin microfibers (Smith *et al.,* 1981). Esterase and proteolytic enzymes (endopeptidase, aminopeptidase, and carboxypeptidase) are produced first (around 24h after inoculation) followed by N-acetylglucosaminidase. Chitinase and lipase were first detected 3-5 days after innoculation. The order of appearance of the enzymes is supported by the sequence of cuticle constituents solubilized, with a rapid release of amino acids. Since chitinase is an inducible enzyme (Smith and Grula, 1983; St. Leger *et al.,*1986c), and cuticular chitin is masked by protein (St. Leger *et al.,*1986a,d), the late appearance of chitinase is presumably a result of induction as chitin eventually becomes available after degradation of the encasing cuticle proteins. The late detection of lipase appears to be due to the fact that the enzyme may be cell bound during initial growth. By testing purified enzymes against locust cuticle *in vitro,* St. Leger *et al.* (1986a,b) showed that pre-treatment or combined treatment with endopeptidase (Pr1; see later) was necessary for high chitinase activity. When locust exuviae (non-digested remains of old cuticle shed at ecdysis) were used as substrate for purified fungal enzymes instead of cuticle from larval sclerites, comparatively little hydrolysis occurred (St. Leger   *et al.,* 1986b). To date peptidases from entomopathogenic fungi have been more thoroughly investigated than chitinases.

Apart from the more obvious role of enzymes in the penetration of the host integument, enzymes may be important in the initiation of the pathogenicity process, during adhesion of conidia to the host cuticle (Boucias and Pendland, 1991), during germination and the colonization process following penetration (Charnley, 2003).

The secretion of enzymes by entomopathogenic fungi has been stated to be a pathogenicity factor and/or virulence factor, especially in the case of the enzymes proven to be actively involved in cuticle degradation during host penetration.

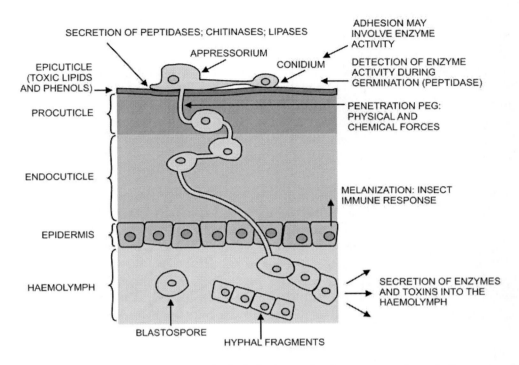

Figure 1. Summary of the events involved in the infection cycle of an entomopathogenic fungus attacking an insect host.

For enzymes considered as virulence factors, speed of colonization related to high expression of cuticle degrading enzymes could be important as well as the possible suppression of immune responses (or over stimulation of immune responses) by the host. Figure 1 shows a summary of the events involved in the infection cycle of an entomopathogenic fungus attacking an insect host via the integument. We will also discuss the possible application of fungal enzymes in crop protection, genetic improvement of fungi using molecular biology techniques and subsequent production of new biological control agents, which may kill their hosts faster and more efficiently. Finally, we will also consider the commercial use fungal enzymes, some of which have already been patented, with diverse biotechnological applications.

4.2. ENZYMES INVOLVED WITH THE ADHESION PROCESS

The infection process of entomopathogenic fungi can be separated into three phases: a) adhesion and germination of the conidia on the insect's cuticle; b) penetration of the conidia into the haemocoel by the secretion of different enzymes such as lipase, chitinase and peptidase and c) development of fungus which results in the death of the host (Amer *et al.*, 2008). The adhesion process is of paramount importance for pathogenic fungi during the colonization of the host. In the first phase, physiochemical properties of the conidia are involved and in the second phase, secretion of enzymes and mucilage are involved in the consolidation of the attachment process and establishing the infection (Boucias *et al.*, 1988).

Some fungi, as *Metarhizium anisopliae*, have evolved mechanisms to adhere to a variety of biological surfaces that initiate and maintain pathogenic and mutualistic interactions. The proteins MAD1 and MAD2, produced by *M. anisopliae*, are responsible for conidia adhesion in insect and plant hosts, respectively. MAD1 cDNA encodes a protein with a molecular mass of 74.6 kDa and MAD2 is a 30.5 kDa protein. Disruption of these genes reduced adherence in approximately 90% (Wang and St. Leger, 2007).

According to Shah *et al.* (2007), the spore-bound Pr1 may help consolidate the initial adhesion of conidia by modifying the cuticle surface, allowing the formation of cross-linkages and providing nutrients for conidial germination.

4.3. THE ROLE OF ENZYMES IN THE GERMINATION PROCESS

Upon landing on a potential host, a fungal propagule initiates a series of steps that may lead to a compatible (infection) or a non-compatible (resistance) reaction. Alternatively, a propagule landing on an insect may elicit no reaction because of the absence of recognition between the fungus and the insect. In a compatible reaction, fungal recognition and attachment proceed to germination on the host cuticle, followed by penetration of the cuticle and colonization of the insect haemocoel (Castrillo *et al.,* 2005).

The infection process involves complex insect host–fungal pathogen interactions that vary among fungi and even among different strains (isolates) of a species for a given host. The initial step in infection, attachment, may be passive and non-specific as has been shown by Boucias *et al.* (1988) for entomopathogenic deuteromycetes such as *B. bassiana, M. anisopliae*, and *Nomuraea rileyi* on host and non-host insects. In these fungi, attachment is due to the hydrophobic interaction between the insect cuticle and the well-organized fascicles of rodlets in the conidia. These rodlets are formed from the self-assembly of hydrophobic proteins or hydrophobins in fungal aerial structures (Kershaw and Talbot, 1998). These hydrophobins were found in aerial conidia of *B. bassiana* but not in blastospores (Bidochka *et al.,* 1995). A mucilaginous coat also permits passive attachment in Entomophthorales and some deuteromycetes (Boucias and Pendland, 1991; Brey *et al.,* 1986). In contrast, attachment was shown to be selective in two *M. anisopliae* strains that were highly specific to their scarabaeid beetle hosts (Vey *et al.,* 1982).

In terrestrial fungi germination is followed by the formation of a germ tube (Boucias and Pendland, 1991) or a germ tube and appressorium (Madelin *et al.,* 1967; Zacharuk, 1970a), which forms a thin penetration peg that breaches the insect cuticle via mechanical (turgor pressure) and/or enzymatic means (e.g., peptidases) (Zacharuk, 1970b). An exocellular mucilage, proposed to enhance binding to the host cuticle, is also secreted by several entomogenous fungi during the formation of infective structures (see review by Boucias and Pendland, 1991). In *M. anisopliae*, appressorium formation and the expression of cuticle-degrading peptidases are triggered by low nutrient levels (St. Leger *et al.,* 1992), demonstrating that the fungus senses environmental conditions or host cues at the initiation of infection. The production of cuticle-degrading enzymes, proteases, chitinases, and lipases, has long been recognized as an important determinant of the infection process in various fungi, facilitating penetration as well as providing nutrients for further development (Charnley, 1984)

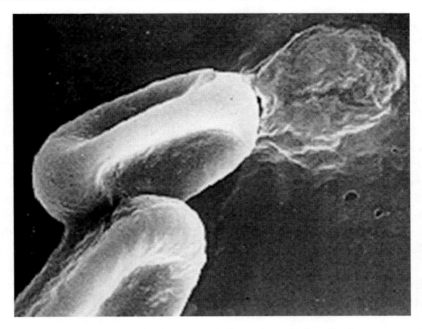

Figure 2. Scanning electron micrograph of appressoria formation with localized enzyme production in the secretions surrounding this structure. (Micrograph kindly supplied by R.J. St. Leger, University of Maryland, USA).

The production of cuticle degrading enzymes by *M. anisoplaie* during appressoria formation has been investigated both *in vitro* and *in vivo*. The transparent wings of blowflies provide an ideal media for combined biochemical and histochemical analysis. Among the first enzymes produced on blowfly wings are endopeptidases and aminopeptidases, which are produced at the same time as the formation of appressoria. N-Acetylglucosaminidase is produced at a later stage when compared to proteolytic enzyme secretion. Chitinase and lipase activities were not detected at this stage. *In situ* histochemical localization confirms that high levels of proteolytic enzymes are secreted by appressoria. The major protein synthesized during appressorium development by *M. anisopliae* is the enzyme Prl. Figure 2 shows an electron-micrograph of appressoria formation on the insect cuticle.

Prl-like peptidases have been found in all deuteromycete and ascomycete entomopathogenic fungi tested to date, although the role of this enzyme is best understood in *Metarhizium*. The addition of either glucose or alanine to conidia germinating *in vitro* repressed both appressorium formation and Prl expression, suggesting regulation by catabolite repression.

Nutrient starvation is therefore likely to be a key environmental signal for the switch from a saprophytic to a pathogenic mode of growth, possibly after depletion of nutrients on the insect surface. Interestingly, isolates of *M. anisoplaie* have been identified in which the differentiation of appresoria is not glucose "repressed". Some of these "catabolite de-repressed" strains were originally isolated from homopteran insects (plant sap suckers), in which nutrients on the cuticle are likely to be supplemented with insect secretions rich in sugars.

4.4. ENZYMES INVOLVED IN PENETRATION OF INSECT INTEGUMENT

Fungi differ from bacteria and viruses in that they are capable of invading the host by penetrating its cuticle. Microbial pathogens apart from nematodes need to be ingested to initiate an infection.

St. Leger and co-workers have shown that entomopathogenic fungi produce a wide range of peptidases (St. Leger *et al.,* 1987c). The high levels of peptidase activity detected in the culture filtrates of *M. anisopliae* (strain ME1), suggested a key role for this type of enzyme in cuticle penetration (St. Leger *et al.,* 1987b). Three major peptidases have so far been purified and characterized from the culture filtrates of *M. anisopliae,* Prl, Pr2 (St. Leger *et al.,* 1987b), and Pr4 (Cole *et al.,* 1993). Prl is a chymoelastase with an active site serine residue, showing strong homology with the Subtilisin class of peptidase (St. Leger *et al.,* 1992) and possessing a broad primary specificity for amino acids with a hydrophobic side group at the second carbon atom (e.g. phenylalanine, methionine and alanine). Prl acts as a general peptidase, hydrolysing a wide range of substrates, e.g. casein, elastin, bovine serum albumen, collagen and most importantly, insect cuticle (St. Leger *et al.,* 1987b). Pr2 is a serine peptidase which apparently occurs as three iso-enzymes (St. Leger *et al.,* 1987b). Only one of the iso-enzymes was characterized and shown to be a classical trypsin-like enzyme with a primary specificity for arginine and lysine residues. Pr2 degrades casein and albumin, but unlike Prl has no activity against elastin. It was also stated to have 4 to 10% of the cuticle-degrading activity of Prl (St. Leger *et al.,* 1987b). However, in a subsequent study, Pr2 was isolated as a single iso-zyme with a pI of 5.4 and had 21% of the cuticle-degrading activity of Prl (Cole *et al.,* 1993). A cysteine peptidase with cuticle degrading activity was characterized from *M. anisopliae* culture filtrates and designated as Pr4 (Cole *et al.,* 1993). Pr4 exhibited trypsin-like specificities, but on the basis of inhibition and activation studies was classified as a cysteine peptidase. Pr4 was in fact more specific in its requirements than Pr2 as it preferred elongated substrates. Pr4 was also more effective in hydrolyzing insect cuticle than Pr2, exhibiting 51% of the activity of Prl. A peptidase with a pH optimum of 5.0-5.5, designated Pr3, was identified by St. Leger and co-workers (1987b). This enzyme, surprisingly, was not characterized further nor its cuticle degrading ability established. Several isolates of *M. anisopliae* produce acidic proteases that occur as multiple iso-enzymes and resemble Prl in their primary specificity, but do not degrade elastin (St. Leger *et al.,* 1987c). Peptidases with similar properties to both Prl and Pr2 have also been isolated from culture filtrates of the entomopathogenic deuteromycetes, *B. bassiana, Verticillium lecanii, Nomuraea rileyi* and *Aschersonia aleyrodris* (St. Leger *et al.,* 1987c). Despite the similarity of the substrate specificities of the chymoelastases from these organisms, the enzymes of all but two strains of *Metarhizium* failed to cross-react (as determined by Ochterlony immunodiffusion) with antibodies raised against Prl from the *M. anisopliae* strain ME1. A single peptidase is produced by *B. bassiana* when grown with gelatin as a sole carbon and nitrogen source. This enzyme, which also had elastolytic activity, was shown to be inhibited by PMSF, has a Mr of 35 kDa and a pH optimum of 8.5 (Bidochka and Khachatourians, 1987). Peptidases with activity *vs* collagen have also been identified in culture filtrates of the entomopathogens *Entomophthora coronata* (Hurien *et al.,* 1977) and *Lagenidium giganteum* (Dean and Domnas, 1983). In contrast to the five species of deuteromycetes described by St. Leger *et al.* (1987c), in which Prl and Pr2-1ike activities are found as separate enzymes, the entomopathogenic entomophthoraceae, *Erynia rhizospora, E. dipterigena* and *E. neoaphidis,* produce single peptidases with activity against tryptic and chymotryptic substrates (Samuels *et al.,* 1990). Prl appears to be a pathogenicity determinant in *M. anisopliae,* first by its ability to degrade cuticle and second by its presence in high levels at the site of fungal penetration (St. Leger *et al.,* 1989; Goettel *et al.,* 1989). Simultaneous applications of fungal conidia and the Prl inhibitor TEl (Turkey egg-white inhibitor) significantly delayed the mortality of *Manduca* larvae, indicating the importance of Prl in the infection process (St. Leger *et al.,* 1988). However, two pathogenic strains of *M. anisopliae* (RS 703 and RS 2134) have been shown to produce very little peptidase activity when grown on insect cuticle *in vitro* (Gupta *et al.,* 1991). The molecular cloning of the Prl gene (St. Leger *et al.,* 1992) and the development of transformation protocols for entomopathogenic fungi (Bernier *et al,* 1989; Daboussi *et al.,* 1989; Goettel *et al.,* 1990) should enable the elucidation of the exact role of Prl in pathogenesis. Transformation of Prl deficient mutants with the Prl gene and a comparison of the pathogenicity of the parental and transformed isolates would provide critical evidence for the role of Prl in cuticle penetration and pathogenesis.

To date 11 subtilisins (Prl) type enzymes have been characterized from *Metarhizium anisopliae* (Bagga *et al.,* 2004), the highest diversity of subtilisins reported for any fungus. Results suggested that each Pr1 paralog contributes to the pathogens fitness. Furthermore, homology modeling predicted differences between the Prl's in their secondary substrate specificities, adsorption properties to cuticle and alkaline stability, indicative of functional differences.

The role of Pr2 has been suggested to be in cellular control mechanisms, catalysing specific proteolytic inactivation and activation processes (St. Leger *et al.,* 1987b; Charnley and St. Leger, 1991). In that case it is surprising to find that Pr2 has been detected on insect cuticle 16hr after inoculation with *M. anisopliae* (St. Leger *et al.,* 1987a). The possibility that the trypsin-like activity detected on infected cuticle was a combination of Pr2 and Pr4 cannot be ruled out as both enzymes degrade similar substrates.

4.4.1. Peptidases Produced by Entomopathogenic Fungi

Aminopeptidase activities have been identified in pathogenic fungi as diverse as *Verticillium cinnabarium, Candida albicans* and *Fusarium* oxysporum. Of the entomopathogenic fungi so far screened for aminopeptidase production, the deuteromycetes and entomophthorales produce a broad spectrum of aminopeptidase activities (Samuels *et al.,* 1990), suggesting an important role for these enzymes in cuticle degradation. St. Leger and his co-workers (1993) partially characterized an aminopeptidase and a X-prolyl dipeptidyl peptidase (Mr 74 kDa, p*I* 4.51) from the culture filtrates of *M. anisopliae.* The aminopeptidase had optimal activity against alanine-β-naphthylamide and was fully inhibited by the metal chelator 1,10-phenanthroline and to a lesser extent by the substrate inhibitor amastatin. The dipeptidyl-peptidase showed a strong preference for substrates having a penultimate proline residue. Inhibition of this enzyme by diprotin A indicated a similarity to mammalian enzymes of this class (Umezama and Aoyagi, 1983). The peptidases from *Metarhizium* had no activity against intact cuticle. However, Charnley and St Leger (1991) state that when combined with Prl they enhanced release of amino acids, although no data was provided. Whilst limited data exits regarding the primary structure of cuticular protein, it is noteworthy that the sequences Ala-Ala-Pro, Ala-Ala-Pro-Ala, Ala-Ala-Val-Ala and Ala-Ala-Ala are repeated many times in several proteins from the migratory locust *Locusta migratoria*. The preferred artificial substrate of Prl is Suc-(Ala)$_3$-Phe-NA, although the substrates Suc-(Ala)$_2$-Pro-Phe-NA and Ac-(AIa)$_3$-NA are also readily hydrolysed (St. Leger *et al.,* 1987b). These substrates have a striking resemblance to the repetitive peptide sequences found in locust cuticle, which would therefore provide excellent substrates for Prl. The action of Prl on cuticle would thus release many alanine- and proline rich peptides, which in turn would be excellent substrates for the aminopeptidase and dipeptidyl peptidase. These enzymes could therefore act synergistically to completely hydrolyze locust cuticle. Aminopeptidase activity has been detected histochemically on mature appressoria and in the surrounding mucilage produced by *M. anisopliae* during infection of blowfly cuticle (St. Leger *et al.,* 1987a).

4.4.2. Chitinases Secreted by Entomopathogenic Fungi

Chitin is present in arthropod exoskeletons and fungal cell walls and these organisms produce chitinases for growth regulation. Pathogens and predators of chitin-containing organisms also produce chitinases, whereas hosts of pathogens that contain chitin (fungi) produce chitinases to defend themselves (Gooday, 1999).

St. Leger *et al.* (1991) purified endochitinase from culture filtrates of *M. anisopliae* grown on 1% ground chitin. The purified enzyme failed to hydrolyze arylglycosides or chitobiose (N-acetylglucosamine dimer), showed only trace activity against chitotriose (trimer), but rapidly degraded chitotetraose (tetramer). Colloidal chitosan (deacetylated form of chitin) and crystalline chitin were degraded to a lesser extent than colloidal chitin, but even so activity was still significant. The chitinase from *M. anisopliae* had many similarities to those produced by other microorganisms (Stirling *et al.,* 1979). These properties include a pH optimum of 5.3, a molecular mass of ca. 33 kDa, and the lack of any requirement for a cofactor. Hydrolysis of crystalline chitin produced only one low-molecular-weight reaction product within 24h; N-acetylglucosamine (NAG). The absence of intermediary oligomers among chitin breakdown products probably means that NAG is released directly from insoluble chitin. Either the chitinase has an exo-acting component or alternatively the reaction proceeds by a single chain progressive mechanism as described for some other endo-acting polysaccharidases (Cooper *et al.,* 1978). This involves the random cleaving of bonds followed by release of monomers or dimers from exposed ends so that a single macromolecule is completely degraded before a new one is attacked. Such a mechanism, especially if it involved simultaneous digestion of several parallel chains, could result in the rapid degradation of chitin fibrils and in addition produce monomers for nutrition and induction for further enzyme synthesis.

Studies have shown the presence of a range of chitinase isoforms (10) in culture filtrates from *M. anisopliae* (St. Leger *et al.,* 1993b). Differences in molecular masses of the enzymes suggest that they are products of different genes rather than post-translational modifications, e.g. glycosylation (St. Leger *et al.,* 1996b). The isoforms produced on cockroach cuticle by *M. anisopliae* sf. *anisopliae* isolate 2575 and *M. anisopliae* sf. *acridum* isolate 324 may be broadly classified by *pI* as basic or acidic. The latter are dominant. Two isoforms 43.5 and 45kDa, *pI* 4.8 have been purified. In this case similarities between N-termini suggest that they may not be separate gene productions but rather result from post-translational modifications, particularly glycosylation. Pinto *et al.* (1997) found a 30-kDa endochitinase from a different isolate of *M. anisopliae* with similar Mr, pH and temperature to the one reported previously by St. Leger *et al.* (1991). However, a 60-kDa endochitinase by Kang *et al.* (1998, 1999) differed from the iso-forms described by St. Leger *et al.* (1991, 1996b) not only in Mr but also protein N-terminus and ORF nucleotide sequence. Bogo *et al.* (1998) isolated a cDNA for a chitinase gene encoding an endochitinase of 58 kDa. The ORF encoded a protein with a predicted final mass of 43 kDa. This compared well with the 45-kDa chitinase purified biochemically by St. Leger *et al.* (1996b). N-Acetylglucosaminidase activity has been partially purified from culture filtrates of *M. anisopliae* grown on 1% ground chitin. The enzyme had substantial activity against p-nitrophenol acetylglucosamine, as well as chitobiose, chitotriose, and chitotetraose, the major product in each case being NAG, showing that the enzyme is a true NAGase rather than a chitobiase (St. Leger *et al.*, 1991). The enzyme had little activity against colloidal or crystalline chitin.

Ultrastructural immunocytochemistry enabled production of CHIT1 (St. Leger *et al.*, 1996a) to be visualized during penetration of host (*Manduca sexta*) cuticle. Chitinase was produced at very low levels by infection structures at the cuticle surface and during the initial penetration of the cuticle. Much greater levels of chitinase accumulated in zones of degradation produced by the peptidases, which suggests that the release of the chitinase is dependent on the accessibility of its substrate (St. Leger *et al.*, 1996a,d). This hypothesis is consistent with the known protective action of protein associated with chitin microfibrils in cuticle (St. Leger *et al.*, 1986a; Hassan and Charnley, 1989) and the induction of chitinases by chitin degradation products that was shown *in vitro* to be facilitated by the unmasking of cuticular chitin (St. Leger *et al.*, 1986b,c). Although it seems likely that proteases initiate cuticle degradation, allowing the chitinase to permeate the cuticle, there is substantial evidence that proteases act synergistically with chitinases in the solubilization of the cuticle (St. Leger *et al.*, 1986a,c). Consistent with this, ultrastructural studies have shown massively enhanced degradation zones and fungal penetration through cuticles of insects treated with an inhibitor of chitin synthesis (Hassan and Charnley, 1989).

Extracellular chitinase activity has been implicated in the pathogenesis of several fungal infections. Following induction with chitin, the insect pathogens *M. anisopliae* sf. *acridum* ARSEF strain 324 and *M. anisopliae* sf. *anisopliae* ARSEF strain 2575 secrete 44-kDa basic and acidic iso-forms of endochitinase, respectively (Screen *et al.*, 2001). These authors found that genetically manipulating these isolates to over-produce chitinase did not result in altered virulence to caterpillars (*Manduca sexta*) compared to the wild-type fungus, suggesting that wild-type levels of chitinase are not limiting for cuticle penetration.

The involvement of chitinase in fungal modification of the procuticle raised the possibility that over-expression of chitinases may provide a means to improve fungal pathogenesis as has been described for hydrolase's from several other systems.

4.4.3. Lipolytic and Esterolytic Enzymes

There has been little work published on entomopathogenic fungal enzymes that are active against lipids and esters. This is perhaps not a serious omission in the role of enzymes in cuticle penetration, because substrates for these enzymes are not important components of the epi- or procuticle. However, hydrocarbons are major constituents of the wax and may be metabolized by germinating fungi (Lecuona *et al.,* 1991). Alternatively, the lipids present in the wax layers may have a role in insect defense against fungal invasion, inhibiting the growth of fungi or inhibiting enzyme activity (for review of integument as a barrier to microbial infection see: St. Leger, 1991). An interesting example of a defensive wax layer can be seen in the scale insects (Hemiptera: Coccoidea). Lipase secretion by isolates of *B. bassiana*, was correlated to virulence against the scale insect *Dactylopius spp.* (Hemiptera: Dactylopiidae), a serious pest of the cactus palm in the NE of Brazil (Samuels, R.I. unpublished results).

Esterase activity produced by *M. anisopliae* in 3-day-old cultures was highest against short- and intermediate-length p-nitrophenol esters with only trace activity occurring above C10, suggesting that lipase is not produced extracellularly by young mycelia. Activity against C14 was detected in older cultures (7-14 days). Expression of extracellular lipase *in vitro* was confirmed using the 'true' lipase substrate olive oil (St. Leger *et al.,* 1986a). Flat-bed IEF, however, revealed 25 distinct esterases (isozymes) from culture filtrates of *M. anisopliae* grown on locust cuticle. On the basis of their reactions with naphthyl esters, the isozymes appeared to have different substrate specificities. However, all the bands were inhibited by PMSF, indicating that they are serine carboxyesterases (esterase B) and not arylesterases (esterase A) that are inhibited by N-ethyl-maleimide. Esterases catalyse many enzymatic reactions though preferentially hydrolysing aliphatic or aromatic esters and amides (Heymann, 1980). The considerable heterogeneity of esterases could account for their collective lack of specificity. Multiple enzyme strategies are believed to play an important role in the ability of an organism to adapt to different environments (Moon, 1975), presumably including that provided by an insect host.

4.5. ENZYME REGULATORY MECHANISMS

The regulation of production of Prl and Pr2 from *M. anisopliae* has also been studied. As with many other fungal proteases, the synthesis of both enzymes is controlled by multiple regulatory circuits. These include carbon and nitrogen de-repression (St. Leger *et al.,* 1988) and induction (Paterson *et al.,* 1993, 1994a). Nuclear run-on experiments have shown that Prl mRNA is absent in rapidly growing cells, but is transcribed within 2 hr of nutrient starvation (St. Leger *et al.,* 1992). Under conditions of carbon and nitrogen starvation Pr2 is induced by a range of proteinaceous substrates; whereas Prl is only induced by insect cuticle (Paterson *et al.,* 1993, 1994a) and specifically by cuticular proteins (Paterson *et al.,* 1994b). The specificity of Prl induction presumably reflects the adaptation of *M. anisopliae* to insect parasitism. St. Leger and co-workers (1987c) have reported that *V. lecanii, B. bassiana, Tolypocladium niveum* and *Paecilomyces farinosus* also produce a Prl-like enzyme during nutrient deprivation. Pr2 is detectable in de-repressed cultures before Prl. This may also be the case in germinating conidia of *M. anisopliae* which initially secrete Pr2 in order to provide nutrients during early saprophytic growth or to activate proteins which are present in zymogenic form in the conidia. Treatment of germinating conidia of *M. anisopliae* with the Pr2 inhibitor, TLCK, selectivity repressed formation of infection structures (R. J. St Leger, pers. comm.).

The cuticle degrading activity of Pr2 may also provide peptides that are capable of inducing Prl production (Paterson *et al.*, 1994b). The cysteine protease Pr4 is not produced under the same conditions as either Prl or Pr2. Pr4 was originally purified from 6-day-old cuticle cultures with little activity detected until day 3, which implies that Pr4 may be produced later in the infection process. There is some evidence to suggest that Pr2 and Pr4 may have a regulatory role as both enzymes facilitate the rapid activation of the pro-Prl zymogen, whereas auto-activation by pure Prl occurs at a much lower rate.

To explore the molecular basis of enzyme regulation, gene expression responses of *Metarhizium* to diverse insect cuticles were surveyed, using cDNA microarrays constructed from an expressed sequence tag (EST) clone collection of 837 genes (Freimoser *et al.*, 2005). During growth in culture containing caterpillar cuticle (*Manduca sexta*), *M. anisopliae* up-regulated 273 genes, representing a broad spectrum of biological functions, including cuticle-degradation (e.g. peptidases), amino acid/peptide transport and transcription regulation. There were also many genes of unknown function. The 287 down-regulated genes were also distinctive, and included a large set of ribosomal protein genes. The response to nutrient deprivation partially overlapped with the response to *M. sexta* cuticle, but unique expression patterns in response to cuticles from another caterpillar (*Lymantria dispar*), a cockroach (*Blaberus giganteus*) and a beetle (*Popilla japonica*) indicate that the pathogen can respond in a precise and specialized way to specific conditions. Comparisons between *M. anisopliae* and published data on *Trichoderma reesei* and *Saccharomyces cerevisiae* identified differences in the regulation of glycolysis-related genes and citric acid cycle/oxidative phosphorylation functions. In particular, *M. anisopliae* has multiple forms of several catabolic enzymes that are differentially regulated in response to sugar levels. These may increase the flexibility of *M. anisopliae* as it responds to nutritional changes in its environment (Freimoser *et al.*, 2005).

4.6. ROLE OF FUNGAL ENZYMES DURING COLONIZATION

Efforts to understand the molecular parameters of fungal virulence have largely focused on the complex array of enzymes they secrete during penetration and colonization as they are thought to be primarily involved in this necrotrophic stage of pathogenesis (Walton, 1996). It is inferred that the production of enzymes that disrupt the physiological integrity of hosts will have a strong selective advantage for pathogens. The possible role of enzymes secreted by the fungi during colonization has been little studied. This may be due to the difficult nature of this type of study when compared to those carried out on the enzymes thought to be involved in cuticle degradation. The strategy of the fungus during this stage may include the production of toxins to reduce the impact of the cellular immune response. However, the acquisition of nutrients during the haemolymph phase is little understood.

Inside the insect haemocoel the fungus switches from filamentous hyphal growth to produce yeast-like hyphal bodies or protoplasts that circulate in the haemolymph and proliferate via budding (Boucias and Pendland, 1982). At a later stage, the fungus switches back to filamentous type growth and begins to invade internal tissues and organs (Mohamed *et al.*, 1978).

The situation in the haemolymph of an insect host under attack is complex. Whilst the fungus is starting to attack its host, the insect's immune responses can mask the actual sequence of events. For example, it is thought that the over-expression of protease (Pr1) by a genetically manipulated *Metarhizium* isolate, may result in the exaggerated immune response to invasion seen in *Manduca sexta*, when the haemolymph became highly melanized (St. Leger *et al.*, 1996c). Melanization of host insects suggests that the recombinants were activating the enzyme (phenoloxidase) responsible for synthesis of melanin, a key component in arthropod immunity and wound healing. In fact, the host haemolymph is the one of the most readily available sources of nutrients for the developing fungi and its depletion of nutrient may actually be responsible for host death.

A large number of phosphorylated compounds have been identified in the haemolymph of insects, including glucose-1-phosphate and many phosphoproteins. However, these phosphorylated compounds cannot be directly utilized by the fungal cells and instead must be hydrolyzed by phosphatase before assimilation. Therefore the fungus must secrete corresponding hydrolytic enzymes. Xia *et al.* (2000) reported that *M. anisopliae* (strain ME1) secretes acid phosphatase (AcP) after invading the haemolymph of the desert locust *Schistocerca gregaria*. Simple sugars (principally the disaccharide trehalose) and phosphorylated sugars (e.g. glucose-1-phosphate) are found in high concentrations in the haemolymph. While repressive for fungal proteases they are a readily available source of carbon for the fungus, though utilization depends on the secretion of appropriate hydrolases. Attempts to identify fungal enzymes in insect haemolymph must take into account the fact that defense related proteins, including enzymes, may be produced by the host in response to the fungal invasion. The insect immune system comprises of a battery of humoral and cellular defenses that act against pathogens (reviewed by Gillespie *et al.,* 1997). The humoral defenses system involves the production of melanin and antimicrobial peptides. Gillespie *et al.* (2000) showed that mycosis of the desert locust with *M. anisopliae* var *acridum* resulted in changes in the properties of the haemolymph that occurred in two stages. During the first stage, (2 days after inoculation), there was an increase in total haemocyte count (mainly due to an increase in coagulocytes), number of nodules and increased pro-phenoloxidase (pPO) activity. This suggests that there is a "signal" which is either a host derived molecule (released from the integument during fungal penetration) or a soluble fungal metabolite that activates the immune system. It is apparent that, whilst the fungus stimulates the immune system, the impact of the cellular host defenses on *M. anisopliae* var *acridum* is minimal since the haemocytes remain unattached to fungal particles and there is no indication that nodules incorporate the fungus. The second stage of the infection process occurred when the fungus had entered the haemocoel and replicated extensively (3–4 days after inoculation). At this time, all parameters measured (apart from pPO) were at levels significantly below those of controls and mycosed locusts in stage 1 of infection. This may be because the immune system has now been overcome by the fungus or fungus-derived metabolites.

In a recent study (Li *et al.,* 2007) investigated the effects of different phosphorous sources (casein, phytic acid, KH_2PO_4 and $C_6H_5Na_2PO_4.2H_2O$) on AcP isoenzymes from fungi, and their results indicate that the activity and AcP isoenzymes are significantly different using different phosphorylated compounds as sole phosphorus source.

Trehalose is the main sugar in the haemolymph of insects and is a key nutrient source for an insect pathogenic fungus. Secretion of trehalose-hydrolysing enzymes may be a pre-requisite for successful exploitation of this resource by the pathogen. An acid trehalase [EC

3.2.1.28] was purified to homogeneity from a culture of a locust-specific pathogen, *M. anisopliae*, and its properties were characterized (Zhao *et al.*, 2006). Results indicated that the acid trehalase may serve as an "energy scavenger" and deplete blood trehalose during fungal pathogenesis. For entomopathogens of the genus *Metarhizium*, growth within the host is confined largely to the haemolymph prior to death. However, the acquisition of nutrients during the haemolymph colonization phase is little understood. Trehalases purified from filamentous fungi have been characterized as "acid" or "neutral" depending on their pH optimum. Neutral trehalases are intracellular proteins involved in the catabolism of internal trehalose, while acid trehalases are extracellular or vacuolar glycoproteins that hydrolyze extracellular trehalose

During haemolymph colonization, fungi also produce secondary metabolites, derivatives from various intermediates in primary metabolism, some of which have insecticidal activities. For entomopathogens producing these toxins, infection has been shown to result in more rapid host death (McCauley *et al.*, 1968) compared to strains that do not produce these metabolites (Samuels *et al.*, 1988; Kershaw *et al.*, 1999).

4.7. APPLICATION OF FUNGAL ENZYMES IN BIOTECHNOLOGY

4.7.1. Crop Protection

Current molecular and genomic methods are now being applied to *M. anisopliae*, the causative agent of green muscardine disease, because of its importance for the biological control of insect pests. It is a very versatile fungus, being able to infect a broad range of insects, 200 species from over 50 insect families and is also adapted to life in the root rhizosphere (Hu and St Leger, 2002). Consistent with its promiscuous nature, an array of expressed sequence tags (ESTs) from *M. anisopliae* strain 2575 identified large numbers of genes dedicated to host interaction and countering insect defenses, as well as regulators for coordinating their implementation (Freimoser *et al.*, 2003). Sequence comparisons and conserved motifs suggest that about 60% of the ESTs of strain 2575, expressed during growth on cuticle, encode secreted enzymes and toxins. Acting collectively, the number and diversity of these effectors may be the key to this pathogen's ability to infect a wide variety of insects. In contrast, ESTs from the specialized locust pathogen *M. anisopliae* sf. *acridum* strain 324 revealed very few toxins (Freimoser *et al.*, 2003). This relates to lifestyles. Strain 2575 kills hosts quickly via toxins, and grows saprophytically in the cadaver. In contrast, strain 324 causes a systemic infection of host tissues before the host dies. This shows that by utilizing ESTs, multiple virulence factors and pathways can be viewed simultaneously, and the different lifestyles that exist in insect–fungus interactions can be understood from a broader perspective.

The involvement of chitinase in fungal modification of the pro-cuticle raises the possibility that over-expression of chitinases may provide a means to improve fungal pathogenesis as described for hydrolases from several other systems. Transformation of *M. anisopliae* with *Prla*, the gene that codes for the major subtilisin protease Prl, under control of a constitutive promoter has improved biological control potential (St. Leger *et al.*, 1996c). Likewise, the biological control potential of *Trichoderma harzianum* (a fungal pathogen of plant pathogens) was improved by over-expression of genes for protease or glucanase under control of native promoters (Flores *et al.,* 1997, Migheli *et al.,* 1998). However, while overexpression of *Ech42* directed by its own promoter did not affect efficacy of *T. harzianum* strains (Carsollo *et al.,* 1999), transformants overexpressing a 33-kDa chitinase under constitutive control showed increased anti-fungal activity (Limon *et al.,* 1999).

Chitinases from mycoparasitic *Trichoderma* species have been characterized and shown to inhibit *in vitro* spore germination and tube elongation in a variety of fungi. The most potent anti-fungal enzyme from *T. harzianum* is the 42-kDa endochitinase (ECH42) that hydrolyzes *Botrytis cinerea* cell walls more efficiently than can plant or bacterial enzymes (Lorito *et al.,* 1993, 1994). Transgenic plants expressing *Ech42* have been shown to be resistant to foliar and soil-borne pathogens (Lorito *et al.,* 1998).

Ultrastructural immunocytochemistry and biochemical evidence suggested that the late appearance of chitinase as compared to protease during the infection processes of *M. anisopliae* sf. *anisopliae* strain 2575 is attributable to the delay caused by the induction process (St. Leger *et al.,* 1996a). However, over-production of CHIT1 by strain 2575 did not enhance virulence of the fungus against *M. sexta,* suggesting that wild-type levels of chitinase and its mode of regulation are not limiting for cuticle penetration, unlike the situation in *Trichoderma* as stated above. Moreover, injection of CHIT1 into larvae had no apparent effect on their mortality, whereas it was previously found that injection of 200 ng of Prl is lethal (St. Leger *et al.,* 1996c). Chitin is only present in insect cuticular structures and is shielded by proteins from enzymolysis, so lack of suitable substrates may explain why injected chitinase is not toxic.

It would be expected that most transformants genes under native control would be regulated in time and space as in the wild type, resulting in the production of enzyme where and when it is normally required, albeit in larger quantities. The results with *T. harzianum* chitinases suggest that the use of regulated promoters that require specific inducers for expression may not be the optimal situation for engineering transformants intended for biological control.

4.7.2. Industrial Applications of Fungal Enzymes

With the increasing expansion of proteomics and genomics technologies, the accuracy of analyses on fungal enzymes has greatly improved in the last years. Microbial enzymes are more frequently used commercially than those derived from plants and animals, as they have the desirable characteristics for biotechnological applications (stability, biochemical diversity, and suitability for genetic manipulation). Rapid microbial growth leads to efficient enzyme production in a very limited cultivation space (Pereira *et al.,* 2007).

Studies on microbial peptidases have increased in the last decade because of their commercial potential. Proteolytic enzymes account for nearly 60% of the industrial enzyme market and are widely used in food industry for cheese ripening, meat tendering, the production of protein hydrolysate and bread making. The use of peptidases as detergent additives in the 1960s stimulated their commercial development and led to a considerable expansion of fundamental research into these enzymes. Since then, there has been renewed interest in the discovery of proteases with novel properties. Fungi as enzymes producers have many advantages, considering that the produced enzymes are normally extra-cellular, simplifying the separation from fermentation media. The use of fungi as enzyme producers is also considered of lower risk than processes using bacteria.

An entomopathogenic fungal trypsin-like enzyme secreted by *M. anisopliae* has been patented in the UK and USA (United States Patent US4987077), although its industrial use has yet to be released. A peptidase from *Paecilomyces lilacinus* with a molecular mass of 20,00 to 200,000 Daltons, an isoelectric point of pH 8 to12, a pH optimum in the range of 9 to 12 and substrate specificity towards surface structures of plant parasitic nematodes, has also been patented in Europe (European Patent EP0623672). This protease may have potential for the control of plant parasitic nematodes.

Microbial *N*-acetylhexosaminidases have potential applications as tools for analysis of complex glycosylation chains on glycoproteins and glycolipids, and in organic chemistry for the preparation of synthetically demanding oligosaccharide structures (Scigelova and Crout, 1999).

Microbial lipases are the major commercial source of this type of enzymes. The biotechnological potential of entomopathogenic fungi for producing lipases is well known. Florczak *et al.* (2007) isolated and purified a lipase synthesized by the Antarctic filamentous fungus *Beauveria* sp., which was found to be an efficient catalyst of enantioselective transesterification of secondary alcohols. A *M. anisopliae* spore surface lipase (MASSL) was recently purified (Silva *et al.,* 2009).

An example of applications of oxygenase-based biocatalysis in industry was reported by Dingler *et al.* (1996) who used a *B. bassiana* isolate as a general biocatalysts to hydroxylate (R)-2-phenoxypropionic acid (POPS) to (R)-2-(4-hydroxyphenoxy) propionic acid (HPOPS). *B. bassiana* resulted in >98% conversion of POPS (1 g litre^{-1}) within three days. *B. bassiana* was also a suitable catalyst for the selective monohydroxylation of other aromatic carboxylic acids. This is of special interest in organic synthesis for industry bioreactors.

CONCLUSION

The most intensively researched enzyme secreted by an entomopathogenic fungus is the peptidase Pr1 (chymo-elastase) primordially produced by *Metarhizium anisopliae*. Other species of entomopathogenic fungi secrete enzymes similar to Pr1 and evidence would lead us to believe that this type of enzyme can be considered a pathogenicity factor as well as a virulence determinant. The trypsin-like peptidase Pr2 also plays an important role during host cuticle penetration as do the chitinases (endo- and exo-chitinases), not only in the disruption of the insects major defensive barrier (integument) but also for obtaining nutrients necessary for fungal development. Fungal enzymes are targets for the production of recombinant microorganisms, with improved biological control potential. Peptidases are also of great industrial importance, therefore entomopathogenic fungi should not be over-looked as a source of novel enzymes. In fact, entomopathogenic fungi are a genetic resource of immense importance, not only as natural controllers of insect populations but also as a highly diverse source of biologically active compounds. More than 700 species of entomopathogenic fungi are currently known to man and with a further intra-species genetic diversity (isolate diversity), the natural habitats of many of these microorganisms, the tropical rain-forests, need to be preserved in order not to lose such a valuable resource.

ACKNOWLEDGMENTS

The authors wish to thank Professors Stuart Eduard Reynolds and Keith A. Charnley (Bath University, England) for their constant help and advice over the last 25 years.

REFERENCES

Amer, M.M., El-Sayed, T.I., Bakheit, H.K., Moustafa S.A. and El-Sayed, Y.A. 2008. Pathogenicity and genetic variability of five entomopathogenic fungi against *Spodoptera littoralis*. *Research Journal of Agricultural and Biological Science*, 4: 354-367.

Bagga, S., Hu, G., Screen, S.E. and St. Leger, R.J. 2004. Reconstructing the diversification of subtilisins in the pathogenic fungus *Metarhizium anisopliae*. *Gene*, 324: 159–169.

Bernier, L., Cooper, R.M., Charnley, A.K. and Clarkson, J.M. 1989. Transformation of the entomopathogenic fungus *Metarhizium anisopliae* to benomyl resistance. *FEMS Microbiology Letters*, 60: 261-266.

Bidochka, M.J. and Khachatourians, G.G. 1987. Purification and properties of an extracellular protease produced by the entomopathogenic fungus *Beauveria bassiana*. *Applied and Environmental Microbiology*, 53: 1679-1684.

Bidochka, M.J., St. Leger, R.J., Joshi, L. and Roberts, D.W. 1995. The rodlet layer from aerial and submerged conidia of the entomopathogenic fungus *Beauveria bassiana* contains hydrophobin. *Mycological Research*, 99: 403–406.

Bogo, M.R., Rota, C.A., Pinto, H., Ocampos, M., Correa, C.T., Vainstein, M.H. and Schrank, A. 1998. A chitinase encoding gene (chitI gene) from the entomopathogen *Metarhizium anisopliae*: Isolation and characterization of genomic and full-length cDNA. *Current Microbiology*, 37: 221 225.

Boucias, D.G. and Pendland, J.C. 1991. Attachment of mycopathogens to cuticle: the initial event of mycosis in arthropod hosts. In: Cole G.T. and Hoch H.C. (Eds.) *The Fungal Spore and Disease Initiation in Plants and Animals*. Plenum Press, New York, pp. 101-128.

Boucias, D.G. and Pendland, J.C. 1982. Ultrastructural studies on the fungus, Nomuraea rileyi, infecting the velvetbean caterpillar, *Anticarsia gemmatalis*. *Journal of Invertebrate Pathology*, 39: 338–345.

Boucias, D.G., Pendland, J.C. and Latgé, J.P. 1988. Nonspecific factors involved in the attachment of entomopathogenic deuteromycetes to host insect cuticle. *Applied Environmental Microbiology*, 54: 1795–1805.

Brey, P.T., Latge, J.P., Prevost, M.C., 1986. Integumental penetration of the pea aphid Acyrthosiphon pisum by *Conidiobolus obscurus*. *Journal of Invertebrate Pathology*, 48, 34–41.

Carsollo, C., Benhamou, N., Haran, S., Cortes, C., Gutierrez, A., Chet, I. and Herrera-Estrella, A. 1999. Role of the *Trichoderma harzianum* endochitinase gene, ech42, in mycoparasitism. *Applied Environmental Microbiology*, 65: 929–935.

Castrillo, L.A., Roberts, D.W. and Vandenberg, J.D. 2005. The fungal past, present, and future: Germination, ramification, and reproduction. *Journal of Invertebrate Pathology*, 89: 46–56

Charnley, A. K. 2003. Fungal Pathogens of Insects: Cuticle Degrading Enzymes and Toxins. *Advances in Botanical Research*, 40: 241-321.

Charnley, A.K. and St. Leger, R.J. 1991. The rote of cuticle-degrading enzymes in fungal pathogenesis of insects. In: Cole G. T. and Hoch H. C. (Eds.) *The Fungal Spore and Disease Initiation in Plants and Animals*. Plenum Press, New York, pp. 129-156.

Charnley, A.K. 1984. Physiological aspects of destructive pathogenesis in insects by fungi: a speculative review. *British Mycological Society Symposium*, 6: 229–270.

Cole, S.C.J., Charnley, A.K. and Cooper, R.M. 1993. Purification and partial characterization of a novel trypsin-like cysteine protease from *Metarhizium anisopliae*. *FEMS Microbiology Letters*, 113: 189-196.

Cooper, R.M., Rankin, B. and Wood, R.K., S. 1978. Cell-wall degrading enzymes of vascular wilt fungi II. Properties and modes of action of polysaccharidases of *Verticillium albo-atrum* and *Fusarium oxysporum* f. sp. *lycopersici*. *Physiological Plant Pathology*, 13: 101-134.

Daboussi, M.J., Djeballi, A., Gerlinger, C., Blaiseau, P.L., Cassan, M., Lebrun, M.H., Parisot, D. and Brygoo, Y. 1989. Transformation of seven species of filamentous fungi using the nitrate reductase gene of *Aspergillus nidulans*. *Current Genetics*, 15: 453 456.

Dean, D,D. and Domnas, A.J. 1983. The extracellular proteolytic enzymes of the mosquito-parasitising fungus *Lagenidium giganteum*. *Experimental Mycology*, 7: 31-39.

Dingler, C., Ladner, W., Krei, G.A., Cooper, B., Hauer, B. 1996. Preparation of (R)-2-(4-Hydroxyphenoxy) propionic Acid by Biotransformation. Pesticide Science, 46: 33-35.

Florczak, T., Makowski, K. and Turkiewicz, M. 2007. The cold-adapted lipase of an Antarctic fungus *Beauveria* sp. P7 as an effective catalyst of enantioselective 1-phenylethanol trasesterification. *Journal of Biotechnology*, 131: S74–S97.

Flores, A., Chet, I., and Herrera-Estrella, A. 1997. Improved biocontrol activity of *Trichoderma harzianum* by overexpression of the proteinase encoding gene prb1. *Current Genetics*, 31: 30–37.

Freimoser, F.M., Screen, S., Bagga, S., Hu, G. and St Leger, R.J. 2003. Expressed sequence tag (EST) analysis of two subspecies of *Metarhizium anisopliae* reveals a plethora of secreted proteins with potential activity in insect hosts. *Microbiology*, 149: 239–247.

Freimoser, F.M., Hu, G., and St Leger, R.J. 2005. Variation in gene expression patterns as the insect pathogen *Metarhizium anisopliae* adapts to different host cuticles or nutrient deprivation in vitro. *Microbiology*, 151: 361–371.

Gabriel, B.P. 1968. Enzymatic activities of some entomopathogenic fungi. *Journal of Invertebrate Pathology*, 11: 70–81.

Gillespie, J.P., Burnett, C. and Charnley, A.K. 2000. The immune response of the desert locust *Schistocerca gregaria* during mycosis of the entomopathogenic fungus, *Metarhizium flavoviride*. *Journal of Insect Physiology*, 46: 429–437.

Gillespie, J.P., Kanost, M.R. and Trenczek, T. 1997. Biological mediators of insect immunity. *Annual Review Entomology*, 42: 611–642.

Goettel, M.S., St Leger, R.J., Bhairi, S., Jung, M. K., Oakley, B.R., Roberts, D.W. and Staples, R.M. 1990. Pathogenicity and growth of *Metarhizium anisopliae* stably transformed to benomyl resistance. *Current Genetics*, 17: 129–132.

Goettel, M.S., St Leger, R.J., Rizzo, N.W., Staples, R.M. and Roberts, D.W. 1989. Uttrastructural localization of a cuticle degrading protease produced by the entomopathogenic fungus *Metarhizium anisopliae* during penetration of the host cuticle. *Journal of General Microbiology*, 135: 2223–2239.

Gooday, G.W. 1999. Aggressive and defensive roles for chitinases. In: Jolles, P. and Muzzarelli, R.A.A. (Eds.), *Chitin and Chitinases*, Birkhauser, Basel. pp. 157–170.

Gupta, S.C., Leathers, T.D., EI-Sayed, G.N. and Ignoffo, C.M. 1991. Production of degradative enzymes by *Metarhizium anisopliae* during growth on defined media and insect cuticle. *Experimental Mycology*, 15: 310–315.

Hassan, A.E.M. and Charnley, A.K. 1989. Ultrastructural-study of the penetration by *Metarhizium anisopliae* through dimilin-affected cuticle of *Manduca sexta*. *Journal of Invertebrate Pathology*, 54: 117–124.

Hepburn, H.R. 1985. Structure of the integument. In Gilbert, L.I. and Kerkut, G. (Eds.), *Comprehensive Insect Physiology, Biochemistry and Pharmacology - Volume 3*. Pergamon Press, Oxford, pp. 1–58.

Heymann, E. 1980. Carboxyesterases and amidases. In: Jakoby, W.B (Ed.), *Enzymatic Basis of Detoxification*, Academic Press, New York, pp. 291–323.

Hu, G. and St Leger, R.J. 2002. Field studies using a recombinant mycoinsecticide (*Metarhizium anisopliae*) reveal that it is rhizosphere competent. *Applied Environmental Microbiology*, 68: 6383–6387.

Hurien, N., Fromentin, H. and Keil, B. 1977. Proteolytic enzymes of *Entomophthora coronata*: Characterization of a collagenase. *Comparative Biochemistry and Physiology*, 56B: 259–264.

Kang, S.C., Park, S. and Lee, D.G. 1998. Isolation and characterization of a chitinase cDNA from the entomopathogenic fungus, *Metarhizium anisopliae*. *FEMS Microbiology Letters*, 165: 267–271.

Kang, S.C., Park, S. and Lee, D.G. 1999. Purification and characterization of a novel chitinase from the entomopathogenic fungus, *Metarhizium anisopliae*. *Journal of Invertebrate Pathology*, 73: 276–281.

Kershaw, M.J. and Talbot, N.J. 1998. Hydrophobins and repellents: proteins with fundamental roles in fungal morphogenesis. *Fungal Genetics and Biology*, 23: 18–33.

Kershaw, M.J., Moorhouse, E.R., Bateman, R., Reynolds, S.E. and Charnley, A.K. 1999. The role of destruxins in the pathogenicity of *Metarhizium anisopliae* for three species of insect. *Journal of Invertebrate Pathology*, 74: 213–223.

Kucera, M. 1980. Proteases from the fungus *Metarhizium anisopliae* toxic for Galleria mellonella larvae. *Journal of Invertebrate Pathology*, 35: 304–310.

Lecuona, R., Riba, G., Cassier, P. and Clement, J.L. 1991. Alterations of insect epicuticular hydrocarbons during infection with *Beauveria bassiana* or *Beauveria brongniartii*. *Journal of Invertebrate Pathology*, 58:10–18.

Leopold, J. and Samsinakova, A. 1970. Quantative estimation of chitinase and several other enzymes in the fungus *Beauvaria bassiana*. *Journal of Invertebrate Pathology*, 15: 34–42.

Li, Z., Wang, Z., Peng, G., Yin, Y., Zhao, H. Cao, Y. and Xia, Y. 2007. Regulation of extracellular acid phosphatase biosynthesis by culture conditions in entomopathogenic fungus *Metarhizium anisopliae* strain CQMa102. *Annals of Microbiology*, 57: 565–570.

Limon, M.C., Pintor-Toro, J.A. and Benitez, T. 1999. Increased anti-fungal activity of *Trichoderma harzianum* transformants that overexpress a 33-kDa chitinase. *Phytopathology*, 89: 254–261.

Lorito, M., Harman, G.E., Hayes, C.K., Broadway, R.M., Woo, S.L. and Pietro, A. 1993. Chitinolytic enzymes produced by *Trichoderma harzianum* II. Antifungal activity of purified endochitinase and chitobiosidase. *Phytopathology*, 83: 302–307.

Lorito, M., Peterbauerg, C., Hayes, C.K., and Harman, G.E. 1994. Synergistic interaction between fungal cell wall degrading enzymes and different antifungal compounds enhances inhibition of spore germination. *Microbiology*, 140: 623–629.

Lorito, M., Woo, S.L., Garcia, I., Colucci, G., Harman, G.E., Pintor-Toro, J.A., Filippone, E., Muccifora, S., Lawrence, C.B., Zoina, A., Tuzun, S., and Scala, F. 1998. Genes from mycoparasitic fungi as a source for improving plant resistance to fungal pathogens. *Proceedings of the National Academy of Sciences USA*, 95: 7860–7865.

Madelin, M.F., Robinson, R.F. and Williams, R.S. 1967. Appressorium-like structures in insect parasitizing deuteromycetes. *Journal of Invertebrate Pathology*, 9: 404–412.

McCauley, V.J.E., Zacharuk, R.Y. and Tinline, R.D. 1968. Histopathology of the green muscardine in larvae of four species of Elateridae (Coleoptera). *Journal of Invertebrate Pathology*, 12: 444–459.

Migheli, Q., Gonzalez-Candelas, L., Dealessi, L., Camponogara, A. and Ramon-Vidal, D. 1998. Transformants of *Trichoderma longibrachiatum* overexpressing the beta-1,4-endoglucanase gene egll show enhanced biocontrol of *Pythium ultimum* on cucumber. *Phytopathology*, 88: 673–677.

Mohamed, A.K.A., Sikorowski, P.P. and Bell, J.V. 1978. Histopathology of *Nomuraea rileyi* in larvae of *Heliothis zea* and in vitro enzymatic activities. *Journal of Invertebrate Pathology*, 31: 345–352.

Moon, T.W. 1975. Temperature adaptation, isozymic function and the maintenance of heterogeneity. In: Marbet, A.P. (Ed.). *Isozymes II. Physiological Function*, Academic Press, New York.

Paterson, I.C., Charnley, A.K., Cooper, R.M. and Clarkson, J.M. 1993. Regulation of production of a trypsin-like protease by the insect pathogenic fungus *Metarhizium anisopliae*. *FEMS Microbiology Letters*, 109: 323–327.

Paterson, I.C., Charnley, A.K., Cooper, R.M. and Clarkson, J.M. 1994a. Partial characterization of specific inducers of a cuticle-degrading protease from the insect pathogenic fungus *Metarhizium anisopliae*. *Microbiology*, 1411: 3153–3159.

Paterson, I.C., Charnley, A.K., Cooper, R.M. and Clarkson, J.M. 1994b. Specific induction of a cuticle-degrading protease of the insect pathogenic fungus *Metarhizium anisopliae*. *Microbiology*, 140: 185–189.

Pereira, J.L., Noronha, E.F., Miller, R.N.G. and Franco, O.L. 2007. Novel insights in the use of hydrolytic enzymes secreted by fungi with biotechnological potential. *Letters in Applied Microbiology*, 44: 573–581.

Pinto, A.D.S., Barreto, C.C., Schrank, A., Ulhoa, C.J. and Vainstein, M.H. 1997. Purification and characterization of an extracellular chitinase from the entomopathogen *Metarhizium anisopliae*. *Canadian Journal of Microbiology*, 43: 322–327.

Samuels, R.I., Charnley, A.K. and St Leger, R.J. 1990. The partial characterization of endoproteases and exoproteases from three species of entomopathogenic entomophthorales and two species of deuteromycetes. *Mycopathologia*, 110: 145–152.

Samuels, R.J., Charnley, A.K. and Reynolds, S.E. 1988. The role of destruxins on the pathogenicity of 3 strains of *Metarhizium anisopliae* for the tobacco hornworm *Manduca sexta*. *Mycopathologia*, 104: 51–58.

Scigelova, M. and Crout, D.H.G. 1999. Microbial b-N acetylhexosaminidases and their biotechnological applications. *Enzyme and Microbial Technology*, 25: 3–14.

Screen, S.E., Hu, G. and St Leger, R.J. 2001. Transformants of *Metarhizium anisopliae* sf. anisopliae overexpressing chitinase from *Metarhizium anisopliae* sf. acridum show early induction of native chitinase but are not altered in pathogenicity to Manduca sexta. *Journal of Invertebrate Pathology*, 78: 260–266.

Shah, F.A., Allen, N., Wright, C.J. and Butt, T.M. 2007. Repeated in vitro subculturing alters spore surface properties and virulence of *Metarhizium anisopliae*. *FEMS Microbiology Letters*, 276: 60–66.

Silva, W.O.B., Santi, L., Berger, M. Pinto, A.F.M., Guimarães, J.A., Schrank, A., and Vainstein, M.H. 2009. Characterization of a spore surface lipase from the biocontrol agent *Metarhizium anisopliae*. *Process Biochemistry*, 44: 829–834.

Smith, R.J., Perkrul, S. and Grula, E.A. 1981. Requirement for sequential enzymatic activities for penetration of the integument of the corn earworm (*Heliothis zea*). *Journal of Invertebrate Pathology,* 38: 335–344.

Smith, R. J. and Grula, E. A. 1983. Chitinase is an inducible enzyme in *Beauveria bassiana*. *Journal of Invertebrate Pathology*, 42: 319-326.

St. Leger, R.J., Butt, T.M., Goettel, M.S., Roberts, D.W. and Staples, R.C. 1989. Synthesis of proteins including a cuticle-degrading protease during differentiation of the entomopathogenic fungus *Metarhizium anisopliae*. *Experimental Mycology*, 13: 253–262.

St. Leger, R.J., Charnley, A.K. and Cooper, R.M. 1986a. Cuticle-degrading enzymes of entomopathogenic fungi: synthesis in culture on cuticle. *Journal of Invertebrate Pathology*, 48: 85–95.

St. Leger, R.J., Charnley, A.K. and Cooper, R.M. 1986b. Cuticle-degrading enzymes of entomopathogenic fungi mechanisms of interaction between pathogen enzymes and insect cuticle. *Journal of Invertebrate Pathology*, 47: 295–302.

St. Leger, R.J., Cooper, R.M. and Charnley, A.K. 1986c. Cuticle-degrading enzymes of entomopathogenic fungi regulation of production of Chitinolytic Enzymes. *Journal of General Microbiology*, 132: 1509–1518.

St. Leger, R.J., Cooper, R.M. and Charnley, A.K. 1987a. Production of cuticle-degrading enzymes by the entomopathogen *Metarhizium anisopliae* during infection of cuticles from *Calliphora vomitoria* and *Manduca sexta*. *Journal of General Microbiology*, 133: 1371–1382.

St. Leger, R.J., Charnley, A.K. and Cooper, R.M. 1987b. Characterization of cuticle-degrading proteases produced by the entomopathogen *Metarhizium anisopliae*. *Archives of Biochemistry and Biophysics*, 253: 221–32.

St. Leger, R.J., Cooper, R.M. and Charnley, A.K. 1987c. Distribution of chymoelastases and trypsin-like enzymes in five species of entomopathogenic Deuteromycetes. *Archives of Biochemistry and Biophysics*, 258: 121–131.

St. Leger, R.J., Cooper, R.M. and Charnley, A.K. 1993. Analysis of aminopeptidase and dipeptidyl peptidase from the entomopathogenic fungus *Metarhiziurn anisopliae*. *Journal of General Microbiology*, 139: 237–243.

St. Leger, R.J., Durrands, P.K., Charnley, A.K. and Cooper, R.M. 1988. The role of extracellular chymoelastase in the virulence of *Metarhizium anisopliae* for *Manduca sexta*. *Journal of Invertebrate Pathology*, 52: 285–294.

St. Leger, R.J. 1991. Integument as a barrier to microbial infections, In: Binnington, K. and Retnakaran, A. (Eds.), *Physiology of the insect epidermis*. Commonwealth Scientific and Industrial Research Organization, Melbourne, Australia. p. 286–308.

St. Leger, R. J., Charnley, A. K. and Cooper, R. M. 1986b. Cuticle-degrading enzymes of entomopathogenic fungi mechanisms of interaction between pathogen enzymes and insect cuticle. *Journal of Invertebrate Pathology* 47: 295-302.

St. Leger, R.J., Charnley, A.K. and Cooper, R.M. 1986c. Cuticle-degrading enzymes of entomopathogenic fungi: synthesis in culture on cuticle. *Journal of Invertebrate Pathology*, 48: 85–95.

St. Leger, R.J., Cooper, R.M. and Charnley, A.K. 1986d. Cuticle-degrading enzymes of entomopathogenic fungi regulation of production of Chitinolytic Enzymes. *Journal of General Microbiology*, 132: 1509–1518.

St. Leger, R.J., Cooper, R.M. and Charnley, A.K. 1991. Characterization of chitinase and chitobiase produced by the entomopathogenic fungus *Metarhizium anisopliae*. *Journal of Invertebrate Pathology*, 58: 415–426.

St. Leger, R.J., Joshi, L., Bidochka, M.J. and Roberts, D.W. 1996c. Construction of an improved mycoinsecticide overexpressing a toxic protease. Proceedings of the National Academy of Sciences USA, 93: 6349–6354.

St. Leger, R.J., Joshi, L., Bidochka, M.J., Rizzo, N.W. and Roberts, D.W. 1996a. Biochemical characterization and ultrastructural localization of two extracellular trypsins produced by *Metarhizium anisopliae* in infected insect cuticles. *Applied and Environmental Microbiology*, 62: 1257–1264.

St. Leger, R.J., Joshi, L., Bidochka, M.J., Rizzo, N.W. and Roberts, D.W. 1996b. Characterization and ultrastructural localization of chitinases from *Metarhizium anisopliae, Metarhizium flavoviride*, and *Beauveria bassiana* during fungal invasion of host (Manduca sexta) cuticle. *Applied and Environmental Microbiology*, 62: 907–912.

St. Leger, R.J., Staples, R.C. and Roberts, D.W. 1993b. Entomopathogenic isolates of *Metarhizium anisopliae, Beauveria bassiana*, and Aspergillus flavus produce multiple extracellular chitinase isozymes. *Journal of Invertebrate Pathology*, 61: 81–84.

St. Leger, R.J., Staples, R.C. and Roberts, D.W. 1992. Cloning and regulatory analysis of starvation-stress gene, ssgA, encoding a hydrophobin-like protein from the entomopathogenic fungus, *Metarhizium anisopliae*. *Gene*, 120: 119–124.

Stirling, J.L., Cook, E.A. and Pope, A.M.S. 1979. Chitin and its degradation. In: Burnett, J.A. and Trinci, C. (Eds.), *Fungal walls and hyphal growth*, Cambridge University Press, Cambridge, pp. 169–188.

Umezama, H. and Aoyagi, T. 1983. Trends in research of low molecular weight protease inhibitors of microbial origin. In: Katanuma, N., Umezawa, H. and Holzer, M. (Eds.), *Protease Inhibitors: Medical and Biological Aspects*, Tokyo and Berlin, Japan Science Society Press and Springer-Verlag, pp. 3–15.

Vey, A., Fargues, J., Robert, P. 1982. Histological and ultrastructural studies of factors determining the specificity of pathotypes of the fungus *Metarhizium anisopliae* for scarabaeid larvae. *Entomophaga*, 27: 387–397.

Walton, J.D. 1996. Host selective toxins: agents of compatibility. *Plant Cell*, 8: 1723–1733.

Wang, C. and St. Leger, R.J. 2007. The MAD1 Adhesin of *Metarhizium anisopliae* links adhesion with blastospore production and virulence to insects, and the MAD2 adhesin enables attachment to plants. *Eukaryot Cell*, 6: 808–816.

Xia, Y., Dean, P., Judge A.J., Gillespie J.P., Clarkson, J.M. and Charnley, A.K. 2000. Acid phosphatases in the haemolymph of the desert locust, *Schistocerca gregaria*, infected with the entomopathogenic fungus *Metarhizium anisopliae*. *Journal of Insect Physiology*, 46: 1249–1257.

Zacharuk, R.Y. 1970a. Fine structure of the fungus *Metarhizium anisopliae* infecting three species of larval Elateridae (Coleoptera). II. Conidial germ tubes and appressoria. *Journal of Invertebrate Pathology*, 15: 81–91.

Zacharuk, R.Y. 1970b. Fine structure of the fungus *Metarhizium anisopliae* infecting three species of larval *Elateridae* (Coleoptera.) III. Penetration of the host integument. *Journal of Invertebrate Pathology*, 15: 372–396.

Zhao, H., Charnley, A.K., Wang, Z., Yin, Y., Li, Z., Li, Y., Cao, Y., Peng, G. and Xia, Y. 2006. Identification of an extracellular acid trehalase and its gene involved in fungal pathogenesis of *Metarhizium anisopliae Journal of Biochemistry*, 140: 319–327.

In: Microbial Insecticides: Principles and Applications
Editors: J. Francis Borgio, K. Sahayaraj, et al.

ISBN: 978-1-61209-223-2
© 2011 Nova Science Publishers, Inc

Chapter 5

PROTEOMIC ANALYSIS OF ENTOMOPATHOGENIC FUNGI SECRETION

*Simoni C. Dias[1] and Octávio L. Franco[1,2] *

[1]Centro de Análises Proteômicas e Bioquímicas, Pós-Graduação em Ciências Genômicas e Biotecnologia UCB, Brasília-DF, Brazil
[2]Universidade Federal de Juiz de Fora, Juiz de Fora-MG, Brazil. Universidade Católica de Brasília, Pós-Graduação em Ciências Gênomicas e Biotecnologia, SGAN 916 – Av W5 – Módulo C, Brasília, DF, Brazil

ABSTRACT

Entomopathogenic fungi are able to synthesize extremely rich enzymes secretions containing chitinases, proteinases and β-glucanases. Moreover, others proteins with different functions also have been found in the same secretions by proteomical techniques, suggesting that those mixtures are much more complexes than researchers believed.

All these proteins can act synergistically, helping fungi to colonize insect body causing devastating mortality. This property could have practical applications to control insect-pests that attack productive crops, offering potential economic advantage to agribusiness. Despite the enormous potential to give contributions to the study of fungi-insect interactions, only in recent times the proteomics approach begun to be applied to this area.

Biological discrepancies and complexity in a situation involving two organisms in close contact are intrinsic challenges for proteomical studies. For these reasons this chapter focuses in the novelties of fungi secretion proteomics studies, describing worst problems and major benefits of different applied approaches. Finally, some perspectives and directions for the future of molecular studies of fungi-insect interactions are also provided.

* Fax number: +556133474797, +556134487220; E-mail: ocfranco@pos.ucb.br

5.1. ENTOMOGENOUS AND ENTOMOPHATOGENIC FUNGI: TWO DIFFERENT COLONIZATION STRATEGIES

The earliest description of entomogenous fungi were originated more than 2.000 years ago, when *Cordyceps* (Ascomycota) infecting lepidopteran larvae was identified in ancient China. This fungus was the subject of the first published report of an entomopathogenic fungus described by Reuamu in 1726 (Cole and Kendrick, 1981). Furthermore, only in the late of 19th century, Metchnikoff firstly described *Metarhizium anisopliae* "green muscardine" infections on the cereal cockchafer (*Melolontha melolontha*), suggesting the utilization of this microorganism as a biological control agent for insects (Zimmermann *et al.*, 1995). With the increasing of different microorganisms discovered with the ability to control insect, a classification was developed dividing fungi in entomogenous and entomopathogenic. Entomogenous fungi include fungi genera which are associated to insects and other host arthropods by different manners such as saprophytic, parasitic or some pathogenic associations. Parasitic entomogenous usually behave as ectoparasites do not penetrate host tegument but some of them can debilitate their host at different degrees. These fungi include a wide number of genera being able to infect all insect orders (Douglas and Pendland, 1991). Despite of their straight pathogenic action against their cognate host, genomic analysis has been show that these fungi can play additional roles in nature. They could live as plant endophytes or plant antagonists, acting as a beneficial factor to rhizosphere-associates and possibly performing as plant growth promoters (Vega *et al.*, 2008, Vega *et al.*, 2009). On the other hand, entomophatogenic fungi are commonly characterized by their capability to attach and penetrate host cuticle and further replicate into host insect body (Tanada and Kaya, 1993). During host colonizing and invading processes, fungi secretions are extremely lethal to insects, and for this specific property, entomophatogenic fungus has been considered an important tool to develop efficient biological control agents (Figure 1).

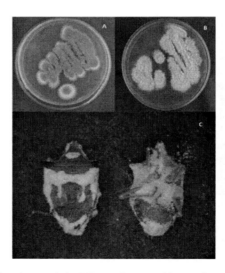

Figure 1. Entomophatogenic fungi growth in BDA culture and insect-host. (A) *Metarhizium anisopliae*, (B) *Beauveria bassiana* and (C) natural infected. stink bug dicehlops melacantus (Hemiptera: pentatomidea) wiht *B. bassiana* collected in soybean crops in Dourados (Brazil).

Entomopathogenic fungi are usually identified according fungal growth observed on insect cadavers. The infective process is unique between the microorganisms, since they cause numerous insect threats during host infection through cuticle. The symptoms caused for fungal disease to insect can vary, depending on the relation with cognate host. For example, insects infected with ectoparasites fungal, which do not penetrate throughout exoskeleton region, may exhibit no signs of disease. Otherwise, endoparasitic and pathogenic infections cause a wide number of symptoms such as: melanization (blackening), appetite loss, restlessness and disorientation. In these cases, host cuticle penetration is the beginning of host body colonization. Fungus adheres to host cuticle (see Figure 1, Douglas and Pendland, 1991), directly interacting with structural and chemical composition surfaces of cuticle, recognizing in summary appropriate chemical signs (Pedrini *et al.,* 2007). At later infection state host insect could change color. Such coloration is useful in order to form a preliminary diagnosis of pathogenic fungi involved (Figure 1, Alves, 1998). Another important additional task to researchers consists to perceive that insect fungus relations are in constantly modification, since both are involved at co-evolution process. In order to overcome insect defenses, entomopathogenic fungi exhibit a diverse array of adaptations. If in one hand, entomopathogenic fungi show the ability to avoid insect immune defenses (Roy *et al.,* 2006), on the other, insect could modify the structure and composition of exoskeleton, synthesize antifungal proteins in hemolymph and produce cellular and humoral defense reactions (Vilcinskas and Gotz, 1999) improving much more the molecular studies of this intriguing interaction. Proteomical studies could help to elucidate at molecular levels the protein compounds involved in both sides of this complex relation, as will cited above. Nevertheless, only entomopathogenic have been focused until now and none entomogenous fungi-insect relation was proteomically elucidated until now.

5.2. *METHARIZIUM ANISOPLIAE* AND *BEAUVERIA BASSIANA*: ENTOMOPATHOGENIC FUNGI PROTOTYPES FOR PROTEOMICAL STUDIES

Despite of seven hundred species of entomopathogenic fungi have been estimated, only approximately 170 products have been developed for insect control, being all of them based in only 12 fungi species. Mostly of commercially fungi produced pertaining to species of *Beauveria, Metarhizium, Lecanicillium* and *Isaria* due to easiness for mass production (Faria and Wraight, 2007). *Beauveria bassiana* and *Metarhizium anisopliae* were world widely evaluated according their virulence toward many insects (Hajek *et al.,* 2001) such as storage bruchidae pests (Cherrya *et al.,* 2005; Murad *et al.,* 2006, Murad *et al.,* 2007), tick species (Fernandes and Bittencourt, 2008), flea species that parasitizes dogs and cats (Melo *et al.,* 2008) and vector of visceral leishmaniasis (Amóra *et al.,* 2009) providing a clear alternative to synthetic chemical insecticides. The genus *Beauveria* (Balsamo) Vuillemin (Ascomycota: Hypocreales) includes several entomophatogenic species, of which the most notable insect controller is *B. bassiana*. *Beauveria* spp. has a cosmopolitan distribution and occupy diverse habitats, including soils and plant phylloplanes (Zimmermann, 2007). For all these reasons *B. bassiana* is an excellent prototype for proteomic studies of insect-fungi interaction. Another model that could be used in order to carry these studies is the green muscardine filamentous fungus *Metarhizium anisopliae*. Once this fungus utilizes a synergistic strategy of secretion of hydrolytic enzymes, such as proteases, chitinases and possibly lipases, associated with mechanical mechanisms to transpose the arthropod's cuticle, proteomical studies could give an obvious contribution in the understanding of mechanistically of parasitic process (Silva *et al.,* 2009). In spite of only few proteomical reports have been conduced in these two species, secreted secondary metabolites with clear importance to their pathogenic action have been studied. These metabolites are often referred as toxins, and several reports describe the properties, production and spatial distribution of the dominant metabolites secreted *in vitro*. There is also increasing evidence of inter- and intra-specific variation in metabolite production, with some strains or species secreting more toxic metabolites than others (Bandani *et al,.* 2000). For example, Beauvericin, a cyclohexadepsipeptide ionophore from the entomopathogen *B. bassiana*, shows antibiotic, antifungal, insecticidal, and cancer cell antiproliferative and antihaptotactic (cell motility inhibitory) activity by using *in vitro* bioassays (Xu *et al.,* 2008). The insecticidal cyclic depsipeptides, destruxins (dtxs), produced by *M. anisopliae* have been suggested to be an important virulence factors to accelerate the demise of infected insects (Wang *et al.,* 2004). Moreover, bassianin and tenellin are yellow pigments which can inhibit erythrocyte membrane ATPases (Jeffs and Khachatourians, 1997) and oosporein is an effective antibiotic against Gram-positive bacteria (Taniguchi *et al.,* 1984), also causing avian gout in broiler chicks and turkeys (Strasser *et al.,* 2000). Moreover, in recent years, *M. anisopliae* has been used as a model to study gene expression during fungal differentiation. These studies have led to related investigations on pathogenicity and identification of the genes involved in insect/fungus interactions (Pedrini *et al.,* 2007; Silva *et al.,* 2009). In resume, all these accumulated data of metabolomics and genomics of both *M. anisopliae* and *B. bassiana* obviously could facilitate and improve proteomical studies, integrating information and help in us to elucidate the molecular mechanisms of entomopathogenic fungi parasitism.

5.3. PROTEOMICAL TECHNOLOGIES APPLIED TO UNDERSTAND ENTOMOPATHOGENIC FUNGI SECRETION

The current expansion of novel technologies such as proteomics and functional genomics, which have been used to complement classical biochemical techniques and enzyme assays, clear improved the sensibility and precision of analyses on fungal proteinaceous secretions in insect-pathogen interactions. Entomopathogenic fungi secretions represent complex open systems, with hundreds of carbohydrates, secondary metabolites and proteins interacting, resulting in combined effects toward insect-pests. Among these compounds, proteins from fungi secretion are responsible for a strictly interaction with their hosts, being in resume, utilized to obtain food sources. But who are these proteins? How they work alone and together? And how can proteomics do to answer these questions? The first step to understand it is to become familiar to the "proteome" concept. The term proteome appeared for the first time in the year of 1995 (as revised by Graves and Haystead, 2002) and it was used to describe all of the proteins expressed in a particular genome. Moreover, proteomics aims to study the whole proteins content of a biological sample in a determinate time point. Since secretions could form a dynamic and complex network interaction, secretomics or the proteomics of secretion, consists in an enormous challenge for researchers from several areas. It's inevitable to start thinking how different kinds of stimulus can affect theses dynamic processes. In fungi, these motivation included food source, oxygen, light, the presence of predators, competition and several other conditions (as revised by Pereira *et al.,* 2007). Although proteomic analysis have completed more than one decade of proteins simultaneous analysis, its technical methods are constantly improving in order to make proteomics a more successful and reproducible method. Besides improving a more efficient way on separating proteins of complexes mixtures, these improvements aim to a better resolution analysis and also for a minimum lost of sample content. Differential secretion proteomic analyses were originally conduced via bidimensional gels, indicating the presence and absence of different proteins, as well the expression rates of each one by image analyses (Murad *et al.,* 2008). In review, two-dimensional gel is based by two independent separation methods. The first, named isoelectric focusing or simply "the first dimension" is characterized to separate proteins with different charges, searching for its charge stability on an immobilized pH gradient (Bjellqvist *et al.,* 1982). This charge stability is intrinsic to each protein and is known as isoelectric point (Figure 2). After charge protein separation, this strip gel is mounted over a second gel (see Figure 2) and resolved on a sodium dodecyl sulfate – polyacrylamide gel (SDS-PAGE), with is able to separated proteins by their native molecular mass, what's called the "second dimension". In summary, by using electricity, proteins migrate according to variants two dimensional variants molecular mass and isoelectric points (Figure 2). Proteins on gels are visualized by staining. Staining can be performed with Coomasie, silver nitrate or a number of new fluorescent dyes. The fluorescent dyes such as SYPRO-Ruby and Flamingo Pink TM, SYPRO Ruby and Krypton TM are very sensitive and compatible with MS technology described above (Deng *et al.,* 2008). Finishing gel analyses, for enhanced data reliability, biological (independently obtained samples) as well technical replicates (the same sample run on a different gel) are compared using a number of software's (Platinum, Bionumerics, etc.). Gel spots deemed differentially expressed based on statistical analysis of gels are excised and further processed.

If by one hand, gels are excellent tools to visualize protein maps and their respective differences, they do not are reliable apparatus for protein identification, being necessary protein sequencing via mass spectrometry methodologies. This technique, associated to 2D gels, contributes enormously to analytical proteomics of global proteinaceous samples, such as secretions derived from complex mixtures, extending knowledge in this area to levels not yet imagined (Nagendran *et al.,* 2009). In MS, the two most commonly used techniques are known as matrix assisted laser desorption / ionization time of flight (MALDI ToF) and electrospray ionisation (ESI) (Fenn *et al.,* 1989). Both MALDI ToF and ESI are soft ionisation techniques, in which ions are formed with low internal energies and thus undergo little fragmentation.

In the first technique cited before (MALDI), samples are co-crystallized with an organic matrix on a metal target. A pulsed laser is utilized to excite the matrix, causing rapid molecular thermal heating and eventually desorption of ions into the gas phase (Guerrera and Kleiner, 2005). The second technique (ESI) is based on spraying an electrically generated fine mist of ions into the inlet of a mass spectrometer at atmospheric pressure, ionizing compounds directly from aqueous solution. This property is extremely important, especially when methods are coupled to an interface that uses liquid partition methods such as liquid chromatography's (van Ulsen *et al.,* 2009).

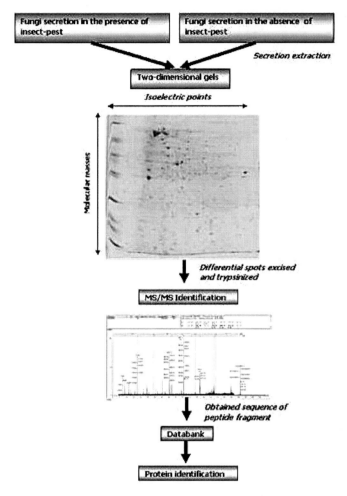

Figure 2. Work flow of protein identification from entomopathogenic fungi secretion by main proteomics techniques.

After ionization by means of these two methods, samples are submitted to a mass analyzer, which separates ions by their mass-to charge (m / z) ratios. Data obtained is submitted to data bank in order to found similar proteins and peptides. In the circumstances of enzymatic secretion studies, MS methods have had an enormous impact in proteome secretion analyses, facilitating the identification of a wide range of fungi enzymes in short period of time. In general, enzymes are isolated from 2D gels and further digested with a genetic modified trypsin with a reduced auto-cleavage, allowing fragments to be sequenced or submitted to a peptide mass fingerprint (Murad *et al.,* 2008). The technique known as peptide mass fingerprinting utilizes the masses of the peptides originated from the spot protein and these masses are further compared by bioinformatics tools to a database containing known protein sequences or the genome of the organism. This software's translate genomes into proteins and then theoretically cut proteins into peptides, calculating the absolute masses of the peptides from each protein. They then compare peptide masses of the unknown target protein to the theoretical peptide masses of each protein added to databanks. Peptide mass fingerprinting is a high throughput technique, however, it will only work if the genomical protein sequence is present in the database utilized. While this latter strategy is extremely quick, the *de novo* sequencing, using a tandem MS/MS methodology, is utilized to obtain the real protein sequence, does not necessity of an associated genomic study. Therefore, many laboratories prefer to use MS/MS to sequence the peptides. The number of such studies has recognized a wide protein range in entomopathogenic fungi is extremely low, especially when secretion is focused (Murad *et al.,* 2008) and much work its necessary to completely elucidate the fungi secretion complexity.

5.4. ELUCIDATING ENTOMOPATHOGENIC FUNGI SECRETION: THE INVOLVEMENT OF MORE THAN HYDROLYTIC ENZYMES

In the last decades, several researchers have been observed that different fungi are able to secrete enzymes in order to predate insects, plants and animals. Among them are commonly found hydrolytic enzymes in fungi secretion, which catalyze the hydrolysis of chemical bonds from diverse substrates (Pereira *et al.,* 2006). This information is completely widespread in literature. For example, the deuteromycetes *B. bassiana, M. anisopliae* and several other entomopathogenic fungi penetrate through the insect cuticle, secreting hydrolytic enzymes such as lipases, chitinases, and proteinases (Fang *et al.,* 2005; Murad *et al.,* 2006; Murad *et al.,* 2007), being these last two extremely efficient to degrade carbon and nitrogen sources. Among enzymes classes cited before, the most common and higher expressed is the chitinase, which are glucosyl hydrolases pertaining from family 18 and 19 (Pereira *et al.,* 2007). Entomopathogenic fungi synthesize a wide diversity of chitinases. This class of enzymes acts synergistically with other enzymes such as β-1,3-glucanases, both of which are strictly involved in insect cuticle cell wall degradation during biological control processes (Punja and Utkhede, 2003). In addition to fungal enzymes that hydrolyze carbohydrates, enzymes that are able to cleave peptide bonds in proteinaceous substrates, were also used in fungal parasitic process (Pereira *et al.,* 2007). While these enzymes have been separated into four different classes (serine-, cysteine, metallo- and aspartyl-) according to a number of properties, including catalytic mechanisms, serine-proteinases are the most common proteolytical enzymes found in entomopathogenic fungal secretions (Zhang *et al.,* 1998).

Despite of chitinases and proteinases have been widely found and further described into entomopathogenic fungi secretion, its impossible to believe that only these two enzymes compose this complex mixture of proteins and other metabolic compounds. Other enzymes such as chitosanases and chitin deacetylases, which converts chitin into its deacetylated form chitosan, have also been described to be produced by *M. anisopliae* in the process of insect pathogenesis (Nahar *et al.,* 2004), but several other compounds are expressed by fungi and further secreted. One of the manners to get a biochemical overview of microorganism's secretion and shed some light over their composition consists in the utilization of proteomical techniques, which are previously described in the topic above. Proteome permits to evaluate several proteins at same time and some interesting experiments were carried before by using *M. anisopliae* and *B. bassiana* under cowpea bruchids. In 2006, Murad and co-workers selected different strains of *M. anisopliae* with variable lethality toward cowpea weevil *C. maculatus.*

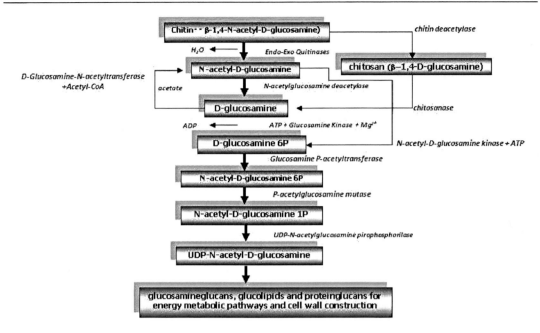

Figure 3. Chitin degradation pathway proposed by proteomical techniques by *M. anisopliae* (adapted from Murad *et al.* 2008). Enzymes are italicized and out of boxes. Substrates and products are inside gray boxes.

The most active fungi was grown in the presence and absence of insect-pest exo-skeleton, being their proteinaceous secretion analyzed by two-dimensional electrophoresis as well by their enzymatic activity which included chitinolytical and proteolytical. One year latter, the same group conduce similar experiments with *B. bassiana*, showing almost identical data (Murad *et al.*, 2007). Summarizing these data, both reports showed the presence of two main hydrolytic enzymes classes (chitinases and proteinases) over-expressed in secretions of entomopathogenic fungi, which was not a real novelty. Otherwise, analyzing the protein maps, it was observed that they completely change in the presence of insect cuticle. Moreover, most proteins do not show any similar isoelectric points or molecular weights to conventional enzymes previously report, suggesting that numerous other proteins may also be produced during insect colonization process. How at this moment, none mass spectrometry identification was carried, this theory was not proved, but the initial idea was provided.

In order to elucidate at least part of this secretion, novel gels were provided and further

MS identification was conduced in proteins differentially expressed in the presence of cowpea weevil exoskeleton (Murad *et al.*, 2008). Surprisingly, in spite of the presence of proteinases, chitinases and glucosidases, most identified proteins do not pertained to these classical enzyme groups. Data reported by Murad *et al.* (2008) indicate a clear nutrient uptake strategy during insect colonization, enabling a proposal of two metabolic pathways involved in degradation of compounds from liquid culture. Figure 3 illustrates the probable chitin degradation cycle proposed for *M. anisopliae*. Initially, enzyme assays indicated the presence of endo/exo chitinases (Murad *et al.*, 2006). Their activity produces N-acetyl-D-Glucosamine that could be phosphorylated by an N-acetyl-D-Glucosamine kinase, identified in fungi secretion (Murad *et al.*, 2008). Moreover, a branch of proposed cycle suggests that a chitin deacetylase may deacetylated chitin (Nahar *et al.*, 2004) (EC 3.5.1.41), and D-glucosamine units may be acetylated by a D-glucosamine N-acetyltransferase (Murad *et al.,* 2008). Moreover D-glucosamine could be further phosphorylated by N-acetyl-D-Glucosamine kinase (Murad *et al.,* 2008) into D-glucosamine-6-phosphate. Furthermore, the phosphorylated product is essential for the synthesis of UDP-N-acetyl-D-Glucosamine which could be utilized for glycogen synthesis, as previously described in fungi (Lomako *et al.,* 2004). N-acetyl-D-glucosamine-6P probably could also be metabolized into UDP-N-acetyl-D-glucosamine, forming a second branch in the described metabolic pathway (Figure 3). This last product, UDP-N-acetyl-D-glucosamine is an important substrate for dolichol metabolic pathway, utilized by fungi in the production of glycosamineglucans, glycolipids and protein glucans, being these compound commonly used for energy storage and cell wall construction (Sorensen *et al.,* 2003). Moreover, same authors proposed that UDP-N-acetyl-D-glucosamine was produced aiming to follow the N-acetyl-D-glucosamine-6P metabolism, being the products utilized for energy metabolism. On the other hand, D-glucosamine can also be transformed into glucose monomer and further used by fungi to construct their own cell wall (Lesage and Bussey, 2006). In resume, this report clear indicates, by proteomical techniques, that *M. anisopliae* is capable to not only hydrolytic enzymes, but also secrete enzymes involved in carbohydrate metabolism during nutritional stress responses. Nevertheless, proteomical secretion elucidation does no showed only enzymes involved at carbohydrate metabolism. Its known that proteolytic enzymes are were produced by *M. anisopliae* during host insect colonization (St-Leger *et al.,* 1994; Murad *et al.,* 2006) and also that numerous fungi utilize different forms of nitrogen capturing processes (Kneip *et al.,* 2007). Since fungi are unable to remove atmospherically nitrogen. Murad *et al.* (2008) suggested how *M anisopliae* are able to process nitrogen obtained from insect's cuticle from other compounds by using proteomical techniques (Figure 4).

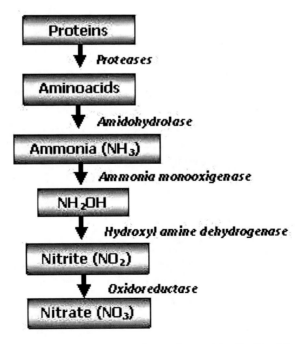

Figure 4. Protein degradation and nitrogen adsorption pathway described by *M. anisopliae* (Murad *et al.* 2008). Gray boxes indicated substrate italicized names represent enzymes.

Firstly, polypeptides were cleaved by serine-preoteinases (St-Leger *et al.,* 1994, Murad *et al.,* 2008), producing free amino acids and small peptides. Moreover, a methionine gamma lyase and amidohydrolase (Figure 4) hydrolyze the amino group found on N-termini and side chains of cationic amino acids. NH_3 becomes a substrate for ammonia monooxygenase, which results in NH_2OH. This compound is then oxidized by hydroxyl amine dehydrogenase to nitrite (NO_2), which is converted by nitrite oxidoreductase to nitrate (NO_3) (Figure 4).

This metabolic pathway is extremely important for nitrogen absorption (Marzluf *et al.,* 1981) and also for reduces high ammonium rates, since this compound is exceedingly deadly for fungi (Hess *et al.,* 2006). Other enzymes such as aminotransferases, α-amylases, ATPases, TRNA synthetases, zinc finger proteins and several others were also identified by proteomical techniques (Murad *et al.,* 2008) suggesting that secretion produced in response to the presence of insect is much more complex that data published until now. This conclusion indicates that we just appreciate the surface of this question and that proteomics could help us to go more deep, obtaining higher information over secretion protein composition.

CONCLUSION

All data and proteomical techniques here described are increasingly appropriate for studying fungi secretions in insect and pathogens biocontrol. A continued and persistent research in this area is extremely necessary, if not essential, in order to enable us to completely elucidate mechanisms of action of fungi enzyme secretion, determine methods to evaluate effectiveness and stability of enzymes and also survival of micro-organisms, promoting their use as potent biotechnological tools, and advancing the knowledge of secretomics. Proteomics techniques here demonstrated (Figure 2) are extremely usual. Nevertheless, other protein chemistry techniques could be applied in the elucidation of fungi secretomic, which included gel free-proteomics. Although the platform based on 2-DE is still the most commonly used, the use of gel-free and second-generation Quantitative Proteomic techniques must be applied to fungi secretomics. In addition, an appropriate experimental design and statistical analysis are essential to provide available information in accordance with the required minimal information about a proteomic experiment standards (Jorrin-Novo *et al.,* 2009). Moreover, several techniques could be coupled to proteomics in order to give enhanced information about secretion composition and activity. Among them we could cited the binding techniques, which was revised by Pereira *et al.* (2007). These binding techniques may be utilized to elucidate interactions between proteins secreted by entomopathogenic fungi and their cognate hosts and comprises two-hybrid system, isothermal titration calorimetry analyses and surface plasmon resonance. This last procedure, as with protein micro-arrays, is in a hot spot in terms of protein-protein interactions and, as the others, was not utilized to study entomopathogenic fungi secretions until now. Another important approach never used to study the interaction of insect-pests and fungi is comparative metagenomics (Tringe *et al.,* 2005). By this technique is possible to find novel microorganism that predate pests, considering these insects as complete environment. By this reason is possible to contrast metabolic activities of different microbial communities using largely unassembled sequence data obtained by next generation DNA sequencing of isolated

and cloned from diverse not-cultivated fungi from insects. Metagenome cloning has become a superior implement to exploit the biocatalytic potential of microbial communities for the discovery of unexplored enzymes. In summary, scientific community has much to learn with entomopathogenic fungi secretion. Compared to other systems as mammalian and yeast, however, entomopathogenic fungi research did not exploit fully the potential of proteomics, in particular its applications to interactomics, and their utilization as a biotechnological tool for medical and crop sciences could be extremely valuable, helping society to solve problems and improving human life quality.

REFERENCES

Alves, S.B. 1998. *Controle microbiano de insetos*. Ed. FEALQ. 1163 pp, Brasília-DF, Brazil.

Amóra, S.S.A., Bevilaqua, M.A.S., Leal, C.M., Feijó, F.M.C., Pereira, R.H.M.A, Silva, S.C., Alves, N.D., Freire, F.A.M., Oliveira, D.M. 2009. Evaluation of the fungus *Beauveria bassiana* (Deuteromycotina: Hyphomycetes), a potential biological control agent of *Lutzomyia longipalpis* (Diptera : Psychodidae). *Biological Control*, 50: 329–335.

Bandani, A. R., Khambay, B. P. S., Faull, J., Newton, R., Deadman, M., Butt, T. M. 2000. Production of efrapeptins by *Tolypocladium* species (Deuteromycotina : Hyphomycetes) and evaluation of their insecticidal and antimicrobial properties. *Mycological Research,* 104: 537-544.

Bjellqvist, B., Ek, K., Righetti, P.G., Gianazza, E., Görg, A., Westermeier, R. and Postel, W. 1982. Iso-electric focusing in immobilized pH gradients: principle, methodology, and some applications. *Journal of Biochemistry and Biophysical Methods*, 6: 317-339.

Douglas, D.G. and Pendland, J.C. 1991. The fungal cell wall and its involvement in the pathogenic process in insect hosts. In: Fungal cell wall and immune response, vol. H 53. Springer-Verlag. Berlin, pp. 303–316.

Cherrya, A.J., Abalob, P. and Hella, K. 2005. A laboratory assessment of the potential of different strains ofthe entomopathogenic fungi *Beauveria bassiana* (Balsamo)Vuillemin and *Metarhizium anisopliae* (Metschnikoff) to control *Callosobruchus maculatus* (F.) (Coleoptera: Bruchidae) in stored cowpea. *Journal of Stored Products Research*, 41: 295-309.

Cole, G.T. and Kendrick, B. 1981. *The fungal spore and disease initiation in plants and animals*. Plenum Press, N.Y.

Deng, Y.Z., Ramos-Pamplona, M. and Naqvi, N.I. 2008. Methods for functional analysis of macroautophagy in filamentous fungi. *Methods Enzymology*, 451: 295-310.

Fang, W., Leng, B., Xiao, Y., Jin, K., Ma, J., Fan, Y., Feng, J., Yang, X., Zhang, Y. and Pei, Y. 2005. Cloning of *Beauveria bassiana* chitinase gene *Bbchit1* and its application to improve fungal strain virulence. *Applied and Environmental Microbiology* 7,1, 363-370.

Faria, M.R. and Wraight, S.P. 2007. Mycoinsecticides and mycoacaricides: a comprehensive list with worldwide coverage and international classification of formulation types. *Biological Control,* 43: 237–256.

Fenn, J.B., Mann, M., Meng, C.K., Wong, S.F. and Whitehouse, C.M. 1989. Electrospray ionization for mass spectrometry of large biomolecules. *Science,* 246: 64–71.

Fernandes, E.K.K. and Bittencourt, V.R.E. 2008. Entomopathogenic fungi against South American tick species. *Experimental and Applied Acarology* 46: 71–93.

Graves, P.R. and Haystead, T.A.J. 2002. Molecular biologist's guide to proteomics. *Microbiologist Molecular Biology Reviews*. 66(1): 39-63.

Guerrera, I.C. and Kleiner, O. 2005. Application of mass spectrometry in proteomics. *Bioscience Reports* 25: 71–93.

Hajek, A.E., Wraight, S.P. and Vandenberg, J.D. 2001. Control of arthropods using pathogenic fungi. In: Bio-exploitation of filamentous fungi: fungal diversity. (Research Series No. 6 - Fungal Diversity), Hong Kong, pp. 309-34.

Hess, D. C., Lu, W., Rabinowitz, J. D. and Botstein, D. 2006. Ammonium toxicity and potassium limitation in yeast. *PloS Biology*, 4(11): e351.

Jeffs, L.B. and Khachatourians, G.G. 1997. Toxic properties of Beauveria pigments on erythrocyte membranes. *Toxicon*, 35(8):1351-6.

Jorrin-Novo, J.V., Maldonado, A.M., Echevarría-Zomeño, S., Valledor, L., Castillejo, M.A., Curto, M., Valero, J., Sghaier, B., Donoso, G. and Redondo, I. 2009. Plant proteomics update (2007-2008): Second-generation proteomic techniques, an appropriate experimental design, and data analysis to fulfill MIAPE standards, increase plant proteome coverage and expand biological knowledge. *Journal of Proteomics*, 72(3): 285-314.

Kneip, C., Lockhart, P., Vob, C. and Maier, U.-G. 2007. Nitrogen fixation in eukaryotes – New models for symbiosis. *BMC Evolutive Biolology* 7, 1-12.

Lesage, G., Bussey, H. 2006. Cell wall assembly in *Saccharomyces cerevisiae. Microbiology Molecular Biology Reviews*, 70: 317-343.

Lomako, J., Lomako, W.M. and Whelan, W.J. 2004. Glycogenin: the primer for mammalian and yeast glycogen synthesis. *Biochim Biophys Acta.*, 1673(1-2): 45-55.

Melo, D.R., Fernandes, E.K.K., Costa, G.L, Scott, F.B. and Bittencourt, V.R.E.P. 2008. Virulence of *Metarhizium anisopliae* and *Beauveria bassiana* to *Ctenocephalides felis felis. Animal Biodiversity and Emerging Diseases: Annual New York Academy of Sciences,* 1149: 388–390.

Marzluf, G.A. 1981. Regulation of nitrogen metabolism and gene expression in fungi. *Microbiol Rev.* 45(3):437-61.

Murad, A.M., Laumann, R.A., Lima T.A., Sarmento, R.B., Noronha, E.F., Rocha, T.L., Valadares-Inglis, M.C. and Franco, O.L. 2006. Screening of entomopathogenic *Metarhizium anisopliae* isolates and proteomic analysis of secretion synthesized in response to cowpea weevil (*Callosobruchus maculatus*) exoskeleton. *Comparative Biochemistry and Physiology Part C Toxicology and Pharmacology,*. 142(3-4): 365-370.

Murad, A.M., Laumann, R.A., Mehta, A., Noronha, E.F. and Franco, O.L. 2007. Screening and secretomic analysis of enthomopatogenic *Beauveria bassiana* isolates in response to cowpea weevil (*Callosobruchus maculatus*) exoskeleton. *Comparative Biochemistry and Physiology Part C Toxicology and Pharmacology,* 145(3): 333-338.

Murad, A.M., Noronha, E.F., Miller, R.N., Costa, F.T., Pereira, C.D., Mehta, A., Caldas, R.A. and Franco, O.L. 2008. Proteomic analysis of *Metarhizium anisopliae* secretion in the presence of the insect pest *Callosobruchus maculatus*. Microbiology USA. 154(12): 3766-3774.

Nagendran, S., Hallen-Adams, H.E., Paper, J.M., Aslam, N. and Walton, J.D. 2009. Reduced genomic potential for secreted plant cell-wall-degrading enzymes in the ectomycorrhizal fungus Amanita bisporigera, based on the secretome of *Trichoderma reesei. Fungal Genetics and Biology*, 46(5):427-35.

Nahar, P., Ghormade, V. and Deshpande, M.V. 2004. The extracellular constitutive production of chitin deacetylase in *Metarhizium anisopliae*: possible edge to entomopathogenic fungi in the biological control of insect pests. *Journal Invertebrate Pathology*, 85(2): 80-88.

Pereira, J.L., Franco, O.L. and Noronha, E.F. 2006. Production and biochemical characterization of insecticidal enzymes from *Aspergillus fumigatus* toward *Callosobruchus maculatus. Current Microbiology*, 52(6): 430-434.

Pereira, J.L., Noronha, E.F., Miller, R.N. and Franco, O.L. 2007. Novel insights in the use of hydrolytic enzymes secreted by fungi with biotechnological potential. *Letters in Applied Microbiology*, 44(6): 573-581.

Pedrini, M. Crespo, R. and Juárez, M.P. 2007. Biochemistry of insect epicuticle degradation by entomopathogenic fungi. *Comparative Biochemistry and Physiology, Part C,* 146 :124–137.

Punja, Z.K. and Utkhede, R.S. 2003. Using fungi and yeasts to manage vegetable crop diseases. *Trends in Biotechnology,* 21, 400–407.

Roy, H.E, Steinkraus, D.C, Eilenberg, J., Hajek, A.E. and Pell J.K. 2006. Bizarre interactions and endgames: entomopathogenic fungi and their arthropod hosts. *Annual Review of Entomology,* 51: 331–357.

Silva, W.O.B., Santi, L., Berger, M., Pinto, A. F.M., Guimarães, J.A., Schrank, A. and Vainstein, M.H. 2009. Characterization of a spore surface lipase from the biocontrol agent *Metarhizium anisopliae. Process Biochemistry,* 44: 829–834.

Sorensen, T. K., Dyera, P. S., Fierroc, F., Laubea, U. and Peberdy, J. F. 2003. Characterization of the gptA gene, encoding UDP N-acetylglucosamine: dolichol phosphate N-acetylglucosaminylphosphoryl transferase, from the filamentous fungus, *Aspergillus niger. Biochimica et Biophysica Acta,* 1619: 89-97.

Strasser, H., Abendstein, D., Stuppner, H. and Butt, T.M. 2000. Monitoring the distribution of secondary metabolites produced by the entomogenous fungus *Beauveria brongniartii* with particular reference to oosporein. *Mycology Research,* 104(10) : 1227-1233.

St-Leger, R. J., Bidochka, M. J. and Roberts, D. W. 1994. Isoforms of the cuticle-degrading Pr1 proteinase and production of a metallo-proteinase by *Metarhizium anisopliae. Archives in Biochemistry Biophysics,* 313: 1-7.

Tanada, Y. and Kaya, H. K. 1993. *Insect Pathology.* Academic Press, San Diego, USA.

Taniguchi, M., Kawaguchi, T., Tanaka, T. and Oi, S. 1984. Antimicrobial and respiration inhibitory activities of oosporein. *Agricultural and Biological Chemistry,* 48: 1065-1067.

Tringe, S.G., Von Mering, C., Kobayashi, A., Salamov, A.A., Chen, K., Chang, H.W., Podar, M., Short, J.M., Mathur, E.J., Detter, J.C., Bork, P., Hugenholtz, P. and Rubin, E.M. 2005. Comparative metagenomics of microbial communities. *Science,* 308, 554–557.

van Ulsen, P., Kuhn, K., Prinz, T., Legner, H., Schmid, P., Baumann, C. and Tommassen, J. 2009. Identification of proteins of *Neisseria meningitidis* induced under iron-limiting conditions using the isobaric tandem mass tag (TMT) labeling approach. *Proteomics,* 9(7): 1771-1781.

Vega, F.E, Goettel M.S. and Blackwell, M. 2009. Fungal entomopathogens: new insights on their ecology. *Fungal Ecology* in press doi:10.1016/j.funeco 2009.05.001.

Vega, F.E, Pousada, F.M. Aime, M.C, Pava-Ripoll, M., Infante, F. and Rehner., S.A. 2008. Entomopathogenic fungal endophytes. *Biological Control,* 46: 72–78.

Vilcinskas, A. and Gotz, P. 1999. Parasitic fungi and their interactions with the insect immune system. *Advances in Parasitology,* 43: 267–313.

Wang, C., Skrobek, A. and Butt, T.M. 2004. Investigations on the destruxin production of the entomopathogenic fungus *Metarhizium anisopliae. Journal of Invertebrate Pathology,* 85:168–174.

Xu, Y., Orozco, R., Wijeratne, E.M.K., Gunatilaka,A.A.L., Stock, P. and Molna, I. 2008. Biosynthesis of the cyclooligomer depsipeptide Beauvericin, a virulence factor of the entomopathogenic fungus *Beauveria bassiana. Chemistry and Biology,* 15: 898–907.

Zhang, Y., Liu, X. and Wang, M. 2008. Cloning, expression, and characterization of two novel cuticle-degrading serine proteases from the entomopathogenic fungus *Cordyceps sinensis. Research in Microbiology,* 159(6): 462-469.

Zimmermann, G., Papierok, B. and Glare, T. 1995. Elias Metschnikoff, Elie Metchnikoff or Ilya Ilich Mechnikov (1845–1916): a pioneer in insect pathology, the first describer of the entomopathogenic fungus *Metarhizium anisopliae* and how to translate a Russian name. *Biocontrol Science and Technology,*5: 527–530.

Zimmermann, G. 2007. Review on safety of the entomopathogenic fungi *Beauveria bassiana* and *Beauveria brongniartii. Biocontrol Science and Technology,* 17: 553–596.

In: Microbial Insecticides: Principles and Applications   ISBN: 978-1-61209-223-2
Editors: J. Francis Borgio, K. Sahayaraj, et al.   © 2011 Nova Science Publishers, Inc

Chapter 6

INNOCUITY AND TOXICITY OF ENTOMOPATHOGENIC FUNGI

*Conchita Toriello[1] *, Armando Pérez-Torres[2], Hortensia Navarro-Barranco[1] and Victor Hernández-Velázquez[3]*

[1]Depto. de Microbiología y Parasitología; Facultad de Medicina, Universidad Nacional Autónoma de México, México D.F. 04360, México
[2]Depto. de Biología Celular y Tisular, Facultad de Medicina, Universidad Nacional Autónoma de México, México D.F. 04360, México
[3]Centro de Investigación en Biotecnología, Universidad Autónoma del estado de Morelos, Cuernavaca 62209, México

ABSTRACT

The characterization of potential hazards and safety concerns of fungal microbial agents for biological control is a pressing need in view of the increased number of entomopathogenic fungi used as non-chemical alternatives for agricultural and medical relevant pest control. The host-damage response framework theory can be applied to fungal microbial agents to understand the innocuity of entomopathogenic fungi when introduced to non-target organisms such as mammals. This chapter is then focused at comparing two models. The first one, the binomial *Metarhizium anisopliae* var. *acridum*/locust (*Schistocerca piceiforns* ssp. *piceifrons*), illustrates how the virulence factors of an entomopathogenic fungus are completely expressed when infecting a susceptible host producing disease and the death of the insect. The second binomial, *M. anisopliae* var. *acridum*/mice, exemplifies that even though the fungus possesses the same virulence factors or attributes, these are not expressed or are ineffective in a non-susceptible host, and is not able to cause damage in the host. These examples may explain the different outcomes of the interaction between a host and a microbe depending on the amount of host damage produced. The final interaction of the first binomial illustrated resulting in the death of the susceptible host, i.e., the insect; and the result of the second binomial exemplified resulting in the innocuity of the fungus to the non-susceptible host, i.e., the mouse.

* E-mail: toriello@servidor.unam.mx; Phone/Fax: (52 55) 56232461

6.1. INTRODUCTION

The use of entomopathogenic fungi as biocontrol agents require a great deal of research concerning their safety for humans, animals, invertebrates and arthropods non target organisms, and their effect on the environment. With the increased interest on available non-chemical alternatives for agricultural and medical relevant pest control, there has been an augmented number of entomopathogenic fungi as biocontrol agents, and therefore an urgent need to characterize the potential hazards and safety concerns of this group of microorganisms.

In this chapter we restrict our discussions to explain how the host-damage response framework theory of Casadevall and Pirovski (1999) can be applied to fungal microbial control agents and provide a probable explanation for the innocuity of fungal entomopathogens when introduced in mammals. This theory may interest all who have safety concerns associated with biocontrol fungi. For ample reviews on the safety of biocontrol fungi see Austwick (1980), Goettel *et al.* (1990, 2001), Prior (1990), and Evans (1998, 2000).

We focused in the pathogenic assays carried on with *Metarhizium anisopliae* var. *acridum* Driver and Milner in locusts, *Schistocerca piceifrons* ssp. *piceifrons* (Walker), as it is being developed as a biopesticide for use against locusts and grasshoppers in Australia (Milner, 1997; Hunter *et al.,* 1999; 2001; Milner and Pereire, 2000), Africa (Lomer *et al.,* 2001), Brazil (Magalhaes *et al.,* 2000), China (Lee *et al.,* 2000) and Mexico (Hernández-Velázquez *et al.,* 2003), among others. The wide application of this fungus for biocontrol over the globe merits an assessment of its innocuity in vertebrate organisms.

6.2. HOST-DAMAGE RESPONSE FRAMEWORK APPLIED TO FUNGAL MICROBIAL CONTROL AGENTS

In order to best explain what we consider the innocuity of fungal entomopathogens for non-target organisms such as mammals and other invertebrates, an integrated theory of microbial pathogenesis has been postulated (Casadevall and Pirofski, 1999) that reflects the outcome of an interaction between a host and a microbe, depending on the amount of host damage that results from this interaction. It is an integrative microbial pathogenesis classification for it involves not only the host but also the microbial contributions, including bacteria, virus, fungi and protozoans (Casadevall and Pirofski, 1999, 2003; Pirofski and Casadevall, 2008). This theory can be extended to microbial agents used in the biological control of plant pests and diseases, for it may explain in a very clear way how a microorganism can in one way be a pathogen for an organism (i.e. entomopathogens) and at the same time be non-pathogenic for another organism (i.e. mammals). It is based on three tenets (Casadevall and Pirofski, 2003): a. Microbial pathogenesis requires two entities, a host and a microbe, and both must interact; b. The host relevant outcome of this interaction is damage to the host; c. Host damage can occur as a result of microbial factors, host factors, or both.

Based on this theory of microbial pathogenesis, fungal entomopathogens may at the same time be pathogens of insects, and non-pathogens for mammals or other non-target organisms. When acting as pathogens of insects, fungi cause damage that includes the breaching of the cuticle of susceptible insects when penetrating into the host, through mechanical and enzymatic activities (Boucias and Pendland, 1991), breaking down the lipoprotein matrix of the epicuticle (Fargues, 1984). Later, the penetrating germ tube will reach the haemocoelic cavity and differentiate into the vegetative growth stage and may produce yeast-like hyphal bodies invading all insect tissues, and lastly, producing new conidia on the dead host (Boucias and Pendland, 1991). An extensive literature on this infection process is available by different authors (St. Leger, 1993; St. Leger and Bidochka, 1996; Boucias and Pendland, 1998; Hajek and St. Leger, 1994; Khachatourians, 1998). Depending on the specific fungus and insect host the outcome of the interaction may result in the elimination of the microbial agent by the host immune response and/or host-mediated behavioral changes, or the death of the host by the fungal multiple insecticide mechanisms used. Here again lethal damage would occur only in the susceptible host.

By the contrary, when a fungal entomopathogen is inoculated in a non-susceptible host, for example, *M. anisopliae* var. *acridum* inoculated intragastrically in mice (Toriello *et al.*, 2009), this microorganism is incapable of causing damage. In this case, the host eliminates the fungus without any problem and the fungal insecticidal pathogenic mechanisms are at not avail.

Therefore, only when host damage exceeds the disease threshold (or the host homeostasis) in a susceptible host there will be disease. The damage-response framework theory views the microbe and the host as the outcome of their interaction. An issue always noted for microbial control agents is if they are pathogens of insects why not for other organisms (including man) (Cook *et al.*, 1996). Therefore, within the context of the damage-response framework theory, the next examples of a fungal entomopathogen may illustrate how the same fungus may be at the same time a pathogen of insects and a non-pathogen for mice (a nontarget organism used for pathogenicity and toxicity safety tests) leading to reassurance in the use of microbial agents for the biological control of plant pests and diseases in view of safety considerations of live microorganisms, regarding the safety issues of pathogenicity and toxicity to nontarget organisms.

Analyzing the pathogenic interaction between *M. anisopliae* var. *acridum* and a locust (Orthoptera: Acrididae) will show all the microbial virulence factors used by the microorganism in order to produce disease in a susceptible host.

6.3. PATHOGENIC INTERACTION OF *METARHIZIUM ANISOPLIAE* VAR. *ACRIDUM* IN A SUSCEPTIBLE HOST: LOCUST, *SCHISTOCERCA PICEIFRONS* SSP. *PICEIFRONS* (ORTHOPTERA: ACRIDIDAE)

A selected virulent isolate (EH-502/8) (García-Gutiérrez *et al.*, 2002) of *M. anisopliae* var. *acridum* (*M. acridum*) with a median lethal time of 6.23 h was deposited (1 x 10^7 conidia) on the pronotum of young adults (after 7 days of the last moult) of the locust *S. piceifrons* (Toriello *et al.*, 2007). The infection process during seven days was studied by photonic and electron microscopy and followed the same pattern as described for other entomopathogenic fungi (Ferron, 1978; Boucias and Pendland, 1991). The conidium of *M. a. acridum* with a thick layered cell wall and an external rodlet layer of hydrophobic nature (Leland *et al.*, 2005) is the infectious propagule which initially attaches to the host cuticle.

This initial attachment may be passive based on hydrophobic interactions, or more permanent mediated through the secretion of mucilage at the cuticle-conidium interface (Boucias and Pendland, 1991). Scanning electron microcopy indicated that conidia of *M. a. acridum* strain EH-502/8 delivered on the dorsal cuticle of the locust *S. piceifrons* were able to adhere through an amorphous extracellular matrix, forming a pit or socket over the superficial epicuticle, mainly on cuticular folds (Figure 1.a). Many conidia germinated within 24-72 h post-inoculation, and produced a single long and slender germination tube across the cuticle surface, which clearly penetrated the cuticle using natural orifices corresponding to dermal gland openings or to pore canals (Figure 1.b). This surprising tropism contrasted with the development of appressoria directly from conidia that exhibited bipolar germination (Figure 1.c) or a single short penetration peg (Figure 1.d). Later, the fungus was frequently associated to amorphous material or debris that could be the result of epicuticle degradation.

Most conidia and appressoria were covered in a thin amorphous mucilage layer also observed for *M. anisopliae* by other researchers previously (Zacharuk, 1970; St. Leger *et al.*, 1989; Jarrold *et al.*, 2007) and also with a mucus-like secretion covering all the structures. These extracellular matrices, produced by many fungi, aid adhesion to the host during pre-penetration growth; the composition of this material is unknown but suggested to be glycoproteins containing β-1,3 glucans in other fungi (Latgé *et al.*, 1987; Stahlman *et al.*, 1992; Carzaniga *et al.*, 2001) and for *M. a. acridum* with a probable β-1,4- component also (Jarrold *et al.*, 2007).

Thus, in the course of breaching the cuticle and then penetrating the host, *M. a. acridum* not only uses the well known enzymatic machinery (St. Leger *et al.*,1986; 1987a; 1987b; 1995; Bidochka and Meltzer, 2000; Freimoser *et al.*, 2003), but also a more simple mechanism using the natural insect openings. Although conidial density was low, attributes or virulence factors of *M. a. acridum* could be related to multiple penetration ways or sites and to a more complex germination and differentiation process of conidia at the cuticular level.

Six days after inoculation, fungal growth and colonization of host cavities and most internal tissues was evident (Figures. 2. a, b). Hyphal bodies, pseudohyphae, and blastospores of *M. a. acridum* were observed in the locust' hemocoel (Figures. 2. c, d).

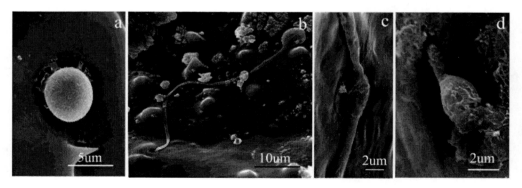

Figure 1. a-d. Scanning electron microscopy of experimental interaction between *M. anisopliae* var. *acridum* and *Schistocerca piceifrons* ssp. *piceifrons*. Initially, the fungus is adhered to the epicuticle through an amorphous extracellular matrix (a); Later, conidia produce germination tubes (b), appresoria with bipolar germination (c), or single short pegs (d), all structures involved in the penetration of the locust's cuticle. Note that dermal gland openings or pore canals are entrance routes for fungal invasion (b).

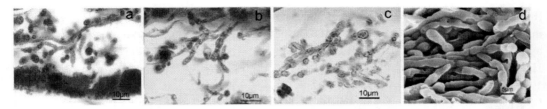

Figure 2. a-d. Histological examination of infected *Schistocerca piceifrons* ssp. *piceifrons* by *M. anisopliae* var. *acridum* six days after inoculation. Fungal colonization with hyphal bodies, pseudohyphae, and blastospores was evident and noteworthy (a-c, HandE stain). Scanning electron microscopy confirmed these findings (d).

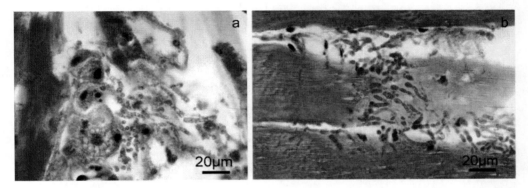

Figure 3. a, b. Histopathology of infected *Schistocerca piceifrons* ssp. *piceifrons* by *M. anisopliae* var. *acridum* 3-6 days after inoculation. Epidermal cells (a) and muscle cells (b) are target cells to the invasion and lesions caused by the fungus. Note that host defense response is almost absent. Many locusts died at this time (HandE stain).

Fungal propagules were dispersed elsewhere in the hemolymph. Epidermal cells (Figure 3. a) and muscle cells (Figure 3. b) are clearly altered. Moreover, muscle cells appeared to be invaded by fungal structures and showed clear signs of lysis and were detached at cuticular insertion. At this stage post-inoculation (or previously) hosts were also probably affected by starvation due to immobilization secondary to muscle damage. The muscle destruction observed could be the cause of the impairment on feeding, movement and flying abilities of the locust as previously reported in locust infection by *Metarhizium* (Moore *et al.*, 1992; Hernández-Velázquez *et al.*, 2007). No evidence of encapsulation of fungal propagules was observed. Clearly, cellular host defense reaction mediated by hemocytes or celomocytes in the locust is very weak to control infection and invasion by *M. acridum*. Although no sites of fungal emergence through host integument were observed directly, conidiophores were formed on the surface of the locust cadaver (Figure 4. a), which also contains chains of conidia (Figure 4. b), free conidia (Figure 4. c), and conidia in clusters and hyphae around seta (Figure4. d). At this time (6 days post-inoculation) all locusts were dead. In fact, 4-5 days after conidia delivery at the surface of *S. piceifrons* many insects were observed floating when they were immersed in fixative solution, indicating a severe damage of internal tissues. This interaction of *M. a. acridum* with *S. piceifrons* demonstrates how the fungus is capable of manifesting its virulence factors to overcome the host's defense to produce disease in the locust. It is evident that the susceptible host response is totally incapable to respond to the fungal aggression, therefore host damage is present in almost all insect tissues, including muscles, an explanation to the lack of the insect feeding as a sign of fungal infection after the application of this biological agent in the field and even in the laboratory (Moore *et al.,* 1992; Hernández Velázquez *et al.,* 2007).

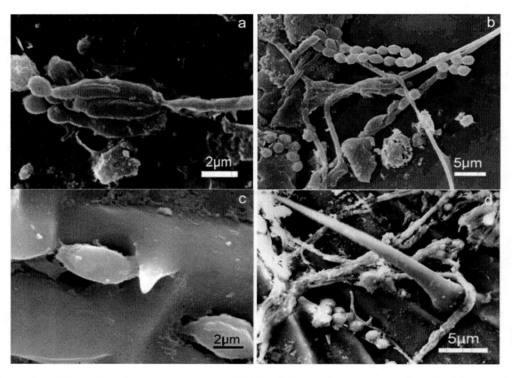

Figure 4. a-d. Scanning electron microscopy of infected *S. piceifrons* ssp. *piceifrons* by *M. anisopliae* var. *acridum* 6 days after inoculation. At this stage of infection many locust died and their surface had a greenish color. Conidiophores (a), catenulated conidia (b) conidium (c) and hypha (d) on the locust surface.

6.4. NON-PATHOGENIC INTERACTION BETWEEN FUNGAL ENTOMOPATHOGEN AND VERTEBRATE HOSTS

Since many years ago, safety studies with *M. anisopliae*, fungal entomopathogen with a wide range of insect host species, showed that this fungus is not pathogenic or toxic to several vertebrate models such as fish, reptiles, birds, rats, rabbits, guinea pigs, and mice (Austwick, 1980; Shadduck *et al.,* 1982; El-Kadi *et al.,* 1983; Tsai *et al.,* 1994; Jevanand and Kannan, 1995; Toriello *et al.,* 1999), and there are also safety studies on other entomopathogenic fungi (Austwick, 1980, Goettel *et al.* 1990; 2001; Prior, 1990; Evans, 1998; 2000).

Therefore these microorganisms are deemed safe and considered an environmentally acceptable alternative to chemical pesticides (Domsch *et al.,* 1980; Zimmermann, 1993) for the biological control of several pests (Butt *et al.,* 2001). More than one century has passed since Metchnikoff (ca. 1879) carried out the first scientific experiments to test entomopathogenic fungi as biological control of pests, and although some cases of human infection and disease by *M. anisopliae* have been reported in both immunocompetent (Cepero de García *et al.*, 1997, Revankar *et al.*, 1999) and immunosuppressed individuals, with one fatality in a child (Burgner *et al.*, 1998), humans seemingly are not a susceptible host for entomopathogens.

Pathogenicity in mammals has not been analyzed. The exposure via used in the safety protocols appear not to be relevant as a virulence factor because inhalation, subcutaneous, intraperitoneal, and intraocular routes (Burges, 1981; Shadduck *et al.*, 1982; El-Kadi *et al.*, 1983; Saik *et al.*, 1990, Toriello *et al.*, 1999), have not demonstrated increased adverse effects, toxicity or pathogenicity.

More recently, strain EH-479/2 of *M. anisopliae* var. *anisopliae* isolated from spittlebugs was inoculated by gavage in mice and showed lack of pathogenicity and toxicity (Toriello *et al.*, 2006). In this study, mycological and histological tests were performed until 21 days after the inoculation.

During this period, none of the animals showed discernible clinical symptoms of illness, positive organ cultures were obtained in 17 of 72 mice, but no inflammatory reactions or tissue damage were identified in the same analyzed organs in all animals. However, one mouse died 2 days after inoculation, before the first necropsy, probably due to the presence of fungal conidia associated to microthrombi and emboli occluding microvasculature, mainly at pulmonary circulation. Moreover, evidence of fungal growth was obtained in this only dead mouse at axillary lymph node and liver, which expressed some degree of damage (Toriello *et al.*, 2006).

The lack of pathogenicity and toxicity of the mycoinsecticide *M. anisopliae* var. *acridum* following acute gastric exposure in mice has been recently reported (Toriello *et al.*, 2009). This fungal entomopathogen infects and is being widely used for microbial control of locust and grasshoppers (Hunter *et al.*, 2001; Lomer *et al.*, 2001; Wraight *et al.*, 2001; Langewald *et al.*, 2002; Milner 2002).

Safety of *M. anisopliae* var. *acridum* has been established in some invertebrate and vertebrate animal models (Stolz 1999; Stolz *et al.*, 2002), such as non-target arthropods (Ball *et al.*, 1994; Peveling and Demba, 1997; Langewald *et al.*, 2002), pheasants (Smits *et al.*, 1999; Johnson *et al.*, 2002), animals belonging to aquatic ecosystems (Milner *et al.*, 2002), and lizards (Peveling and Demba, 2003).

Mycological and histological results in mice inoculated with a high dose of viable or heat-killed conidia of *M. a. acridum* showed that, while the fungus can persist temporarily (3 days post-inoculation), it does not seem to grow in the host, and no inflammatory acute reactions (Figure 5. a-c) were observed in several analyzed organs (Toriello *et al.*, 2009).

Gross pathology did not reveal abnormalities associated with toxicity or fungal infection and growth in tissue organs. However, splenomegaly was observed in this study and has also been reported in another two similar protocols of fungal entomopathogens inoculation in mice, *Paecilomyces fumosoroseus* (Mier *et al.*, 2005) and *M. anisopliae* var. *anisopliae* (Toriello *et al.*, 2006). In all of these cases, enlarged spleens showed congestive red pulp with numerous and large megakaryocytes but without an inflammatory infiltrate, which seems to be a non-specific response because it has been observed in mice exposed to vanadium (Fortoul *et al.*, 2008; Piñón-Zárate *et al.*, 2008) and some solvents as paint thinner (A. Cárabez, Mexico City, personal communication).

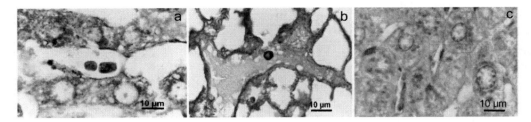

Figure 5. a-c. Histological finding in mice inoculated by gavage with viable conidia of *M. anisopliae* var. *acridum*. Fungus can persist 3 days post-inoculation but does not induce inflammatory response by the host at liver (a) and lung (b). Note the similar aspect of liver from inoculated (a) and control mice (c). Modified Gomori´s methenamine silver nitrate procedure (a and b), and PAS stain (c).

Our group has also demonstrated previously the lack of pathogenicity/toxicity of other entomopathogenic fungi such as *Erynia neoaphidis* and *Conidiobolus major* (Toriello *et al.,* 1986), *Hirsutella thompsonii* (Mier *et al.,* 1989), and *Verticillium lecanii* (Mier *et al.,* 1994), in guinea pigs and mice.

All these works show that even though fungi have the attributes of virulence to be able to cause disease in susceptible hosts, when introduced to a non-susceptible host, the same attributes of virulence are not manifested and therefore become incapable of causing damage. In this regard, there is much yet to study but with all the new molecular techniques available there is no doubt that further on we will be able to know the precise biological or physical factors that may switch-on or switch-off the genes involved in the complete pathogenesis mechanisms, and therefore have the knowledge to eliminate the possible fungal entomopathogen's damage to vertebrate and invertebrate non-target hosts.

CONCLUSIONS

With the increased concern in regard for a better and non-polluted environment, different measures have been taken. Among them, the efforts to substitute chemical pesticides by more environment friendly pesticides such as microbial agents for biological control. However, scientists and the general population are now aware of the problems that resulted because of the lack of knowledge and testing in regard to chemical pesticides safety half a century before. Nowadays, there is concern for safety procedures of microbial agents used in biological control and therefore the need to test them to minimize any possible risk or hazard for humans, animals, plants and the environment. The example given in this work of how the same fungus may be at the same time a pathogen of insects and a non-pathogen for mice taking into account that the entomopathogenic fungus needs a susceptible host to cause infection and/or death, may lead to the reassurance in the use of microbial agents for the biological control of plant pests and diseases in view of safety considerations of live microorganisms. Actually, with fungal specific molecular probes available to study the detail of host/microbe interactions, and other aspects of ecological safety procedures, it will be feasible to minimize the possible risks for humans and the environment, bearing in mind Burges (1981): "A no risk situation does not exist, certainly not with chemical pesticides, and even with biological agents, one cannot absolutely prove a negative. Registration of a chemical is essentially a statement of usage in which risks are acceptable, and the same must be applied to biological agents".

ACKNOWLEDGMENTS

The authors would like to thank Biol. Armando Zepeda for his technical assistance with SEM processes. This work was supported by the Consejo Nacional de Ciencia y Tecnología (CONACYT) de México, grant G-31451-B.

REFERENCES

Austwick, P.K.C. 1980. The pathogenic aspects of the use of fungi: The need for risk analysis and registration of fungi. In: Ludholm, B. and Stackerud, M. (Eds.), *Environmental Protection and Biological Forms of Control of Pest Organisms*. Volume 31. *Ecology Bulletin,* Stockholm, pp.91-102.

Ball, B.V., Pye, B.J., Carreck, N.L., Moore, D. and Bateman, R.P. 1994. Laboratory testing of a mycopesticide on non-target organisms: of an oil formulation of *Metarhizium flavoviride* applied to *Apis mellifera*. *Biocontrol Science and Technology,* 4: 289-296.

Bidochka, M.J. and Meltzer, M.J. 2000. Genetic polymorphisms in three subtilisin-like protease isoforms (Pr1, Pr1B, and PrC) from *Metarhizium* strains. *Canadian Journal of Microbiology,* 46: 1138-1144.

Boucias, D.G. and Pendland, J.C. 1991. The fungal cell wall and its involvement in the pathogenic process in insect hosts. In: Latgé, J.P. and Boucias, D.G. (Eds.), *Fungal cell wall and immune response*. NATO ASI Series Volume 53. Springer Verlag, Berlin, pp. 303-316.

Boucias, D.G. and Pendland, J.C. 1998. *Principles of insect pathology*. Kluwer Academic Publishers, Boston, MA, USA.

Burges, H.D. 1981. Safety, safety testing and quality control of microbial pesticides. In: Burges, H.D. (Ed.), *Microbial Control of Pests and Plant Diseases 1970-1980*. Academic Press, London, pp. 737-767.

Burgner, D., Eagles, G., Burgess, M., Procopis, P., Rogers, M., Muir, D., Pritchard, R., Hocking, A. and Priest, M. 1998. Disseminated invasive infection due to *Metarhizium anisopliae* in an immunocompromised child. *Journal of Clinical Microbiology,* 36: 1146-1150.

Butt, T.M., Jackson, C. and Magan, N. 2001. Introduction-Fungal biological control agents: Problems and potential. In: Butt, T.M., Jackson, C. and Magan, N. (Eds.), *Fungi as Biocontrol Agents*. CAB International, Wallingford, pp. 1-9.

Casadevall, A. and Pirofski, L. 1999. Host-pathogen interactions: redefining the basic concepts of virulence and pathogenicity. *Infection and Immunity,* 67: 3703-3713.

Casadevall, A. and Pirofski, L. 2003. The damage-response framework of microbial pathogenesis. *Nature Reviews of Microbiology*, 1: 17-24.

Carzaniga, R., Bowyer, P.O. and O' Connell, R.J. 2001. Production of extracellular matrices during development of infection structures by the downy mildew *Peronospora parasitica*. *New Phytologist*, 149: 83-93.

Cepero de García, M.C., Arboleda, M.L., Barraquer, F. and Grose, E. 1997. Fungal keratitis caused by *Metarhizium anisopliae* var. *anisopliae*. *Journal of Clinical Microbiology*, 35: 361-363.

Cook, R.J., Bruckart, W.L., Coulson, J.R., Goettel, M.S., Humber, R.A., Lumsden, R.D., Maddox, J.V., McManus, M.L., Moore, L., Meyer, S.F., Quimby, P.C., Stack, J.P. and Vaughn, J.L. 1996. Safety of microorganisms intended for pest and plant disease control. A framework for scientific evaluation. *Biological Control,* 7: 333-351.

Domsch, K.H., Gams, W. and Anderson, T.H. 1980. *Compendium of Soil Fungi*. Academic Press, London.

El-Kadi, M.K., Xará, L.S., De Matos, P.F., Da Rocha, J.V.N. and De Oliveira, D.P. 1983. Effects of the entomopathogen *Metarhizium anisopliae* on guinea pigs and mice. *Environmental Entomology*, 12: 37-42.

Evans, H.C. 1998. The safe use of fungi for biological control of weeds. *Phytoprotection*, 79: S67-S74.

Evans, H.C. 2000. Evaluating plant pathogens for biological control of weeds; an alternative view of pest risk assessment. *Australian Plant Pathology*, 29: 1-14.

Fargues, J. 1984. Adhesion of the fungal spore to the insect cuticle in relationship to pathogenicity. In: Roberts, D.W. and Ais, J.R. (Eds.), *Infection processes of fungi*. Rockefeller Foundation Report, pp. 90-110.

Ferron, P. 1978. Biological control of insect pests by entomopathogenic fungi. *Annual Review of Entomology*, 23: 409-442.

Fortoul, T.I., Piñón-Zárate, G., Díaz-Bech, M.E., González-Villalva, A., Mussali-Galante, P., Rodríguez-Lara, V., Colín–Barenque, L., Martínez-Pedraza, M. and Montaño, L.F. 2008. Spleen and bone marrow megakaryocytes as targets for inhaled vanadium. *Histology and Histopathology,* 23: 1321-1326.

Freimoser, F.M., Screen, S., Bagga, G.H. and St. Leger, R.J. 2003. Expressed sequence tag (EST) analysis of two subspecies of *Metarhizium anisopliae* reveals plethora of secreted proteins with potential activity in insect hosts. *Microbiology,* 149: 239-247.

García-Gutiérrez, K., Castellanos-Moguel, J., Cano-Ramírez, C., Mier, T., Hernández-Velázquez, V.M. and Toriello, C. 2002. Actividad de proteasas y tiempo letal medio en langosta (Orthoptera: Acrididae) de aislados de *Metarhizium anisopliae* var. *acridum* de México. In: Báez-Sañudo, R. and Juvera Bracamontes, J.J. (Eds.), *Actas del XXV Congreso Nacional de Control Biológico,* November 14-15, 2002, Hermosillo, Sonora, México. pp. 70-72.

Goettel, M.S., Poprawski, T.J., Vandenberg, J.D., Li, Z. and Roberts, W.D. 1990. Safety to non-target in vertebrates of fungal biocontrol agent. In: Laird, M., Lacey, L.A. and Davidson, E.W. (Eds.), *Safety of Microbial Insecticides.* CRC Press, Boca Raton, Florida, pp. 209-232.

Goettel, MS., Hajek, A.E., Siegel, J.P. and Evans, HC. 2001. Safety of fungal biocontrol agents. In: Butt T.M., Jackson, C.W. and Magan, N. (Eds.), *Fungi as Biocontrol Agents*. Cabi Publishing, Wallingford, pp. 347-375.

Hajek, A.E. and St. Leger, R.J. 1994. Interactions between fungal pathogens and insect hosts. *Annual Review of Entomology,* 39: 293-322.

Hernández-Velázquez, V.M., Hunter, D.M., Barrientos-Lozano, L., Lezama-Gutiérrez, R. and Reyes-Villanueva, F. 2003. Susceptibility of *Schistocerca piceifrons* (Orthopera: Acrididae) to *Metarhizium anisopliae* var. *acridum* (Deuteromycotina: Hyphomycetes): laboratory and field trials. *Journal of Orthoptera Research,* 12: 89-92.

Hernández-Velázquez, V.M., Berlanga, A., and Toriello, C. 2007. Reduction of feeding by *Schistocerca piceifrons piceifrons* (Orthoptera: Acrididae), following infection by *Metarhizium anisopliae* var. *acridum. Florida Entomologist,* 90(4): 786-789.

Hunter, D.M., Milner, R.J., Scanlan, J.C. and Spurgin, P.A. 1999. Aerial treatment of the migratory locust, *Locusta migratoria* (L.) (Orthopthera: Acrididae) with *Metarhizium anisopliae* (Deuteromycotina: Hyphomycetes) in Australia. *Crop Protection,* 18: 699-704.

Hunter, D.M., Milner, R.J. and Spurgin, P.A. 2001. Aerial treatment of the australian plague locust, *Chortoicetes terminifera* (Orthoptera: Acrididae) with *Metarhizium anisopliae* (Deuteromycotina: Hyphomycetes). *Bulletin of Entomological Research,* 91: 93-99.

Jarrold, S.l., Moore, D., Potter, U. and Charnley, A.K. 2007. The contribution of surface waxes to pre-penetration growth of an entomopathogenic fungus on host cuticle. *Mycological Research,* 111: 240-249.

Jevanand, H.R. and Kannan, N. 1995. Evaluation of *Metarhizium anisopliae* as a biocontrol agent for coconut pest *Oryctes rhinoceros* and its mammalian toxicity test on rats. *Journal of Ecotoxicology and Environmental Monitoring,* 5: 51-57.

Johnson, D.L., Smits, J.E., Jaronski, S.T. and Weaver, D.K. 2002. Assessment of health and growth of ring-necked pheasants following consumption of infected insects or conidia of entomopathogenic fungi, *Metarhizium anisopliae* var. *acridum* and *Beauveria bassiana,* from Madagascar and North America. *Journal of Toxicology and Environmental Health-Part A,* 65 (24): 2145-2162.

Khachatourians, G.G. 1998. Biochemistry and molecular biology of entomopathogenic fungi. In: Howard, D.H. and Miller, J.D. (Eds.), *The Mycota,* Volume VI: *Human and Animal Relationships.* Springer Verlag, Berlin, pp. 331-363.

Langewald, J., Stolz, I., Everts, J. and Peveling, R. 2002. Toward the registration of microbial insecticides in Africa: non target arthropod testing on Green Muscle[TM], a grasshopper and locust control product based on the fungus *Metarhizium anisopliae* var. *acridum.* In: Neuenschwander. P. (Ed.), *Biological Control in IPM Systems in Africa.* CABI Publishing, Wallingford, pp. 207-225.

Latgé, J.P., Sampedro, L., Brey, P. and Diaquin, M. 1987. Aggresiveness of *Conidiobolus obscurus* against the pea aphid-influence of cuticular extracts on ballistospore germination of aggressive and non-aggressive strains. *Journal of General Microbiology,* 133: 1987-1997.

Lee, B.I., Bateman, R., Guoyou, L.I., Meng, L. and Zheng, Y.A. 2000. Field trial on the control of grasshoppers in mountain grassland by oil formulation of *Metarhizium flavoviride. Chinese Journal of Biological Control,* 16: 145-147.

Leland, J.E., Mullins, D.E., Vaughan, L.J. and Warren, H.L. 2005. Effects of media composition on submerged culture spores of the entomopathogenic fungus, *Metarhizium anisopliae* var. *acridum,* Part 1: Comparison of cell wall characteristics and drying stability among three spore types. *Biocontrol Science and Technology,* 15(4): 379-392.

Lomer, C.J., Bateman, R.P., Johnson, D.L., Langewald, J. and Thomas, M. 2001. Biological control of locusts and grasshoppers. *Annual Review of Entomology,* 46: 667-702.

Magalhaes, B.P., Lecoq, M., Defaria, M.R., Schmidt, F.G.V. and Guerra, W.D. 2000. Field trial with the entomopathogenic fungus *Metarhizium anisopliae* var. *acridum* against bands of the grasshopper, *Rhammatocerus schistocercoides* in Brazil. *Biocontrol Science and Technology,* 10: 427-441.

Mier, T., Pérez, J., Carrillo-Farga, J. and Toriello, C. 1989. Study on the innocuity of *Hirsutella thompsonii*. I. Infectivity in mice and guinea pigs. *Entomophaga,* 34: 105-110.

Mier, T., Rivera, F., Rodríguez-Ponce, M.P., Carrillo-Farga, J. and Toriello, C. 1994. Infectividad del hongo entomopatógeno *Verticillium lecanii* en ratones y cobayos. *Revista Latino-Americana de Microbiologia,* 36: 107-111.

Mier, T., Olivares-Redonda, G., Navarro-Barranco, H., Pérez-Mejía, A., Lorenzana, M., Pérez-Torres, A. and Toriello, C. 2005. Acute oral intragastric pathogenicity and toxicity in mice of *Paecilomyces fumosoroseus* isolated from whiteflies. *Antonie van Leeuwenhoek,* 88: 103-111.

Milner, R.J. 1997. Insect pathogens - how effective are they against soil insect pests? In: Allsopp, P.G., Rogers, D.J. and Robertson, L.N. (Eds.), *Soil Invertebrates in 1997.* Bureau of Sugar Experiment Station, Brisbane Paddington, Australia, pp. 63-67.

Milner, R.J., and Pereire, R.M. 2000. Microbial control of urban pests - cockroaches, ants and termites. In: Lacey, L.A., and Kaya, H.K. (Eds.), *Field Manual of Techniques in Invertebrate Pathology.* Kluwer Academic Publisher, Boston, pp. 721-740.

Milner, R.J. 2002. Green Guard®. *Pesticide Outlook.* 13: 20-24.

Milner, R.J., Lim, R.P. and Hunter, D.M. 2002. Risks to the aquatic ecosystem from the application of *Metarhizium anisopliae* for locust control in Australia. *Pest Management Science,* 58: 718-723.

Moore, D., Reed, M., Le Patourel, G., Abraham, Y.J. and Prior, C. 1992. Reduction of feeding by desert locust, *Schistocerca gregaria*, after infection with *Metarhizium flavoviride*. *Journal of Invertebrate Pathology,* 60: 304-307.

Peveling, R. and Demba, S.A. 1997. Virulence of the entomopathogenic fungus *Metarhizium flavoviride* Gams and Rozypal and toxicity of diflubenzuron, fenitrothion-esfenvalerate and profenofos-cypermethrin to nontarget arthropods in Mauritania. *Archives of Environmental Contamination and Toxicology,* 32: 69-79.

Peveling, R. and Demba, S.A. 2003. Toxicity and pathogenicity of *Metarhizium anisopliae* var. *acridum* (Deuteromycotina: Hyphomycetes) and fipronil to the fringe-toed lizard *Acanthodactylus dumerili* (Squamata: Lacertidae). *Environmental Toxicology and Chemistry,* 22: 1437-1447.

Piñón-Zárate, G., Rodríguez-Lara, V., Rojas-Lemus, M., Martínez-Pedraza, M., González-Villalva, A., Mussali-Galante, P., Fortoul, T.I., Barquet, A., Masso, F. and Montaño, L.F. 2008. Vanadium pentoxide inhalation provokes germinal center hyperplasia and suppressed humoral immune responses. *Journal of Immunotoxicology,* 5: 115-122.

Pirofski, L.A. and Casadevall, A. 2008. The damage-response framework of microbial pathogenesis and infectious diseases. In: Huffnagle, G.B. and Noverr, M.C. (Eds.), *GI Microbiota and Regulation of the Immune System.* Springer, New York, pp. 135-146.

Prior, C. 1990. The biological basis for regulating the release of micro-organisms, with particular reference to the use of pest control. *Aspect of Applied Biology,* 24: 231-238.

Revankar, S.G., Sutton, D.A., Sanche, S.E., Rao, J., Zervos, M., Dashti, F. and Rinald, M.G. 1999. *Metarhizium anisopliae* as a cause of sinusitis in immunocompetent host. *Journal of Clinical Microbiology,* 37: 195-198.

Saik, J.E., Lacey, L.A. and Lacey, C.M. 1990. Safety of microbial insecticides to vertebrate-domestic animals and wildlife. In: Laird, M., Lacey, L.A. and Davidson, E.W. (Eds.), *Safety of Microbial Insecticides*. CRC Press, Boca Raton, Florida, pp.115-134.

Shadduck, J.A., Roberts, D.W. and Lause, S. 1982. Mammalian safety tests of *Metarhizium anisopliae*: preliminary results. *Environmental Entomology,* 80: 189-192.

Smits, J.E., Johnson, D.L. and Lomer, C. 1999. Pathological and physiological responses of ring-necked pheasant chicks following dietary exposure to the fungus *Metarhizium flavoviride*, a biocontrol agent for locusts in Africa. *Journal of Wildlife Diseases,* 35: 194-203.

Stahlman, K.P., Pielken, P., Schimz, K.L. and Sahm, H. 1992. Degradation of extracellular beta-(1,3)(1,6)-D-glucans by *Botrytis cinerea*. *Applied and Environmental Microbiology*, 58: 3347-3354.

St. Leger, R.J., Cooper, R.M. and Charnley, A.K. 1986. Cuticle degrading enzyme of entomopathogenic fungi: cuticle degrading *in vitro* by enzymes from entomopathogens. *Journal of Invertebrate Pathology,* 47: 167-177.

St. Leger, R.J., Charnley, A.K. and Cooper, M. 1987a. Characterization of cuticle-degrading proteases produced by the entomopathogen *Metarhizium anisopliae*. *Archives of Biochemistry and Biophysics*, 253-221-232.

St. Leger, R.J., Charnley, A.K. and Cooper, M. 1987b. Distribution of chymoelastases and trypsin-like enzymes in five species of entomopathogenic deuteromycetes *Archives of Biochemistry and Biophysics*, 258: 123-131.

St. Leger, R.J., Butt, T.M., Goettel, M.S., Staples, R.C. and Roberts, D.W. 1989. Production *in vitro* of appressoria by the entomopathogenic fungus *Metarhizium anisopliae*. *Experimental mycology,* 13: 274-288.

St. Leger, R.J. 1993. Biology and mechanisms of insect-cuticle invasion by deuteromycetous fungal pathogens. In: Beckage, N.C., Thompson, S.N. and Federici, B.A. (Eds.), *Parasites and pathogens of insects* –Volume 2. Academic Press, New York, pp. 211-229.

St. Leger, R.J. 1995. The role of cuticle-degrading proteases in fungal pathogenesis of insects. *Canadian Journal of Botany*, 73: 1119-1125.

St. Leger, R.J. and Bidochka M.J. 1996. Insect-fungal interactions. In: Soderhall, K, Iwanaga S. and Vasta, G.R. (Eds.), *New directions in invertebrate immunology*. SOS Publications, New Jersey, pp. 443-479.

Stolz, I. 1999. The effect of *Metarhizium anisopliae* (Metsch.) Sorokin (=*flavoviride*) Gams and Rozypal var. *acridum* (Deuteromycotina: Hyphomycetes) on non-target Hymenoptera. Ph.D. Dissertation, Universität Basel, Switzerland.

Stolz, I., Nagel, P., Lomer, C. and Peveling, R. 2002. Susceptibility of the Hymenopteran parasitoids *Apoanagyrus* (=*Epidinocarsis*) *lopezi* (*Encyrtidae*) and *Phanerotoma* sp. (*Braconidae*) to the entomopathogenic fungus *Metarhizium anisopliae* var. *acridum* (Deuteromycotina. Hyphomycetes). *Biocontrol Science and Technology,* 12: 349-360.

Toriello, C., Hernández-Ibañez, J.M., López-Martínez, R., Martínez, A., López- González, L., Mier, T., Carrillo, J. and Latgé, J.P. 1986. The pathogenic fungi of the spittlebug in México. III. Innocuity of *Erynia neoaphides* and *Conidiobolus major* in experimental animals. *Entomophaga,* 31: 317-376.

Toriello, C., Navarro-Barranco, H., Martínez-Jacobo, A. and Mier, T. 1999. Seguridad en ratones de *Metarhizium anisopliae* (Metsc.) Sorokin aislado de *Aeneolamia* sp. (Homoptera: Cercopidae) en México. *Revista Mexicana de Micología*, 15: 123-125.

Toriello, C., Pérez-Torres, A., Burciaga-Díaz, A., Navarro-Barranco, H., Pérez-Mejía, A., Lorenzana-Jiménez, M. and Mier, T. 2006. Lack of acute pathogenicity and toxicity in mice of *Metarhizium anisopliae* var. *anisopliae* from spittlebugs. *Ecotoxicology and Environmental Safety,* 65: 278-287.

Toriello, C., Jiménez-Gutiérrez, P., Pérez-Torres, A., Navarro-Barranco, H., Berlanga-Padilla, A. and Hernández-Velázquez, V. 2007. Microscopía electrónica de barrido en la infección experimental de *Schistocerca piceifrons piceifrons* (Orthoptera: Acrididae) por *Metarhizium anisopliae* var. *acridum. Memorias XXX Congreso Nacional de Control Biológico.* Simposio del Organismo Internacional de Control Biológico. Mérida, Yucatán, México. November 11-15, 2007, pp. 109-112.

Toriello, C., Pérez-Torres A., Vega-García, F., Navarro-Barranco, H., Pérez-Mejía, A., Lorenzana-Jiménez, M., Hernández-Velázquez, V. and Mier, T. 2009. Lack of pathogenicity and toxicity of the mycoinsecticide *Metarhizium anisopliae* var. *acridum* following acute gastric exposure in mice. *Ecotoxicology and Environmental Safety*, 72: 2153-2157.

Tsai, S.F., Liao, J.W., Hung, W.K. and Wang, S.H. 1994. Acute pulmonary toxicity, infectivity and pathogenicity of *Metarhizium anisopliae*, on rats. *Plant Protection Bulletin* (Taichung), 36: 65-73.

Wraight, S.P., Jackson, M.A. and de Kock, S.L. 2001. Production, stabilization and formulation of fungal biocontrol agents. In: Butt, T.M., Jackson, C., and Magan, N. (Eds.), *Fungi as Biocontrol Agents*. CAB International, United Kingdom, pp. 253-287.

Zacharuk, R.Y. 1970. Fine structure of the fungus *Metarhizium anisopliae* infecting 3 species of larval *Elateridae Coleoptera*. Part 2. Conidial germ tubes and appressoria. *Journal of Invertebrate Pathology*, 15: 81-91.

Zimmermann, G. 1993. The entomopathogenic fungus *Metarhizium anisopliae* and its potential as a biocontrol agent. *Pesticide Science,* 37: 375-379.

In: Microbial Insecticides: Principles and Applications   ISBN: 978-1-61209-223-2
Editors: J. Francis Borgio, K. Sahayaraj, et al.   © 2011 Nova Science Publishers, Inc

Chapter 7

PHYLOGENETIC ANALYSIS OF ENTOMOPATHOGENIC FUNGI

*Milton A. Typas[1,] * and Vassili N. Kouvelis[2]*

[1] Department of Genetics and Biotechnology, Faculty of Biology, University of Athens, Panepistimiopolis, Kouponia, Athens 157 01, Greece
[2] Department of Genetics and Biotechnology, Faculty of Biology, University of Athens, Panepistimiopolis, Kouponia, Athens 157 01, Greece

ABSTRACT

Entomopathogenic fungi (EPF) are cosmopolitan insect pathogens that produce several biologically active metabolites and some of them have already been commercially used as Biological Control Agents. The most common molecular techniques that have been developed lately for phylogenetic studies like RAPDs, RFLPs, AFLPs, microsatellite analysis, telomeric fingerprinting, direct sequencing and analysis of particular genomic regions and genes, alone or in combination with other genes (multi-gene approach), have been applied in studies of phylogeny and consequently classification, evolution and taxonomy of EPFs. Each method has advantages and disadvantages that are analysed. The best results for EPF phylogenetic studies have been obtained from the sequences of several genes not only within taxa but also within genera, orders and subphyla where EPF belong. A number of phylogenetic studies based on sequences of the nuclear rRNA gene-complex, housekeeping genes like *ben*A, *tef*1, *rpb*1 and *rpb*2, or mitochondrial genomes, revealed the taxonomic status of many EPFs and helped in the revision of the best known genera like *Beauveria, Metarhizium, Paecilomyces* and *Lecanicillium* (former *Verticillium lecanii*). The generally accepted notion that insect hosts or EPF geographic location are related to certain fungal genotypes has also been tested in a number of phylogenetic studies. However, the results obtained were contradictory since some support a correlation of fungal genetic loci to their insect hosts or origins, while others show complete lack of association. Studies based on gene clusters of different evolutionary origin (nuclear and/or mitochondrial), were the most informative and provided all the necessary data for establishing well-supported and accurate fungal phylogenetic relationships.

* Tel.: +302107274633, Fax: +302107274318, E-mail: matypas@biol.uoa.gr

7.1. INTRODUCTION

Entomopathogenic or entomophagous fungi are the fungi that are capable of infecting/parasitizing insects and cause their death. Surprisingly enough, they were known since the time of Aristotle (384-322 B.C.), who first described the "disease" of honeybees by a fungus. However, it was only in mid 19th century when the Italian entomologist Agostino Bassi identified the first entomopathogen fungus as the cause of the muscardine disease of domesticated silkworms (1835 A.C.), later to be named after his discoverer as *Beauveria bassiana* (Bals.-Criv.) Vuill., and almost 50 years later, Metschnicoff discovered and named another fungal species, *Metarhizium anisopliae* (Metschn.) Sorokin, after the insect species it was originally isolated, namely the beetle *Anisopliae austriaca* (Metschnicoff, 1879). The revolution of chemistry in the 20th century diminished the importance of such natural biological control agents (BCAs) and for many years, the application of vast amounts of chemicals was imposed as a one-way choice for the protection of agriculture crops, at a level that became dangerous to human and animal health and polluted air and water reservoirs. This has created an urgent need for environmentally more friendly agricultural practices and since the successful application of *Bacillus thurungiensis* and its registration as a commercial product in the U.S.A. in 1961 (Registration Eligibility Decision Document, 1998), it led to the rediscovery of BCAs. It also prompted insect pathologists express strong optimism during recent years for the potential of fungal entomopathogens as BCAs, hoping to use them alone or complementary to chemicals for sustainable pest control (Inglis *et al.,* 2001; Butt, 2002).

Entomopathogenic fungi (EPF) can be found mainly in the phyla Ascomycota, Zygomycota but also the Chytridiomycota and Oomycota (Samson *et al.,* 1988; Eilenberg, 2002; Shah and Pell, 2003). However, the best known and currently investigated for their potential use as BCAs are primarily those that belong to the class Hyphomycetes in the Deuteromycota and, to a lesser extent, those of the class Entomophthorales in the Zygomycota. Entomopathogenic hyphomycetes include genera with cosmopolitan members of haploid, asexual, soil-borne fungi of significance for their role as insect pathogens and the production of biologically active metabolites (Butt *et al.,* 2001). They are opportunistic pathogens that share common habitats with those of the insect hosts and, in general, synchronize their life cycles with insect host stages and co-evolve with them. Unlike bacterial and viral pathogens of insects, hyphomycetes infect the insect with contact and do not need to be consumed by their host to cause infection. As soon as conidia of these fungi land on the cuticle of a suitable insect host, they adhere and form germ tubes or appresoria, with the help of which and the concomitant secretion of an array of enzymes, penetrate the external insect skeleton, invade the insect body and the haemolymph, completely overwhelming host defence mechanisms and causing insect death (Clakson and Charnley, 1996; Roberts and St. Leger, 2004). Death caused by hyphomycetes is invariably associated with the production of several enzymes and secondary metabolites (Clakson and Charnley, 1996; Butt *et al.,* 2001). Usually, once the fungus has killed its host, it grows back out through the softer portions of the cuticle, covering the insect with a layer of millions of spores that are released to the environment, in which they persist for long periods of time and can infect new insects (Fegan *et al.,* 1993). Entomophtorales, on the contrary, infect insects and gradually colonize tissues, parasitizing the insect for long periods of time before causing host death and making use of little, if none, of toxins (Humber, 2008). Some species even sporulate while the host is still alive and mobile (e.g., members of the genus *Strongwellsea*; Eilenberg, 2002). Since several species of Entomophthorales are known only from a single host or relatives of that host, a high degree of host-parasite specialization is suggested. However, in spite of their potential, these fungi still remain the least studied fungal BCAs.

In accordance with a generally accepted notion in fungi, species and often isolates within a species, have different genetic properties which –depending on the infection strategies the fungi have developed- allow them to behave differently towards their hosts or to the various environmental conditions. Such genetic differences enable them either to infect a broad range of insect hosts, as is the case for *Beauveria bassiana*, which is capable of infecting more than 750 different insect hosts (Uribe and Khachatourians, 2004) or to specialize more at a genus or family level (e.g., *Beauveria brongniartii*, found only in species of *Melolontha*, Strasser *et al.*, 2004; *Metarhizium acridum* stat. nov., found only in members of acrididae, Bischoff *et al.*, 2009). Clearly, these genetic properties directly correlate to the entomopathogenic lifestyle that these fungi have adopted through the course of evolution and also reflect to the "plasticity" of their genome. Entomopathogenicity in hyphomycetes has been suggested to have arisen or have lost, multiple times in many independent lines of fungal evolution, from common soil saprophytic ancestors and following inter-kingdom jumps in insect, animals, plants and fungi to have extensively radiated into different host groups (Nikoh and Fukatsu, 2000; Humber, 2008). The Clavicipitaceae (Ascomycota, Hypocreales) contain more than 33 genera and approximately 800 species, including the most prominent families of EPF like *Beauveria*, *Metarhizium*, *Lecanicillium* (former *Verticillium lecanii*), *Nomurea*, *Paecilomyces* and *Trichoderma*, but at the same time plant parasites, endophytes and epiphytes (Eriksson, 2006; Index Fungorum http://www.speciesfungorum.org /Names/Names.asp). Moreover, an increasing number of recent reports indicate that EPF themselves play additional roles in nature, including endophytism, antagonism or pathogenicity to plant pathogens, and possibly act as plant growth promoting agents (Harman *et al.*, 2004; Posada and Vega, 2005; Quesada-Moraga *et al.*, 2006). This underlines the necessity to understand in depth the functional ecology of EPF and their true phylogenetic relationships, in a way that combined knowledge provides answers on the evolution, role(s) and putative exploitation of these fungi (Goettel, 2008). In addition, the deliberate introduction of fungal BCAs in the environment and the possible effects on non-target hosts creates social concerns and accentuates the need for efficient methods capable of monitoring the establishment and spread of the released fungi in the field (Strasser *et al.*, 2000; Butt *et al.*, 2001).

The characterization and classification of asexual fungi for many years was based on traditional morphological taxonomic criteria (Tulloch, 1976; Rombach *et al.*, 1987; Glare and Inwood, 1998) and later - in the 80s and early 90s- was strengthened by the analysis of enzyme profiles (Riba *et al.*, 1986; St. Leger *et al.*, 1992; Tigano-Milani *et al.*, 1995a). However, morphological characteristics have only limited potential to distinguish entomopathogenic species and enzyme synthesis can vary significantly during fungal growth (Driver *et al.*, 2000). Thus, the advent of new molecular methods during the last two decades provided valuable tools that enable us to resolve the genetic, taxonomic and phylogenetic status of the fungal species under examination. In this chapter an effort will be made to present the progress made through recent years in molecularly characterising, identifying and monitoring inter- and intra- specific phylogenetic relations of entomopathogenic hyphomycete fungi.

7.2. MOLECULAR TOOLS MOST COMMONLY USED FOR PHYLOGENETIC ANALYSES OF EPF

The basic principle behind any molecular approach that aims at the differentiation or clarification of the phylogenetics of members of a family, genus, species or even isolates is to seek for polymorphisms in the genetic material of these organisms. In the beginning methods were based on random differences (mutations) all over the genome and later on specific regions of the genetic material that provided clear advantages in handling, namely multiple copies of a gene (nuclear rRNA gene complex, repetitive regions like microsatellites, transposable elements, direct or inverted repeats, etc, and the mitochondrial DNA).

Gradually, it became evident that nucleotide sequences of a particular gene were necessary to establish adequate number of informative characters for phylogenetic analyses but also that no single gene alone or single gene region alone could provide reliable information for phylogenetic analyses. Thus, we have ended up with analyses of nucleotide sequences from several genes -multi-gene approaches- and the most recent developments fully support this view [e.g., the National Science Foundation project undertaken by several research groups "The Assembling of the Fungal Tree of Life (AFTOL)" (http://aftol.org/) aiming to molecularly analyse as many as 5,000 different fungal taxa and establish their phylogenetic relations and course of evolution (Hibbett *et al.,* 2007), and entire issues appearing in scientific Journals like "Mycologia" (2007 vol. 98), "Studies in Mycology" (2007 vol. 57) and numerous independent research groups –too many to mention here-].

The most common molecular approaches in studies of phylogenesis and consequently classification, evolution and taxonomy of EPFs are: (a) Random Amplification of Polymorphic DNA (RAPDs), (b) Restriction Fragment Length Polymorphisms (RFLPs) and Amplified Fragment Length Polymorphisms (AFLPs), (c) microsatellites and telomeric fingerprinting, and (d) direct sequencing and analysis of particular genomic regions and genes, alone or in combination with other genes (multi-genic approach). A brief description of these methods and exemplar articles concerning EPFs are given below:

7.2.1. Random Amplification of Polymorphic DNA (RAPD)

RAPD is generally considered as a powerful diagnostic tool for the differentiation of several EPF when a variety of 10-mer primers are used for the PCR amplification of total genomic DNA. The technique utilizes short primers of arbitrary sequence that anneal to multiple target sequences producing diagnostic patterns (Williams *et al.,* 1990). It does not require prior knowledge of target site sequence and, therefore, can be easily adapted to study various entomogenous fungi, even those with poorly studied genomes.

For instance, Tigano-Milani *et al.,* (1995b) showed that arbitrarily primed PCR with the OPA and *trn* primers generated a finer level of resolution in *Paecilomyces fumosoroseus* (Wise) A.H.S. Br. and G. Sm., in comparison with respective results based on morphological and isozyme data analyses.

Similarly, studies on *B. bassiana* showed that RAPD markers could give better resolution than isozymes between strains (Castrillo and Brooks, 1998) and in the case of *Metarhizium* helped place previously unidentified strains of the genus into *M. anisopliae* or *Metarhizium flavoviride* W. Gams and Rozsypal (Bidochka *et al.,* 1994; Bridge *et al.,* 1997). However, the main disadvantage of this method is that reproducibility is not always guaranteed since even the slightest changes in experimental conditions may alter results and the electrophoretic patterns obtained are based on bands of unknown origin and function (Figure 1).

7.2.2. Restriction Fragment Length Polymorphisms (RFLPs) and Amplified Length Polymorphisms (AFLPs)

Digestion of total genomic DNA from an EPF strain/isolate can be performed either using a number of common restriction endonucleases (recognizing and cleaving a target sequence of six nucleotides) until some prove to be discriminative or by restriction endonucleases (RE) that recognize four-, eight- or even twelve- nucleotide targets. The differences obtained in electrophoretic patterns of digested DNA are either due to small genetic changes of the restricted genomic DNA (single point mutations or small deletions/additions) that affect the composition of the RE target sequence or to genetic alterations of larger extend (deletions, duplications, inversions, translocations, insertion of mobile elements like introns and transposons, and replicons like plasmids and dsRNA viruses). Thus, depending on the RE applied, the effects can be substantially different. For example, tetra-cutter REs are usually chosen to restrict regions rich in AT or GC composition, obviously for contrasting reasons. The mitochondrial DNA (mtDNA) of fungi has a rich in AT composition –over 70%- whereas the nuclear DNA of the same organism has, in general, an even distribution of AT and GC –around 50%- in chromosomes and only islands of GC-rich regions. Consequently, in the case of REs like *Hae*III (GGCC), *Hpa*II (CCGG) and *Cfo*I (GCGC) the chromosomal DNA will be digested into fragments smaller than 1.6 kb and will reveal bands larger than this molecular size, all of which will be predominantly of mt origin (Figure 2). On the contrary, REs that recognize nucleotide sequences rich in AT restrict mtDNA into very small fragments and provide a clearer picture of nuclear DNA bands.

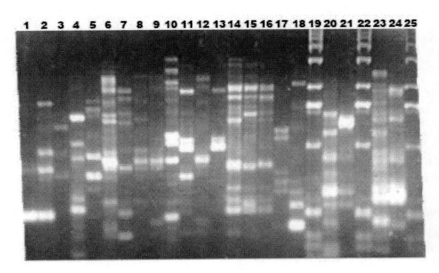

Figure 1. RAPD patterns obtained from 18 *Beauveria bassiana* and 7 unclassified *Beuveria* sp. isolates (lanes 1-18 and 19-25, respectively) with the use of random decamer primers (unpublished data).

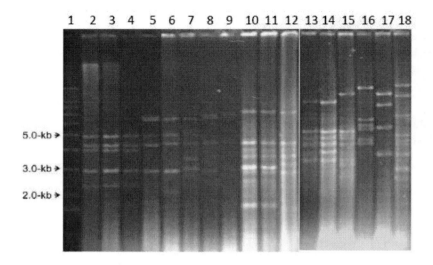

Figure 2. RFLP patterns of total genomic DNA from 17 isolates of *Lecanicillium* (lanes 2-18; lane 1=1-kb ladder Invitrogen) digested with restriction endonuclease *Hae*III.

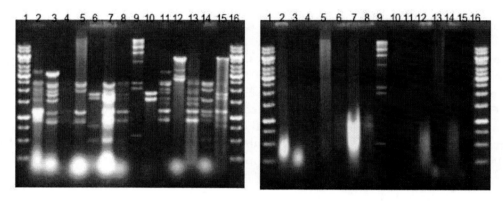

Figure 3. Agarose gel with the bands of total nucleic acids (i) extracted from 14 *B. bassiana* strains (lanes 2-15; lanes 1 and 16= 1-Kb ladder Fermentas). All bands disappearing following incorporation of RNase in the gel (ii) are dsRNA virus-like particles (Kotta-Loizou and Typas, 2009 – unpublished data).

In cases that a fungal strain contains dsRNA viruses (Giménez-Pecci *et al.,* 2002; Dalzoto *et al.,* 2006), the viral RNA will not be restricted by any RE and the electrophoretic band(s) corresponding to its M.W. will be clearly detectable (Figure 3). Providing that appropriate probes are available, the origin of these bands (i.e., nuclear rRNA repeat, genomic or mitochondrial, plasmid or viral) can be verified by DNA hybridization and this allows a very good differentiation of species and/or isolates of a given EPF species (Figure 4; Walsh *et al.,* 1990; Typas *et al.,* 1992; Mavridou and Typas, 1998). An alternative variable of RFLP is the Amplified Length Polymorphism (AFLP) technique which is based on RE-digestions of PCR amplification products of a particular gene. This method is widely used especially for amplified parts of the nuclear rRNA gene complex, mt genes and specific genes of phylogenetic interest as described in the project Assembling of the Fungal Tree of Life (AFTOL)" (http://aftol.org/).

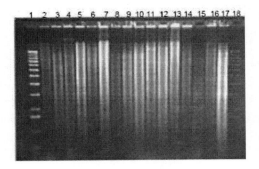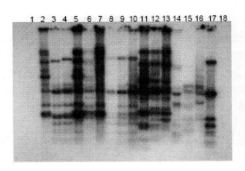

Figure 4. (a) Total genomic DNA from 16 isolates of *Lecanicillium* digested with restriction endonuclease *Eco*RI (lanes 2-17; lanes 1 and 18: 1-kb ladder Invitrogen, quantity in lane 18 1/10 of lane 1) and (b) hybridised with the total mt DNA probe of *L. muscarium* strain C42 (unpublished data).

From the numerous of works employing RFLPs or AFLPs, only a few examples are given below for the three more popular EPF species: (It should be noted that both methods provide credible results but they are costly and time consuming).

(i) In *B. bassiana*, a weak correlation between fungal isolates and their origin and/or habitat was recorded with the use of AFLPs (Aquino de Muro *et al.,* 2003), whereas in other cases, no association with either geographic origin or habitat was observed, however, cryptic phylogenetic species were emerged from the fungal populations studied (Devi *et al.,* 2006). RFLPs of *Beauveria* mt genomes proved ideal for the discrimination of species and further for *B. bassiana* isolates (Uribe and Khachatourians, 2004).

(ii) In *M. anisopliae*, AFLPs gave highly reproducible results and also revealed a high level of genotypic diversity among the fungal populations examined. *M. anisopliae* isolates could be distinguished from other fungal taxa by AFLP studies and could be also grouped into distinct groups (Leal *et al.,* 1997; Inglis *et al.,* 2008; Enkerli *et al.,* 2009). RFLPs of mt origin were again an excellent tool for the genetic fingerprinting and discrimination of *Metarhizium* species and *M. anisopliae* isolates (Mavridou and Typas, 1998).

(iii) Equally good discrimination between species and isolates was achieved by AFLPs with amplicons of nuclear (b-tubulin, rDNA) and mt genes (*rns*) in population studies of the formerly *Verticillium lecanii* (Zimm.) Viégas -now known as *Lecanicillium*- (Sugimoto *et al.,* 2003). RFLP studies also provided valuable data for the discrimination of *V. lecanii* isolates and helped for their taxonomic revision in the newly introduced *Lecanicillium* and *Simplicillium* species (Kouvelis *et al.,* 1999; Zare and Gams, 2001a).

7.2.3. Microsatellites and Telomeric Fingerprinting

Microsatellite markers are considered as a powerful tool for EPF species or strain differentiation and may also contribute substantial information for phylogenetic studies. They allow differentiation with high resolution, are easy to handle and also are, amenable to automated high-throughput analyses. A perfect example of their value in studies with EPFs is that of *M. anisopliae*, where the variability obtained with microsatellites primers and probes led not only to the phylogenetic analysis of the isolates examined but also to the development of molecular tools for tracking and monitoring specific BCA strains in the field (Enkerli *et al.*, 2001; Enkerli *et al.*, 2005; Oulevey *et al.*, 2009). The main disadvantage of this approach is that analysis of large numbers of markers in large populations of strains remains costly (Oulevey *et al.*, 2009).

The telomeric fingerprinting, a method in which total genomic DNA is digested with several REs and then hybridized with a probe consisting of a telomeric sequence (e.g., the repeated hexanucleotide sequence [TTAGGG]$_{18}$), appears as an additional option for the molecular phylogeny and taxonomy of EPF species (Viaud *et al.*, 1996; Boucias *et al.*, 2000; Padmavathi *et al.*, 2003). The telomeric sequences serve as an effective multicopy probe for detecting band variation among fungal isolates and polymorphisms recorded reflect to the variation among chromosome(s) telomere-associated regions of the different isolates. The technique has successfully been used with isolates of *Nomuraea rileyi* (Farl.) Samson that were grouped into two different clusters and *Beauveria* isolates collected from different host insects that showed several distinct fingerprints (Boucias *et al.*, 2000). However, it must be noted that in a previous report (Viaud *et al.*, 1996) *Beauveria* isolates collected from the same hosts produced similar fingerprints.

7.2.4. Direct Sequencing and Analysis of Specific Gene Regions and Genes, Alone or as Part of a Multi-Genic Approach

Today, it is generally accepted that the best results for inferring phylogenetic and evolutionary relationships are obtained from the full exploitation of DNA sequences (as nucleotide or amino acid datasets). The trees produced with Maximum Likelihood (ML) and Maximum Parsimony (MP) analyses allow a view in the deep phylogeny of the microorganisms studied.

For studies concerning phylogeny within a genus or a species, a distant method, like the Neighbour-Joining (NJ), is much faster compared to the other methods and results in equally credible phylograms. Lately, Bayesian inference (BI) is capturing the attention of all researchers as it can be applied in all studies having the advantages of a maximum likelihood approach but requiring less time in producing the expected results. Since the phylogeny of EPFs follows the rules and approaches applied in general for all fungi, no further details are provided in this section, other than the molecules used for these studies.

The most popular genetic region used in the phylogeny of almost all eukaryotic microorganisms and consequently of EPFs is the nuclear rRNA gene complex (Figure 5).

The "ITS" domain consists of the internal transcribed spacers 1 and 2 (ITS1 and ITS2) flanking the 5.8S rRNA gene. It is the region used as the basis for every phylogenetic analysis of an EPF almost by every research group that has worked with these fungi (e.g., Driver *et al.*, 2000; Zare and Gams, 2001a; Rehner and Buckley, 2005).

The small (SSU or 18S) and large (LSU or 28S) ribosomal subunits are both coding for the respective rRNA gene products and are the second most popular regions for phylogenetic and taxonomical analyses of EPFs. Since these genes often harbour group-I introns they provide additional information for phylogenetic analyses. Interestingly enough, these introns are inserted not only at exactly the same point of insertion in all fungi but also after highly conserved target sequences that are flanking the intron (namely, 516, 943, 989 and 1199 of *Eschericia coli* for the 18S rRNA gene, and 1921, 2066, 2449 and 2563 of *E. coli* for 28S rRNA; Figure 5). Thus, nucleotide sequences of SSU or LSU, group-I introns as nucleotide sequences or as drawn secondary structures (Figure 6), each one alone or in any possible combination, can provide valuable information for convincing phylogenetic and taxonomic analyses of EPFs (e.g., for the 28S rRNA: Neuveglise *et al.,* 1997; Mavridou *et al.,* 2000; Pantou *et al.,* 2003; Wang *et al.,* 2003a; Marquez *et al.,* 2006; Sung *et al.,* 2007; and for the 18S rRNA: Nikoh and Fukatsu, 2001; Pantou *et al.,* 2003; Luangsa-ard and Samson, 2004; Yokoyama *et al.,* 2006). Finally, perhaps the most informative molecule, especially in cases where differentiation between isolates of a given fungal species is the problem to solve, is the intergenic spacer region (IGS; see Figure 5, examples Pantou *et al.,* 2003; Hughes *et al.,* 2004). However, due to its large size (often larger than 3,000 nt; Pantou *et al.,* 2003), the difficulty in designing "universal" primers and the interference of its secondary structures in obtaining good PCR products makes it the less used molecule of the rRNA repeat.

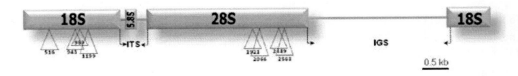

Figure 5. Schematic representation of the fungal nuclear rRNA gene complex. Dotted triangles represent introns and numbers underneath triangles indicate the exact insertional position in the respective gene sequence of *Escherichia coli*.

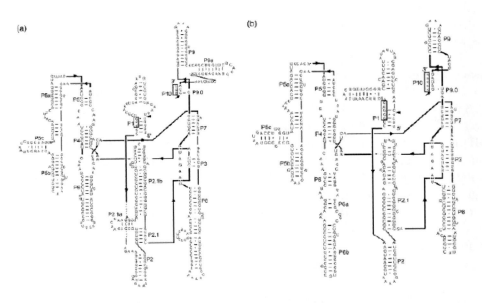

Figure 6. Predicted RNA secondary structures of two *Metarhizium anisopliae* var. *anisopliae* introns: (a) Ma-int4 and (b) Ma-int5 (Pantou *et al.,* 2003).

Besides the rDNA matrix, several single genes, again alone or as part of a multi-gene approach, have been used in analyses of EPFs, following fungal phylogeny methodologies mentioned above. More specifically, genes like those coding for b-tubulin (benA), translation elongation factor 1-a (tef1), the RNA polymerase II subunits 1 and 2 (rpb1 and rpb2) are usually included in molecular analyses as they demonstrate levels of polymorphism that can be exploited with good discriminatory results (e.g., Sung et al., 2007; Chaverri et al., 2008; Liu et al., 2009). In the case of EPFs, pathogenicity-associated genes like proteases or chitinases hold a prominent position and have extensively been used in phylogenetic studies, providing useful information on fungal phylogeny, but more particularly offering better insights in the evolution of these genes/gene families (Bidochka et al., 1999; Bidochka et al., 2001; Bagga et al., 2004; Hu and St. Leger, 2004; Sanz et al., 2004; Seidl et al., 2005; Wang et al., 2009). Lately, mt genes or even entire mt genomes are increasingly used to examine genetic diversity within fungal populations and to establish phylogenetic relationships (e.g., Bullerwell et al., 2003; Tambor et al., 2006; Pantou et al., 2008). MtDNA has several advantages that make it an attractive molecule in phylogenetics because it evolves faster than nuclear DNA (Burger et al., 2003), contains introns and mobile elements (Paquin et al., 1997) and exhibits extended polymorphisms (Hegedus and Khachatourians, 1993; Mavridou and Typas, 1998; Sugimoto et al., 2003; Pantou et al., 2006). Moreover, it is present in many copies per cell, has highly conserved gene functions and gene content (see a comparison of mtDNA coding product sizes from representatives of all fungal phyla and human as an outgroup in Table I) and in many cases has extended synteny in genome organization. For example, the complete mtDNAs of the Hypocreales B. bassiana, B. brongniartii, L. muscarium, M. anisopliae, Fusarium oxysporum, Giberella zeae and Hypocrea jecorina maintain four fully conserved syntenic units [namely $rnl-trn_{(11-12)}-nad2-nad3$, $nad4L-nad5-cob-cox1$, $nad1-nad4-atp8-atp6$ and $rns-trn_{(1-5)}-cox3-trn_{(1-5)}-nad6-trn_{(2-5)}$] and these units have been suggested to be equally conserved –though with some minor genetic rearrangements- in all Sordariomycetes (Pantou et al., 2006; Pantou et al., 2008). On the other hand, the variable number of introns, mobile elements and long-variable intergenic regions that EPF mtDNAs contain (Paquin et al., 1997; Kouvelis et al., 2004), were exploited in phylogenetic studies using RFLPs, AFLPs and RAPDs (Kouvelis et al., 1999; Sugimoto et al., 2003; Uribe and Khachatourians, 2004). Compared with nucleotide datasets from standard nuclear single genes, mt genes like the small rRNA gene (rns), and the NADH dehydrogenase subunits 1(nad1) and 3 (nad3) often provide more informative / discriminatory results (Kouvelis et al., 2008a,b; Sosa-Gomez et al., 2009). Moreover, particular mt intergenic regions, like for example that of nad3-atp9, could resolve phylogenetic uncertainties in Metarhizium and Beauveria that couldn't be resolved with nuclear datasets (Ghikas et al., 2006; Kouvelis et al., 2008b). Finally, when the sequences of all 14 essential mt genes (amino acid or nucleotide) were concatenated and the datasets were used in phylogenetic studies, they helped to resolve the exact phylogenetic status of several EPFs within the kingdom of Fungi (Kouvelis et al., 2004; Ghikas et al., 2006; Pantou et al., 2008). However, in spite of the great value of the latter approach in phylogenetics because it reflects the evolution of a complete small genome that co-evolves with the nuclear genome of the organism, its obvious disadvantage is that it requires complete mtDNA nucleotide sequences (usually 24-50 kb in size). Yet, with the fast progress of deep sequencing and the continuous lowering of cost, this may prove to be one of the powerful phylogenetic tools in the near future.

Table 1. Mitochondrial protein sizes (in a.a.) of complete fungal mt genomes from all known Ascomycetes and representatives from the other subphyla. Shaded cells show the size of EPF mt proteins. The highest and lowest values of identity are shown underlined and in bold, while "–" and "X" denote the lack and missed characterisation of the respective mt proteins

Phylum	Group	Class	Order	Protein (Fungi)	Cob	Cox1	Cox2	Cox3	Nad1	Nad2	Nad3	Nad4	Nad4L	Nad5	Nad6	Atp6	Atp8	Atp9
Ascomycota	Pezizomycotina	Sordariomycetes	Hypocreales	B.b	391	531	249	269	368	567	139	498	89	661	212	263	48	74
				B.br	391	538	249	269	368	568	139	498	89	660	212	263	48	74
				M.a	390	526	249	269	371	559	137	485	90	686	225	261	55	74
				L.m	385	533	249	269	375	562	141	492	89	642	214	262	53	74
				F.ox	390	530	249	269	369	554	137	507	89	662	223	266	48	74
				H.jec	375	523	236	269	380	594	137	489	99	692	250	259	48	67
			Phyl	V.dah	390	540	248	270	368	554	136	495	88	667	222	264	57	74
			Sorda-riales	N.cr	385	557	250	269	371	583	147	543	89	715	213	261	54	74
				P.an	387	541	250	269	367	556	137	519	89	652	221	264	51	X
				C.gl	393	536	249	269	377	562	149	505	89	695	224	262	55	X
		Doth	Pleos	P.nod	399	525	248	269	371	572	148	488	94	658	198	258	X	69
		Leot	Hel	B.fuc	391	559	227	269	383	565	153	489	89	665	231	259	48	80
				S.scl	398	530	252	268	385	557	153	489	89	664	231	259	48	64
		Eurot	Onyg	E.floc	400	528	264	269	358	565	134	487	89	653	199	255	48	74
				T.rub	386	528	237	269	343	563	138	497	89	653	203	255	48	74
			Eurot	P.mar	386	561	251	269	359	580	135	487	89	658	194	256	48	74
				A.n	387	535	252	269	352	X	136	488	X	657	228	256	48	74
	Sacchar	Sacc		Y.lip	385	535	242	268	341	469	128	486	89	655	185	255	48	76
				P.can	386	535	247	269	351	567	148	511	93	644	207	256	48	76
				S.cer	385	534	251	269	-	-	-	-	-	-	-	259	48	76
	Schiz	Schiz		S.jap	389	530	250	269	-	-	-	-	-	-	-	256	52	75
Basid	Het	Trem		C.neo	385	528	251	258	336	499	126	483	88	662	203	277	48	72
Basid	Hom	Agar		S.com	383	527	252	268	337	599	142	505	88	686	215	262	52	73
Chytr	Mon			Harp	366	487	233	269	297	428	109	457	87	580	203	237	49	74
Zygom	Muc			R.or	386	528	255	265	325	523	124	475	88	655	214	259	48	74
Chytr	Blast			A.mac	382	536	249	274	332	477	114	482	99	641	204	262	47	74

Table 1. (Continued)

Protein	Cob	Cox1	Cox2	Cox3	Nad1	Nad2	Nad3	Nad4	Nad4L	Nad5	Nad6	Atp6	Atp8	Atp9
H. sapiens	380	513	227	261	318	347	115	459	98	603	174	226	68	-
Size Difference %	8.5	12.9	14	5.8	22.8	27.9	26.8	10.6	12.1	10.3	19.9	14.4	17.5	20

B.b: B.bassiana, B.br: B.brongniartii, M.a: M. anisopliae, L.m: Lecanicillium muscarium, F.ox: Fusarium oxysporum, H.jec: Hypocrea jecorina, V.dah: Verticillium dahliae, N.cr: Neurospora crassa, P.an: Podospora anserina, C.gl: Chaetomium globosum, P.nod: Phaeosphaeria nodorum, B.fuc: Botryotinia fuckeliana, S.scl: Sclerotinia sclerotiorum, E.floc: Epidermophyton floccosum, T.rub: Trichophyton rubrum, P.mar: Penicillium marneffei, A.n: Aspergillus nidulans, P.can: Pichia canadensis, S.cer: Saccharomyces cerevisiae, S.jap: Schizosaccharomyces japonicus, C.neo: Cryptococcus neoformans, S.com: Schizophyllum commune, Harp: Harpochytrium sp., R.or: Rhizopus oryzae, A.mac: Allomyces macrogynus, Basid: Basidiomycota, Chytr: Chytridiomycota, Zygom: Zygomycota, Sacchar: Saccharomycetes, Schiz: Schizosaccharomycetes, Het: Heterobasidiomycetes, Hom: Homobasidiomycetes, Mon: Monoblepharidales, Muc: Mucorales, Blast: Blastocladiales, Doth: Dothideomycetes, Leot: Leotiomycetes, Eurot: Eurotiomycetes, Sacc: Saccharomycetales, Sch: Schizosaccharomycetales, Trem: Tremalles, Agar: Agaricales, H. sapiens: Homo sapiens

7.3. PHYLOGENETIC STUDIES REVEALING THE TAXONOMIC STATUS OF EPF

The taxonomy of several species or genera of asexual fungi has for long created many ambiguities and contraversions, and it is only with the aid of molecular techniques during the last 15 years that major revisions of hyphomycete EPF genera or species have taken place and resolved such problems.

Perhaps the most comprehensive report on the taxonomic status of EPFs and their associations with morphological traits, entomopathogenic or entomoparasitic nutritional properties are those recently provided by Sung *et al.* (2007) and Humber (2008).

In brief, EPFs can be found in the:

(a) Blastocladiomycota (formerly: Chytridiomycota; *Coelomomyces* spp, *Coelomycidium simulii* Debais),

(b) Entomophthoromycotina (formerly: Zygomycota; with entomopathogens in five of its six families and with the great majority of genera being obligatory entomopathogenic),

(c) Kickxellomycotina (formerly: Zygomycota; endocommensal *Harpellales* and *Asellariales*),

(d) Eurotiomycetes (Phylum Ascomycota, Subphylum Pezizomycotina; *Ascosphaera* and other genera),

(e) Laboulbeniomycetes (Phylum Ascomycota, Subphylum Pezizomycotina),

(f) Dothideomycetes (Phylum Ascomycota, Subphylum Pezizomycotina; *Myriangium*),

(g) Sordariomycetes (Phylum Ascomycota, Subphylum Pezizomycotina; mostly in order Hypocreales) and

(h) Pucciniomycetes (Phylum Basidiomycota, Subphylum Pucciniomycotina; *Septobasidium* and its relatives).

Additional species in the Zoopagomycotina, Entomophthoromycotina, Orbiliomycetes, and Sordariomycetes have obligatory parasitic relationships with micro-invertebrates outside the Arthropoda such as nematodes, tardigrades, amoebae, rotifers, and several other classes. However, as stated previously in this chapter the emphasis is placed only on the best studied hyphomycetes for which extended phylogenetic analyses have been performed.

7.3.1. Phylogeny of Entomophthoralean Fungi

The entomophthoralean fungi belong to the previously known phylum of Zygomycota according to the latest taxonomy. They occupy a basal position among zygomycetes with the two major entomopathogenic families (Entomophthoraceae and Neozygitaceae) being so divergent that they represent separate lineages within this order.

Until today, only a few studies exploited molecular data to such an extent that safe conclusions can be drawn for Entomophthorales and the species included in this order (Hodge *et al.,* 1995; Jensen *et al.,* 1998; Freimoser *et al.,* 2001; Nielsen *et al.,* 2001; Tymon *et al.,* 2004; James *et al.,* 2006). ITS and 18S rRNA gene sequences provided good species differentiation within EPF genera (Jensen *et al.,* 1998; Freimoser *et al.,* 2001), but failed to provide information that could be used for diagnostic purposes of *Pandora neoaphidis* (Remaud. and Hennebert) Humber isolates (Tymon *et al.,* 2004).

In a similar way, "universal" primers used for the amplification of the *rns* and IGS (White *et al.,* 1990; Anderson and Stasovski, 1992) failed to amplify reproducible products from any of the *P. neoaphidis* isolates tested, strongly indicating extended nucleotide variation within the regions of these primers (Tymon *et al.,* 2004). Thus, additional molecular data is still needed for members of Entomophthorales before more detailed phylogenetic relationships within species of the order can be inferred.

7.3.2. Phylogeny of Entomopathogenic Ascomycetes Within Pezizomycotina

In contrast to the rest of EPFs, phylogenetic data on entomopathogenic members of Ascomycota is abundant with most of the studies concerning members of the order Hypocreales.

Reconstruction of broad phylogenies of ascomycetes, based on rDNA (Berbee, 1996; Berbee *et al.,* 2000; Tehler *et al.,* 2003) or nuclear protein-coding genes (e.g., Keeling *et al.,* 2000; Reeb *et al.,* 2004) were extensively used to resolve fungal taxonomy. These investingations have identified several monophyletic lineages and are, in general, the most accepted by the scientific community.

However, they could not resolve all higher-order relationships with statistical significance, a task that was fulfilled with the help of alternative approaches that incorporate several concatenated gene sequences or even entire mt genomes in phylogenetic studies. These approaches fully resolved species-trees, with maximum support and without incongruences, and established the phylogenetic relationships of several genera and species within the subphylum of Pezizomycotina (selected references: Bullerwell *et al.,* 2003; Lutzoni *et al.,* 2004; Fitzpatrick *et al.,* 2006; Robbertse *et al.,* 2006; Spatafora *et al.,* 2006; Hibbett *et al.,* 2007).

The most relevant studies along these lines for EPFs are those on populations of *L. muscarium* (Kouvelis *et al.,* 2004), *M. anisopliae* (Ghikas *et al.,* 2006), and *B. bassiana* and *B. brongniartii* (Ghikas, Kouvelis and Typas –send for publication). It is worth noticing that the proposed phylogeny is identical to that of the most complete studies based on nuclear ribosomal regions (Berbee *et al.,* 2000) or other nuclear genes (Keeling *et al.,* 2000) and, moreover, the support of all topologies of the orders examined is the highest possible [Parsimonial Bootstrap support (BP) and Posterior Probabilities (PP) =100%], no matter which method (Maximum Likelihood, Maximum Parsimony and Bayesian Inference) is implemented or what data set (nucleotide or amino acid) is used (Figure 7). Thus, dealing with complete mt genomes as a phylogenetic approach, makes it possible to surpasse the two major inefficiencies of standard molecular phylogenetic studies, namely

    (a) phylogenies based on single or few genes,
    (b) selection of the most appropriate data matrix which may limit the accurate recording of evolutionary history of the species under examination.

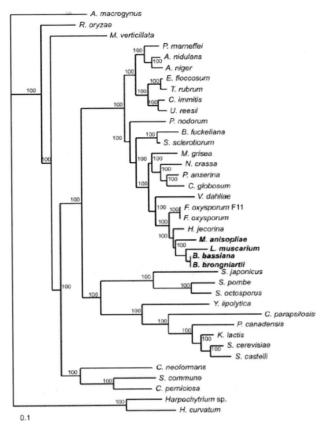

0.1

Figure 7. The single phylogenetic tree constructed from unambiguously aligned portions of concatenated protein sequences of 14 mt genes as produced by Bayesian analysis [and in accordance (100%) to the tree of maximum likelihood (ML) and maximum parsimony (MP) analyses]. Clade credibility using MrBayes (numbers) is shown. GenBank sequences used: *Allomyces macrogynus* (NC_001715), *Aspergillus nidulans* (http://www.megasun.bch.umontreal.ca/People/ lang/species/asp/), *Aspergillus niger* (NC_007445), *Beauveria bassiana* (EU100742), *Beauveria brongniartii* (NC_011194), *Candida parapsilosis* (NC_005253), *Chaetomium globossum* (AAFU01000232, AAFU01000427), *Coccidioides immitis* (AAEC01000203, AAEC01000204), Cryptococcus neoformans (NC_004336), Crinipellis perniciosa (NC_005927), *Epidermophyton floccosum* (NC_007394), *Fusarium oxysporum* (AY945289), *Harpochytrium* sp. JEL105 (NC_004623), *Hyaloraphidium curvatum* (NC_003048), Hypocrea jecorina (NC_003388), *Kluyveromyces lactis* (NC_006077), *Lecanicillium muscarium* (NC_004514), *Metarhizium anisopliae* (AY884128), *Mortierella verticillata* (NC_006838), *Phaeosphaeria nodorum* (AAGI01000412), *Pichia canadensis* (NC_001762), *Penicillium marneffei* (NC_005256), *Podospora anserina* (NC_001329), *Rhizopus oryzae* (NC_006836), *Saccharomyces castelli* (NC_003920), *Saccharomyces cerevisiae* (AJ011856), *Schizophyllum commune* (NC_003049), *Schizosaccharomyces japonicus* (NC_004332), *Schizosaccharomyces octosporus* (NC_004312), *Schizosaccharomyces pombe* (NC_001326), *Trichophyton rubrum* (Y18476, X65223, and X88896), *Verticillium dahliae* (DQ351941) and *Yarrowia lipolytica* (NC_002659). Protein sequences of *Neurospora crassa*, were downloaded from http://www.broad.mit.edu/cgibin/annotation/fungi/neurospora_crassa_7/download_license.cgi.

Sequences of *Botryotinia fuckeliana*, *Magnaporthe grisea*, *Sclerotinia sclerotiorum* and *Uncinocarpus reesii* were determined from the draft released sequences of the Broad Institute of Harvard and MIT sequencing projects (http://www.broad.mit.edu). The translation of mt proteins for species *N. crassa*, *H. jecorina* and *Y. lipolytica* was interpreted and corrected as previously described (Kouvelis *et al.*, 2004). The mtDNA of entomopathogenic fungi are shown in bold.

7.3.3. Phylogeny of Entomopathogenic Ascomycetes at Genus and/or Species Level

Most of the important ascomycete EPFs are asexual species, like *Beauveria*, *Metarhizium*, *Lecanicillium*, *Nomuraea* and *Paecilomyces*. In such cases, molecular approaches help immensely in resolving ambiguities which sole morphological criteria cannot sort out.

(a) For the cosmopolitan EPF genus *Beauveria* morphological characteristics were combined with molecular data based on ITS1-5.8S-ITS2 and EF-1a sequences from 86 exemplar isolates and the phylogenetic study assigned these isolates to six major clades (A-F), where all known *Beauveria* species were included (Rehner and Buckley, 2005). *B. bassiana* isolates were grouped into two unrelated and morphologically indistinguishable clades (Clades A and C), while *B. brongniartii* formed a third sister clade to the other two (designated as Clade B). A new species, *Beauveria malawiensis* S.A. Rehner and Aquino de Muro, was later introduced and placed as sister to clade E (Rehner *et al.*, 2006a), and several other *B. bassiana* isolates pathogen to coffee berry borer from Africa and the Neotropics were added to Clades A and C, comprising *B. bassiana s. l.* and "pseudobassiana" isolates, respectively (Rehner *et al.*, 2006b). Our results based on mt intergenic regions are in full agreement with and further strengthen the phylogeny proposed (Ghikas, Kouvelis and Typas – un published data).

(b) The case of *Metarhizium* is of particular interest since neither morphological characters nor simple molecular data could initially provide sufficient information to differentiate species or isolates in the most abundant species *M. anisopliae* var. *anisopliae*. Unlike most other mitosporic fungi which demonstrate considerably high levels of polymorphisms in their nuclear rRNA gene complex, all population studies with strains of *M. anisopliae* var. *anisopliae* isolated from different hosts and geographic locations, showed that its ITS1-5.8S-ITS2, 18S and 28S sequences were highly conserved and failed to provide information for the safe differentiation of isolates or for phylogenetic studies (Mavridou and Typas, 1998; Driver *et al.*, 2000; Pantou *et al.*, 2003; Bidochka *et al.*, 2005). Differentiation was achieved by analysing IGS sequences that revealed motifs specific for three principal groups (Pantou *et al.*, 2003; Hughes *et al.*, 2004) and/or mt intergenic region sequences that allowed differentiation of *M. anisopliae* var. *anisopliae* populations at strain-specific level (Ghikas *et al.*, 2006). Finally, a multi-locus approach based on the nuclear genes EF-1a, RPB1, RPB2, b-tubulin and morphological characteristics of *Metarhizium* species led to the recognition of nine species belonging to genus *Metarhizium* (Bischoff *et al.*, 2009).

(c) In phylogenetic studies of former *Verticillium* EPFs, the ITS1-5.8S-ITS2 sequences, alone or in combination with data from the small and large nuclear rRNA subunits, b-tubulin genes and mtDNA regions were examined (*V. lecanii*; Zare *et al.*, 1999, 2000) leading to the re-classification of *Verticillium* sect. *Prostrata*, with the introduction of the EPF genera *Lecanicillium*, *Haptocillium* and *Simplicillium* (Zare *et al.*, 2000; Gams and Zare, 2001; Sung *et al.*, 2001; Zare and Gams, 2001a,b). However, as many fungi with ambiguous taxonomic status still exist within the old

Verticillium group the exploitation of other genetic loci is necessary to clarify with certainty the phylogenetic relationships newly introduced and remaining in the group species (Sugimoto *et al.,* 2003; Kouvelis *et al.,* 2008a).

(d) *Cordyceps* is the most diverse genus in the family *Clavicipitaceae* in terms of number of species and host range, with more than 400 species and a broad range of hosts that belong to ten different orders of arthropods. Several phylogenetic studies using the nuclear rRNA complex have been conducted to test and refine the classification of this genus (Sung *et al.,* 2001, Stensrud *et al.,* 2005), but limited taxon sampling and the inadequate resolution power of rRNA, resulted in restricted conclusions regarding the phylogeny and the systematics of the genus. Recent phylogenetic studies (Spatafora *et al.,* 2006, Sung *et al.,* 2007) based on multiple independent loci, [namely the nuclear 18S and 28S rRNA genes, the elongation factor 1α (*tef*1), the largest and the second largest subunits of RNA polymerase II (*rpb*1 and *rpb*2), β-tubulin (*ben*A), and mitochondrial ATP6 (*atp6*)], provided a greater level of resolution and support, and revealed that neither *Cordyceps* nor the family *Clavicipitaceae* is monophyletic. Three major well supported clades of clavicipitaceous fungi were introduced by these phylogenetic analyses, with Clade A comprising the entomopathogenic *Hypocrella* subspecies and the anamorph genera of *Aschersonia*, *Metarhizium* and *Pochonia* among others, Clade B including the *Paecilomyces lilacinus* clade and the anamorphs *Haptocillium*, *Hirsutella* and *Tolypocladium,* and Clade C with the *Cordyceps* subclade and anamorphic genera like *Beauveria*, *Isaria*, *Lecanicillium* and *Simplicillium* (for detailed taxonomic revisions and classification see Sung *et al.,* 2007).

Several other studies based on the methodologies described have been recently published presenting significant results for phylogenetic relationships of EPF species with potential as biological control agents [e.g., for genera *Hypocrella* (Mongkolsamrit *et al.,* 2009), *Moelleriella* (Liu *et al.,* 2009), *Samuelsia* and their Aschersonia-like anamorphs (Chaverri *et al.,* 2008)] but are not analysed further here due to space limitations.

7.3.4. Phylogenetic Studies of EPFs and Their Association to Their Hosts

A generally accepted notion that insect hosts are related to certain genotypes of EPFs has been tested in several phylogenetic studies in the past for fungal species that already have been used as BCAs.

In the case of the two most important species of the genus *Beauveria*, *B. bassiana* and *B. brongniartii*, there are some reports that suggest a host – fungal genotype specificity, like in the case of *B. brongniartii* and *Melolontha melolontha* and *M. hippocastani* (Cravanzola *et al.,* 1997) and *Hoplochelus marginalis* (Neuveglise *et al.,* 1994), or the cases of *B. bassiana* for which a common genotype was found for isolates of *Ostrinia nubilalis* (Viaud *et al.,* 1996) and *Alphitobius diaperinus* (Castrillo *et al.,* 1999). However, more often, *B. bassiana* isolates collected from the same insect species were genetically dissimilar (Coates *et al.,* 2002; Urtz and Rice, 1997) or showed cross-infectivity (Glare and Inwood, 1998). Similarly, fungal isolates derived from different insect species/families/orders clustered together (Gaitan *et al.,* 2002). This indicates that *B. bassiana* is a generalized insect pathogen, which is easily understandable due to its world-wide distibution, the vast variety of hosts from which it has been isolated and its entomopathogenic and/or endophytic characteristics (Quesada-Moraga *et al.,* 2006; Rehner and Buckley, 2005). Similarly, studies on the other two most widely used as BCAs genera, *Metarhizium* and *Lecanicillium*, supported by various gene phylogenies, showed only loose correlation of species and host specificity (Driver *et al.,* 2000; Zare and Gams, 2001a; Pantou *et al.,* 2003; Kouvelis *et al.,* 2008a). The use of RAPDs and isozyme markers indicated that there is persistence of particular fungal genotypes in specific locations and that in some instances RAPD groupings correlate with insect hosts, leading to the conclusion that *M. anisopliae* may contain a number of cryptic species (Bidochka *et al.,* 1993; Fegan *et al.,* 1993; Fungaro *et al.,* 1996; Bridge *et al.,* 1997). [The latter was recently confirmed by a multigenic phylogenetic approach (Bischoff *et al.,* 2009)]. Also, phylogenetic studies based on rRNA gene complex sequences demonstrated that the correlation of *Metarhizium* species with their hosts is only limited (Driver *et al.,* 2000; Pantou *et al.,* 2003; Hughes *et al.,* 2004) and, similarly, analyses based on rRNA and mt genes in *Lecanicillium*, showed that there is a loose association between fungal isolates and their host (Kouvelis *et al.,* 2008a). Host specificity and habitats were previously used as general additional criteria for species classification in *Lecanicillium* (Zare and Gams, 2001a). It was found that *L. lecanii* (Zimm.) Zare and W. Gams contained isolates found mainly in tropical regions (like Sri Lanka, Indonesia, Peru) and with a host specificity to soft scale insects (like Coccidae or Lecanidae), whereas *L. muscarium* isolates were more common in temperate climates, and their hosts comprised not only insects of orders such as Thysanoptera and Lepidoptera, but also acari or rusts. The "failure" to establish a strong genetic, and in extent phylogenetic, correlation with the species-host association, in almost all known EPFs, may be attributed to the nutritional needs of these fungi, as they have to adapt to every nutritional source available under the continuous changing of natural selective forces.

7.3.5. Phylogenetic Studies of EPFs and Their Geographic Origin

An increasing number of recent studies point towards a broad correlation of fungal EPFs with their place of origin and/or habitats [e.g., for *B. bassiana* (Bidochka *et al.,* 2002; Wang *et al.,* 2003b; Aquino de Muro *et al.,* 2005; Kouvelis *et al.,* 2008b), for *M. anisopliae* (Bidochka *et al.,* 2001), and for *Lecanicillium* spp. (Zare and Gams, 2001a)]. On the contrary, other reports show no clear genetic/phylogenetic relationship of EPFs and their place of origin or habitats (Pantou *et al.,* 2003; Hughes *et al.,* 2004; Kouvelis *et al.,* 2008a). Obviously, many different factors can influence fungal population structures, like climate conditions, the range of temperatures in which the various isolates can grow in nature, humidity levels, the doses of daily UV exposure, habitat type, cropping system and soil properties (Bidochka *et al.,* 2002; Wang *et al.,* 2003b; Quesada-Moraga *et al.,* 2007). Even the molecular marker used for the phylogenetic study may be a misleading factor as its evolutionary rate may be different and under different constraints. Thus, although the description of the effects of a single variable on the population of EPF fungi in a habitat can give significant and useful ecological and agronomical information, there may be relationships among the different variables that must be studied in detail to adequately understand the source of genetic variability in these fungi before coming to safe conclusions based on the phylogenetic analyses (Bidochka *et al.,* 2002; Quesada-Moraga *et al.,* 2007).

CONCLUSION

Phylogenetic studies with EPF show a continuous increase in numbers and are associated not only with fungal taxonomy and evolution but also with intra- and inter- species differentiation, genetic fingerprinting of isolates and host and origin/habitat interactions. Until now, the majority of these studies have focused on the entomopathogenic members within the order of Hypocreales (Ascomycota, Pezizomycotina), but studies with fungal parasites on insect from other orders are appearing timidly. Undoubtedly, the rRNA gene complex still is the most commonly used domain for phylogenetic studies of most fungi, including EPF. Yet, it is universally accepted that a single gene or a single gene region alone cannot faithfully represent the phylogeny of a fungus and a multi-gene approach is necessary. The recent employment of sequences from a number of nuclear housekeeping genes, like β-tubulin (*tub*-a, *tub*-b), the small and large subunits of the RNA polymerase II (*rpb*1, *rpb*2) and the elongation factor (*tef*1) along with the traditional SSU, LSU and ITS1-5.8S-ITS2, and independent mt genes, for a higher-level phylogenetic classification of Fungi (Hibbett *et al.,* 2007), has set the standards for such studies, and these inevitably must be also applied to EPF. Mt genes seem to play an important role in phylogenetic studies providing good statistical significance and resolving higher-order relationships that otherwise remain unclear with analysis of traditional nuclear gene sequences. Even more, entire mt genomes -although still few at present for general studies- show a great potential in resolving phylogenetic ambiguities found in nuclear-gene based analyses. Finally, genes associated with entomopathogenicity like proteases (Pr-like) and chitinases (*chit*) genes, are certainly of great importance in phylogenetic studies, because they are directly associated with functional properties of the EPF that determine their very identity. Thus, a multi-gene phylogenetic approach that combines sequence information from Pr-like and *cht* genes with that of selected nuclear housekeeping genes like benA and *rpb*1, and mt genes like *nad*1 and *rms*, is envisaged to provide the best data for resolving many phylogenetic, taxonomic and evolutionary ambiguities which arise from less powerful criteria (morphological and biochemical properties). In the era of genomics and proteomics, with transcriptomics and metabolomics also evolving very fast and being applied to several fungi of economic importance, providing that the cost of these approaches will continue to drop drastically, as it has been done during the past decade, it seems inevitable that the future phylogenetic approaches for fungi, including EPF, will be based on these methodologies. Phylogenetic analyses based on gene clusters and chromosome or whole genome syntenies will efficiently and beyond any doubt provide all the necessary proof for establishing the correct fungal relationships in a group with scientific, biotechnological and ecological importance as EPF.

REFERENCES

Anderson, J.B. and Stasovski, E. 1992. Molecular phylogeny of Northern Hemisphere species of *Armillaria*. *Mycologia*, 84: 505-516.

Aquino de Muro, M., Mehta, S. and Moore, D. 2003. The use of amplified fragment length polymorphism for molecular analysis of *Beauveria bassiana* isolates from Kenya and other countries and their correlation with host and geographic origin. *FEMS Microbiology Letters*, 229: 249-257.

Aquino de Muro, M., Elliott, S., Moore, D., Parker, B.L., Skinner, M., Reid, W. and El Bouhssini, M. 2005. Molecular characterisation *of Beauveria bassiana* isolates obtained from overwintering sites of Sunn Pest (*Eurygaster* and *Aelia* species). *Mycological Research*, 109: 294-306.

Bagga, S., Hu, G., Screen, S.E. and St. Leger, R.J. 2004. Reconstructing the diversification of subtilisins in the pathogenic fungus *Metarhizium anisopliae. Gene*, 324: 159-69.

Berbee, M.L. 1996. Loculoascomycete origins and evolution of filamentous ascomycete morphology based on 18S rRNA gene sequence data. *Molecular Phylogenetics and Evolution,* 13: 462-470.

Berbee, M.L., Carmean, D.A. and Winka, K. 2000. Ribosomal DNA and resolution of branching order among the Ascomycota: how many nucleotides are enough? *Molecular Phylogenetics and Evolution*, 17: 337-344.

Bidochka, M.J., McDonald, M.A., St Leger, R.A. and Roberts, D.W. 1993. Differentiation of species and strains of entomopathogenic fungi by random amplified polymorphic DNA (RAPD). *Current Genetics*, 21: 107-113.

Bidochka, M.J., Mc Donald, M.A., St. Leger, R.J. and Roberts, D.W. 1994. Differentiation of species and strains of entomopathogenic fungi by random amplification of polymorphic DNA (RAPD). *Current Genetics*, 25: 107-113.

Bidochka, M.J., St Leger, R.J., Stuart, A. and Gowanlock, K. 1999. Nuclear rDNA phylogeny in the fungal species *Verticillium* and its relationship to insect and plant virulence, extracellular proteases and carbohydrases. *Microbiology*, 145: 955-63.

Bidochka, M.J., Kamp, A.M., Lavender, T.M., Dekoning, J. and De Croos, J.N. 2001. Habitat association in two genetic groups of the insect-pathogenic fungus *Metarhizium anisopliae*: uncovering cryptic species? *Applied and Environmental Microbiology*, 67: 1335-1342.

Bidochka, M.J., Menzies, F.V. and Kamp, A.M. 2002. Genetic groups of the insect-pathogenic fungus *Beauveria bassiana* are associated with habitat and thermal growth preferences. *Archives of Microbiology*, 178: 531-537.

Bidochka, M.J., Small, C.-L.N. and Spironello, M. 2005. Recombination within sympatric cryptic species of the insect pathogenic fungus *Metarhizium anisopliae. Environmental Microbiology*, 7: 1361-1368.

Bischoff, J.F., Rehner, S.A. and Humber, R.A. 2009. A multilocus phylogeny of the *Metarhizium anisopliae* lineage. *Mycologia*, 101: 512-530.

Boucias, D.G., Tigano, M.S., Sosa-Gomez, D.R., Glare, T.R. and Inglis, P.W. 2000. Genotypic properties of the entomopathogenic fungus *Nomurea rileyi. Biological Control*, 19: 124-138.

Bridge, P.D., Prior, C., Sagbohan, J., Lomer, C.J., Carey, M. and Buddie, A. 1997. Molecular characterization of isolates from *Metarhizium* from locusts and grasshoppers. *Biodiversity and Conservation*, 6: 177-189.

Bullerwell, C.E., Forget, L. and Lang, B.F. 2003. Evolution of monoblepharidalean fungi based on complete mitochondrial genome sequences. *Nucleic Acids Research*, 31: 1614-1623.

Burger, G., Gray, M.W. and Lang, B.F. 2003. Mitochondrial genomes: anything goes. *Trends in Genetics*, 19: 709-716.

Butt, T.M., Jackson, C.W. and Magan, N. 2001. Introduction—fungal biological control agents: progress, problems and potential. In: Butt, T.M., Jackson, C.W. and Magan, N. (Eds.), *Fungi as biocontrol agents*. CABI Publishing, Walingford, U.K., pp. 1-8.

Butt, T.M. 2002. Use of entomogenous fungi for the control of insect pests. In: Kempken, F. (Ed.), *The Mycota XI. Agricultural applications*. Springer-Verlag, Berlin, Heidelberg, Germany, pp. 111-134.

Castrillo, L.A. and Brooks, W.M. 1998. Identification of *Beauveria bassiana* isolates from the darkling beetle, *Alphitobius diaperinus*, using isozyme and RAPD analyses. *Journal of Invertebrate Pathology*, 72: 190-194.

Castrillo, L.A., Wiegmann, B.M. and Brooks, W.M. 1999. Genetic variation in *Beauveria bassiana* populations associated with the darkling beetle, *Alphitobius diaperinus*. *Journal of Invertebrate Pathology*, 73: 269-275.

Chaverri, P., Liu, M. and Hodge, K.T. 2008. A monograph of entomopathogenic genera *Hypocrella*, *Moelleriella*, and *Samuelsia* gen. nov. (*Ascomycota*, *Hypocreales*, *Clavicipitaceae*), and their aschersonia-like anamorphs in the Neotropics. *Studies in Mycology*, 60: 1-66.

Clakson, J.M. and Charnley, A.K. 1996. New insights into the mechanisms of fungal pathogenesis in insects. *Trends in Microbiology*, 4: 197-203.

Coates, B.S., Hellmich, R.L. and Lewis, L.C. 2002. *Beauveria bassiana* haplotype determination based on nuclear rDNA internal transcribed spacer PCR-RFLP. *Mycological Research*, 106: 40-50.

Cravanzola, F., Piatti, P., Bridge, P.D. and Ozino, O.I. 1997. Detection of genetic polymorphism by RAPD-PCR in strains of the entomopathogenic fungus *Beauveria brongniartii* isolated from the European cockchafer (*Melolontha* spp.). *Letters in Applied Microbiology*, 25: 289-294.

Dalzoto, P.R., Glienke-Blanco, C., Kava-Cordeiro, V., Ribeiro, J.Z., Kitajima, E.W. and Azevedo, J.L. 2006. Horizontal transfer and hypovirulence associated with double-stranded RNA in *Beauveria bassiana*. *Mycological Research*, 110: 1475-1481.

Devi, U.K., Reineke, A., Reddy, N.N.R., Rao, U.M.C. and Padmavathi, J. 2006. Genetic diversity, reproductive biology, and speciation in the entomopathogenic fungus *Beauveria bassiana* (Balsamo) Vuillemin. *Genome*, 49: 495-504.

Driver, F., Milner, R.J. and Trueman, J.W.H. 2000. A taxonomic revision of *Metarhizium* based on phylogenetic analysis of rDNA sequence data. *Mycological Research*, 104: 134-150.

Eilenberg, J. 2002. Biology of fungi from the order Entomophtorales with emphasis on the genera *Entomophthora*, *Strongwellsea* and *Eryniopsis*: A contribution to insect pathology and biological control. Department of Ecology, Royal Vetinary and Agricultural University, Copenhagen, Denmark, p. 407.

Enkerli, J., Widmer, F., Gessler, C. and Keller, S. 2001. Strain-specific microsatellite markers in the entomopathogenic fungus *Beauveria brongniartii*. *Mycological Research*, 105: 1079-1087.

Enkerli, J., Kölliker, R., Keller, S. and Widmer, F. 2005. Isolation and characterization of microsatellite markers from the entomopathogenic fungus *Metarhizium anisopliae*. *Molecular Ecology Notes*, 5: 384-386.

Enkerli, J., Ghormade, V., Oulevey, C. and Widmer, F. 2009. PCR-RFLP analysis of chitinase genes enables efficient genotyping of *Metarhizium anisopliae* var. *anisopliae*. *Journal of Invertebrate Pathology,* 102: 185-188.

Eriksson, O.E. 2006. Outline of Ascomycota — 2006. *Myconet*, 12: 1-82.

Fegan, M., Manners, J.M., Maclean, D.J., Irwin, J.A., Samuels, K.D., Holdom, D.G. and Li, D.P. 1993. Random amplified polymorphic DNA markers reveal a high degree of genetic diversity in the entomopathogenic fungus *Metarhizium anisopliae* var. *anisopliae*. *Journal of General Microbiology*, 139: 2075-2081.

Fitzpatrick, D.A., Logue, M.E., Stajich, J.E., Butler, G. 2006. A fungal phylogeny based on 42 complete genomes derived from supertree and combined gene analysis. *BMC Evolutionary Biology,* 6: 99.

Freimoser, F.M., Jensen, A.B., Tuor, U., Aebi, M. and Eilenberg, J. 2001. Isolation and *in vitro* cultivation of the aphid pathogenic fungus *Entomophthora planchoniana*. *Canadian Journal of Microbiology*, 47: 1082-1087.

Fungaro, F.H.P., Vieira, M.L.C., Pizzirani-Kleiner, A.A. and Azevedo, de J.L. 1996. Diversity among soil and insect isolates of *Metarhizium anisopliae* var. *anisopliae* detected by RAPD. *Letters in Applied Microbiology*, 2: 389-392.

Gaitan, A., Valderrama, A.M., Saldarriaga, G., Velez, P. and Bustillo, A. 2002. Genetic variability of *Beauveria bassiana* associated with the coffee berry borer *Hypothenemus hampei* and other insects. *Mycological Research*, 106: 1307-1314.

Gams, W. and Zare, R., 2001. A revision of *Verticillium* sect. *Prostrata*. III. Generic classification. *Nova Hedwigia*, 72: 329-337.

Ghikas, D.V., Kouvelis, V.N. and Typas, M.A. 2006. The complete mitochondrial genome of the entomopathogenic fungus *Metarhizium anisopliae* var. *anisopliae*: gene order and trn gene clusters reveal a common evolutionary course for all Sordariomycetes. *Archives of Microbiology*, 185: 393-401.

Giménez-Pecci, M., Bogo, M.R., Santi, L., Moraes, C.K., Corrêa, C.T., Henning Vainstein, M. and Schrank, A. 2002. Characterization of mycoviruses and analyses of chitinase secretion in the biocontrol fungus *Metarhizium anisopliae*. *Current Microbiology*, 45: 334-349.

Glare, T.R. and Inwood, A.J. 1998. Morphological characterization of *Beauveria* spp. from New Zealand. *Mycological Research*, 102: 250-256.

Goettel, M.S. 2008. Are entomopathogenic fungi only entomopathogens? A permeable. *Journal of Invertebrate Pathology*, 98: 255.

Harman, G.E., Howell, C.R., Viterbo, A., Chet, I. and Lorito, M. 2004. *Trichoderma* species – opportunistic, avirulent plant symbionts. *Nature Reviews*, 2: 43-56.

Hegedus, D.D. and Khachatourians, G.G. 1993. Identification of molecular variants in mitochondrial DNAs of members of the genera *Beauveria*, *Verticillium*, *Paecilomyces*, *Tolypocladium* and *Metarhizium*. *Applied and Environmental Microbiology*, 59: 4283-4288.

Hibbett, D.S., Binder, M., Bischoff, J.F., Blackwell, M., Cannon, P.F., Eriksson, O.E., Huhndorf, S., James, T., Kirk, P.M., Lücking, R., Thorsten Lumbsch, H., Lutzoni, F., Brandon Matheny, P., McLaughlin, D.J., Powell, M.J., Redhead, S., Schoch, C.L., Spatafora, J.W., Stalpers, J.A., Vilgalys, R., Aime, M.C., Aptroot, A., Bauer, R., Begerow, D., Benny, G.L., Castlebury, L.A., Crous, P.W., Dai, Y.-C., Gams, W., Geiser, D.M., Griffith, G.W., Gueidan, C., Hawksworth, D.L., Hestmark, G., Hosaka, K.,

Humber, R.A., Hyde, K.D., Ironside, J.E., Kõljakg, U., Kurtzman, C.P., Larsson, K.-H., Lichtward, R., Longcore, J., Miądlikowska, J., Miller, A., Moncalvo, J.-M., Mozley-Standridges, S., Oberwinkler, F., Parmasto, E., Reeb, V., Rogers, J.D., Roux, C., Ryvarden, L., Sampaio, J.P., Schüßler, A., Sugiyama, J., Thorn, R.G., Tibell, L., Untereiner, W.A., Walker, C., Wang, Z., Weir, A., Weiss, M., White, M.M., Winka, K., Yao, Y.-J. and Zhang, N. 2007. A higher-level phylogenetic classification of the Fungi. *Mycological Research*, 111: 509-547.

Hodge, K.T., Sawyer, A.J. and Humber, R.A. 1995. RAPD-PCR for identification of *Zoophthora radicans* isolates in biological control of potato leafhopper. *Journal of Invertebrate Pathology*, 65: 1-9.

Hu, G. and St. Leger, R.J. 2004. A phylogenomic approach to reconstructing the diversification of serine proteases in fungi. *Journal of Evolutionary Biology*, 17: 1204-1214.

Hughes, W.O.H., Thomsen, L., Eilenberg, J. and Boomsma, J.J. 2004. Diversity of entomopathogenic fungi near leaf-cutting ant nests in a neotropical forest, with particular reference to *Metarhizium anisopliae* var. *anisopliae*. *Journal of Invertebrate Pathology*, 85: 46-53.

Humber, R.A. 2008. Evolution of entomopathogenicity in fungi. *Journal of Invertebrate Pathology*, 98: 262-266.

Inglis, G.D., Goettel, M.S., Butt, T.M. and Strasser, H. 2001. Use of hyphomycetous fungi for managing insect pests. In: Butt, T.M., Jackson, C. and Magan, N. (Eds.), *Fungi as Biocontrol Agents: Progress, Problems and Potential*. CABI Publishing, Wallingford, U.K., pp. 27-69.

Inglis, G.D., Duke, G.M., Goettel, M.S. and Kabaluk, T.J. 2008. Genetic diversity of *Metarhizium anisopliae* var. *anisopliae* in southwestern British Columbia. *Journal of Invertebrate Pathology,* 98: 101-113.

James, T.Y., Kauff, F., Schoch, C., Matheny, P.B., Hofstetter, V., Cox, C.J., Celio, G., Geuidan, C., Fraker, E., Miadlikowska, J., Lumbsch, H.T., Rauhut, A., Reeb, V., Arnold, A.E., Amroft, A., Stajich, J.E., Hosaka, K., Sung, G.-H., Johnson, D., O'Rourke, B., Crockett, M., Binder, M., Curtis, J.M., Slot, J.C., Wang, Z., Wilson, A.W., Schüßler, A., Longcore, J.E., O'Donnell, K., Mozley-Standridge, S., Porter, D., Letcher, P.M., Powell, M.J., Taylor, J.W., White, M.M., Griffith, G.W., Davies, D.R., Humber, R.A., Morton, J.B., Sugiyama, J., Rossman, A., Rogers, J.D., Pfister, D.H., Hewitt, D., Hansen, K., Hambleton, S., Shoemaker, R.A., Kohlmeyer, J., Volkmann-Kohlmeyer, B., Spotts, R.A., Serdani, M., Crous, P.W., Hughes, K.W., Matsuura, K., Langer, E., Langer, G., Untereiner, W.A., Lócking, R., Bódel, B., Geiser, D.M., Aptroot, A., Diederich, P., Schmitt, I., Schultz, M., Yahr, R., Hibbett, D.S., Lutzoni, F., McLaughlin, D.J., Spatafora, J.W., Vilgalys, R., 2006. Reconstructing the early evolution of Fungi using a six-gene phylogeny. *Nature*, 443: 818-822.

Jensen, A.B., Gargas, A., Eilenberg, J. and Rosendahl, S., 1998. Relationships of the insect pathogenic order Entomophthorales (Zygomycota, Fungi) based on phylogenetic analyses of nuclear small subunit ribosomal DNA sequences (SSU rDNA). *Fungal Genetics and Biology*, 24: 325-334.

Keeling, P.J., Luker, M.A., and Palmer, J.D. 2000. Evidence from beta tubulin phylogeny that microsporidia evolved from within the fungi. *Molecular Biology and Evolution*, 17: 23-31.

Kouvelis, V.N., Zare, R., Bridge, P.D. and Typas, M.A. 1999. Differentiation of mitochondrial subgroups in the *Verticillium lecanii* species complex. *Letters in Applied Microbiology*, 28: 263-268.

Kouvelis, V.N., Ghikas, D.V. and Typas, M.A. 2004. The analysis of the complete mitochondrial genome of *Lecanicillium muscarium* (synonym *Verticillium lecanii*) suggests a minimum common gene organization in mtDNAs of Sordariomycetes: phylogenetic implications. *Fungal Genetics and Biology*, 41: 930-940.

Kouvelis, V.N., Sialakouma A. and Typas, M.A. 2008a. Mitochondrial gene sequences alone or combined with ITS region sequences provide firm molecular criteria for the classification of *Lecanicillium* species. *Mycological Research*, 112: 829-844.

Kouvelis, V.N., Ghikas, D.V., Edgington, S., Typas, M.A. and Moore, D. 2008b. Molecular characterization of isolates of *Beauveria bassiana* obtained from overwintering and summer populations of Sunn Pest (*Eurygaster integriceps*). *Letters in Applied Microbiology*, 46: 414-420.

Leal, S.C.M., Bertioli, D.J., Butt, T.M., Carder, J.H., Burrows, P.R. and Peberdy, J.F. 1997. Amplification and restriction endonuclease digestion of the Pr1 gene for the detection and characterization of *Metarhizium* strains. *Mycological Research* 101: 257-265.

Liu, M., Milgroom, M.G., Chaverri, P. and Hodge, K.T. 2009. Speciation of a tropical fungal species pair following transoceanic dispersal. *Molecular Phylogenetics and Evolution*, 51: 413-426.

Luangsa-ard, J.J. and Samson, R.A. 2004. The polyphyletic nature of *Paecilomyces sensu lato* based on 18S-generated rDNA phylogeny. *Mycologia*, 96: 773-780.

Lutzoni, F., Kauff, F., Cox, C.J., McLaughlin, D., Celio, G., Dentinger, B., Padamsee, M., Hibbett, D., James, T.Y., Baloch, E., Grube, M., Reeb, V., Hofstetter, V., Schoch, C., Arnold, A.E., Miadlikowska, J., Spatafora, J., Johnson, D., Hambleton, S., Crockett, M., Shoemaker, R., Sung, G.-H., Lücking, R., Lumbsch, T., O'Donnell, K., Binder, M., Diederich, P., Ertz, D., Gueidan, C., Hansen, K., Harris, R.C., Hosaka, K., Lim, Y.-W., Matheny, B., Nishida, H., Pfister, D., Rogers, J., Rossman, A., Schmitt, I., Sipman, H., Stone, J., Sugiyama, J., Yahr, R. and Vilgalys, R. 2004. Assembling the fungal tree of life: progress, classification, and evolution of subcellular traits. *American Journal of Botany*, 91: 1446-1480.

Marquez, M., Iturriaga, E.A., Quesada-Moraga, E., Santiago-Álvarez, C., Monte, E. and Hermosa, R. 2006. Detection of potential valuable polymorphisms in four group I intron insertion sites at the 3'-end of the LSU rDNA genes in biocontrol isolates of *Metarhizium anisopliae*. *BMC Microbiology*, 6: 77

Mavridou, A. and Typas, M.A. 1998. Intraspecific polymorphism in *Metarhizium anisopliae* var. *anisopliae* revealed by analysis of rRNA gene complex and mtDNA RFLPs. *Mycological Research*, 102: 1233-1241.

Mavridou, A., Cannone, J. and Typas, M.A. 2000. Identification of group-I introns at three different positions within the 28S rDNA gene of the entomopathogenic fungus *Metarhizium anisopliae* var. *anisopliae*. *Fungal Genetics and Biology*, 31: 79-90.

Mongkolsamrit, S., Luangsa-ard, J.J., Spatafora, J.J., Sung, G.-H. and Hywel-Jones, N.L. 2009. A combined ITS rDNA and β-tubulin phylogeny of Thai species of *Hypocrella* with non-gragmenting ascospores. *Mycological Research*, 113: 684-699.

Neuveglise, C., Brygoo, Y., Vercambre, B. and Riba, G. 1994. Comparative analysis of molecular and biological characteristics of *Beauveria brongniartii* isolated from insects. *Mycological Research*, 98: 322-328.

Neuveglise, C., Brygoo, Y. and Riba, G. 1997. 28S rDNA group-I introns: a powerful tool for identifying strains of *Beauveria brongniartii*. *Molecular Ecology*, 6: 373-381.

Nielsen, C., Sommer, C., Eilenberg, J., Hansen, K.S. and Humber, R.A. 2001. Characterization of aphid pathogenic species in the genus *Pandora* by PCR techniques and digital image analysis. *Mycologia*, 93: 864-874.

Nikoh, N. and Fukatsu, T. 2000. Interkingdom host jumping underground: phylogenetic analysis of entomoparasitic fungi of the genus *Cordyceps*. *Molecular Biology and Evolution*, 17: 629-638.

Nikoh, N. and Fukatsu, T. 2001. Evolutionary dynamics of multiple group I introns in nuclear ribosomal RNA genes of entomoparasitic fungi of the genus *Cordyceps*. *Molecular Biology and Evolution*, 18: 1631-1642.

Oulevey, C., Widmer, F., Kölliker, R. and Enkerli, J. 2009. An optimized microsatellite marker set for detection of *Metarhizium anisopliae* genotype diversity on field and regional scales. *Mycological Research*, 113: 1016-1024.

Padmavathi, J., UmaDevi, K., Rao, C.U. and Reddy, N.N. 2003. Telomere fingerprinting for assessing chromosome number, isolate typing and recombination in the entomopathogen *Beauveria bassiana*. *Mycological Research*, 107: 572-580.

Pantou, M.P., Mavridou, A. and Typas, M.A. 2003. IGS sequence variation, group-I introns and the complete nuclear ribosomal DNA of the entomopathogenic fungus *Metarhizium*: excellent tools for isolate detection and phylogenetic analysis. *Fungal Genetics and Biology*, 38: 159-174.

Pantou, M.P., Kouvelis, V.N. and Typas, M.A. 2006. The complete mitochondrial genome of the vascular wilt fungus *Verticillium dahliae*: a novel gene order for *Verticillium* and a diagnostic tool for species identification. *Current Genetics*, 50: 125-136.

Pantou, M.P., Kouvelis, V.N. and Typas, M.A. 2008. The complete mitochondrial genome of *Fusarium oxysporum*: insights into fungal mitochondrial evolution. *Gene*, 419: 7-15.

Paquin, B., Laforest, M.-J., Forget, L., Roewer, I., Wang, Z., Longcore, J. and Lang, B.F. 1997. The fungal mitochondrial genome project: evolution of fungal mitochondrial genomes and their gene expression. *Current Genetics*, 31: 380-395.

Posada, F. and Vega, E.F. 2005. Establishment of the fungal entomopathogen *Beauveria bassiana* (Ascomycota: Hypocreales) as an endophyte in cocoa seedlings (*Theobroma cacao*). *Mycologia*, 97: 1195-1200.

Quesada-Moraga, E., Landa, B.B., Muñoz-Ledesma, J., Jiménez-Diáz, R.M. and Santiago-Alvarez, C. 2006. Endophytic colonization of opium poppy, *Papaver somniferum*, by an entomopathogenic *Beauveria bassiana* strain. *Mycopathologia*, 161: 323-329.

Quesada-Moraga, E., Navas-Cortés, J.A., Maranhao, E.A.A., Ortiz-Urquiza, A. and Santiago-Álvarez, C. 2007. Factors affecting the occurrence and distribution of entomopathogenic fungi in natural and cultivated soils. *Mycological Research*, 111: 947-966.

Reeb, V., Lutzoni, F. and Roux, C. 2004. Contribution of RPB2 to multilocus phylogenetic studies of the euascomycetes (Pezizomycotina, Fungi) with special emphasis on the lichen-forming Acarosporaceae and evolution of polyspory. *Molecular Phylogenetics and Evolution*, 32: 1036-1060.

Registration Eligibility Decision Document: *Bacillus thurungiensis*; EPA-738-R-98-004; U.S.A. Environmental Protection Agency, Office of Pesticide Programs, U.S. Government Printing Office: Washington, D.C., March 1998.

Rehner, S.A. and Buckley, E.P. 2005. A *Beauveria* phylogeny inferred from nuclear ITS and EF1-α sequences: evidence for cryptic diversification and links to *Cordyceps* teleomorphs. *Mycologia*, 97: 84-98.

Rehner, S.A., Aquino de Muro, M. and Bischoff, J.F. 2006a. Description and phylogenetic placement of *Beauveria malawiensis* sp. nov. (Clavicipitaceae, Hypocreales). *Mycotaxon*, 98: 137-145.

Rehner, S.A., Posada, F., Buckley, E.P., Infante, F., Castillo, A. and Vega, F.E. 2006b. Phylogenetic origins of African and Neotropical *Beauveria bassiana s. l.* pathogens of the coffee berry borer, *Hypothenemus hampei. Journal of Invertebrate Pathology*, 93: 11-21.

Riba, G., Soares, G.G.Jr, Samson, R.A., Onillon, J. and Caudal, A. 1986. Isoenzyme analysis of isolates of the entomogenous fungi *Tolypocladium cylindrosporum* and *Tolypocladium extinguens* (Deuteromycotina; Hyphomycetes). *Journal of Invertebrate Pathology*, 48: 362-367.

Roberts, D.W. and St. Leger, R.J. 2004. *Metarhizium* spp., Cosmopolitan insect-pathogenic fungi: mycological aspects. *Advances in Applied Microbiology*, 54: 1-70.

Robbertse, B., Reeves, J.B., Schoch, C.L. and Spatafora, J.W. 2006. A phylogenomic analysis of the Ascomycota. *Fungal Genetics and Biology*, 43: 715-725.

Rombach, M.C., Humber, R.A. and Evans, H.C. 1987. *Metarhizium album* a fungal pathogen of leaf- and plant-hoppers of rice. *Transactions of British Mycological Society*, 37: 37-45.

Samson, R.A., Evans, H.C. and Latge, J.P. 1988. *Atlas of Entomopathogenic Fungi*. Spring-Verlag, Berlin, Germany.

Sanz, L., Montero, M., Grondona, I., Viccaino, J.A., Hermosa, R. and Monte, E. 2004. Cell wall-degrading isoenzyme profiles of *Trichoderma* biocontrol strains show correlation with rDNA taxonomic species. *Current Genetics*, 46: 277-286.

Seidl, V., Huemer, B., Seiboth, B. and Kubicek, C.P. 2005. A complete survey of *Trichoderma* chitinases reveals three distinct subgroups of family 18 chitinases. *FEBS Journal*, 272: 5923-5939.

Shah, P.A. and Pell, J.K. 2003. Entomopathogenic fungi as biological control agents. *Applied Microbiology and Biotechnology*, 61: 413-423.

Sosa-Gomez, D.R., Humber, R.A., Hodge, K.T., Binnek, E. and Silva-Brandao, K.L. 2009. Variability of the mitochondrial ssu rDNA of *Nomurea* species and other entomopathogenic fungi from Hypocreales. *Mycopathologia*, 167: 145-154.

Spatafora, J.W., Sung, G.H., Johnson, D., Hesse, C., O' Rourke, B., Serdani, M., Spotts, R., Lutzoni, F., Hofstetter, V., Miadlikowska, J., Reeb, V., Gueidan, C., Fraker, E., Lumbsch, T., Lücking, R., Schmitt, I., Hosaka, K., Aptroot, A., Roux, C., Miller, A.N., Geiser, D.M., Hafellner, J., Hestmark, G., Arnold, A.E., Büdel, B., Rauhut, A., Hewitt, D., Untereiner, W.A., Cole, M.S., Scheidegger, C., Schultz, M., Sipman, H. and Schoch, C.L. 2006. A five-gene phylogeny of Pezizomycotina. *Mycologia,* 98: 1018-1028.

St. Leger, R.J., May, B., Allee, L.L., Frank, D.C., Staples, R.C. and Roberts, D.W. 1992. Genetic differences in allozymes and information of infection structures among isolates of the entomopathogenic fungus *Metarhizium anisopliae. Journal of Invertebrate Patholology*, 60: 89-101.

Stensrud, O., Hywel-Jones, N.L. and Schumacher, T. 2005. Towards a phylogenetic classification of *Cordyceps*: ITS nrDNA sequence data confirm divergent lineages and paraphyly. *Mycological Research*, 109: 41-56.

Strasser, H., Vey, A. and Butt, T.M. 2000. Are there any risks in using entomopathogenic fungi for pest control, with particular reference to the bioactive metabolites of *Metarhizium, Tolypocladium* and *Beauveria* species? *Biocontrol and Science Technology*, 10: 717-735.

Strasser, H., Langle, T., Pernfuss, B. and Seger, C. 2004. Experiences with the entomopathogenic fungus *Beauveria brongniartii* for the biological control of the common cockchafer Melolontha melolontha. In: Elad, Y., Pertot, I. and Enkegaard, A. (Eds.), *Management of Plant Diseases and Arthropod Pests by BCAs and their Integration in Agricultural Systems. IOBC wprs Bulletin*, 27: 131-132.

Sugimoto, M., Koike, M., Hiyama, N. and Nagao, H. 2003. Genetic, morphological, and virulence characterization of the entomopathogenic fungus *Verticillium lecanii. Journal of Invertebrate Pathology*, 82: 176-187.

Sung, G.-H., Spatafora, J.W., Zare, R., Hodge, K.T. and Gams, W. 2001. A revision of *Verticillium* sect. *Prostrata*. II. Phylogenetic analyses of SSU and LSU nuclear rDNA sequences from anamorphs and teleomorphs of the Clavicipitaceae. *Nova Hedwigia*, 72: 311-328.

Sung, G.-H., Hywel-Jones, N.L., Sung, J.-M., Luangsa-ard, J., Shretha, B., and Spatafora, J. 2007. Phylogenetic classification of *Cordyceps* and the clavicipitaceous fungi. *Studies in Mycology*, 57: 5-59.

Tambor, J.H., Guedes, R.F., Nobrega, M.P. and Nobrega, F.G. 2006. The complete DNA sequence of the mitochondrial genome of the dermatophyte fungus *Epidermophyton floccosum. Current Genetics*, 49: 302-308.

Tehler, A., Little, D.P. and Farris, J.S. 2003. The full-length phylogenetic tree from 1551 ribosomal sequences of chitinous fungi, Fungi. *Mycological Research*, 107: 901-916.

Tigano-Milani, M.S., Carneiro, R.G., Faria, M.R., Frazão, H.S. and Mc Coy, C.W. 1995a. Isozyme characterization and pathogenicity of *Paecilomyces fumosoroseus* and *Paecilomyces lilacinus* to *Diabotrica speciosa* (Coleoptera: Chrysomelidae) and *Meloidogyne javanica* (Nematoda: Tylenchidae). *Biological Control*, 5: 378-382.

Tigano-Milani, M.S., Honeycutt, R.J., Lacey, L.A., Assis, R., McClelland, M. and Sobral, W.S. 1995b. Genetic variability of *Paecilomyces fumosoroseus* isolates revealed by molecular markers. *Journal of Invertebrate Pathology*, 65: 274-282.

Tulloch, M. 1976. The genus *Metarhizium. Transactions of British Mycological Society*, 66: 407-411.

Tymon, A.M., Shah, P.A. and Pell, J.K. 2004. PCR-based molecular discrimination of *Pandora neoaphidis* isolates from realted entomopathogenic fungi and development of species-specific diagnostic primers. *Mycological Research*, 108: 419-433.

Typas, M.A., Griffen, A.M., Bainbridge, B.W. and Heale, J.B. 1992. Restriction fragment length polymorphisms in mitochondrial DNA and ribosomal RNA gene complexes as an aid to the characterization of species and sub-species populations in the genus *Verticillium. FEMS Microbiology Letters*, 95: 157-162.

Uribe, D. and Khachatourians, G.G. 2004. Restriction fragment length polymorphisms of mitochondrial genome of the entomopathogenic fungus *Beauveria bassiana* reveals high intraspecific variation. *Mycological Research*, 108: 1070-1078.

Urtz, B.E., and Rice, W.C. 1997. RAPD-PCR characterization of *Beauveria bassiana* isolates from the rice water weevil *Lissorhoptrus oryzophilus*. *Letters in Applied Microbiology*, 25: 405-409.

Viaud, M., Couteaudier, Y., Levis, C. and Riba, G. 1996. Genome organization in *Beauveria bassiana* electrophoretic karyotype, gene mapping, and telomeric fingerprinting. *Fungal Genetics and Biology*, 20: 175-183.

Walsh, S.R.A., Tyrell, D., Humber, R.A. and Silver, J.C. 1990. DNA restriction fragment length polymorphisms in the rDNA repeat unit of *Entomophaga*. *Experimental Mycology*, 14: 381-392.

Wang, C., Li, Z. Typas, M.A. and Butt, T.M. 2003a. Nuclear large subunit rDNA group I intron distribution in a population of *Beauveria bassiana* strains: phylogenetic implications. *Mycological Research*, 107: 1189-1200.

Wang, C., Shah, F.A., Patel, N., Li, Z. and Butt, T.M. 2003b. Molecular investigation on strain genetic relatedness and population structure of *Beauveria bassiana*. *Environmental Microbiology*, 5: 908–915.

Wang, B., Liu, X., Wu, W., Liu, X. and Li, S. 2009. Purification, characterization, and gene cloning of an alkaline serine protease from a highly virulent strain of the nematode-endoparasitic fungus *Hirsutella rhossiliensis*. *Microbiological Research*, doi:10.1016/j.micres.2009.01.003.

White, T.J., Bruns, T.D., Lee, S. and Taylor, J. 1990. Analysis of phylogenetic relationship by amplification and direct sequencing of ribosomal RNA genes. In: Innis, M.A., Gelfand, D.H., Sninsky, J.J. and White, T.J. (Eds.), *PCR Protocols: A Guide to Methods and Applications*, Academic Press, San Diego, pp. 315-322.

Williams, J.G., Kubelik, A.R., Livak, K.J., Rafalski, J.A. and Tingey, S.V. 1990. DNA polymorphisms amplified by arbitrary primers are useful genetic markers. *Nucleic Acid Research* 18: 6531-6535.

Yokoyama, E., Arakawa, M., Yamagishi, K. and Hara, H. 2006. Phylogenetic and structural analyses of the mating-type loci in *Clavicipitaceae*. *FEMS Microbiology Letters*, 264: 182-191.

Zare, R., Kouvelis, V.N., Bridge, P.D. and Typas, M.A. 1999. Presence of a 20bp insertion/deletion in the ITS1 region of *Verticillium lecanii*. *Letters in Applied Microbiology*, 28: 258-262.

Zare, R., Gams, W. and Culham, A. 2000. A revision of *Verticillium* sect. *Prostrata* I. Phylogenetic studies using ITS sequences. *Nova Hedwigia*, 71: 465-480.

Zare, R. and Gams, W. 2001a. A revision of *Verticillium* section *Prostrata*. IV. The genera *Lecanicillium* and *Simplicillium* gen. nov. *Nova Hedwigia*, 73: 1-50.

Zare, R. and Gams, W. 2001b. A revision of *Verticillium* sect. *Prostrata*. VI. The genus *Haptocillium*. *Nova Hedwigia*, 73: 271-292.

PART II:
ENTOMOPATHOGENIC NEMATODES

In: Microbial Insecticides: Principles and Applications
Editors: J. Francis Borgio, K. Sahayaraj, et al.

ISBN: 978-1-61209-223-2
© 2011 Nova Science Publishers, Inc

Chapter 8

ECOLOGY OF ENTOMOPATHOGENIC NEMATODES

*Ho Yul Choo[1], Sang Myeong Lee[2], Dong Woon Lee[3] and Hyeong Hwan Kim[4, *]*

[1]Department of Applied Biology and Environmental Sciences, College of Agriculture and
Life Sciences, Gyeongsang National University,
Jinju, Gyeongnam, 660-701, Republic of Korea
[2]Korea Forest Research Institute, Southern Forest Research Center,
Jinju, 660-300, Gyeongnam, Republic of Korea
[3]Department of Applied Biology, Kyungpook National University,
Sangju, 742-711, Kyungpook, Republic of Korea
[4]Horticultural Environmental Division, National Horticultural Research Institute, RDA,
Suwon, 441 – 440, Gyeonggi, Republic of Korea

ABSTRACT

This chapter provides an overview on the ecology of entomopathogenic nematodes
with emphasis on recent developments. These nematodes, represented by the genera
Steinernema and *Heterorhabditis*, are important natural mortality agents of soil insects.
Since the 1980s, great strides have been made in understanding their ecology, but much
more work needs to be done as most studies have focused only on a few model nematode
species. In spite of this, major accomplishments have been achieved in understanding the
infection ecology such as their foraging behavior and understanding their active and
passive dispersal and the effects of abiotic factors on nematode survival, dispersal and
infectivity. In foraging behavior, the infective nematode stage was classified as being
an ambusher (sit-and-wait forager), cruiser (widely forager) or intermediate (having
characteristics of both types of foragers). With abiotic factors, there is a large body of
literature with a good understanding on the effects of temperature and moisture extremes
on the model nematode species. Effects of soil texture on the nematodes are less
understood as soil is complex and various factors such as the amount of sand, silt, and
clay and other factors as temperature, moisture, biotic organisms, soil chemistry, etc.
affect the nematodes' survival, dispersal and infectivity. Field studies have focused on

* E-mail: hychoo@gnu.ac.kr. Tel.: 82-55-751-5444 Fax: 82-55-758-5110

biogeography, but our knowledge base in population ecology and epizootiology of the nematodes and their hosts has only scratched the surface and much more studies in this area are warranted.

8.1. INTRODUCTION

Entomopathogenic nematodes represented by two families, Steinernematidae and Heterorhabditidae, are important natural mortality factors of soil insects and some nematode species are commercially produced and applied as biological control agents against insect pests (Grewal *et al.,* 2005). The family, Steinernematidae, has two genera, *Neosteinernema* containing one species and *Steinernema* containing more than 61 species, whereas the family, Heterorhabditidae, has only one genus, *Heterorhabditis*, with more than 14 species (http://nematology.ifas.ufl.edu/nguyen/morph/kbnstein.htm). New nematode isolates from both families are continually being found, and many of them are new species that are being described or need to be described. Recently described species include *Steinernema khoisanae* (Nguyen *et al.,* 2006; Ansari *et al.,* 2004) and *Heterorhabditis safricana* (Malan *et al.,* 2008) from South Africa and *H. sonorensis* from Sonoran desert in Mexico (Stock *et al.,* 2009).

These nematodes are obligate, parasitic soil animals that are mutualistically associated with bacteria (Kaya and Gaugler, 1993; Gaugler, 2002). With respect to the mutualistic relationship, the steinernematid nematodes are associated with bacteria in the genus *Xenorhabdus* and the heterorhabditid nematodes are associated with bacteria in the genus *Photorhabdus*. The bacteria provide the pathogenic quality to the nematodes; hence, the word "entomopathogenic" is used with these nematodes. Thus, in this mutualistic relationship, the nematode benefits from the relationship because the bacterium does the following: (1) kills its insect host quickly, (2) creates a suitable environment in the cadaver for the nematode by producing antibiotics that suppress competing microorganisms, (3) transforms the host tissues into a food source, and (4) serves as a food resource. The bacterium benefits from the association because the nematode does the following: (1) provides protection to the bacterium from the harsh external environment as it cannot survive well in the soil, (2) serves as a vector for the bacterium into the host's hemocoel, and (3) inhibits the host's antibacterial proteins (Griffin *et al.,* 2005).

The life cycle of entomopathogenic nematodes is initiated by the third-stage infective (often called dauer) juvenile which houses the mutualistic bacterium in its intestine and is the only free-living stage (Kaya and Stock, 1997; Koppenhöfer, 2007) that is morphologically and physiologically adapted to remain in the soil environment for a prolonged period (Poinar, 1990). It searches for an insect host in the soil and enters it through natural openings (mouth, anus, or spiracles) and penetrates into the hemocoel. In some species, especially heterorhabditids, the infective juvenile has an anterior tooth that allows it to penetrate directly into the hemocoel through soft cuticle. The infective juvenile voids the bacterial cells from its intestine in the insect's hemocoel killing its host by septicemia within 48 hours. The nematodes feed on the bacterial cells and host tissues, produce two to three generations, and emerge from the host as infective juveniles to search for new hosts (Kaya and Gaugler, 1993; Griffin *et al.,* 2005).

The publication of the book "Silent Spring" by Rachel Carson in 1962 launched, in part, the environmental movement (Carson, 1962). Because of the problems with ground water contamination with chemical pesticides, human health concerns, the need to find alternative control agents for soil-inhabiting insects, and some encouraging results with entomopathogenic nematodes against soil insect pests (Klein, 1990), there was great interest in commercializing these nematodes. In the early 1980s, a number of entomopathogenic nematode species began to be commercially produced for use against soil insect pests (Kaya et al., 2006; Georgis et al., 2006). However, many of the early field studies had poor results (Georgis and Gaugler, 1991), in a large part, because the ecology and behavior of the nematodes were not understood at the time of application (Lewis, 2002). When field tests were successful, the reasons for the nematodes' success of the soil pests were not known. Even today, the reasons for success or failure of a nematode application against a soil insect pest are often unknown (Georgis et al., 2006). Consequently, there is a need to understand the ecology and behavior of these nematodes to make them better biological control agent, and a major effort has been made to better understand the ecological and behavioral factors that affect the nematodes. Once commercialization of these nematodes begin in earnest in the early 1980s (Friedman, 1990), they readily became available for researchers to do more ecological and behavioral studies.

Although there have many attempts to use these nematodes to control insect pests in habitats other than the soil with varying degrees of success (Georgis et al., 2006), such habitats are unnatural for them. Accordingly, the focus of our chapter is to provide ecologically useful information on these entomopathogenic nematodes in the soil environment for the effective control of soil-inhabiting insect pests in the field. Furthermore, most ecological and behavioral research has focused on the infective juvenile of these nematodes. Once infective juveniles are applied into the host's environment or they naturally emerge from a host cadaver in the environment, they are subject to abiotic and biotic factors that affect their survival. In addition, the developing nematodes within a host cadaver are subject to the same ecological parameters, but less is known about this aspect.

8.2. BIOGEOGRAPHY

Entomopathogenic nematodes have a global distribution and are essentially ubiquitous (Hominick, 2002). Some species such as *S. carpocapsae* and *S. feltiae* are widely distributed in temperate and subtropical regions, *Heterorhabditis bacteriophora* is common in regions with continental and Mediterranean climates, and *H. indica* occurs in much of the tropical and subtropical regions of the world. However, these species are not restricted to specific temperate, subtropical or tropical regions. Moreover, some of these species, especially since the early 1990s, have been used as biological control agents (e.g., *S. carpocapsae*, *S. feltiae*, *S. glaseri*, *S. scapterisci*, *H. bacteriophora*, and *H. megidis*) and have been introduced into new areas, and if they have become established, their biogeographic distribution is expanded.

Some nematode species such as *S. kushidai*, *S. ritteri* and *H. argentinensis* seem to have a more restricted distribution. As more surveys are conducted in various parts of the world, these species may be found more widely distributed than previously thought. For example, *S. glaseri* was considered to be restricted to North America, but surveys have shown that it can be isolated from Korea (Stock *et al.,* 1997) and Europe (de Doucet and Gabarra, 1994). In Korea, *S. glaseri,* as far as we know was never introduced into the country, yet have been isolated there (Stock *et al.,* 1997), and *S. websteri,* initially described from China (Cutler and Stock, 2003), has been isolated from the Andean region of Colombia (Lopez-Nuñez *et al.,* 2007). In another example, *H. megidis* was initially described from a population in Ohio (USA) by Poinar *et al.* (1987), but subsequently, this species has been difficult to isolate there, but is a common species that occurs throughout Europe (Hominick *et al.,* 1995; Miduturi *et al.,* 1996, 1997; Steiner, 1996; Menti *et al.,* 1997; Griffin *et al.,* 1999; Mráček *et al.,* 1999a; Sturhan, 1999; Kramer *et al.,* 2000).

At a local level, some habitats seem to have a greater propensity for entomopathogenic nematodes to be found. In part, preference habitat may have the proper hosts as well as the right conditions for long-term survival. Habitat preference is likely to display the distribution of suitable hosts and may explain the coexistence of species (Hominick, 2002). Each nematode species will be physiologically and behaviorally adapted to survive in a particular habitat. Some nematode species clearly show a habitat preference, whereas other species can be found in many different habitats. Even though a species may show a preference for a given habitat, it does not preclude them from being isolated in other habitats. On tropical and subtropical islands, *H. indica* is found along coastal areas in calcareous sand (Hara *et al.,* 1991; Constant *et al.,* 1998; Mauléon *et al.,* 2006). In fact, it seems that heterorhabditid species are commonly found near coastal areas (see Mauléon *et al.,* 2006; Griffin *et al.,* 1991, 1994, 2000; Hara *et al.,* 1991; Poinar, 1993; Amarasinghe *et al.,* 1994; Hominick *et al.,* 1995; Mason *et al.,* 1996; Constant *et al.,* 1998; Yoshida *et al.,* 1998), although they do occur in many inland areas as well (Hominick, 2002). Several hypotheses have been proposed for occurrence of heterorhabditids along the coastal areas including their introduction from ship ballast (Akhurst and Bedding, 1986), the evolution of heterorhabditids from marine nematodes (Poinar, 1993), the availability of insect hosts (Hominick *et al.,* 1996), and suitable conditions of the soil type (Mauléon *et al.,* 2006). In temperate regions, *S. affine* prevails in grassland and *S. intermedium* and *S. kraussei* are isolated primarily from forest areas (Sturhan, 1999; Spiridonov and Moens, 1999). *S. affine* and *S. feltiae* are commonly found in hedgerows in Europe (Hominick *et al.,* 1995), *S. feltiae* is clearly ubiquitous in many different habitats in tropical and temperate regions of the world (Hominick, 2002).

8.3. INFECTION ECOLOGY

Host finding by the infective juveniles requires a number of ecological steps for successful infection. Griffin *et al.* (2005) separated in the steps into four phases including dispersal, foraging strategies, host discrimination, and host infection, whereas Campbell and Lewis (2002) first discussed foraging behavior followed by separating the host-finding steps into three phases including host-habitat selection, host finding, and host acceptance. Lewis (2002) provided a detail examination of the host-finding behavior of the infective juvenile including dispersal and location within the soil profile, foraging strategy, host discrimination, and infection dynamics. On the other hand, Lewis *et al.* (2006) described host-finding behaviors and strategies and then took a mechanistic approach at how the infective juveniles recognize their hosts and how they infect the hosts. As pointed out by Lewis (2002) and

Lewis *et al.* (2006), entomopathogenic nematodes are represented by a diversity of species, but the ecological behaviors of only a few nematode species have been studied in detail. Accordingly, these few nematode species have served as model organisms and form the bulk of our knowledge on foraging behavior. In our chapter, we will discuss locomotory behavior, foraging strategies, and host finding from an ecological perspective.

8.3.1. Locomotory Behavior

Like most nematode species, infective juveniles crawl by sinusoidal movement on the soil substrate using the water film to propel them forward or backward (Croll, 1970). While crawling, the nematodes may scan for environmental cues such as chemical and temperature gradients that influence their behavior (Croll, 1970; Dusenbery, 1980; Lewis *et al.*, 2006) and serve as a means of dispersal or to find hosts. In addition to crawling, the infective juveniles of some entomopathogenic nematode species exhibit standing and jumping behavior that enables them to find and infect their hosts. Most nematode species can raise the anterior portion of their body off the substrate and wave it back and forth (body waving). In some *Steinernema* species, the infective juvenile can elevate more than 95% of their body off the substrate, balance on a bend in their tail (*i.e.*, "stand"), and do the body wave (Reed and Wallace 1965; Ishibashi and Kondo, 1990; Campbell and Gaugler, 1993; Campbell and Kaya, 2002). This standing behavior has been called "nictation" (Ishibashi and Kondo, 1990) or more recently as "standing" (Campbell and Kaya, 1999a, b).

Standing behavior by the infective juveniles of some *Steinernema* species is a prerequisite for jumping. However, not all species of nematodes that stand can jump. To jump, a standing infective juvenile forms a loop with its body and when released from the loop, it propels itself many times its body length through the air (Campbell and Kaya, 1999a, b; 2000).

For example, the forces generated by the jumping mechanism of *S. carpocapsae* can propel an infective juvenile an average distance of nine times or an average height of seven times its body length (558 µm) (Campbell and Kaya, 1999a, b). The frequency of jumping, like standing behavior, varies among species of *Steinernema* and may be used for dispersal or attaching to an insect host (Campbell and Kaya, 2002). More detailed functions of these crawling, standing and jumping behaviors are discussed below under foraging strategies.

8.3.2. Foraging Strategies and Host Finding

An organism's foraging behavior is a critical component of its life history because acquisition of resources is closely linked with fitness, and foraging mode can be correlated with suites of ecological, behavioral, physiological and morphological traits (Campbell and Lewis, 2002).

Foraging behaviors of entomopathogenic nematodes are divided into two broad categories, cruise (widely foraging) and ambush (sit-and-wait) that represent end points on a continuum. Those nematode species that fall between these two end points have an intermediate strategy (Campbell and Kaya, 2000; Lewis *et al.*, 2006). This classification is based on differences in how foraging time is allocated to motionless scanning versus moving through the environment (see Lewis *et al.,* 2006).

Cruise foragers (cruisers) allocate their foraging time to scanning for resource-associated cues when crawling through the environment or during short pauses, whereas ambush foragers (ambushers) scan during long standing pauses that are interrupted by repositioning bouts (crawling) of relatively short duration (Lewis *et al.,* 2006). Lewis (2002), Lewis and Campbell (2002) and Campbell *et al.* (2003) sorted some *Steinernema* based on behavioral tests. Thus, *S. arenarium, S. cubanum, S. glaseri, S. karii, S. longicaudum,* and *S. puertoricense* and *Heterorhabditis* spp. are cruisers, *S. carpocapsae, S. scapterisci,* and *S. siamkayai* are ambushers, and *S. abbasi, S. affine, S. ceratophorum, S. feltiae, S. intermedium, S. monticolum, S. oregonense,* and *S. riobrave* are intermediate foragers. It is important to remember, however, that this classification does not mean that all individuals in a population response in the same manner. Individuals within a species that are classified as ambushers or intermediate foragers, for example, may response as cruisers.

Some comparisons have been made on the response among foraging nematodes to various cues. For example, the cruiser, *H. bacteriophora,* has superior host finding ability to the ambusher, *S. carpocapsae* (Choo *et al.,* 1989). The cruiser, *S. glaseri,* strongly responded to host volatiles compared with the ambusher *S. carpocapsae* and intermediate forager *S. feltiae.* Comparisons between an ambusher and an intermediate forager showed that more infective juveniles of *S. feltiae* moved towards a host than *S. carpocapsae* (Lewis *et al.,* 1995). Electrical current is another factor for host finding of nematodes where the infective juveniles respond differently to electrical current. Thus, the cruiser, *S. glaseri,* moved to a higher electric potential, whereas the ambusher, *S. carpocapsae,* moved to a lower electric potential (Shapiro-Ilan *et al.,* 2009). That is, a higher number of *S. carpocapsae* moved to the cathode relative to the anode, whereas *S. glaseri* moved in the opposite direction. With another cue, Torr *et al.* (2004) reported that *S. carpocapsae, S. feltiae,* and *H. megidis* responded positively to vibration. These types of comparisons show the differential or similar response of the nematodes to various cues.

8.3.2.1. Cruise Foragers and Host Finding

Cruise foraging infective juveniles move sinusoidally using relatively linear patterns that are typical of ranging movements in the absence of host-associated cues (Lewis *et al.,* 1992). This ranging movement is believed to increase the search area for hosts, and when the infective juvenile detects a source of volatile cues, it switches to localized search after contacting the host-associated cues (Lewis *et al.,* 2006).

For example, *S. glaseri* responded positively to volatile cues from an insect host (Lewis *et al.,* 1993) and switched to localized search after contacting host-associated cues such as cuticle and feces (Lewis *et al.,* 1992). Other cruise foragers in *Steinernema* had similar levels of response to volatile cues from insects (Grewal *et al.,* 1993a; Campbell *et al.,* 2003) as did two species of *Heterorhabditis* (Grewal *et al.,* 1994b).

In addition, chemical signals from injured plant roots on which an insect host feed have been implicated in host finding of weevil (van Tol *et al.,* 2001) and corn rootworm larvae (Rasmann *et al.,* 2005; Köllner *et al.,* 2008[) by *H. megidis.* In corn plants, the volatile cue that served as an attractant for *H. megidis* was identified as (*E*)-β-caryophyllene (Rasmann *et al.,* 2005; Köllner *et al.,* 2008).

The positive response to volatile and host-contact cues assist the cruise foragers in finding sedentary host. The host-contact cues maximize localized search and increases the chance that a searcher will either remain in a potentially profitable area or reestablish contact with a host that was lost.

8.3.2.2. Ambush Foragers and Host Finding

Campbell and Kaya (2002) recognized two behavioral modes –standing and jumping - which are associated with the ambush foraging strategy. The first behavioral mode is standing that includes body waving which is common among ambushers and also occurs among intermediate foragers. This standing behavior allows the infective juvenile to maintain a straight and immobile posture with interspersed periods of body waving for extended periods of time, especially for ambushers and less so for intermediate foragers (Campbell and Kaya, 2002; Lewis *et al.,* 2006). Standing and body-waving behavior may waste the nematode's energy but can be an efficient trait to encounter hosts because the infective juvenile can easily attach itself to a passing insect host (Campbell and Gaugler, 1993).

The second behavioral mode among ambushers and intermediate foragers is jumping. Jumping (leaping) was first reported in infective juveniles of *S. carpocapsae* by Reed and Wallace (1965) and can be an important behavior in response to information about insect proximity to nematode attack (Campbell and Kaya, 1999a). Infective juveniles that are standing respond differently depending on the information they receive from the environment. Thus, air movement influences the direction of the jump, whereas insect-associated volatile cues are important activators for the jumping behavior (Campbell and Kaya, 1999a). The body-waving and jumping behavior of some *Steinernema* species enhances host searching for mobile hosts, and these nematode species tend to be or at or near the soil surface.

The response of ambushing infective juveniles to volatile cues has been studied in some detail. When ambushers like *S. carpocapsae* infective juveniles are placed on a smooth substrate, they use a ranging search behavior typical of a cruiser, but they are not attracted to host volatile cues such as CO_2 (Lewis *et al.,* 1993). Moreover, ambush foragers do not switch their searching behavior in response to contact with host cues such as cuticle and feces like *S. glaseri* (Lewis *et al.,* 1992) because their response is fundamentally different from cruisers. Thus, *S. carpocapsae* infective juveniles that are crawling on the smooth substrate are not attracted to host volatile cues until after contact with host cuticle which suggests that they respond to cues only when the cues are presented in the appropriate sequence (Lewis *et al.,* 1995).

In another finding, ambushers will respond to volatile cues prior to host contact, but their responses are expressed only when nematodes are standing and not when they are crawling. In contrast to cruisers that scan while moving through the soil environment or during short pauses, ambushers scan the environment during long standing pauses. Campbell and Kaya (1999a, 2000) observed that ambushing infective juveniles when presented with host-associated cues during a standing bout will wave back and forth and jump toward the source of the cue. Such behavior appears when two types of cues occur which are volatile chemical ones and air movement. This response increases the area of attack surrounding a standing infective juvenile and the probability of attaching to a host. It suggests that the standing infective juveniles are actively scanning the environment and increasing the probability of host encounter when host cues are present (Lewis *et al.,* 2006). On the other hand, not all nematode species that can stand respond to host cues when standing (Campbell and Kaya,

2002). *S. siamkayai*, for example, appears to use the standing behavior as a means to jump for dispersal rather than attaching to a host.

8.3.2.3. Intermediate Foragers and Host Finding

Intermediate forager responses to host-associated cues are more variable than for cruising or ambushing species. Campbell *et al.* (2003) found that many intermediate foragers are more similar to cruisers than ambushers. That is, they are attracted to host volatiles and switch to localized search in response to contact with host cuticle, but unlike true cruisers, most intermediate foraging species can stand and jump. However, jumping is not triggered by the sudden introduction of host cues and they do not to jump toward the source of cues (Campbell and Kaya, 2000). Accordingly intermediate foragers seem to use host cues in a manner consistent with cruisers, but because they can stand or jump, they can also attach to moving hosts. Significantly more information is available on the foraging behavior of cruisers and ambushers than intermediate foragers. We encourage more studies on the behavioral ecology of these intermediate foragers as they are represented by a number of important species that can be used in biological control programs.

8.4. DISPERSAL

Kaya (1990) and Downes and Griffin (1996) reviewed active and passive dispersal by infective juveniles. Those reviews were done before there was a greater understanding of foraging and host finding behavior of the infective juveniles. In retrospect, foraging behavior can now explain much of the older literature on active dispersal from subsequent observations made from laboratory studies. In this section, we examine active and passive dispersal of these nematodes.

We define active dispersal as movement by the nematodes using their own locomotory power, whereas passive dispersal is through the action of another agent (Kaya, 1990). Previously, we have discussed jumping behavior and response of the cruiser and ambusher nematodes to various cues. These behaviors encompass active dispersal but will not be repeated here. In addition, we recognize that both passive and active dispersal may be involved in nematode movement. When entomopathogenic nematodes are applied as biopesticides, it is highly recommended that the nematodes be irrigated into the soil after application to get them into the soil environment and provide adequate moisture for active dispersal [see various chapters in Lacey and Kaya (2007) concerning application techniques for entomopathogenic nematodes].

Dispersal by nematodes is a prerequisite to host finding. Generally, the location of insect host in soil is usually within 10 cm of the soil surface. For example, the black cutworm, *Agrotis ipsilon,* and the scarab, *Ectinohoplia rufipes*, are distributed within the upper 3 cm of soil (Kim *et al.,* 2001; Choo *et al.,* 2002b), and the oriental beetle, *Exomala orientalis*, is within 3 – 7 cm from the soil surface (Choo *et al.,* 2002b). The fungus gnat, *Bradysia difformis*, damages seedling roots in 5 cm deep tray cell in the nursery (Kim *et al.,* 2003), but the European chafer, *Rhizotrogus majalis*, moves down to a depth of 5 – 25 cm to pupate (Brandenburg and Villani, 1995). Thus, in using entomopathogenic nematodes for biological control of these soil insect pest species, the dispersal ability (*i.e.,* foraging behavior) must be taken into consideration. The nematode species needs to possess the foraging ability to infect the susceptible insect host in the soil. That is, it is critical to match the right nematode species with the target insect pest to have an effective biological control program.

8.5. ABIOTIC FACTORS

8.5.1. Temperature

8.5.1.1. Infectivity, Development and Reproduction

In common with almost all poikilotherms, the rate of nematode activity varies with temperature increasing over a range from 5 to 40°C (Croll, 1970). Thus, temperature affects entomopathogenic nematode's infectivity, development, reproduction, mobility, and survival. The temperature range for infection and host-killing is species and isolate dependent (Mason and Hominick, 1995, Menti *et al.,* 2000). The maximum temperature causing host death by *H. bacteriophora* is 32.3°C and the minimum 11 – 12°C with the optimum temperature for activity being approximately 28°C (Blackshaw and Newell, 1987). Extreme temperature range may be different depending on isolate. The maximum temperature of *H. bacteriophora* Hamyang strain was 35°C while the optimum was 24°C (Chung, 2001).

There are different temperature ranges affecting nematode infection between genera or among species within a genus. For example, Molyneux (1986) demonstrated that *Steinernema* spp. usually have wider activity range than *Heterorhabditis* spp. *Steinernema* spp. are more active at lower temperatures and infect the sheep blowfly (*Lucilia cuprina*) larvae over a greater temperature range, whereas maximal infection by *Heterorhabditis* spp. occurred within a narrower temperature range. The steinernematid species were motile at temperatures below 9°C, whereas the heterorhabditid species were inactive below 9°C. In another example, *S. riobrave* has wider temperatures for infection than *S. feltiae*. The range of temperatures for infection for *S. riobrave* is between 10 and 39°C (Grewal *et al.,* 1994a), while it is 8 to 30°C for *S. feltiae* (Hazir *et al.,* 2001) and 10 to 37°C in *S. glaseri* (Grewal *et al.,* 1994a). *S. carpocapsae* and *H. bacteriophora* also have wide temperature ranges and can infect hosts between 1 and 32°C (Grewal *et al.* 1994a). As expected, infection generally takes longer time al low temperatures. At 10°C, infection of the insect host, *Galleria mellonella* by the Canadian isolate of *S. kraussei* increased with time of exposure to the infective juveniles and ranged from 80% at 28 h to 100% at 265 h (Mráček *et al.,* 1999b). With *S. rarum*, it infected *G. mellonella* larvae from 15 to 33°C with an optimum temperature of 25°C, but it did not infect at 10°C and only 29% at 35°C (Koppenhöfer and Kaya, 1999). Time to kill is its host is associated with activity of the entomopathogenic nematodes. Host death by *S. feltiae* occurs at 12 day at 8°C, 4 to 5 days at 15°C, 3 days at 20°C, and 2 days at 25°C (Hazir *et al.,* 2001).

The effects of fluctuating temperatures on the development and infectivity of *S. carpocapsae* showed that arrested development occurred at low temperatures. For example, Bornstein-Forst *et al.* (2005) conducted a series of experiments and showed that when nematode-killed hosts were incubated for 2 days 23°C and then subjected to lower temperatures of −10, 4, 10 or 14°C, respectively, from days 3 to 36 post-infection, the nematodes remained as adults. When 2-day-old nematode-killed hosts were held a 23°C and then subjected to the cold temperatures for 7 days followed by incubation at 23°C for another

14 days, a limited number of the host cadavers that had been held at 10 and 14°C produced infective juvenile. However, the emergent numbers were significantly lower than those of control infections incubated continuously at 23°C. When the nematode-killed hosts were held at for 4 days at 23°C and then subjected to cold temperatures followed by incubation at 23°C, the survival of first- and second-stage juveniles was reduced by at least 95% or more at the lower temperatures compared with controls.

Nematode infection is not always in accord with nematode development and bacterium growth. *H. bacteriophora* and symbiotic bacterium *P. luminescens* killed *G. mellonella* larva at 12 to 28°C. Although nematode development is inhibited at 12 and 30°C, *Photorhabdus* can cause mortality from 12 to 33°C (Milstead, 1981). *S. carpocapsae* Pocheon strain killed its host at all tested temperatures from 13 to 35°C with the optimal temperature at 24°C, but it did not developed at 13 or 35°C (Choo *et al.*, 2002a). Optimum temperatures for *S. longicaudum* Gongju strain infection were from 24 to 30°C, and it cannot produce progenies at 13°C (Choo *et al.*, 1999).

8.5.1.2. Cold Tolerance

Entomopathogenic nematodes have been isolated from a wide variety of temperature regimes ranging from the sub-Arctic to tropical climates (Glazer, 1996, 2002). Studies conducted with these isolates demonstrate that the nematodes are usually acclimatized to the temperature regimes from which they were isolated.

A number of nematode species have been categorized as being cold tolerant because they can survival at low temperatures. For example, some steinernematids and heterorhabditis have been isolated from cold temperate sites from northern Europe and Canada and from the far north such as Moscow, Russia or from alpine sites in the Swiss Alps (Glazer, 1996; Lee, 2002). Species known to be cold tolerant include *S. anomali*, *S. feltiae*, *S. kraussei*, and *H. megidis,* and some isolates of *H. bacteriophora*. The lower lethal temperatures are -22, -19, and -14°C for the infective juveniles of *S. feltiae*, *H. bacteriophora*, and *S. anomali*, respectively. Survival after prolonged freezing at -4°C was 6, 5, and 3 days for *S. feltiae*, *H. bacteriophora*, and *S. anomali*, respectively (Brown and Gaugler, 1996). Acclimation to lower temperatures increases freezing tolerance. The freezing tolerance of *H. bacteriophora* increased under a stepwise acclimation regime, whereas *S. feltiae* is better under a direct acclimation regime (Brown and Gaugler, 1996).

Some species and/or isolates not considered to be cold tolerant can tolerate cold temperatures. Freezing of the infective juveniles for a short time does not affect their pathogenicity, but freezing them for long time significantly reduces their survival and pathogenicity (Brown and Gaugler, 1996, 1998). *S. carpocapsae* and *H. bacteriophora* resuscitated from freezing retained their pathogenicity. Other studies with *S. riobrave*, *S. carpocapsae*, and *S. glaseri* showed that these nematodes could survive prolonged exposure to freezing at -4°C, although survival was decreased with time. *S. riobrave* and *S. carpocapsae* were still pathogenic after 6 days after freezing but *S. glaseri* showed poor pathogenicity after 4 days of freezing (Brown and Gaugler, 1998). Addition of 20% glycerol for 48 h prior to freezing at -20°C enhanced survival and pathogenicity of *S. carpocapsae* (Brown and Gaugler, 1998).

The sheath (second-stage cuticle that is retained and surrounds infective juveniles) also plays an important role in nematode response to freezing temperature. The infective juvenile sheath of *H. zealandica* may prevent protect it from freezing, allowing extensive supercooling to -32°C, whereas exsheathed infective juveniles freeze above -6°C (see Glazer, 2002). According to some cryopreservation studies, entomopathogenic nematodes can be stored indefinitely in liquid nitrogen (Lee, 2002). When infective juveniles were frozen in sand, *S. feltiae* and *S. carpocapsae* had about 50% survival after 1 week of freezing at -8°C which represented a significant reduction in survival compared to after 1 day of freezing. On the other hand, *S. glaseri* and *H. bacteriophora* suffered complete mortality after 1 week of freezing. These species had reduced survival after 2 days of freezing (Lewis and Shapiro-Ilan, 2002). In general, nematodes can survive cold stress in the soil because the nematodes are buffered from extremely cold temperatures by moving deeper in the soil profile.

8.5.1.3. Heat Tolerance

Soil temperatures from the sun's radiation change with depth and the time of day (Miller and Gardiner, 2001) but rarely reach the lethal point for most organisms in the most ecosystems. The soil is protected from sunlight by insulators such as vegetation or organic coverings; yet, the temperature can reach levels affecting nematodes directly. At high temperatures, nematode metabolism is likely to be reduced and they become quiescent and exhibit heat stupor followed by heat coma (Lee, 2002). High temperatures (above 32°C) have an adverse effect on the reproduction, growth and survival of many organisms including nematodes (Glazer, 1996). However, the effects of high soil temperature on entomopathogenic nematodes under natural conditions are not known. Most studies on the effect of high temperature have been done under laboratory conditions.

Thermal niche breadths generally do not differ among conspecific populations isolated from different localities (Hazir *et al.,* 2003). However, different species isolated from the same locality do show differences (Grewal *et al.,* 1994a). For example, *S. glaseri* and *S. longicaudum* do not produce progenies at 35°C (Hang *et al.,* 2007), and *S. riobrave* showed higher temperate tolerant (10 - 39 °C) than *S. feltiae* (8 - 30°C) (Grewal *et al.,* 1994a; Hazir *et al.,* 2003). Thermal niche breadth for reproduction was 12 - 32 °C for *S. glaseri,* 20 - 30 °C for *S. carpocapsae,* 20 - 32 °C for *S. scapterisci,* and 20 - 35 °C for *S. riobrave* (Grewal *et al.,* 1994a). Although heterorhabditids are endemic to warmer climates, the upper thermal limits and temperature for reproduction of *H. bacteriophora* and *H. megidis* were cooler than that of some of the steinernematids from South America and the Caribbean (Grewal *et al.,* 1994a).

Judging from isolation of *Steinernema* and *Heterorhabditis* from such warm regions as Sri Lanka (Amarasinghe *et al.,* 1994), *Heterorhabditis* spp. from semiarid region of Israel (Glazer *et al.,* 1996), *S. anatoliense, S. carpocapsae, S. feltiae,* and *H. bacteriophora* from Jordan (Stock *et al.,* 2008), or *H. sonorensis* from Sonoran desert in Mexico (Stock *et al.,* 2009) suggests that they are heat tolerant because nematodes inhabiting high-temperature habitats must exhibit the capacity to adapt with life and activity occurring at a range of temperatures higher than that of nematodes from more moderate habitats (Lee, 2002). *Heterorhabditis* sp. IS-5 isolated from under the canopy of fruit trees in the arid "Negev" region of Israel is highly tolerant to temperatures above 30 °C (Glazer *et al.,* 1996). Two hours at 37 °C killed 79% of infective juveniles of *H. bacteriophora* HP88 but only 18% of *Heterorhabditis* sp. IS-5. One hour at 40 °C killed all infective juveniles of the HP88 isolate but only 28% of IS-5 (Glazer *et al.,* 1996).

The ability of nematodes to survive at temperatures up to the thermal death point, and the temperature at which death occurs, may be increased by sublethal exposure to high temperature by heat shock response (Lee, 2002). Although more than 90% of infective juveniles of *H. bacteriophora* were inactivated upon exposure to 40 °C within 2 hours, preconditioning of *H. bacteriophora* for 3 hours at 35 °C with a subsequent latency period of 2 hours at 25°C, extended their survival from less than 1 to 6 hours at 40 °C . In addition, infectivity and establishment of preconditioned *H. bacteriophora* were significantly higher than the untreated *H. bacteriophora* at 30, 35, and 40 °C (Selvan *et al.,* 1996). Heat shock treatment for 2 hours at 37°C before exposure to 40°C also enhanced the survival of *Heterorhabditis* sp. IS-5 juveniles to 43% compared with non heat-shocked control (Glazer *et al.,* 1996). When this IS-5 population is reared at 30 and 25°C for several passages through *Galleria mellonella*, nematode survival at 40°C after 8, 10, 12 passages was greater in the IS-5 strain than *H. bacteriophora* HP88 strain regardless of selection pressure (Shapiro *et al.,* 1996). IS-5 nematodes reared at 30°C had greater survival abilities than those reared at 25°C, but HP88 strain did not exhibit any increase in survival ability when reared at 30°C (Shapiro *et al.,* 1996). Moreover, progeny from both IS-5 populations had significantly greater survival rates at 40°C than the HP88 strain (Shapiro *et al.,* 1996). The IS-5 strain survived significantly longer at 25°C than at 10°C, whereas HP88 survived longer at 10°C (Shapiro *et al.,* 1996).

Survival of hybrid transferred heat tolerance of *H. bacteriophora* IS-5 to *H. bacteriophora* HP88 and treated for 2 hours at 40°C was significantly greater than survival of the HP88 strain and similar to the survival of the IS-5 strain. At 32°C, the IS-5 and hybrid strains caused mortality of *G. mellonella* at a faster rate than the HP88 strain. Similar to IS-5 strains, the hybrids exhibited sensitivity to cold (Shapiro *et al.,* 1997).

Grewal *et al.* (1996) selected the reproductive thermal niche breadth extremes of 15 and 30°C for *H. bacteriophora* and *S. anomali* for 12 passages (equal to 24 – 36 generations) by repeated passage through *G. mellonella*. The selected *H. bacteriophora* extended the upper thermal limit for infection from 32 to 35 °C and from 32 to 37°C with selection at 15 and 30°C, respectively. On the other hand, *S. anomali* extended the lower thermal virulence niche breadth at the selection of 15°C, but the upper and lower limits were extended at the selection of 30°C (Grewal *et al.,* 1996). Thus, the temperature-specific virulence and thermal niche breadth of entomopathogenic nematodes are malleable and adaptation to warm temperature enhanced fitness in a warm environment (Grewal *et al.,* 1996).

8.5.2. Moisture

Nematodes are hydrophilic animals, and water governs the life of nematodes to accomplish their normal activities in soil. For these soil-dwelling nematodes including entomopathogenic ones, a film of water is needed for them to move through soil pores. Water potential, the potential energy of water per unit mass of water in the system and measured as MPa (megapascals) or kPa (kilopascals), and pore size can differentially affect the activity of various organisms including nematodes (Barbercheck, 1992). However, not all nematode studies report soil water potential and use percent relative humidity or percent soil moisture. As with temperature, the nematodes have a certain optimum range of moisture levels.

In the oak forest near Ceske Budejovice, Czech Republic, Půža and Mráček (2007) demonstrated that the abundance of infective juveniles of entomopathogenic nematodes is strongly correlated with soil moisture. They studied the effect of drought on population dynamics of *S. affine* at soil temperatures of 17.5 - 18°C and found that infective juvenile numbers were positively correlated with soil moisture. They also found that infective juvenile numbers rose immediately after an increase in soil moisture, suggesting that the infective juveniles remained within a host at low ambient moisture instead of emerging.

In laboratory studies, entomopathogenic nematodes show different emergence from cadavers and survival in the cadavers at different relative humidities (Brown and Gaugler, 1997). Increasing relative humidity significantly increased cadavers with *H. bacteriophora* or *S. feltiae* with infective juveniles emerging, whereas cadavers with *S. glaseri* and *S. carpocapsae* showed no significant decrease (Brown and Gaugler, 1997). At 75% relative humidity, *S. feltiae* and *H. bacteriophora* showed significant decrease in emergence, whereas *S. glaseri* and *S. carpocapsae* did not. All the cadavers from which no nematodes emerged at 75% relative humidity contained nematodes. The infective juveniles of *S. carpocapsae* in cadavers had more than 75% survival of nematodes, but *S. glaseri* had less than 25% survival, *S. feltiae* less than 5% survival, and *H. bacteriophora* no survival.

Low moisture is probably a more critical factor affecting nematode survival. However, infective juveniles can survive for considerable lengths of time within desiccating host cadavers in dry soil, and different nematode species are affected in different ways by the low soil moisture. When survival of *S. glaseri*, *S. carpocapsae*, *S. riobrave*, and *H. bacteriophora* within host cadavers was investigated from -500 MPa (very dry) to -0.006 MPa (moist), no infective juveniles emerged from cadavers at -500 MPa and only few *S. glaseri* and *S. carpocapsae* emerged at -40 MPa. Large numbers of infective juveniles of *S. carpocapsae*, *S. glaseri*, and *H. bacteriophora* emerged from all cadavers at $\geq$-5 MPa, but no *S. riobrave* emerged. All nematode species emerged in large numbers at $\geq$0.3 MPa. The first day of infective juvenile emergence was delayed by low moisture in *S. carpocapsae*, *S. riobrave*, and *H. bacteriophora* but not in *S. glaseri*. At the lowest soil moisture *S. glaseri* and *S. carpocapsae* emerged for a significantly period while *H. bacteriophora* emergence was not affected. On the other hand, *S. riobrave* emergence was significantly shorter at the highest soil moisture. Survivability of entomopathogenic nematodes at different moisture levels is species dependent.

Relationship between log of percent survival and days after emerging from wax moth cadavers is linear in *S. carpocapsae* and *S. riobrave*, but is quadratic in *H. bacteriophora* held in desiccators at 75% relative humidity and 25°C (Perez *et al.*, 2003). *H. bacteriophora* TTO1 strain is highly susceptible to desiccation stress and does not survive beyond 72 hours at 85% relative humidity, whereas the survival percentages of *S. feltiae* IS-6 and Carmiel strain, *S. carpocapsae* Mexican strain, and *S. riobrave* were between 10 and 40% at that humidity (Somvanshi *et al.*, 2008). *S. carpocapsae* survives 4 days exposure to a relative humidity of 48.4%, which corresponds to very dry soil with more than 80% of nematodes surviving (Simons and Poinar, 1973). *S. carpocapsae* and *H. bacteriophora* persisted up to 12 weeks at 20 - 40% moisture of nursery (bed) soil (Kim *et al.*, 2003). In addition, Korean isolates of *S. glaseri* and *H. bacteriophora* persisted up to 21 weeks in the rough of golf courses (Choo *et al.*, 2002b). However, extreme moisture condition in a soil is usually unfavorable to nematodes, especially a long span of saturation or drying. Saturation results in deficiency of oxygen and leads to nematode death by hypoxia.

Molyneux and Bedding (1984) observed that infection of *L. cuprina* by *Heterorhabditis* sp. D1 and *S. glaseri* did not occurred at low water potential in the sand and clay loam. However, both nematode species migrated at low moisture potentials with high mortality in the loamy sand. At high moisture potentials that approached saturation, infection was much higher in the fine sand than in the loamy sand and clay loam. In another study, the infective juveniles of *S. riobrave* showed some degree of positive geotaxis and were detected most frequently in the upper 12.7 cm within a water potential range of -40 to -0.0055 MPa (2 to 14% moisture) (Gouge *et al.,* 2000). The optimum soil moisture for *S. riobrave* is 2 to 4% moisture although this nematode survives from 0.5% to 12% moisture (Duncan *et al.,* 1996). Gouge *et al.* (2000) also demonstrated that *S. riobrave* moved down and aggregated between 15 and 23 cm deep in sand columns that were gradually dehydrating. The nematodes redistributed themselves over time and were found within a water potential range of -0.1 to -0.012 MPa (5.2 to 9.5% moisture).

Infective juveniles of *S. carpocapsae* actively nictate (=body-waving) on the surface of soil particles and invade the last instar of the common cutworm, *S. litura* at 25 to 40% relative humidity. However, infective juveniles did not infect the insects at 10, 15, or 50% relative humidity in spite of the higher rates of persistence of infective juveniles than at 25 to 40% relative humidity (Kondo and Ishibashi, 1985). Infectivity is adversely affected by extreme soil moisture. Too dry or too wet soil affects nematode activity and decreases infectivity of nematodes. Koppenhöfer and Fuzy (2007) reported that infectivity of *H. bacteriophora*, *H. zealandica*, *S. scarabaei*, and *S. glaseri* to the Japanese beetle, *Popillia japonica,* was the highest at moderate soil moisture (-10 to -100kPa), and tended to be lower in wet (-1 kPa) and moderately dry (-1000 kPa) soil.

Consistently, optimum soil moisture provides favorable environment for nematode survival and this induces high nematode infectivity. Accordingly, watering is recommended for effective control of insects when entomopathogenic nematodes are applied in the field. Watering moves the nematodes into the soil closer to their hosts, prevents desiccation, avoids the negative effects of sunlight, and prolongs survival in soil.

8.5.3. Soil Texture

Nematodes must move through soil pore spaces to find their hosts. Soil texture is classified into clay, silt, and sand based on the proportions of each of the soil types. A soil that does not exhibit the dominant physical properties of any of three groups is loam (Miller and Gardiner, 2001). Diameter ranges of soil are 0.25 mm to 2.0 mm in sand, 0.002 mm to 0.05 mm in silt, and <0.002 mm in clay, respectively. Texture is an important characteristic affecting water intake rates, water storage in the soil, and soil aeration which are associated with nematode movement in soil. As stated previously, moisture and soil pore space work together to allow nematode movement and provide suitable conditions for the nematode to infect their host. Sandy soils usually have larger pores, i.e., they are coarse-textured, so there are more channels through which the nematodes can move, providing that adequate moisture is present. In addition, aeration is higher which provides better conditions for nematode survival.

The influence of soil texture on infection of entomopathogenic nematodes has not received as much attention as temperature or moisture. Generally, soil texture affects the infectivity of entomopathogenic nematodes. Entomopathogenic nematode occurrence is correlated significantly, although weakly, with sand and silt but not with clay in turfgrass. High sand and moderate silty soils are positive correlated with nematode presence. *H. bacteriophora* was recovered from sandy loam soils with pH 5.1 – 6.8, whereas *S. carpocapsae* and *S. glaseri* were recovered from silty loam to sandy clay loam soils with pH 4.6 – 6.2. Mráček *et al.* (2005) collected more steinernematids from light soils such as sand/silt soil rather than heavy soils with low or high organic content.

Soil texture influences the successful control of soil-dwelling insects by entomopathogenic nematodes because infective juveniles must move through soil pore space. Size of pore spaces depends on the size of soil particles. Movement is impossible if the pore spaces are too small for the nematodes to squeeze through (Taylor and Sasser, 1978). In terms of soil particles associated with nematode movement, a nematode of given body length and width cannot move between soil particles when the pore diameters are less than the nematode width. The optimum particle size for nematode movement is 1 : 3, the ratio of particle diameter to nematode length (Wallace, 1963). The size of infective juvenile of nematode has a close relationship with soil texture. The lengths of infective juveniles in the genus *Steinernema* ranged from 360 μm in *S. asiaticum* to 1580 μm in *S. arenarium* and those in the genus *Heterorhabditis* ranged from 332 μm in *H. tayseare* to 800 μm in *H. megidis*. In addition, the body diameter of the infective juvenile is not always proportioned to the body length. The range of body diameter is from 11 μm in *S. rarum* to 54 μm in *S. puertoricense* and from 18 μm in *H. downesi* to 32 μm in *H. marelatus* (Nguyen and Hunt, 2007). Therefore, nematode movement with a given soil texture may be different depending on nematode species and is probably more hindered in clay soils because of the smaller pore size.

Soils with high sand content are more favorable to the motility and survival of entomopathogenic nematodes than soils with high clay content (Barbercheck and Kaya, 1991a; Kung *et al.,* 1990a). Thus, *H. bacteriophora* infection towards an insect host was significantly greater in humus and sand than in loam and clay (Choo and Kaya, 1991). This nematode infection of hosts is probably influenced by soil pore size as the largest pores occur in humus, followed by sand, loam, and clay.

In the four different soil type (pure silica sand, sandy loam, silty clay loam, and clay), vertical migration of nematode decreased as the percentage of clay and silt increased (Georgis and Poinar, 1983a; 1983b). The greatest dispersal occurred in pure silica sand and coarse sandy loam. *S. carpocapsae* remained within 2 cm of the surface, but some penetrated to a depth of 10 cm in pure silica sand and coarse sandy loam. In addition, *S. carpocapsae* movement was least in clay soil and limited in silty clay loam soil. Migration of infective juveniles tends to decrease as the proportion of silt and clay increase in soil.

In using entomopathogenic nematodes as biopesticides against soil insect pests, soil texture needs to be considered. Assuming that the insect pests are susceptible to the nematodes, one of the best situations for their use are in nurseries where the sandy soil or soil made of bark and other plant sources are used. Other situations include turfgrass, especially in golf courses such as the greens where sand is used to grow the grass or where the soil is naturally sandy (Susurluk, 2006).

8.5.4. Ultraviolet Light

Ultraviolet light (uv) from the sun for naturally occurring entomopathogenic nematodes are not a concern as they are protected from that danger in the soil. Uv light has been shown to be detrimental to the infective juveniles of *S. carpocapsae* (Gaugler and Boush, 1978). To minimize the effects of uv light, nematodes applied as biopesticides during sunny days can be watered in immediately after application.

8.5.5. Agrochemicals

Commercially produced entomopathogenic nematodes are often applied to systems/substrates that are regularly treated with many other agents, including chemical or biorational pesticides, soil amendments, and fertilizers. These nematodes appear to be compatible with many, but not all, herbicides, fungicides, acaricides, insecticides, and even some nematicides (Rovesti and Deseö, 1990; Ishibashi, 1993; Georgis and Kaya, 1998). They are also compatible with azadirachtin (Stark, 1996), and pesticidal soap (Kaya *et al.,* 1995). Negative effects of various pesticides on the infective juveniles have been also documented (*e.g.*, Rovesti and Deseö, 1990; Patel and Wright, 1996). On the other hand, synergistic or additive interaction between entomopathogenic nematodes and other control agents has been observed for various chemical insecticides (*e.g.*, Nishimatsu and Jackson, 1998; Koppenhöfer *et al.,* 2000a, b, c, 2002; Koppenhöfer and Fuzy, 2008). In view of the diversity of available chemical and biorational insecticides, a generalization on pesticide-nematode compatibility cannot be made. Moreover, newer chemical pesticides with different mode of actions are being manufactured. Therefore, the compatibility of each chemical pesticide and nematode species should be evaluated on a case-by-case basis (Susurluk, 2008).

8.6. BIOTIC FACTORS

8.6.1. Natural Enemies

Nematodes have their own suite of natural enemies including microorganisms and invertebrate predators (Poinar and Jansson, 1988a, b). Kaya and Koppenhöfer (1996) and Kaya (2002) have reviewed the various biotic factors (*i.e.*, natural enemies) that affect nematode survival. Entomopathogenic nematodes have been found infected with microsporidian parasites (Poinar, 1988). *S. glaseri* from a cerambycid beetle was infected with a microsporidium. Recently, several field populations of the well-studied free-living nematode, *Caenorhabditis elegans*, was found infected with microsporidian parasites (Troemel *et al.,* 2008) suggesting that these parasites may be common in nematodes in general.

Nematophagous fungi have long been considered as potential biological control agents of plant-parasitic nematodes. These fungi have also been shown to be "predators" or pathogens of entomopathogenic nematodes. These fungi include the trapping fungi such as *Monacrosporium* spp., *Arthrobotrys* spp. (Poinar and Jansson, 1986; Fowler and Garcia,

1989; Jaffee *et al.,* 1996; Kaya and Koppenhöfer, 1996; Koppenhöfer *et al.,* 1996; Kaya 2002,; El-Borai *et al.,* 2009) and *Gamsylella gephyropaga* (El-Borai *et al.,* 2009) that can capture and consume entomopathogenic nematodes, and the endoparasitic fungi, *Hirsutella rhossiliensis*, *Catenaria* sp., and *Myzocytium* sp., can infect these nematodes (El-Borai *et al.,* 2009). Timper *et al.* (1991) found that more conidia of *H. rhossiliensis* adhered to the cuticle *S. glaseri* and *H. bacteriophora* compared to the cuticle of *S. carpocapsae*. The cruiser, *S. glaseri*, was more susceptible to the fungus because of its inability to retain the second-stage cuticle and its foraging activity in soil allowing it to come into contact with more conidia. Although *H. bacteriophora* is also a cruiser, it retains its second-stage cuticle making it refractory to fungal infection (Timper and Kaya, 1989; Timper *et al.,* 1991). Differential susceptibility of entomopathogenic nematodes to nematophagous fungi was also observed in the citrus soil under laboratory conditions by El-Borai *et al.* (2009). The nematode-trapping fungus, *G. gephyropaga*, reduced 56 - 92% of the five nematode species tested including *S. diaprepesi*, *S. glaseri*, *S. riobrave*, *H. zealandica* and *H. indica*, whereas three different species of *Arthrobotrys* did not affect numbers of *S. diaprepesi* and *S. glaseri*. However, *Arthrobotrys musiformis* reduced numbers of *S. riobrave*, *H. zealandica* and *H. indica*, and *A. oligospora* reduced numbers of *S. riobrave* and *H. zealandica*. Both endoparasitic fungi, *Catenaria* sp., and *Myzocytium* sp., reduced all tested nematode species except *H. indica* by 82%. El-Borai *et al.* (2009) concluded that the entomopathogenic fungi may play a role in the abundance and diversity of the naturally occurring entomopathogenic nematodes in citrus soil which in turn regulates the abundance of the citrus root weevil, *Diaprepes abbreviatus* a major pest of citrus roots in Florida, USA.

Entomopathogenic nematodes are a minor part of the soil community, but they can occur in high numbers near a nematode-killed insect. Jaffee and Strong (2005) showed that there is a numerical response of nematophagous fungi near an insect cadaver releasing infective juveniles into the soil. Adding animal manure mulch, especially of chicken, to soil can decrease the prevalence of the nematode-trapping fungi (Duncan *et al.,* 2007). These mulched soils were augmented with applications of *S. riobrave*. Interestingly, the mulched soils showed an increase in free-living bacterial feeding nematodes which reduced the presence of *S. riobrave* because the cadavers containing *S. riobrave* were invaded by the free-living bacterial feeding nematodes. *S. riobrave* could not effectively recycle in the presence of the bacterial feeding nematodes (Duncan *et al.,* 2003a).

Augmenting raw citrus soil with *S. diaprepesi*, *S. riobrave*, *H. zealandica* or no entomopathogenic nematodes and *S. diaprepesi* was added 5 days later, the survival of both *S. diaprepesi* and total of other entomopathogenic nematode species was greater in soil that received no pretreatment than soil pretreated with *S. riobrave* (El-Borai *et al.,* 2007). On the other hand, pretreatment with *H. zealandica* or *S. diaprepesi* had no effect on *S. diaprepesi* or total entomopathogenic nematode numbers. Placing the infective juveniles of *S. diaprepesi* on water agar showed a higher *S. diaprepesi* mortality by endoparasitic and nematode trapping fungi in soils pretreated with steinernematids than in soils with no nematode treatment or with *H. zealandica*. The data suggest that augmenting soils with entomopathogenic nematodes, especially *Steinernema* species, have the potential to reduce natural populations of other entomopathogenic nematode species by nematophagous fungi.

8.6.2. Competition

Competition, defined as a mutually negative interaction between organisms of the same species (intraspecific) or different species (interspecific) that does not involve predation and results in lowered fitness of one by the presence of another (Pianka, 1999; Wootton, 1994; Stuart *et al.,* 2006). When coexistence between species occurs, they may share limiting resources resulting in competition. To avoid competition, the coexisting species may use different hosts, be separated spatially in the same habitat, have different foraging strategies, or have different times (i.e., temporal) of activity.

8.6.2.1. Coexistence of Nematode Species

A number of field studies have shown that multiple species of entomopathogenic nematode species from a single site can occur (Akhurst and Brooks, 1984; Duncan *et al.,* 2003b; Stuart and Gaugler, 1994). Coexistence occurs when behavioral differences such as foraging strategies and variability in environmental factors enable niche separation and avoidance of competition. This type of coexistence was documented in laboratory and greenhouse studies by Koppenhöfer and Kaya (1996a, b). Differences in foraging behaviors apparently reduced competition for hosts between two entomopathogenic nematode species, a cruiser and an ambusher, which permitted coexistence. In field studies conducted in turfgrass, Campbell *et al.* (1996) demonstrated that natural nematode spatial separation occurred between the cruiser, *H. bacteriophora*, and the ambusher, *S. carpocapsae*. With three nematode species coexisting in a corn field, Millar and Barbercheck (2001) showed that the native population of *H. bacteriophora* was found deepest in the soil profile, the native *S. carpocapsae* remained near the soil surface, and the introduced *S. riobrave*, an intermediate foraging strategist, occurred at intermediate depths.

In a bush lupine plant ecosystem, two entomopathogenic nematode species, *H. marelatus* and *S. feltiae* coexist (Gruner *et al.,* 2009). When both nematode species were provided with same the naive host (*G. mellonella*), *S. feltiae* dominated over *H. marelatus* confirming earlier work by Alatorre-Rosas and Kaya (1990, 1991). When *S. feltiae* and *H. marelatus* were provided to the native, root- and stem-boring ghost moth caterpillar, *Hepialus californicus*, *S. feltiae* infected this host at a low frequency and showed a lower reproductive fitness than *H. marelatus* (Gruner *et al.,* 2009). The researchers hypothesized that host resistance to *S. feltiae* may provide a mechanism for coexistence of both nematode species. The high field prevalence and rapid life cycle of *S. feltiae* suggest that it uses abundant, small-bodied hosts and indicates a lack of direct competition with *H. marelatus* in the *Hepialus*-bush lupine trophic cascade (see trophic cascade below).

8.6.2.2. Interspecific Competition among Entomopathogenic Nematodes

Interspecific competition can be direct meaning that competition entails direct interference between two species or indirect meaning that competition is due to a wide range of effects which are mediated by the presence of a third species or by a change in the chemical or physical environment (Pianka, 1999; Wootton, 1994; Stuart *et al.,* 2006). Moreover, indirect competition may be positive or negative. There are at least five simple types of indirect effects that have been demonstrated in ecological communities (Wootton, 1994) including (1) exploitative competition, (2) trophic cascade, (3) apparent competition,

(4) indirect mutualism and commensalism, and (5) higher order interactions. In each type of indirect effects, we will use nematode examples.

In exploitative competition, one nematode species indirectly reduces a second nematode species by diminishing the availability of a shared resource. Generally, an insect host is rarely infected by more than one species of entomopathogenic nematode. In laboratory experiments, however, Kondo (1989) showed that *Spodoptera litura* co-infected with *S. glaseri* and *S. feltiae* produced mixed progeny. Similarly, when *S. carpocapsae* and *S. glaseri* co-infect the same host, both nematode species produced progeny, but *S. glaseri* was less negatively affected compared to *S. carpocapsae* (Koppenhöfer *et al.,* 1995b). The reason for this result may be that *S. glaseri* has a faster development and has the ability to use the bacterial symbiont of *S. carpocapsae*. On the other hand, interference when a heterorhabditid and steinernematid species co-infected the same host at the same time, the steinernematid developed and the heterorhabditid did not (Alatorre-Rosas and Kaya, 1990, 1991). Because of the involvement of the two symbiotic bacterial species, *Xenorhabdus* and *Photorhabdus*, there is an indirect effect in which one bacterial species indirectly reduces a second bacterial species by reducing the abundance of a shared resource. Thus, *Xenorhabdus* species are known to produce bacteriocins that kill *Photorhabdus* species (Boemare, 2002).

In a trophic cascade, there is an indirect effect mediated through a series of consumer-resource interactions. That is, trophic cascades "are predator-prey interactions that indirectly alter the abundance, biomass or productivity of a community across more than one trophic link in a food web" (Denno *et al.,* 2008). Denno *et al.* (2008) conducted a meta-analysis (a statistical method that combines results from independently conducted experiments) of 35 published studies to test the effect of entomopathogenic nematodes on lower trophic levels. A key element in meta-analysis is to examine "effect size" which provides information about how much change is evident across all studies and for subsets of studies. Thus, this analysis looked at effect sizes to quantitatively assess the impact of these nematodes on herbivore/pest density or mortality, herbivore damage or plant growth biomass, survival or yield and the strength of the correlation of these two factors. This meta-analytical approach integrated the role that entomopathogenic nematodes play as drivers in food-web dynamics and biological control. Denno *et al.* (2008) found that effect sizes for insect hosts as the result of application of entomopathogenic nematodes were consistently negative and indirect effects on plants were consistently positive. The positive impact on plant responses increased as the negative effect of the nematodes on the insect host increased which provides strong support for the mechanism of trophic cascades.

In a specific example of trophic cascade, the entomopathogenic nematode, *H. marelatus,* is dynamically linked with populations of the root- and stem-boring ghost moth caterpillars, *H. californicus,* and its bush lupine host plant (Strong, 2002; Strong *et al.,* 1996, 1999; Preisser, 2003; Preisser *et al.,* 2006; Ram *et al.,* 2008). *H. californicus* caterpillars cause heavy root damage by their boring activity and can kill bush lupines, but *H. marelatus* causes high mortality of the ghost moth caterpillars, and the spatial distribution of *H. marelatus* is positively correlated with long-term fluctuations in the local distribution of lupines (Preisser *et al.,* 2006). By also protecting bush lupine, *H. marelatus* may mediate additional positive effects in the ecosystem (Preisser, 2003).

In apparent competition, two prey species share a common natural enemy with an increase in one prey resulting in an increase in the natural enemy and a decline in the second prey (Stuart *et al.,* 2006). When one entomopathogenic nematode species occurs naturally at a site and another entomopathogenic nematode species is applied, the overall abundance of these nematodes may cause a numerical response, for example, in predatory mites resulting in a reduction of both nematode species. The best example of apparent competition is with some soil mites where a numerical response occurred when fed entomopathogenic nematodes (Walter *et al.,* 1986).

With indirect mutualism and commensalism, there are no concrete examples with entomopathogenic nematodes but can be inferred from either exploitative (indirect) or interference (direct) competition. For instance, certain ant species might preferentially prey on steinernematid-killed insects compared to heterorhabditid-killed ones. This preferential predation on a competitive dominant nematode species would allow an inferior sympatric competitor to increase in abundance (Baur *et al.,* 1998; Zhou *et al.,* 2002).

Lastly, in higher order interactions, non-additive effects may occur between groups of species or individuals. That is, one species may modify the interaction between two other species. As with direct mutualism and commensalism, there are no good examples of this occurring with entomopathogenic nematodes. We can speculate that food plants may modify the susceptibility of an insect herbivore to one entomopathogenic nematode species but not to another (Barbercheck, 1993; Barbercheck *et al.,* 1995). In this scenario, nematode progeny production was highest in the rootworm larvae, *Diabrotica undecimpunctata howardi,* fed corn, followed by peanut, and then squash. Nematode progeny production was lower in bitter squash compared to non-bitter squash, and this result was attributed the effects by cucurbitacians and plant primary metabolites (Barbercheck *et al.,* 1995). It is possible that one nematode species could develop better in insect hosts that fed on bitter squash compared to another nematode species, giving the first nematode species a competitive edge.

Interspecific competition among entomopathogenic species is of particular interest because of the potential effects that applications of commercial non-indigenous (exotic) or indigenous (native) nematode species might have on the native nematode communities. These native entomopathogenic nematodes are widespread in nature in both natural and managed habitats and are considered to be a major natural mortality factor in the soil (Hominick, 2002). However, applications of large numbers of either non-indigenous or indigenous nematodes may have far reaching ecological consequences (Dillon *et al.,* 2008) as has been observed in a citrus grove (Duncan *et al.,* 2003a, b). Application of an exotic nematode, *S. riobrave,* provided levels of control of the citrus root weevil, *D. abbreviatus,* higher than those caused by the native nematodes (*S. diaprepesi, H. bacteriophora, H. indica,* and *H. zealandica*) in untreated plots but only during months of treatment while providing less control during non-treatment months (Duncan *et al.,* 2003b). *S. riobrave* partially displaced the native nematodes, but it reproduced poorly in the weevil host, partly because a native bacterial-feeding nematode, *Pellioditis* sp., also invaded the cadaver (Duncan *et al.,* 2003a). Whether this phenomenon occurs in other situations with other nematodes, insect hosts, and soil or plant communities remains to be studied.

8.6.2.3. Intraspecific Competition

For successful entomopathogenic nematode infection to occur, a certain number of infective juveniles are needed to overcome the host defenses (Gaugler *et al.,* 1994; Wang *et al.,* 1994) and to guarantee mating for steinernematids (Kaya and Gaugler, 1993). If too many infective juveniles infect a host (superinfection), nematode development, survival, and reproduction are adversely affected (Selvan *et al.,* 1993; Zervos *et al.,* 1991). Can superinfection be avoided because a large number of infective juveniles emerge from a single nematode-killed host? If a susceptible host is available when the infective juveniles emerge from the cadaver, superinfection could occur. Various strategies could have evolved to regulate the dispersal and infectivity of emerging nematodes to minimize intraspecific competition and inbreeding. These strategies include: (1) at some point after death of a nematode-killed host, the cadaver containing the nematodes is no longer attractive to other infective juveniles (Glazer, 1997); (2) infective juveniles show staggered patterns of infectivity (Bohan and Hominick, 1995, 1996, 1997; Hominick and Reid, 1990; Kaya and Koppenhöfer, 1996), but infective juveniles of some steinernematid species are all infectious at the same time (Campbell *et al.,* 1999); (3) some heterorhabditid species seem to show phase infectivity where all the infective juveniles are not infectious at the same time; and (4) later emerging nematodes are more mobile and less responsive to host cues than early emerging nematodes (Lewis and Gaugler, 1994), suggesting that later emerging nematodes might be adapted for greater dispersal than early emerging nematodes. However, Fujimoto *et al.* (2007) demonstrated that a significantly greater proportion of early-emerging *S. carpocapsae* infective juveniles from a cadaver dispersed compared with late-emerging infective juveniles from a cadaver. Interestingly, infective juveniles that dispersed were less infectious that those that did not disperse.

By examining the early events of the infection process, determination of how intraspecific competition can be minimized can be assessed. The first infective juveniles to successfully invade a host insect and reach the adult stage are likely to have the advantage in competing for mates and contributing to the overall reproductive output from the cadaver than are late arriving infective juveniles that infect the host. Yet, the early invading infective juveniles are probably at risk since the early invaders could suffer high mortality from host defenses (Gaugler *et al.,* 1994; Peters and Ehlers, 1994; Wang *et al.,* 1994). Fushing *et al.* (2008) using a mathematical model and biological data with *S. feltiae* determined that there are "risk-prone" and "risk-averse" infective juveniles. The "risk-averse" individuals only infect after the "risk-prone" individuals have infected the hosts.

Experiments indicate that there are optimal invasion times for the infective juveniles to infect a host. Initial infections by *S. feltiae* facilitate subsequent infections (Hay and Fenlon, 1997) providing further basis for the mathematical model by Fushing *et al.* (2008). Once the insect host is infected, the infective juveniles do not continue to invade the host. Rather, in ongoing infections by *S. carpocapsae, S. feltiae* and *S. riobrave,* there is a release of a chemical(s) that deters further infection (Fairbairn *et al.,* 2000; Glazer, 1997). It is unclear, however, how constrained optimal invasion times might be regulated. With *S. glaseri,* Stuart *et al.* (1998) found that the infective juveniles invaded the insect host, *G. mellonella,* for up to at least 14 h after the first nematode penetrated. These data suggest that the optimal invasion intervals could be quite plastic and depends on a number of factors including the rate of host invasion and dynamics of the interaction between a nematode species and its hosts.

8.6.2.4. Competition with Organisms Other than Entomopathogenic Nematodes
The nematode-killed insect hosts and infective juveniles of entomopathogenic nematodes occur in the soil, and therefore, compete with other organisms that live in this habitat. These include other nematode species such as free-living bacterivorous, fungivorous, predatory and omnivorous nematodes that are important components of decomposition and nutrient cycling food webs (Ferris *et al.,* 2004). Duncan *et al.* (2003a) studied the interactions with the

application of *S. riobrave* to the native S. *diaprepesi* and a native, free-living bacterial feeding nematode, *Pellioditis* sp., and the mortality of the citrus root weevil larvae in a citrus grove. The application of *S. riobrave* increased the number of *Pellioditis* sp. that developed in *S. riobrave*-killed weevils, reducing the number of *S. riobrave* infective juvenile progeny. There was no detectable interaction between *Pellioditis* sp. and *S. diaprepesi*. Thus, the native bacterial feeding nematode had a negative effect on *S. riobrave* by competing for the same resource. Application of *S. carpocapsae* or *S. glaseri* to soil caused a temporary increase in predatory and free-living rhabditid nematodes (Ishibashi and Kondo, 1986, 1987), but this is not always the case, because Grewal *et al.* (1997) found no effects of application of *S. carpocapsae* or *S. glaseri* on free-living nematodes. Entomopathogenic nematodes can compete with plant-parasitic nematodes such as the root-knot nematode complex and can reduce their populations and associated plant damage (Ishibashi and Kondo, 1986; Jagdale *et al.*, 2002; Lewis *et al.*, 2001; Somasekhar *et al.*, 2002; Fallon *et al.*, 2002, 2004).

Entomopathogenic nematodes may compete with other entomopathogens for resources. The nematodes will infect and develop in nucleopolyhdrovirus-infected insects, but the insect cadavers have a fragile integument that easily ruptures (Kaya and Burlando, 1989). When the integument ruptures, it exposes the developing nematodes to the environment resulting in reduced infective juvenile progeny production (Kaya and Burlando, 1989). In contrast, when the nematode infects a granulovirus-infected host, the integument is not fragile, but the infective juvenile progeny production is less compared to an uninfected host (Kaya and Brayton, 1978). It may be the granulovirus-infected insect has fewer resources than an uninfected insect. The infective juvenile from virus-infected insects will sequester the viral occlusion bodies which remain infectious to other insects (Kaya and Brayton, 1978; Kaya and Burlando, 1989).

S. carpocapsae can infect *Bacillus thuringiensis*-(BT)-infected insects, and the development of the nematode is normal if the nematode is given first and BT is given 24 hours later (Kaya and Burlando, 1989; Poinar *et al.,* 1990). The development of the nematode is poor or non-existent and no infective juvenile progeny are produced if BT is given to the insects 24 hours prior to being exposed to the nematode (Kaya and Burlando, 1989). If nematode progeny are produced in BT-infected insects, they are smaller and have less food reserves compared with progeny from a non-BT infected insect (Poinar *et al.,* 1990). Apparently, when BT is given first followed by the nematode, BT prevents the symbiotic bacterium from developing. On the other hand, combinations of entomopathogenic nematodes and BT can have an additive or synergistic effect increasing the mortality of scarab grubs (Koppenhöfer *et al.,* 1999). With the milky disease bacterium, *Paenibacillus* (= *Bacillus*) *popilliae* in the scarab, *Cyclocephala hirta*, Thurston *et al.* (1993, 1994a) showed that the scarab larvae infected with the bacterium were more susceptible to *H. bacteriophora* and *S. glaseri* than uninfected scarabs. The presence of *P. popilliae* in the scarab did not hinder nematode reproduction and the *P. popilliae* spores survived the nematode/bacterium complex invasion and were available to infect other scarab larvae.

Armer *et al.* (2004) demonstrated that *H. marelatus* and its bacterial symbiont, *Photorhabdus temperata*, killed the Colorado potato beetle, *Leptinotarsa decemlineata*, but the nematode was unable to complete its life cycle. Blackburn *et al.* (2007, 2008) conducted research to show that enteric bacteria in the genera, *Pantoea, Enterobacter, Pseudomonas, Serratia,* and *Bacillus* that were isolated from the Colorado potato beetle inhibited the growth of *P. temperata* which adversely affected the development of the nematode. This type of natural, bacterial interaction may be prevalent with other insect hosts where the symbiotic bacterium is prevented from developing and ultimately affecting the development of the nematode.

Entomopathogenic nematodes may compete with entomopathogenic fungi such as *Beauveria bassiana* and *Metarhizium anisopliae* as they commonly infect soil insects (Tanada and Kaya, 1993). With entomopathogenic fungus, *B. bassiana*, Barbercheck and Kaya (1990, 1991b) showed that when *S. carpocapsae* or *H. bacteriophora* was applied simultaneously with the fungus, the nematodes developed normally. Lezama-Gutiérrez *et al.* (1996) found similar results with *H. bacteriophora* and the fungus, *Nomuraea rileyi*. However, when the insects were co-infected with *B. bassiana* and the nematode, the nematode usually out-competed the fungus but this result was influenced by temperature and the relative time of infection (Barbercheck and Kaya, 1990, 1991b). For example, if *B. bassiana* is given a head start of 3 - 4 days at 30°C, 1 - 2 days at 22°C and 1 day at 15°C, *B. bassiana* will exclude the nematode. The data suggest that the fungus out-competed the nematode for the same insect resource. Interestingly, when the nematode was given a choice between uninfected and fungal-infected hosts in the soil environment, the nematodes tended to avoid the fungal-infected host suggesting that this nematode behavior minimized the antagonistic interaction with fungus (Barbercheck and Kaya, 1991b). On the other hand, in a laboratory study *M. anisopliae* can act synergistically with entomopathogenic nematodes against the barley chafer, *Coptognathus curtipennis,* providing that larvae are exposed to the fungus 3 weeks prior to the nematode (Anbesse *et al.,* 2008). In a field study, the fungus *B. brongniartii* plus *S. carpocapsae* significantly enhanced the efficacy over the application of *B. brongniartii* alone against oriental beetle, *Exomala orientalis* (Choo *et al.,* 2002b).

8.7. PERSISTENCE AND RECYCLING

8.7.1. Persistence

The persistence of applied infective juveniles in the field is generally short-lived (Smits, 1996). Baur and Kaya (2001) summarized much of the nematode persistent data available up to 1996. A number of abiotic factors such as extreme temperatures (especially at the soil surface), lack of soil moisture, osmotic stress, soil texture, relative humidity, and uv radiation and biotic factors such as antibiosis, competition, and natural enemies (Glazer, 1996; Smits, 1996; Baur and Kaya, 2001; Kaya and Koppenhöfer, 1996; Kaya, 2002; Susurluk and Ehlers, 2008) have been given as the primary extrinsic reasons for poor infective juvenile survival. Even though 100,000 to 350,000 infective juveniles can emerge from a single waxworm cadaver (Dutky *et al.,* 1964), naturally occurring entomopathogenic nematodes are usually at a low level suggesting that few infective juveniles survive.

Different species and/or strains of entomopathogenic nematodes can survive different temperature regimes with some species surviving temperatures up to 40°C (Grewal *et al.,* 1994a). Extended exposure to temperature extremes (below 0°C or above 40°C) is lethal to most species of entomopathogenic nematodes (Brown and Gaugler, 1996), but in the soil environment, infective juveniles are normally buffered from temperature extremes. Although uv can kill nematodes within minutes (Gaugler and Boush, 1978), it is not a major mortality except when the infective juveniles are applied as biological control agents to surfaces exposed to sunlight.

Soil texture affects infective juvenile survival with the poorest occurring in clay soils (at the same water potentials). The poor survival rate in clay soils is probably due to the lower oxygen levels in the smaller soil pores (Kaya, 1990). Oxygen is also a limiting factor in water-saturated soils and soils with high organic matter content. However, Koppenhöfer and Fuzy (2006) studied persistence of *S. scarabaei*, *S. glaseri*, *H. zealandica* and *H. bacteriophora* in seven different soil types and found that both *Heterorhabditis* species and *S. glaseri* had the shortest persistence in the potting soil (highly organic) and no clear differences among the six other soil types. Persistence of *S. scarabaei* was high in all soil types but over time, its persistence did decline in clay loam soil. They point out that generalizations on nematode persistence in different soil types must be made carefully because other variables may affect the results.

Soil salinity and pH have a negligible effect on infective juvenile survival. Infective juveniles can survive even at salinities above the tolerance levels of most crop plants (Thurston   *et al.,* 1994b). Seawater has no negative effects on survival of several *Heterorhabditis* species/strains (Griffin   *et al.,* 1994) as they have been frequently isolated from soils near the coastal areas.

Perhaps, the most critical factor in entomopathogenic nematode survival in soil in the absence of a host is moisture. A number of published reports state that soil moisture is a key factor for nematode survival (Koppenhöfer *et al.* 1995a; Koppenhöfer and Fuzy, 2007; Grant and Villiani, 2003; Preisser   *et al.,* 2005, 2006). In one large-scale laboratory study, Shapiro-Ilan *et al.* (2006) demonstrated that different species and strains of entomopathogenic nematodes have different longevity when held under the same conditions. Nematode species with the greatest survival over 42 to 56 days were *S. carpocapsae* Sal strain and *S. diaprepesi*, followed by *H. bacteriophora* Lewiston strain, *H. megidis*, *S. feltiae*, and *S. riobrave* 3 - 3 and 355 strains, and then by *H. bacteriophora* Hb, HP88 and Oswego strains, *S. glaseri*, and *S. rarum*. The poorest survival occurred with *H. marelatus* and *H. mexicana*. Koppenhöfer and Fuzy (2007) found that *H. zealandica* and *H. bacteriophora* persisted poorly at -10 kPa in pasteurized soil and persisted the best at -3000 kPa after 56 days. *S. glaseri* persistence was also lowest at -10 kPa but it did not differ at -100 to -3000 kPa, and interestingly, *S. scarabaei* persistence was not affected by soil moisture over the 140 day exposure period.

Preisser *et al.* (2005) conducted a 1-year field study with *H. marelatus* infective juveniles. The experiment was done with unpasterurized soil moistened with 0.2 ml water/g of soil placed in 50 ml plastic centrifuge tubes with the bottom of the removed and covered with a plastic mesh. The tubes were placed in the coastal grassland and destructively sampled during the course of the year, baited with wax moth larvae and examined for *H. marelatus* infection. The results showed that some *H. marelatus* can survive in the soil for up to one year. In a follow up study, Preisser *et al.* (2006) showed that dry-season survival of *H. marelatus* infective juveniles occurred because of their proximity to the moist soil around the taproots of the bush lupine. Moreover, the ghost moth larvae which bore into the roots and stems of the lupine bush are the host for *H. marelatus*. Infection of the ghost moth larvae provide a strong top-down effect on lupine survival, and this action enables a below ground trophic cascade. In comparison, infective juveniles went to extinct in the grasslands during the dry season.

The mechanism for infective juvenile survival under low moisture conditions can be accomplished by lowering the rate of metabolism. Soil usually has a gradual drying regime and under these conditions, the infective juveniles can adapt to the desiccating conditions (Patel *et al.,* 1997; Solomon *et al.,* 1999). They may also survive desiccating conditions by remaining with the host cadaver until the soil moisture situation improves (Brown and Gaugler, 1997). The gradual removal of water from the infective juveniles lowers their rate of metabolism.

8.7.2. Recycling

Recycling of entomopathogenic nematodes occurs in their insect hosts, but only few studies have examined the dynamics of nematode populations and the factors affecting them. Obviously, these nematodes do recycle as they occur naturally in many soils throughout the world (Hominick, 2002), and long-term persistence occurs in the presence of suitable hosts (Klein and Georgis, 1992; Parkman *et al.,* 1993; Shields *et al.,* 1999; Koppenhöfer and Fuzy, 2009; McGraw and Koppenhöfer, 2009). A number of abiotic and biotic factors play a key role in recycling including seasonal fluctuations, foraging strategy, host population dynamics, alternate hosts, etc. (Campbell *et al.,* 1996, 1998; Strong *et al.,* 1996: Stuart and Gaugler, 1994). McGraw and Koppenhöfer (2009) showed that a distinct seasonality occurred with high levels of *S. carpocapsae* and *H. bacteriophora* appearing in soil at a golf course following high densities of first generation of the annual bluegrass weevil, *Listronotus maculicollis.*

Recycling is desirable attribute after an application of entomopathogenic nematodes because it can provide additional and prolonged control of an insect pest. Because they are obligate pathogens, a key to recycling of the nematodes is the availability of suitable hosts. Recycling is more apt to occur after nematode application if the pest or pest complex is present throughout much of the year, the pest has a high economic threshold and is moderately susceptible to the nematode, and soil conditions are favorable for nematode persistence (Kaya, 1990). Although recycling is rather common (Kaya, 1990) after nematode application, it is usually not sufficient for prolonged host suppression and the nematode may have to be reapplied to maintain adequate control of soil insect pests.

8.8. POPULATION ECOLOGY

Research into various aspects of the biology of entomopathogenic nematodes has moved forward at an increasingly rapid rate. We have a better understanding of their foraging behavior, diversity and distribution, and abiotic and biotic factors that affect their survival and can produce them on a large scale in vitro for commercial use (Susurluk, 2009). Their importance as biological control agents of soil insect pests in natural and cultivated ecosystems has gained recognition. However, an area that has lagged behind is the investigations into the structure and dynamics of their population biology and ecology. There is a lot of interest in the population ecology of these nematodes, but the inherent difficulties of sampling, manipulating, and studying organisms in the soil environment have affected the fundamental progress in this area (Brown and Gange, 1990). In this section, we address the spatial distributions of populations and the various biotic and abiotic factors that influence these distributions.

CONCLUSION

Since the 1980s, great strides have been made in understanding the ecology of entomopathogenic nematodes. Most studies were focused in the laboratory and examined individuals or a small group of nematodes as to their response to abiotic factors such as temperature, moisture, soil types, etc. in terms of survival, dispersal, and infectivity. Major accomplishments have been achieved in understanding infection ecology. Such aspects as foraging behavior and determining their response to host cues were elucidated. A few laboratory studies examined the impact of natural enemies such as nematophagous fungi and predaceous mites of these nematodes. Field studies were focused on biogeography and their diversity and distribution in soil. However, only a few nematode species have been studied in depth and there is still much to learn about the ecology of other species.

We noted that there is more intensity in studying the population dynamics of these nematodes. Our knowledge base in population ecology has only scratched the surface. Scientists are being to study natural enemy complexes in the field and looking more at the population level rather than at individuals. The impact of various abiotic factors, especially soil moisture, is being examined. However, the soil is not an easy medium in which to do long-term field studies.

An area in which very little information exists is the epizootiology of nematode diseases. Nematode epizootics do occur in soil (Kaya, 1987), but data to understand the dynamics associated with these epizootics are lacking. As molecular tools are developed, perhaps more population ecology studies will be initiated such that these nematodes can be not only be used in biological control programs, but provide a fundamental understanding of population ecology of these important mortality factors of soil insects.

ACKNOWLEDGMENT

We thank Dr. Harry K. Kaya, University of California, Davis for his comments on the initial draft of the manuscript and for providing some of the recent references associated with entomopathogenic nematode ecology.

REFERENCES

Akhurst, R.J. and Bedding, R.A. 1986. Natural occurrence of insect pathogenic nematodes (Steinernematidae and Heterorhabditidae) in soil of Australia. *Journal of the Australian Entomological Society*, 25: 241 – 243.

Akhurst, R.J. and Brooks, W.M. 1984. The distribution of entomophilic nematodes (Heterorhabditidae and Steinernematidae) in North Carolina. *Journal of Invertebrate Pathology*, 44: 140 – 145.

Alatorre-Rosas, R. and Kaya, H.K. 1990. Interspecific competition between entomopathogenic nematodes of the genera *Heterorhabditis* and *Steinernema* for an insect host in sand. *Journal of Invertebrate Pathology*, 55: 179 – 188.

Alatorre-Rosas, R. and Kaya, H.K. 1991. Interaction between two entomopathogenic nematode species in the same host. *Journal of Invertebrate Pathology*, 57: 1 – 6.

Amarasinghe, L.D., Hominick, W.M., Briscoe, B.R., and Reid, A.P. 1994. Occurrence and distribution of entomopathogenic nematodes (Rhabditida: Heterorhabditidae and Steinernematidae) in Sri Lanka. *Journal of Helminthology*, 68: 277 - 286.

Anbesse, S.A., Adge, B.J., and Gebru, W. M. 2008. Laboratory screening for virulent entomopathogenic nematodes (*Heterorhabditis bacteriophora* and *Steinernema yirgalemense*) and fungi (*Metarhizium anisopliae* and *Beauveria bassiana*) and assessment of possible synergistic effects of combined use against grubs of the barley chafer *Coptognathus curtipennis*. *Nematology*, 10: 701 – 709.

Ansari, M.A., Shah, F.A. and Butt, T.M. 2010. The entomopathogenic nematode *Steinernema kraussei* and *Metarhizium anisopliae* work synergistically in controlling overwintering larvae of the black vine weevil, *Otiorhynchus sulcatus*, in strawberry growbags. *Biocontrol Science and Technology*, 20(1): 99-105.

Armer, C.A., Rao, S., Berry, R.E. 2004. Insect cellular and chemical limitations to pathogen development: the Colorado potato beetle, the nematode *Heterorhabditis marelata* and its symbiotic bacteria. *Journal of Invertebrate Pathology*, 87: 114 – 122.

Aydin, H. and Susurluk, A. 2005. Competitive Abilities of Entomopathogenic Nematodes, *Steinernema feltiae* and *Heterorhabditis bacteriophora* in the Same Host at Different Temperatures. *Turk. J. of Biol.* 29 (1): 35-39.

Barbercheck, M.E. 1992. Effect of soil physical factors on biological control agents of soil insect pests. *Florida Entomologist*, 75: 539 – 548.

Barbercheck, M.E. 1993. Tritrophic level effects on entomopathogenic nematodes. *Environmental Entomology*, 22: 1166 – 1171.

Barbercheck, M.E. and Kaya, H.K. 1990. Interactions between *Beauveria bassiana* and the entomogenous nematodes, *Steinernema feltiae* and *Heterorhabditis heliothidis*. *Journal of Invertebrate Pathology*, 55: 225 – 234.

Barbercheck, M.E. and Kaya, H.K. 1991a. Effect of host condition and soil texture on host finding by the entomogenous nematodes *Heterorhabditis bacteriophora* (Rhabditida: Heterorhabditidae) and *Steinernema carpocapsae* (Rhabditida: Steinernematidae). *Environmental Entomology*, 20: 582 - 589.

Barbercheck, M.E. and Kaya, H.K. 1991b. Competative interactions between entomopathogenic nematodes and *Beauveria bassiana* (Deuteromycotina: Hyphomycetes) in soilborne larvae of *Spodoptera exigua* (Lepidoptera: Noctuidae). *Environmental Entomology*, 20: 707 – 712.

Barbercheck, M.E., Wang, J., and Hirsh, I.S. 1995. Host plant effects on entomopathogenic nematodes. *Journal of Invertebrate Pathology*, 66: 169 – 177.

Baur, M.E. and Kaya, H.K. 2001. Persistence of entomopathogenic nematodes. In: Baur, M.E., Kaya, H.K., and Fuxa, J. (Eds.), *Environmental persistence of entomopathogens and nematodes. Southern Cooperative Series Bulletin 398. Oklahoma Agricultural*

*Experiment St*ation, Stillwater, Oklahoma, USA, p.47. Also available at: http://web.archive.org/web/20030516054055/www.agctr.lsu.edu/s265/baur.htm

Baur, M.E., Kaya, H.K., and Strong, D.R. 1998. Foraging ants as scavengers of entomopathogenic nematode-killed insects. *Biological Control*, 12: 231 – 236.

Blackburn, M.B., Farrar, R.R.Jr., Gundersen-Rindal, D.E., Lawrence, S.D., and Martin, P.A.W. 2007. Reproductive failure of *Heterorhabditis marelatus* in the Colorado potato beetle: evidence of stress on the nematode symbiont *Photorhabdus temperata* and potential interference from the enteric bacteria of the beetle. *Biological Control*, 42: 207 – 215.

Blackburn, M.B., Gundersen-Rindal, D.E., Weber, D.C., Martin, P.A.W., and Farrar, R.R.Jr., 2008. Enteric bacteria of field-collected Colorado potato beetle larvae inhibit growth of the entomopathogens *Photorhabdus temperata* and *Beauveria bassiana*. *Biological Control* 46: 434 – 441.

Blackshaw, R.P and Newell, C.R. 1987. Studies on temperature limitations to *Heterorhabditis heliothidis* activity. *Nematologica*, 33: 180 - 185.

Boemare, N. 2002. Biology, taxonomy and systematics of *Photorhabdus* and *Xenorhabdus*. In: Gaugler, R. (Ed.), *Entomopathogenic nematology*. CABI Publishing, CABI Publishing, New York, NY, USA, pp. 35-56.

Bohan, D.A. and Hominick, W.M. 1995. Intra-population infectious structure and temporal variation in *Steinernema feltiae*. In: Griffin, C.T., Gwynn, R.L., and Masson, J.P. (Eds.), *Ecology and transmission strategies of entomopathogenic nematodes*. European Commission, Luxembourg, pp. 83–94.

Bohan, D.A. and Hominick, W.M. 1996. Investigations on the presence of an infectious proportion amongst populations of *Steinernema feltiae* (site 76 strain) infective stages. *Parasitology*, 112: 113 – 118.

Bohan, D.A. and Hominick, W.M. 1997. Long-term dynamics of infectiousness within the infective-state pool of the entomopathogenic nematode *Steinernema feltiae* (site 76 strain) Filipjev. *Parasitology*, 114: 301 – 308.

Bornstein-Forst, S., Kiger, H., and Rector, A. 2005. Impacts of fluctuaring temperature on the development and infectivity of entomopathogenic nematode *Steinernema carpocapsae* A10. *Journal of Invertebrate Pathology*, 88: 147 – 153.

Brown, I. M. and Gaugler, R. 1998. Survival of steinernematid nematodes exposed to freezing. *Journal of Thermal Biology*, 23: 75 – 80.

Brown, I.M. and Gaugler, R. 1996. Cold tolerance of steinernematid and heterorhabditid nematodes. *Journal of Thermal Biology*, 21: 115 – 121.

Brown, I.M. and Gaugler, R. 1997. Temperature and humidity influence emergence and survival of entomopathogenic nematodes. *Nematologica*, 43: 363 - 375.

Brown, V.K. and Gange, A.C. 1990. Insect herbivory below ground. Advanced Ecological

Campbell, J.F. and Gaugler, R. 1993. Nictation behavior and its ecological implications in the host search strategies of entomopathogenic nematodes (Heterorhabditidae and Steinernematidae). *Behaviour*, 126: 155 – 169.

Campbell, J.F. and Kaya, H.K. 1999a. How and why a parasitic nematode jumps. *Nature*, 397: 485 - 486.

Campbell, J.F. and Kaya, H.K. 1999b. Mechanism, kinematic performance, and fitness consequences of entomopathogenic nematode (*Steinernema* spp.) jumping behavior. *Canadian Journal of Zoology*, 77: 1947 – 1955.

Campbell, J.F. and Kaya, H.K. 2000. Influence of insect associated cues on the jumping behavior of entomopathogenic nematodes (*Steinernema* spp.). *Behaviour*, 137: 591 – 609.

Campbell, J.F. and Kaya, H.K. 2002. Variation in entomopathogenic nematode (Steinernematidae : Heterorhabditidae) infective-stage jumping behaviour. *Nematology*, 4: 471 - 482.

Campbell, J.F. and Lewis, E.E. 2002. Entomopathogenic nematode host-search strategies. In: Lewis, E.E., Campbell, J.F. and Sukhdeo, M.V.K. (Eds.), *The behavioural ecology of parasites*. CAB International, Wallingford, UK, pp.13-38.

Campbell, J.F., Koppenhöfer, A.M., Kaya, H.K., and Chinnasri, B. 1999. Are there temporarily non-infectious dauer stages in entomopathogenic nematode populations: a test of the phased infectivity hypothesis. *Parasitology*, 118: 499 – 508.

Campbell, J.F., Lewis, E.E, Yoder, F., and Gaugler, R. 1996. Entomopathogenic nematode (Heterorhabditidae : Steinernematidae) spatial distribution in turfgrass. *Parasitology*, 113: 473 - 482.

Campbell, J.F., Lewis, E.E., Stock, S.P., Nadler, S., and Kaya, H.K. 2003. Evolution of host search strategies in entomopathogenic nematodes. *Journal of Nematology*, 35: 142 – 145.

Choo, H.Y. and Kaya, H.K. 1991. Influence of soil texture and presence of roots on host finding by *Heterorhabditis bacteriophora*. *Journal of Invertebrate Pathology*, 58: 279 – 280.

Choo, H.Y., Kaya, H.K., Burlando, T.M., and Gaugler, R. 1989. Entomopathogenic nematodes: Host-finding ability in the presence of plant roots. *Environmental Entomology*, 18: 1136 - 1140.

Choo, H.Y., Kaya, H.K., Lee, H, J., Kim, D.W., Lee, H.H., Lee, S.M. and Choo, Y.M. 2002b. Entomopathogenic nematodes (*Steinernema* spp. and *Heterohabditis bacteriophora*) and a fungus *Beauveria brongniartii* for biological control of the white grubs, *Ectinohoplia rufipes* and *Exomala orientalis*, in Korean golf courses. *BioControl*, 47: 177 - 192.

Choo, H.Y., Lee, D.W., Ha, P.J., Kim, H.H., Chung, H.J., and Lee, S.M. 1999. Temperature and dose-size effects on entomopathogenic nematode, *Steinernema longicaudum* Gongju strain. The Korean *Journal of Pesticide Science*, 3: 60 - 68.

Choo, H.Y., Lee, D.W., Yoon, H.S., Lee, M.S., and Hang, D.T. 2002a. Effect of temperature and nematode concentration on pathogenicity and reproduction of entomopathogenic nematode, *Steinernema carpocapsae* Pocheon strain (Nematode: Steinermatidae). *Korea Journal of Applied Entomology*, 41: 269 – 277.

Chung, H.J. 2001. Effects of temperature and dose on infection and reproduction of heterorhabditids in *Galleria mellonella* larvae. MSc dissertation, Gyeongsang National University, Jinju, Gyeongnam, Korea.

Constant, P., Marchay, L., Fischer-Le-Saux, M., Briand-Panoma, S. and Mauleon, H. 1998. Natural occurrence of entomopathogenic nematodes (Rhabditida: Steinernematidae and Heterorhabditidae) in Guadeloupe islands. *Fundamental and Applied Nematology*, 21: 667 – 672.

Croll, N.A. 1970. The behaviour of nematodes. Edward Arnold (Publishers) Ltd, London, UK.

Cutler, C.G. and Stock, S.P. 2003. *Steinernema websteri* n. sp. (Rhabditida: Steinernematidae), a new entomopathogenic nematode from China. *Nematologia Mediterranea*, 31: 215 – 224.

de Doucet, M.M.A. and Gabarra, R. 1994. On the occurrence of *Steinernema glaseri* (Steiner, 1929) (Steinernematidae) and *Heterorhabditis bacteriophora* Poinar, 1976 (Heterorhabditidae) in Catalogne, Spain. *Fundamental and Applied Nematology*, 17: 441 – 443.

Denno, R.F., Gruner, D.S., and Kaplan, I. 2008. Potential for entomopathogenic nematodes in biological control: a meta-analytical synthesis and insights from trophic cascade theory. *Journal of Nematology*, 40: 61 – 72.

Dillon, A.B., Rolston, A.N., Meade, C. V., Downes, M.J. and Griffin, C.T. 2008. Establishment, persistence, and introgresson of entomopathogenic nematodes in a forest ecosystem. *Ecological Applications*, 18: 735 – 747.

Downes, M.J. and Griffin, C.T. 1996. Dispersal behavior and transmission strategies of the entomopathogenic nematodes *Heterorhabditis* and *Steinernema*. *Biocontrol Science and Technology*, 6: 347 – 356.

Duncan, L.W., Dunn, D.C., and McCoy, C.W. 1996. Spatial patterns of entomopathogenic nematodes in microcosms: implications for laboratory experiments. *Journal of Nematology*, 28: 252 – 258.

Duncan, L.W., Dunn, D.C., Bague, G., and Nguyen, K. 2003a. Competition between entomopathogenic and free-living bactivorous nematodes in larvae of the weevil *Diaprepes abbrreviatus*. *Journal of Nematology*, 35: 187 – 193.

Duncan, L.W., Graham, J.H., Dunn, D.C., Zeller, J., McCoy, C.W., and Nguyen, K. 2003b. Incidence of entomopathogenic nematodes following application of *Steinernema riobrave* for control of *Diaprepes abbreviates*. *Journal of Nematology*, 35: 178 – 186.

Duncan, L.W., Graham, J.H., Zellers, J., Bright, D., Dunn, D.C., El-Borai, F.E., and Poraazinska, D.L. 2007. Food web responses to augmenting the entomopathogenic nematodes in bare and animal manure-mulched soil. *Journal of Nematology*, 39: 176 – 189.

Dusenbery, D.B. 1980. Behavior of free-living nematodes. In: Zuckerman, B.M. (Ed.), *Nematodes as biological models*. Volume 1. Academic Press, New York, USA, pp.127–196.

Dutky, S.R., Thompson, J.V., and Cantwell, G.E. 1964. A technique for the mass propagation of the DD-136 nematode. *Journal of Insect Pathology*, 6: 417 – 422.

El-Borai, F.E., Brentu, C.F., and Duncan, L.W. 2007. Augmenting entomopathogenic nematodes in soil from a Florida citrus orchard: non-target effects of a trophic cascade. *Journal of Nematology*, 39: 203 – 210.

El-Borai, F.E., Bright, D. B., Graham, J.H., Stuart, R.J., Cubero, J., and Duncan, L.W. 2009. Differential susceptibility of entomopathogenic nematodes to nematophagous fungi from Florida citrus orchards. *Nematology*, 11: 231 – 241.

El-Borai, F.E., Duncan, L.W., and Preston, J.F. 2005. Bionomics of a phoretic association between *Paenibacillus* sp. and the entomopathogenic nematode *Steinernema diaprepesi*. *Journal of Nematology*, 37: 18 – 25.

Fairbairn, J.P., Fenton, A., Norman, R., and Hudson, P.J. 2000. Re-assessing the infection strategies of the entomopathogenic nematode *Steinernema feltiae* (Rhabditida; Steinernematidae). *Parasitology*, 121: 211 – 216.

Fallon, D. J., H. K. Kaya, R. Gaugler, and B. S. Sipes. 2002. Effects of entomopath-ogenic nematodes on *Meloidogyne javanica* on tomatoes and soybeans. *Journal of Nematology*, 34: 239 - 245.

Fallon, D. J., H. K. Kaya, R. Gaugler, and B. S. Sipes. 2004. Effect of *Steinernema feltiae-Xenorhabdus bovienii* insect pathogen complex on *Meloidogyne javanica*. *Nematology*, 6: 671 - 680.

Ferris, H., Venette, B., and Scow, K.M. 2004. Soil management to enhance bacterivore and fungivore nematode populations and their nitrogen mineralization function. *Applied Soil Ecology*, 24: 19 – 35.

Fowler, H.G. and Garcia, C.R. 1989. Parasite-dependent protocooperation. *Naturwissenschaften*, 76: 26 – 27.

Friedman, M.J. 1990. Commercial production and development. In: Gaugler, R. and Kaya, H.K. (Eds.), *Entomopathogenic nematodes in biological control*. CRC Press, Boca Raton, Florida, USA, pp. 153-172.

Fujimoto, A., Lewis, E.E., Cobanoglu, G., and Kaya, H.K. 2007. Dispersal, infectivity and sex ratio of early- or late-emerging infective juveniles of the entomopathogenic nematode *Steinernema carpocapsae*. *Journal of Nematology*, 39: 333 – 337.

Fushing, H., Zhu, L., Shapiro-Ilan, D.I., Campbell, J.F., and Lewis, E.E. 2008. State-space based mass event-history model I: many decision-making agents with one target. *The Annuals of Applied Statistics*, 2: 1503 – 1522.

Gaugler, R. 2002. Entomopathogenic nematology. CABI Publishing, New York, NY, USA.

Gaugler, R. and Boush, G.M. 1978. Effects of ultraviolet radiation and sunlight on the entomogenous nematode *Neoaplectana carpocapsae*. *Journal of Invertebrate Pathology*, 32: 291 - 296.

Gaugler, R., Wang, Y., and Campbell, J.F. 1994. Aggressive and evasive behaviors in *Popillia japonica* (Coleoptera: Scarabaeidae) larvae: defences against entomopathogenic nematode attack. *Journal of Invertebrate Pathology*, 64: 193 – 199.

Georgis, R. and Gaugler, R. 1991. Predictability in biological control using entomopathogenic nematodes. *Journal of Economic Entomology*, 84: 713 – 720.

Georgis, R. and Kaya, H.K. 1998. Formulation of entomopathogenic nematodes. In: Burges, H.S. (Ed.), *Formulation of microbial biopesticides, beneficial microorganism, nematodes, and seed treatments*. Kluwer Academic Press, Dordrecht, The Netherlands, pp. 289–308.

Georgis, R. and Poinar, G.O. Jr. 1983a. Effect of soil texture on the distribution and infectivity of *Neoaplectana carpocapsae* (Nematoda: Steinernematidae). *Journal of Nematology,* 15(2): 308 - 311.

Georgis, R. and Poinar, G.O., Jr. 1983b. Vertical migration of *Heterorhabditis bacteriophora* and *H. heliothidis* (Nematoda: Heterorhabditidae) in sandy loam soil. *Journal of Nematology*, 15: 652 – 654.

Georgis, R., Koppenhöfer, A.M., Lacey, L.A., Belair, G., Duncan, L.W., Grewal, P.S., Samish, M., Tan, L., Torr, P., van Tol, R.W.H.M. 2006. Successes and failures in the use of parasitic nematodes for pest control. *Biological Control*, 38: 103 – 123.

Glazer, I. 1996. Survival mechanisms of entomopathogenic nematodes. *Biocontrol Science and Technology*, 6: 373 - 378.

Glazer, I. 1997. Effects of infected insects on secondary invasion of steinernematid entomopathogenic nematodes. *Parasitology*, 114: 597 – 604.

Glazer, I. 2002. Survival biology. In: Gaugler, R. (Ed.), *Entomopathogenic nematology*. CABI Publishing, New York, NY, USA, pp. 169-187.

Glazer, I., Kozodoi, E., Salame, L., and Nestel, D. 1996. Spatial and temporal occurrence of natural populations of *Heterorhabditis* spp. (Nematoda: Rhabditida) in a semiarid region. *Biological Control*, 6: 130 - 136.

Gouge, D.H., Smith, K.A., Lee, L.L., and Henneberry, T.J. 2000. Effect of soil depth and moisture on the vertical distribution of *Steinernema riobrave* (Namatoda: Steinernematidae). *Journal of Nematology*, 32: 223 - 228.

Grant, J.A. and Villani, M.G. 2003. Soil moisture effects on entomopathogenic nematodes. *Environmental Entomology*, 32: 80 – 87.

Grewal, P.S., Gaugler, R., and Selvan, S. 1993a. Host recognition by entomopathogenic nematodes: behavioral response to contact with host feces. *Journal of Chemical Ecology*, 19: 1219 - 1231.

Grewal, P.S., Gaugler, R., and Shupe, C. 1996. Rapid changes in thermal sensitivity of entomopathogenic nematodes in response to selection at temperature extremes. *Journal of Invertebrate Pathology*, 68: 65 – 73.

Grewal, P.S., Lewis, E.E., Gaugler, R., and Campbell, J.F. 1994b. Host finding behavior as a predictor of foraging strategy in entomopathogenic nematodes. *Parasitology*, 108: 207 – 215.

Grewal, P.S., Martin, W.R., Miller, R.W., and Lewis, E.E. 1997. Suppression of plant-parasitic nematode populations in turfgrass by application of entomopathogenic nematodes. *Biocontrol Science and Technology*, 7: 393 – 399.

Grewal, P.S., Selvan, S., and Gaugler, R. 1994a. Thermal adaptation of entomopathogenic nematodes: niche breadth for infection, establishment, and reproduction. *Journal of Thermal Biology*, 19: 245 - 253.

Grewal, R., Ehlers, R.-U., and Saphiro-Ilan, D.I. 2005. Nematodes as biocontrol agents. CABI Publishing, Cambridge, MA, USA.

Griffin, C.T., Boemare, N.E., and Lewis, E.E. 2005. Biology and behavior. In: Grewal, P.S., Ehlers, R.-U. and Shapiro-Ilan, D.I. (Eds.), *Nematodes as biocontrol agents*. CABI Publishing, New York, NY, USA, pp. 47-64.

Griffin, C.T., Chaerani, R., Fallon, D., Reid, A.P., and Downes, M.J. 2000. Occurrence and distribution of the entomopathogenic nematodes *Steinernema* spp. and *Heterorhabditis indica* in Indonesia. *Journal of Helminthology*, 74: 143 – 150.

Griffin, C.T., Dix, I., Joyce, S.A., Burnell, A.M., and Downes, M.J. 1999. Isolation and characterisation of *Heterorhabditis* spp. (Nematoda: Heterorhabditidae) from Hungary, *Estonia and Denmark. Nematology*, 1: 321 – 332.

Griffin, C.T., Joyce, S.A., Dix, I., Burnell, A.M., and Downes, M.J. 1994. Characterisation of entomopathogenic nematodes *Heterorhabditis* (Nematoda: Heterorhabditidae) from Ireland and Britan by molecular and cross-breeding techniques, and the occurrence of the genus in these islands. Fundam*ental and Applied Nematology*, 17: 245 – 253.

Griffin, C.T., Moore, J.F., and Downes, M.J. 1991. Occurrence of insect-parasitic nematodes (Steinernematidae, Heterorhabditidae) in the Republic of Ireland. *Nematologica*, 37: 92 – 100.

Gruner, D.S., Kolekar, A., McLaughlin, J.P., and Strong, D.R. 2009. Host resistance reverses the outcome of competition between microparasites. *Ecology*, 90: 1721 – 1728.

Hang, D.T., Choo, H.Y., Lee, D.W., Lee, S.M., Kaya, H.K., and Park, C.G. 2007. Temperature effects on Korean entomopathogenic nematodes, *Steinernema glaseri* and *S.*

longicaudum and their symbiotic bacteria. *Journal of Microbiology and Biotechnology*, 17: 420 - 427.

Hanski, I. 1998. Metapopulation dynamics. *Nature*, 396: 41 – 49.

Hanski, I. 1999a. Habitat connectivity, habitat continuity, and metapopulations in dynamic landscapes. Oikos, 87: 209 – 219.

Hanski, I. 1999b. Metapopulation ecology. Oxford University Press, Oxford, UK.

Hanski, I. 2001. Spatially realistic theory of metapopulation ecology. *Naturwissenschaften*, 88: 372 – 381.

Hanski, I. and Simberloff, D. 1997. The metapopulation approach, its history, conceptual domain and application to conservation. In: Hanski, I. and Gilpin, M.E. (Eds.), *Metapopulation biology: ecology, genetics, and evolution*. Academic Press, London, UK, pp. 5-26.

Hara, A.H., Gaugler, R., Kaya, H.K., and Lebeck, L.M. 1991. Natural populations of entomopathogenic nematodes (Rhabditida: Heterorhabditidae, Steinernematidae) from the Hawaiian Islands. *Environmental Entomology*, 20: 211 – 216.

Harrison, S. and Taylor, A.D. 1997. Empiracle evidence for metapopulation dynamics. In: Hanski, I. and Gilpin, M.E. (Eds.), *Metapopulation biology: ecology, genetics, and evolution*. Academic Press, London, UK, pp. 27–42.

Hay, D.B. and Fenlon, J.S. 1997. A modified binomial model that describes the infection dynamics of the entomopathogenic nematode *Steinernema feltiae* (Steinernematidae; Nematoda). *Parasitology*, 111: 627 – 633.

Hazir, S., Stock, S.P., Kaya, H.K., Koppenhofer, A.M., and Keskin, N. 2001. Developmental temperature effects on five geographic isolates of the entomopathogenic nematode *Steinernema feltiae* (Namatoda: Steinernematidae). *Journal of Invertebrate Pathology*, 77: 243 – 250.

Hazir, S., Keskin, N., Stock, S.P., Kaya, H.K., Özcan, S. 2003. Diversity and distribution of entomopathogenic nematodes (Rhabditida: Steinernematidae and Heterorhabditidae) in Turkey. *Biodiversity and Conservation*, 12: 375 - 386.

Hominick, W. M. 2002. Biogeography. In: Gaugler, R. (Ed.), *Entomopathogenic nematology*. CABI Publishing, New York, USA, pp. 115-143.

Hominick, W. M. and Ried, A.P. 1990. Perspectives on entomopathogenic nematology. In: Gaugler, R. and Kaya, H.K. (Eds.), *Entomopathogenic nematodes in biological control*. CRC Press, Boca Raton, FL, USA, pp. 327-345.

Hominick, W.M., Reid, A.P. and Briscoe, B.R. 1995. Prevalence and habitat specificity of steinernematid and heterorhabditid nematodes isolated during soil surveys of the UK and the Netherlands. *Journal of Helminthology*, 69: 27 – 32.

Hominick, W.M., Reid, A.P., Bohan, D.A., and Briscoe, B.R. 1996. Entomopathogenic nematode: biodiversity, geographical distribution and the convention on biological diversity. *Biocontrol Science and Technology*, 6: 317 – 331.

Ishibashi, N. 1993. Integrated control of insect pests by *Steinernema carpocapsae*. In: Bedding, R.A., Akhurst, R.J., and Kaya, H.K. (Eds.), *Nematodes and the biological control of insects*. CSIRO Publications, East Melbourne, Australia, pp. 105-113.

Ishibashi, N. and Kondo, E. 1986. *Steinernema feltiae* (DD-136) and *S. glaseri*: persistence in soil and bark compost and their influence on native nematodes. *Journal of Nematology*, 18: 310 - 316.

Ishibashi, N. and Kondo, E. 1987. Dynamics of the entomogenous nematode *Steinernema feltiae* applied to soil with and without nematicide treatment. *Journal of Nematology*, 19: 404 – 412.

Ishibashi, N. and Kondo, E. 1990. Behavior of infective juveniles. In: Gaugler, R. and Kaya, H.K. (Eds.), *Entomopathogenic nematodes in biological control*. CRC Press, Boca Raton, Florida, USA, pp. 139-150.

Jaffee, B.A. and Strong, D.R. 2005. Strong bottom-up and weak top-down effects in soil: nematode-parasitized insects and nematode-trapping fungi. *Soil Biology and Biochemistry*, 37: 1011 – 1021.

Jaffee, B.A., Strong, D.R., and Muldoon, A.E. 1996. Nematode-trapping fungi of a natural shrubland: test for food chain involvement. *Mycologia*, 88: 554 – 564.

Jagdale, G.B., Somasekhar, N., Grewal, P.S., and Klein, M.G. 2002. Suppression of plant-parasitic nematodes by application of live and dead infective juveniles of an entomopathogenic nematode, *Steinernema carpocapsae*, on boxwood (*Buxus* spp.). *Biological Control*, 24: 42 – 49.

Kaya, H.K. 1987. Diseases caused by nematodes. In: Fuxa, J.R. and Tanada, Y. (Eds.), *Epizootiology of insect diseases*. John Wiley and Sons, New York, USA, pp. 453–470.

Kaya, H.K. 1990. Soil ecology. In: Gaugler, R. and Kaya, H.K. (Eds.), *Entomopathogenic nematodes in biological control*. CRC Press, Boca Raton, Florida, USA, pp. 93-115.

Kaya, H.K. 2002. Natural enemies and other antagonists. In: Gaugler, R. (Ed.), *Entomopathogenic nematology*. CABI Publishing, New York, USA, pp. 189-202.

Kaya, H.K. and Brayton, M.A. 1978. Interaction between *Neoaplectana carpocapsae* and a granulosis virus of the armyworm *Pseudaletia unipuncta*. *Journal of Nematology*, 10: 350 – 354.

Kaya, H.K. and Burnando, T.M. 1989. Development of *Steinernema feltiae* (Rhabditida: Steinernematidae) in diseased insect hosts. *Journal of Invertebrate Pathology*, 53: 164 – 168.

Kaya, H.K. and Gaugler, R. 1993. Entomopathogenic nematodes. *Annual Review of Entomology*, 38: 181 - 206.

Kaya, H.K. and Koppenhöfer, A.M. 1996. Effects of microbial and other antagonistic organism and competition on entomopathogenic nematodes. *Biocontrol Science and Technology*, 6: 357 – 371.

Kaya, H.K. and Stock, S.P. 1997. Techniques in insect nematology. In: Lacey, L.A (Ed.), *Manual of techniques in insect pathology*. Academic Press, London, UK, pp. 281-324.

Kaya, H.K., Aguillera, M.M., Alumai, A., Choo, H.Y., de la Torre, M., Foder, A., Ganguly, S., Hazir, S., Lakatos, T., Pye, A., Wilson, M., Yamanaka, S., Yang, H., and Ehlers, R.-U. 2006. Status of entomopathogenic nematodes and their symbiotic bacteria from selected countries or regions of the world. *Biological Control*, 38: 134 – 155.

Kaya, H.K., Burlando, T.M., Choo, H.Y., and Thurston, G.S. 1995. Integration of entomopathogenic nematodes with *Bacillus thuringiensis* or pesticidal soap for control of insect pests. *Biological Control*, 5: 432 – 441.

Kim, H.H., Choo, H.Y., Lee, D.W., Lee, S.M., Jeon, H.Y., Cho, M.R., and Yiem, M.S. 2003. Control efficacy of Korean entomopathogenic nematodes against fungus gnat, *Bradysia agrestis* (Diptera: Sciaridae) and persistence in bed soil. *Korean Society for Horticultural Science*, 44: 393 – 401.

Kim, K.W., Ohh, M.H., Kim, J.H., and Baek, C.W. 2001. Damage and control of cutworms in the tobacco fields. *Korean Journal of Applied Entomology*, 40: 315 - 319.

Klein, M.G. 1990. Efficacy against soil-inhabiting insect pests. In: Gaugler, R. and Kaya, H.K. (Eds.), *Entomopathogenic nematodes in biological control*. CRC Press, Boca Raton, Florida, USA, pp. 195-214.

Klein, M.G. and Georgis, R. 1992. Persistence of control of Japanese beetle (Coleoptera: Scarabaeidae) larvae with steinernematid and heterorhabditid nematodes. *Journal of Economic Entomology*, 85: 727 – 730.

Köllner, T.G., Held, M., Lenk, C., Hiltpold, I., Turlings, T.C.J., Gershenzon, J., and Degenhardt, J. 2008. A maize (*E*)-β-caryophyllene synthase implicated in indirect defense responses against herbivores is not expressed in most American maize varieties. *The Plant Cell*, 20: 482 – 494.

Kondo, E. 1989. Studies on the infectivity and propagation of entomogenous nematodes, *Steinernema* spp. (Rhabditida: Steinernematidae) in the common cutworm, *Spodoptera litura* (Lepidoptera: Noctuidae). Bulletin Faculty of Agriculture, Saga University, 67: 1 – 88.

Kondo, E. and Ishibashi, N. 1985. Effects of soil moisture on the survival and infectivity of the entomogenous nematode, *Steinernema feltiae* (DD-136). Proceedings of Association Plant Protection, Kyushu, Japan, 31: 186 - 190.

Koppenhöfer A.M. and Fuzy E.M. 2008. Effects of chlorantraniliprole on *Heterorhabditis bacteriophora* (Rhabditida: Heterorhabditidae) efficacy against and reproduction in white grubs (Coleoptera: Scarabaeidae). *Biological Control*, 45: 93-102.

Koppenhöfer, A.M. 2007. Nematodes. In: Lacey, L.A. and Kaya, H.K. (Eds.) *Field manual of techniques in invertebrate pathology* (2nd Ed.). Springer Publishing Company, New York, USA, pp. 249-264.

Koppenhöfer, A.M. and Fuzy, E.M. 2006. Effect of soil type on infectivity and persistence of the entomopathogenic nematodes *Steinernema scarabaei*, *S. glaseri*, *Heterorhabditis zealandica*, and *H. bacteriophora*. *Journal of Invertebrate Pathology*, 92: 11 – 22.

Koppenhöfer, A.M. and Fuzy, E.M. 2007. Soil moisture effects on infectivity and persistence of the entomopathogenic nematodes *Steinernema scarabaei*, *S. glaseri*, *Heterorhabditis zealandica*, and *H. bacteriophora*. *Applied Soil Ecology*, 35: 128 - 139.

Koppenhöfer, A.M. and Fuzy, E.M. 2009. Long-term effects and persistence of *Steinernema scarabaei* applied for suppression of *Anomala orientalis* (Coleoptera: Scarabaeidae). Biological Control, 48: 63 – 72.

Koppenhöfer, A.M. and Kaya, H.K. 1996a. Coexistence of two steinernematid nematode species (Rhabditida: Steinernematidae) in the presence of two host species. *Applied Soil Ecology*, 4: 221 - 230.

Koppenhöfer, A.M. and Kaya, H.K. 1996b. Coexistence of entomopathogenic nematode species (Steinernematidae and Heterorhabditidae) with different foraging behavior. *Fundamental and Applied Nematology*, 19: 175 – 183.

Koppenhöfer, A.M. and Kaya, H.K. 1999. Ecological characterization of *Steinernema rarum*. *Journal of Invertebrate Pathology*, 73: 120 - 128.

Koppenhöfer, A.M., Brown, I.M., Gaugler, R., Grewal, P.S., Kaya, H.K., and Klein, M.G. 2000c. Synergism of entomopathogenic nematodes and imidacloprid against white grubs: greenhouse and field evaluation. *Biological Control*, 19: 245 – 251.

Koppenhöfer, A.M., Choo, H. Y., Kaya, H.K., Lee, D.W., and Gelernter, W.D. 1999. Increased field and greenhouse efficacy against scarab grubs with a combination of an entomopathogenic nematode and *Bacillus thuringiensis*. *Biological Control*, 14: 37 – 44.

Koppenhöfer, A.M., Ganguly, S., and Kaya, H.K. 2000a. Ecological characterization of *Steinernema monticolum*, a cold-adapted entomopathogenic nematode from Korea. *Nematology*, 2: 407 – 416.

Koppenhöfer, A.M., Jaffee, B.A., Muldoon, A.E., Strong, D.R., and Kaya, H.K. 1996. Effect of nematode-trapping fungi on an entomopathogenic nematode originating from the same field site in California. *Journal of Invertebrate Pathology*, 68: 246 – 252.

Koppenhöfer, A.M., Kaya, H.K., and Taormino, S. 1995a. Infectivity of entomopathogenic nematodes (Rhabditida: Steinernematidae) at different soil depths and moistures. *Journal of Invertebrate Pathology*, 65: 193 – 199.

Koppenhöfer, A.M., Kaya, H.K., Shanmugam, S., and Wood, G.L. 1995b. Interspecific competition between steinernematid nematodes within an insect host. *Journal of Invertebrate Pathology*, 66: 99 – 103.

Kramer, I., Hirschy, O. and Grunder, J.M. 2000. Survey of baited insect parasitic nematodes from the Swiss lowland. In: Griffin, C.T., Burnell, A.M., Downes, M.J. and Mulder, R. (Eds.). *COST 819 Development in entomopathogenic nematode/bacterial research*. European Commission, DG XII, Luxembourg, pp. 172-176.

Kung, S.P., Gaugler, R., and Kaya, H.K. 1990a. Soil types and entomopathogenic nematode persistence. *Journal of Invertebrate Pathology*, 55: 401 - 406.

Lacey, L. and Kaya, H.K. 2007. Field manual of techniques in invertebrate pathology. Springer Publishing Company, New York, NY, USA.

Lee, D.L. 2002. The biology of nematodes. Taylor and Francis, London, UK.

Lewis, E.E. 2002. Behavioral ecology. In: Gaugler, R. (Ed.), *Entomopathogenic nematology*. CABI Publishing, New York, NY, USA, pp. 205-223.

Lewis, E.E. and Campbell, J.F. 2002. Entomopathogenic nematode host-search strategies. In: Lewis, E.E., Campbell, J.F., and Sukhdeo, M.V.K. (Eds.), *The behavioural ecology of parasites*. CAB International, Wallington, UK, pp. 13-38.

Lewis, E.E. and Gaugler, R. 1994. Entomopathogenic nematode (Rhabditida: Steinernematidae) sex ratio relates to foraging strategy. *Journal of Invertebrate Pathology*, 64: 238 – 242.

Lewis, E.E. and Shapiro-Ilan, D.I. 2002. Host cadavers protect entomopathogenic nematodes during freezing. *Journal of Invertebrate Pathology*, 81: 25 - 32.

Lewis, E.E., Campbell, J., Griffin, C., Kaya, H.K., and Peters, A. 2006. Behavioral ecology of entomopathogenic nematodes. *Biological Control*, 38: 66 - 77.

Lewis, E.E., Gaugler, R., and Harrison, R. 1992. Entomopathogenic nematode host finding: response to contact cues by cruise and ambush foragers. *Parasitology*, 105: 309 – 315.

Lewis, E.E., Gaugler, R., and Harrison, R. 1993. Response of cruiser and ambusher entomopathogenic nematodes (Steinernematidae) to host volatile cues. *Canadian Journal of Zoolology*, 71: 765 - 769.

Lewis, E.E., Grewal, P.S., and Gaugler, R. 1995. Hierarchical order of host cues in parasite foraging: a question of context. *Parasitology*, 110: 207 – 213.

Lewis, E.E., Grewal, P.S., and Sardanelli, S. 2001. Interactions between *Steinernema feltiae-Xenorhabdus bovienii* insect pathogen complex and root-knot nematode *Meloidogyne incognita*. *Biological Control*, 21: 55 – 62.

Lezama-Gutiérrez, R., Alatorre-Rosas, R., Arenas-Vargas, M., Bojalil-Jaber, L.F., Melina-Ochoa, J., Gonzalez-Ramirez, M., Rebolledo-Dominguez, O. 1996. Dual infection of *Spodoptera frugiperda* (Lepidoptera: Noctuidae) by the fungus *Nomuraea rileyi* (Deuteromycotina: Hyphomycetes) and the nematode *Heterorhabditis bacteriophora* (Rhabditida: Heterorhabditidae). Vedalia, 3: 41 – 44.

Lopez-Nuñez, J.C., Cano, L. Gongora-Botero, C.E., and Stock, S. P. 2007. Diversity and evolutionary relationships of entomopathogenic nematodes (Steinernematidae and Heterorhabditidae) from the central Andean region of Colombia. Nematology, 9: 333 – 341.

Mason, J.M. and Hominick, W.M. 1995. The effect of temperature on infection, development and reproduction of heterorhabditids. *Journal of Helminthology*, 69: 337 - 345.

Mason, J.M., Razak, A.R., and Wright, D.J. 1996. The recovery of entomopathogenic nematodes from selected areas within Peninsular Malaysia. *Journal of Helminthology*, 70: 303 – 307.

Mauléon, H., Denon, D., and Briand, S. 2006. Spatial and temporal distribution of *Heterorhabditis indica* in their natural habitats of Guadeloupe. *Nematology*, 8: 603 – 617.

McCauley, D.E. 1995. Effects of population dynamics on genetics in mosaic landscape. In: Hansson, L., Fahrig, L., and Merriam, G. (Eds.), *Mosaic landscapes and ecological processs*. Chapman and Hall, London, UK, pp. 178 – 198.

McGraw, B.A. and Koppenhöfer, A.M. 2009. Population dynamics and interactions between endemic entomopathogenic nematodes and annual bluegrass weevil populations in golf course turfgrass. *Applied Soil Ecology*, 41: 77 – 89.

Menti, H., Wright, D.J., and Perry R.N. 2000. Infectivity of populations of the entomopathogenic nematodes *Steinernema feltiae* and *Heterorhabditis megidis* in relation to temperature, age and lipid content. *Nematology*, 2: 515 - 521.

Menti, H., Wright, D.J., and Perry, R.N. 1997. Dessication survival of populations of the entomopathogenic nematodes *Steinernema feltiae* and *Heterorhabditis megidis* from Greece and the UK. *Journal of Helminthology*, 71: 41 – 46.

Miduturi, J.S., Moens, M., Hominick, W.M., Briscoe, B.R., and Reid, A.P. 1996. Naturally occurring entomopathogenic nematodes in the province of West Flanders, Belgium. *Journal of Helminthology*, 70: 319 – 327.

Miduturi, J.S., Waeyenberge, L., and Moens, M. 1997. Natural distribution of entomopathogenic nematodes (Heterorhabditidae and Steinernematidae) in Belgian soils. *Russian Journal of Nematology*, 5: 55 – 65.

Millar, L.C. and Barbercheck, M.E. 2001. Interaction between endemic and introduced entomopathogenic nematodes in conventional-till and no-till corn. *Biological Control*, 22: 235 – 245.

Miller, R.W. and Gardiner, D.T. 2001. Soils in our environment. Prentice Hall, Upper Saddle River, New Jersey, USA.

Milstead, J.E. 1981. Influence of temperature and dosage on mortality of seventh instar larvae of *Galleria mellonella* (Insecta: Lepidoptera) caused by *Heterorhabditis bacteriophora* (Nematoda: Rhabditoidea) and its bacterial associate *Xenorhabdus luminescens*. *Nematologica*, 27: 167 – 171.

Molyneux, A.S. 1986. *Heterorhabditis* spp. and *Steinernema* (= *Neoaplectana*) spp.: temperature, and aspects of behavior and infectivity. *Experimental Parasitology*, 62: 169 – 180.

Molyneux, A.S. and Bedding, R.A. 1984. Influence of soil texture and moisture on the infectivity of *Heterorhabditis* sp. D1 and *Steinernema glaseri* for larvae of the sheep blowfly, *Lucila cuprina*. *Namatologica*, 30: 358 - 365.

Mráček, Z., Bečvář, S., and Kindlmann, P. 1999a. Survey of entomopathogenic nematodes from the families Steinernematidae and Heterorhabditidae (Nematoda: Rhabditida) in the Czech Republic. *Folia Parasitologica*, 46: 145 – 148.

Mráček, Z., Bečvář, S., Kindlmann P. and Webster J.M. 1999b. Factors influencing the low temperature infectivity of a *Steinernema kraussei* (Nematoda:Steinernematidae) Canadian isolate in laboratory experiments. *Journal of Invertebrate Pathology*, 73: 243 - 247.

Mráček, Z., Bečvář, S., Kindlmann, P., and Jersakova, J. 2005. Habitat preference for entomopathogenic nematodes, their insect hosts and new faunistic records for the Czech Republic. *Biological Control*, 34: 27 - 37.

Nguyen, K.B. and Hunt, D.V. 2007. Entomopathogenic nematodes: systematics, phylogeny and bacterial symbionts. Brill, Leiden, The Netherlands.

Nguyen, K.B., Mallan, A.P., and Gozel, U. 2006. *Steinernema khoisanae* n. sp. (Rhabditida: Steinernematidae), a new entomopathogenic nematode from South Africa. *Nematology*, 8: 157 – 175.

Nishimatsu, T. and Jackon, J.J. 1998. Interaction of insecticides, entomopathogenic nematodes, and larvae of the western corn rootworm (Coleoptera: Chrysomelidae). *Journal of Economic Entomology*, 91: 410 – 418.

Parkman, J.P., Hudson, W.G., Frank, J.H., Nguyen, K.B., and Smart, G.C.,Jr. 1993. Establishment and persistence of *Steinernema scapterisci* (Rhabditida: Steinernematidae) in field populations of *Scapteriscus* spp. mole crickets (Orthoptera: Gryllotalpidae). *Journal of Entomological Science*, 28: 182 – 190.

Patel, M.N. and Wright, D.J. 1996. The influence of neuroactive pesticides on the behavior of entomopathogenic nematodes. *Journal of Helminthology*, 70: 53 – 61.

Patel, M.N., Perry, R.N., and Wright, D.J. 1997. Dessication survival and water contents of entomopathogenic nematodes, *Steinernema* spp. (Rhabditida: Steinernematidae). *International Journal for Parasitology*, 27: 61 – 70.

Perez, E.E., Lewis, E.E., and Shaporo-Ilan, D.I. 2003. Impact of the host cadaver on survival and infectivity of entomopathogenic nematodes (Rhabditida: Steinernematidae and Heterorhabditidae) under desiccating conditions. *Journal of Invertebrate pathology*, 82: 111 - 118.

Peters, A. and Ehlers, R. U. 1994. Susceptibility of leather jackets (*Tipula paludosa* and *Tipula oleracea*; Tipulidae; Nematocers) to the entomopathogenic nematode *Steinernema feltiae*. *Journal of Invertebrate Pathology*, 63: 163 – 171.

Pianka, E.P. 1999. Evolutionary ecology 6th ed. Harper and Row, New York, NY, USA.

Poinar, G.O., Jr. 1988. A microsporidian parasite of *Neoaplectana glaseri* (Steinernematidae: Rhabditida). *Revue de Nematologie*, 11: 359 – 360.

Poinar, G.O., Jr. 1990. Biology and taxonomy of Steinernematidae and HJeterorhabditidae. In: Gaugler, R. and Kaya, H.K. (Eds.), *Entomopathogenic nematodes in biological control*. CRC Press, Boca Raton, Florida, USA, pp. 23-61.

Poinar, G.O., Jr. 1993. Origins and phylogenetic relationships of the entomophilic rhabditids, *Heterorhabditis* and *Steinernema*. *Fundamental and Applied Nematology* 16: 332 – 338.

Poinar, G.O., Jr. and Jansson, H.B. 1988a. Diseases of nematodes. Vol. I. CRC Press, Boca Raton, FL. USA.

Poinar, G.O., Jr. and Jansson, H.B. 1988b. Diseases of nematodes. Vol. II. CRC Press, Boca Raton, FL. USA.

Poinar, G.O., Jr., Thomas, G.M., and Lighthart, B. 1990. Bioassay to determine the effect of commercial preparations of *Bacillus thuringiensis* on entomogenous rhabditoid nematodes. *Agriculture, Ecosystems and Environment*, 30: 195 – 202.

Poinar, G.O., Jr. and Jasson, J.B. 1986. Infection of *Neoaplectana* and *Heterorhabditis* (Rhabditida: Nematoda) with the predatory fungi, *Monacrosporium ellipsosporum* and *Arthrobotrys oligospora* (Moniliales: Deuteromycetes). *Revue de Nematlogie*, 9: 241 – 244.

Preisser, E.L. 2003. Field evidence for a rapidly cascading underground food web. *Ecology*, 84: 869 – 874.

Preisser, E.L., Dugaw, C. J., Dennis, B., and Strong, D.R. 2005. Long-term survival of the entomopathogenic nematode *Heterorhabditis marelatus*. *Environmental Entomology*, 34: 1501 – 1506.

Preisser, E.L., Dugaw, C. J., Dennis, B., and Strong, D.R. 2006. Plant facilitation of a belowground predator. *Ecology*, 87: 1116 – 1123.

Půža, V. and Mráček, Z. 2007. Natural population dynamics of entomopathogenic nematode *Steinernema affine* (Steinernematidae) under dry conditions: possible nematode persistence within host cadavers? *Journal of Invertebrate Pathology*, 96: 89 – 92.

Ram, K., Gruner, D.S., McLaughlin, J.P., Preisser, E.L., and Strong, D.R. 2008. Dynamics of a subterranean trophic cascade in space and time. *Journal of Nematology*, 40: 85 – 92.

Rasmann, S., K., Köllner, T.G., Degenhardt, J., Hiltpold, I., Toepfer, S., Kuhlmann, U., Gershenzon, J., and Turlings, T.C. 2005. Recruitment of entomopathogenic nematodes by insect-damaged maize roots. *Nature*, 434: 732 – 737.

Reed, E.M. and Wallace, H.R. 1965. Leaping locomotion by an insect-parasitic nematode. *Nature*, 206: 210 - 211.

Rovesti, L. and Deseö, K.K. 1990. Compatibility of chemical pesticides with entomopathogenic nematodes, *Steinernema carpocapsae* Weiser and *S. feltiae* Filipjev (Nematoda: Steinernematidae). *Nematologica*, 36: 237 – 245.

Selvan, S., Campbell, J.F., Gaugler, R. 1993. Density-dependent effects on entomopathogenic nematodes (Heterorhabditidae and Steinernematidae) within an insect host. *Journal of Invertebrate Pathology*, 62: 278 – 284.

Selvan, S., Grewal, P.S., Leustek, T., and Gaugler, R. 1996. Heat shock enhances thermatolerance of infective juvenile insect-parasitic nematodes *Heterorhabditis bacteriophora* (Rhabditida: Heterorhabditidae). *Experimentia*, 52: 727 – 730.

Shapiro, D.I., Glazer, I., and Segal, D. 1996. Trait stability and fitness of the heat tolerant entomopathogenic nematode *Heterorhabditis bacteriophora* IS5 strain. *Biological Control*, 6: 238 – 244.

Shapiro, D.I., Glazer, I., and Segal, D. 1997. Genetic improvement of heat tolerance *Heterorhabditis bacteriophora* through hybridization. *Biological Control*, 8: 153 – 159.

Shapiro-Ilan, D.I., Campbell, J.F., Lewis, E.E., Elkon, J.M., and Kim-Shapiro, D.B. 2009. Directional movement of steinernematid nematodes in response to electrical current. *Journal of Invertebrate Pathology*, 100: 134 - 137.

Shapiro-Ilan, D.I., Cottrell, T.E., Brown, I., Gardner, W.A., Hubbard, R.K., and Wood, B.W. 2006. Effect of soil moisture and a surfactant on entomopathogenic nematode suppression of the pecan weevil, *Curculio caryae*. *Journal of Nematology* 38: 474 – 482.

Shields, E.J., Testa, A., Miller, J.M., and Flanders, K.L. 1999. Field efficacy and persistence of the entomopathogenic nematodes *Heterorhabditis bacteriophora* 'Oswego' and *H. bacteriophora* 'NC' on Alfalfa snout beetle larvae (Coleoptera: Curculionidae). *Environmental Entomology*, 28: 128 – 136.

Simons, W.R. and Poinar, G.O., Jr 1973. The ability of *Neoaplectana carpocapsae* (Steinernematidae: Nematodea) to survive extended periods of desiccation. *Journal of Invertebrate Pathology*, 22: 228 – 230.

Smits, P.H. 1996. Post-application persistence of entomopathogenic nematodes. *Biocontrol Science and Technology*, 6: 379 – 387.

Solomon, A., Paperna, I., and Glazer, I. 1999. Dessication survival of the entomopathogenic nematode *Steinernema feltiae*: induction of anhydrobiosis. In: Glazer, I., Richardson, P., Boemare, N. and Coudert, F. (Eds.), *Survival strategies of entomopathogenic nematodes*. EUR 18855 EN Report, pp. 83-98.

Somasekhar, N., Grewal, P.S., DeNardo, E.A.B., and Stinner, B.R. 2002. Non-target effects of entomopathogenic nematodes on the soil nematode community. *Journal of Applied Ecology*, 39: 735 – 744.

Somvanshi, V.S., Koltai, H., and Glazer, I. 2008. Expression of different desiccation-tolerance related genes in various species of entomopathogenic nematodes. *Molecular and Biochemical Parasitology*, 158: 65 - 71.

Spiridonov, S.E. and Moens, M. 1999. Two previously unreported species of steinernematids from woodlands in Belgium. *Russian Journal of Nematology*, 7: 39 – 42.

Stark, J.D. 1996. Entomopathogenic nematodes (Rhabditida: Steinernematidae): toxicity of neem. *Journal of Economic Entomology*, 89: 68 – 73.

Steiner, W.A. 1996. Distribution of entomopathogenic nematodes in the Swiss Alps. *Revue Suisse de Zoologie*, 103: 439 – 452.

Stock, P.S., Choo, H.Y., and Kaya, H.K. 1997. First record of *Steinernema glaseri* Steiner, 1929 (Nematoda: Steinernematidae) in Asia, with notes on intraspecific variation. *Nematologica*, 43: 377 – 381.

Stock, S. P., Rivera-Orduno, B., and Flores-Lara, Y. 2009. *Heterorhabitis sonorensis* n. sp. (Nematoda: Heterorhabditidae), a natural pathogen of the seasonal cicada *Diceroprocta ornea* (Walker) (Homoptera: Cicadidae) in the Sonoran desert. *Journal of Invertebrate Pathology,* 100: 175 – 184.

Stock, S.P., Banna, L.A., Darwish, R. and Katbeh, A. 2008. Diversity and distribution of entomopathogenic nematodes (Nematoda: Steinernematidae, Heterorhabditidae) and their bacterial symbionts (γ-Proteobacteria: Enterobacteriaceae) in Jordan. *Journal of Invertebrate Pathology,* 98: 228 - 234.

Strong, D.R. 2002. Populations of entomopathogenic nematodes in foodwebs. In: Gaugler, R. (Ed.), *Entomopathogenic nematology*, CABI, Wallingford, UK.

Strong, D.R., Kaya, H.K., Whipple, A.V., Child, A.L., Kraig, S., Bondonno, M., and Maron, J.L. 1996. Entomopathogenic nematodes : natural enemies of root-feeding caterpillars on bush lupine. *Oecologia*, 108: 85 – 92.

Strong, D.R., Kaya, H.K., Whipple, A.V., Child, and Dennis, B. 1999. Model selection for a sunterraean trophic cascade: root-feeding caterpillars and entomopathogenic nematodes. *Ecology*, 80: 2750 – 2761.

Stuart, R.J. and Gaugler, R. 1994. Patchiness in population of entomopathogenic nematodes. *Journal of Invertebrate Pathology*, 64: 39 – 45.

Stuart, R.J., Babercheck, M.E., Grewal, P.S., Taylor, R.J., and Hoy, C.W. 2006. Population biology of entomopathogenic nematodes: concepts, issues, and models. *Biological Control*, 38: 80 – 102.

Stuart, R.J., Hatab, A. and Gaugler, R. 1998. Sex ratio and the infection process in entomopathogenic nematodes: are males the colonizing sex? *Journal of Invertebrate Pathology*, 72: 288 – 295.

Sturhan, D. 1999. Prevalence and habit specificity of entomopathogenic nematodes in Germany. In: Gwynn, R.L., Smits, P.H., Griffin, C., Ehlers, R.-U., Boemare, N., and Masson, J.-P. (Eds.). *COST 819 – Entomopathogenic nematodes – Application and persistence of entomopathogenic nematodes*. European Commission, Luxembourg, DG XII, pp. 123 – 132.

Susurluk, A. 2006. Effectiveness of the Entomopathogenic Nematodes, *Heterorhabditis bacteriophora* and *Steinernema feltiae* against *Tenebrio molitor* (Yellow Mealworm) Larvae at Different Temperature and Soil Types. *Turk. Journal of Biology*, 30 (4): 199-205).

Susurluk, A. and Ehlers, R.-U. 2008. Field persistence of the entomopathogenic nematode *Heterorhabditis bacteriophora* in different crops. *BioControl*, 53: 627–641.

Susurluk, I. A. 2009. Seasonal and vertical distribution of the entomopathogenic nematodes, *Heterorhabditis bacteriophora* (TUR-H2) and *Steinernema feltiae* (TUR-S3), in turf and fallow areas. *Nematology*, Vol. 11 (2): 309–315.

Susurluk, I.A. 2008. Effects of various agricultural practices on persistence of the inundative applied entomopathogenic nematodes, Heterorhabditis bacteriophora and *Steinernema feltiae* in the field. *Russian Journal of Nematology*, 16 (1): 23–32.

Tanada, Y. and Kaya, H.K. 1993. Insect pathology. Academic Press, San Diego, California, USA.

Taylor, A.L. and Sasser, J.N. 1978. Biology, identification and control of root-knot nematodes. North Carolina State University Graphics., Raleigh, North Carolina, USA.

Thurston, G. S., Kaya, H.K, and Gaugler, R. 1994a. Characterizing the enhanced susceptibility of milky disease-infected scarabaeid grubs to entomopathogenic nematodes. *Biological Control*, 4: 67-73.

Thurston, G. S., Kaya, H.K, Burlando, T.M., and Harrison, R.E. 1993. Milky disease bacterium as stressor to increase susceptibility of scarabaeid larvae to an entomopathogenic nematode. *Journal of Invertebrate Pathology*, 61: 167-172.

Thurston, G.S., Ni, Y. and Kaya, H.K. 1994b. Influence of salinity on survival and infectivity of entomopathogenic nematodes. *Journal of Nematology*, 26: 345 - 351.

Timper, P. and Kaya, H.K. 1989. Role of the 2nd –stage cuticle of entomogenous nematodes in preventing infection by nematophagous fungi. *Journal of Invertebrate Pathology*, 54: 314 – 321.

Timper, P., Kaya, H.K. and Jaffee, B.A. 1991. Survival of entomogenous nematodes in soil infected with the nematode-parasitic fungus *Hirsutella rhossiliensis* (Deuteromycotina: Hyphomycetes). *Biological Control*, 1: 42 – 50.

Torr, P., Heritage, S., and Wilson, M.J. 2004. Vibration as novel signal for host location by parasitic nematodes. *International Journal for Parasitology*, 34: 997 – 999.

Troemel E.R, Félix M.A, Whiteman N.K, Barrière A, and Ausubel F.M. 2008. Microsporidia are natural intracellular parasites of the nematode *Caenorhabditis elegans*. *PLoS Biology*, 6: 736 - 2752.

van Tol, R.W.H.M., van der Sommen, A.T.C., Boff, M.I.C., van Bezooijen, J., Sabelis, M.W. and Smits, P.H. 2001. Plants protect their roots by alerting the enemies of grubs. *Ecology Letters,* 4: 292 – 294.

Wallace, H.R. 1963. The biology of plant parasitic nematodes. Edward Arnold Publishers Ltd, London, UK.

Walter, H.R., Hudgens, R.A., and Freckman, D.W. 1986. Consumption of nematodes by fungivorous mites, *Tyrophagus* spp. (Acarina: Astimata: Acaridae). *Oecologia*, 70: 357 – 361.

Wang, Y., Gaugler, R., and Cui, L. 1994. Variation in immune response of *Popillia japonica* and *Acheta domesticus* to *Heterorhabditis bacteriophora* and *Steinernema* species. *Journal of Nematology*, 26: 11 – 18.

Wootton, J.T. 1994. The nature and consequences of indirect effects in ecological communities. *Annual Review of Ecology and Systematics*, 25: 443 – 466.

Yoshida, M., Reid, R.P., Briscoe, B.R., and Hominick, W.M. 1998. Survey of entomopathogenic nematodes (Rhabditida: Steinernematidae and Heterorhabditidae) in Japan. *Fundamental and Applied Nematology*, 21: 185 – 198.

Zervos, S., Johnson, S.C., and Webster, J.M. 1991. Effect of temperature and inoculums size on reproduction and development of *Heterorhabditis heliothidis* and *Steinernema glaseri* (Nematoda: Rhabditoidea) in *Galleria mellonella*. *Canadian Journal of Zoology*, 69: 1261 – 1264.

Zhou, X.S., Kaya, H.K., Heungens, K., and Goodrich-Blair, H. 2002. Response of ants to a deterrent factor(s) produced by the symbiotic bacteria of entomopathogenic nematodes. *Applied and Environmental Microbiology*, 68: 6202 – 6209.

In: Microbial Insecticides: Principles and Applications
Editors: J. Francis Borgio, K. Sahayaraj, et al.

ISBN: 978-1-61209-223-2
© 2011 Nova Science Publishers, Inc

Chapter 9

ISOLATION, PRESERVATION, CHARACTERIZATION AND IDENTIFICATION OF ENTOMOPATHOGENIC NEMATODES

I. Alper Susurluk[*]

Uludag University, Agriculture Faculty, Plant Protection Department,
16059 Nilufer-Bursa, TURKEY

ABSTRACT

Entomopathogenic nematodes (EPNs) in the families Heterorhabditidae and Steinernematidae have considerable potential as biological control agents of soil-inhabiting insect pests. EPNs can be one alternative to chemical insecticide for control of some important pests. EPNs containing the mutualistic bacteria are extraordinarily lethal to many insect soil-dwelling pests. Infective stage juvenile nematodes that also called dauer larvae (DL) enter insects through the mouth, anus, spiracles, or areas of thin cuticle. After penetrating to the haemocoel the nematodes release their bacteria, which quickly multiply and overwhelm the hosts, usually within 24 to 48 hr. In order to obtain current and important studies related with EPNs, some important methods have to be correctly done. Correct and complete application of the methods has a key role on success of a study. Therefore, in this chapter, the important methods of EPN experiments; isolation, preservation, characterization and identification of EPNs were summarised with many details, photographs and examples. It is thought that this chapter can give very useful and basic information to BSc, MSc or PhD students and scientists studying with EPNs.

9.1. INTRODUCTION

Entomopathogenic nematodes (EPNs) of the genera *Heterorhabditis* Poinar, 1975 and *Steinernema* Travassos, 1927 (Rhabditida) are safe biocontrol agents (Ehlers and Hokkanen,

---

[*] E-mail: susurluk@uludag.edu.tr, Phone: + 90 224 294 15 79, Fax: +90 224 294 14 02

1996) used to control soil-borne insect pests (Ehlers, 1996). They are symbiotically associated with bacteria of the genera *Photorhabdus* and *Xenorhabdus*, respectively (Forst and Nealson, 1996). Like parasitoids or predators, EPNs have chemoreceptors and are motile. Like pathogens, they are highly virulent, killing their hosts quickly and they can be cultured easily *in vivo* or *in vitro* (Ehlers, 2001).

In addition to above attributes, they are safe to most non-target organisms and to the environment. No evidence exists for a mammalian pathogenicity (Ehlers and Peters, 1995; Boemare *et al.*, 1996; Ehlers and Hokkanen, 1996). There are also no difficulties to apply EPNs as they are easily sprayed using standard equipment and can be combined with almost all chemical control compounds (Georgis and Mamweiler, 1994; Georgis and Kaya, 1998). EPNs have been used against many different pests in the soil, in cryptic habitats and on the foliage (Begley, 1990). Soil is the natural reservoir of EPNs (Akhurst, 1986; Gaugler, 1988) offering excellent conditions for nematode survival and activity. The opportunity to use EPN is promising more than 90% of insect pests spend part of their life cycle in the soil. Their life cycle consists of a free-living phase of the infective juvenile (IJ), also called dauer juvenile (DJ), occurring in the soil and the propagative phase, which occurs inside of the insect host body. The term dauer (German word for enduring) was introduced by Fuchs (1915). The IJ stage is the 3rd stage juvenile and it is morphologically and physiologically adapted for long-term survival and formed as a response to depleting food resource and adverse environmental conditions in the soil. The IJs are morphologically distinct from propagative J3 stages (Johnigk and Ehlers, 1999). Due to their ability to survive without food they can be used for insect control purposes.

Nowadays, several species after safety and effectiveness test are used in marketing against several economical important insect pests. EPNs have very high control potential especially on soil-dwelling insect pests. The nematodes can persist in applied areas till 2 years (Susurluk and Ehlers, 2008).

In Europe and the US, larvae of the black vine weevil, *Otiorhynchus sulcatus* (F.) (Coleoptera: Curculionidae), are controlled in strawberries, cranberries, ornamentals and tree nurseries with *Heterorhabditis* spp., *Steinernema feltiae*, or *S. carpocapsae* (Georgis, 1992; Ehlers, 1996).

Moreover, turf insect pests, sciarid fly larvae (Diptera: Sciaridae), *Frankliniella* spp. (Thysanoptera: Thripidae) and white grubs (Scarabaeidae) can be successfully controlled by the using of EPNs. On the other hand, nowadays, effectiveness of EPNs on some endemic pests is still being studied in many countries (Ehlers, 1996; Hui and Webster 2000). Therefore, interest in entomopathogenic nematodes (EPNs) has increased rapidly in recent years and research with these beneficial organisms is being conducted in many laboratories worldwide.

Several new nematode species have recently been discovered (Hominick *et al.*, 1996); however, little is known about their control potential in variable environmental conditions. Thus, firstly, detected species should be identified, and then some ecological tests of the species should be performed.

Successfully using of an EPN species in biological control depends on its isolation, identification and preservation during experiments. Each step is very important and can negatively effect to next. Therefore, each part is clearly explained in this chapter. Moreover, general or common methods related with these parts are expressed.

9.2. ISOLATION, PRESERVATION AND PROPAGATION OF ENTOMOPATHOGENIC NEMATODES

9.2.1. Soil Sampling

It is well known that most nematodes live into soil. Soil is natural habitat for the nematodes. However, only IJs (=3rd Juvenile stage) of EPNs present into soil, they need insect body for completing their life spans (Figure 1). IJs can present very long time into soil without any hosts.

First step of the EPN survey studies is soil sampling; suitable soil collection is very important step for reaching to EPNs into soil. Each soil sample should be taken between from 0 and 20 cm deep of the soil. However, most IJs present 0-10 cm soil deep. Since IJs move to deeper into soil in winter months.

Therefore, if soil sampling is done in winter, soil sample should be collected deeper than in summer. Moreover, if sampling is occurred in fallow areas or areas without any plants, it is also recommended deeper soil sampling. Since most IJs prefer remain near the root system of plants (Susurluk, 2009).

Soil sample collections can be made by using of a shovel or a soil core. Moreover, soil sampling area should not be very dry, optimal soil humidity level is nearly 7 %. Lower humidity level is more risky than higher humidity level (Susurluk *et al.*, 2001; Glazer, 2002). Taken samples should be placed in a plastic bag and transferred to laboratory. The samples were kept at 4 °C until analyzed. After collection, a label must be fixed on the plastic bag.

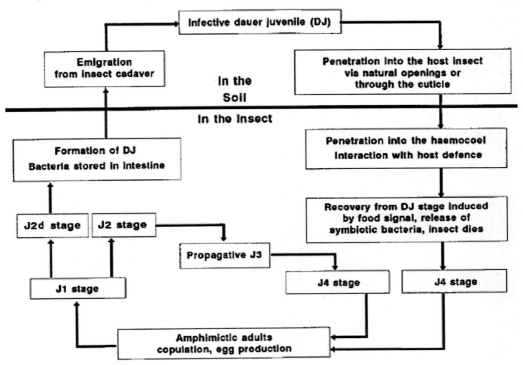

Figure 1. Life cycle of *Steinernema* spp. into soil and into insect (Ehlers, 1996).

9.2.2. Detection of Entomopathogenic Nematodes

Several methods for detection of EPNs from soil are available and these are compared in Table 1. The live bait method is most commonly used. The advantage of this method is that it is simple and selective towards EPNs. The disadvantage, however, is that it is less quantitative, because it depends on successful infection of the bait larvae. This can be improved by counting the number of invading nematodes by dissection. Baermann extractions and flotation methods are not selective towards EPNs and both, nematological experience and special equipment are required. Although these methods are regarded as quantitative, the extraction efficacy can be various. Examples of this were given by Barbercheck and Kaya (1991), and Kung *et al.* (1990). Sturhan and Mracek (2000) compared the methods, *Galleria mellonella* L. (Lepidoptera: Galleriidae) baiting technique and direct extraction method for recovering IJs of *Steinernema* from soil collected in Germany and the Czech Republic. The results of the study show that all *Steinernema* species were recovered with both methods, but the baiting technique was generally less effective and mixtures of species were frequently undetected. The direct extraction method provided quantitative estimates of IJs density, but no information on their infectivity or morphological characters of adult nematodes because cultures can not be established (nematodes can not survive the extraction process). Hass *et al.* (1999) stated further that the most efficient extraction method of IJs from soil was the centrifugal flotation using the Baermann funnel and dissection of bait insect.

Insect baiting technique:

- Collected soil is homogenized by hand.
- Each soil sample (50-150 g) is placed in a plastic box with two last instar larvae of *G. mellonella* (Figure 2).
- The boxes are turned round.
- Stored at 25 ± 2 °C.
- Three days later, dead larvae are gently dissected with Pasteur pipette in Ringer solution for nematode counting and recording.
- To detect new generation IJs from infected larvae (cadaver), infected larvae are transferred to White trap (Kaya and Stock, 1997).
- Incubated for 3-4 weeks at 25 °C (Figures 3 and 4).

Figure 2. Collected soil samples in plastic bags and baiting with *Galleria mellonella* larvae (left) (Photo: A. Susurluk, Bursa-Turkey) and insect bait technique with 2 last instar larvae of *G. mellonella* in 100 g soil in each box (middle and right) (Photo: A. Susurluk, Raisdorf-Germany).

Table 1. Comparison of methods used to detect EPNs in soil samples

Method	Selective towards EPN	Quantitative	Qualitative [a]	Method possible without		References
				special equipment	nematological experience	
Live-bait (*Galleria* trap)	Yes	(yes)[b]	yes	yes	(yes)[c]	1
Dissection of bait insects	Yes	yes	yes	yes	(yes)[c]	2, 3, 4
Baermann extraction	No	yes	yes	no	no	3, 5, 6
Extraction by flotation	No	yes	no	no	no	3, 5, 6

[a] The ability of the nematodes to migrate through the soil (Baermann, bait) and to infect an insect (bait)
[b] Depends on the number of bait larvae used
[c] The selectivity of the method supports identification.
[1] Bedding and Akhurst (1975), [2] Curran and Heng (1992), [3] Bednarek (1998), [4] Koppenhöfer *et al.* (1998), [5] Saunders and All (1982), [6] Sturhan (1995).

Figure 3. Infected *Galleria mellonella* larvae in White traps. Left infected with *Heterorhabditis bacteriophora* and right infected with *Steinernema feltiae* (Photo: A. Susurluk, Bursa-Turkey).

Figure 4. Many IJs on *Galleria mellonella* larvae in White trap (Photo: A. Susurluk, Raisdorf, Germany).

9.2.3. Preservation and Storage of Isolated Entomopathogenic Nematodes

Isolated EPNs from soil are need to storage in order to use very long time. During storage, they remain live and active enough to insect infection. After White trap process, IJs are harvested and transferred into storage plastic box that contains Ringer solution. Ringer solutions protect IJs from other saprophyte microorganisms. Different Ringer solution contents are given below.

Ringer's Solution (Laboratory Standard):   9.0 g   NaCl
    0.42 g KCl
    0.37 g $CaCl_2$ * 2 H2O
    0.20 g $NaHCO_3$
    add 1 l aqua dest.

Ringer's Solution (Commercially Prepared): 8 Ringer Tablets (Merck)
    add 1 l aqua dest.

Insect Ringer's Solution: 7.5 g   NaCl
    0.35 g KCl
    0.21 g $CaCl_2$ * 2 H2O
    add 1 l aqua dest.

Ringer solution should be added not so much, just 2 or 3 mm depth into plastic box. Otherwise, IJs can rapidly inactive because of lacking oxygen. After addition IJs into box, the box should be preserved at 4 °C (Figure 5). In this storage, IJs can active until 3 or 4 months. Afterwards, IJs should be propagated in the last instar larvae of *G. mellonella*.

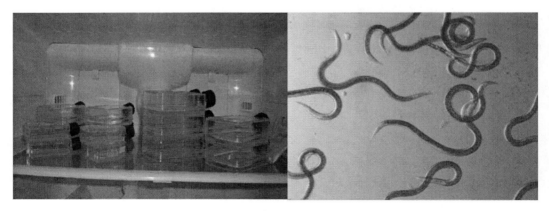

Figure 5. Some IJs with Ringer's solution in boxes at 4 °C (Photo: A. Susurluk, Bursa-Turkey) and IJs of *Steinernema scarabaei* in a box (Photo: P. Hyrsl, Raisdorf-Germany).

9.2.4. *Galleria mellonella* Culture Media

IJs culture should be propagated maximum 3 or 4 months as *in vivo*. For this reason, last instar of *G. mellonella* is used as host. The greater wax moth *G. mellonella* is the most frequently used for research scale propagation of EPNs. It is easy to maintain, relatively big and has a very low resistance to EPNs due to its evolutionary history. For culturing of *G. mellonella* currently two different media are used.

Galleria mellonella simple medium (modified according to Glazer, personal correspondence):

200 g	Honey
183 g	Glycerol
47 g	Yeast extract
4 g	Nipagine (fungicide)

Mix ingredient in a water bath at 60 °C. When softened, add

320 g	Wheat bran (crushed oats in the original medium)

Mix until medium is almost homogenous.
Remark: *G. mellonella* pupate very fast when being fed with this diet.

Galleria mellonella rich medium (according to Wiesner, 1993):

22 %	Corn-grouts
22 %	Wheat-flour or bruised-rain wheat
11 %	Milk powder
11 %	Honey
11 %	Glycerol
5.5 %	Yeast powder
17.5 %	Bee-wax

Preparation:

- Mix honey, glycerol and yeast powder at 80 °C until mixture is homogenous.
- Melt bee wax separately at 80 °C.
- Add cereals and milk powder to the honey-glycerol-yeast mixture.
- Then add melted bee wax.
- Mix ingredients until is almost homogenous.

Remark: This medium is more laborious in preparation, but *G. mellonella* larvae do not pupate as fast as in the simple medium described above.

Rearing of *Galleria mellonella:*

- The insect culture is reared in 1.500 ml volume glass containers (11 cm diameter and 15 cm height) at 30-32 °C on one of the artificial mediums.

- The glass containers are closed with filter paper and a metal screen (Figure 6).
- Females laid eggs on the filter paper from where they are collected and transferred into fresh medium.
- The eggs hatch within 3-4 days.
- Larvae are fed weekly.
- After 5-6 weeks, larvae reach the last instar and are collected to be used in the experiments.
- Some larvae are left in the containers to pupate.
- Two weeks later, the adult females emerge and lay eggs.

Figure 6. Culturing of *Galleria mellonella* into rich medium in glasses containers and incubated at 30-32 °C (Photo: A. Susurluk, Bursa-Turkey).

9.2.5. *In Vivo* Propagation of Entomopathogenic Nematodes in *Galleria mellonella*

In vivo propagation of *Galleria mellonella:*

- Each insect larva is inoculated with 50 IJs for *Steinernema* spp. and 100 IJs for *Heterorhabditis* spp.
- 24 Well-plate is used for assay.
- One larva is placed into each well.
- Each well contains moist sand (10 % relative humidity).
- Incubate for 3-5 days at 25 °C in dark.
- Infected larvae are transferred into White trap (Figure 7).

Figure 7. After 10 days, IJs of *S. carpocapsae* TUR-S4 that exited from larvae *Dorcadion pseudopreissi* (Coleoptera: Cerambycidae) in Ringer's solution in White trap. Milky appearance is mass of IJs (Photo: A. Susurluk, Bursa- Turkey).

9.3. CHARACTERIZATION AND IDENTIFICATION OF ENTOMOPATHOGENIC NEMATODES

Today, new species and strains of EPNs are always being discovered. Unfortunately, when new isolates are discovered, their identification is not always straightforward. Indeed, there has been much confusion over the nomenclature of nematodes with many isolates being misidentified by standard morphological criteria (Reid and Hominick, 1993). For this reason, many scientists turned to molecular techniques for their identifications. The Polymerase Chain Reaction (PCR)-Restriction Fragment Length Polymorphism (RFLP) of the ribosomal DNA (r DNA) repeat unit is widely used for the identification of EPNs. It is an ideal choice for identification purpose, because it is present as a multi-copy tandem repeat in the genome of most organisms. On the other hand, genus specification of EPNs can distinctively be made by color of cadaver. According to symptomatical observation of cadaver based on the color of infected larvae, *Steinernema* spp. appear yellowish, and *Heterorhabditis* spp. appear reddish color. In addition to this color difference of infected larvae, there are some important morphological, biological and phylogenetical distinguishing features between families of Heterorhabditidae and Steinernematidae. The first generation adults of *Heterorhabditis* are hermaphroditic and the IJs carry the symbiotic bacteria in the midgut of intestine. First generation *Steinernema* are amphimictic males and females, and the bacteria are carried in a specialized intestinal vesicle. The other differences are summarized in Table 2 and tails of first generation males of *Heterorhabditis* and *Steinernema* are shown in Figure 7. Moreover, geographical distribution of natural populations of EPNs is little bit different that *S. carpocapsae* and *S. feltiae* are widely distributed in temperate regions; *H. bacteriophora* is common in regions with continental and Mediterranean climates and *H. indica* is found throughout the tropics and subtropics. For some species, the known distributions are much more restricted for example; *S. cubanum* and *S. kushidai* are so far known only in Cuba and Japan, respectively (Griffin, *et al.,* 2005).

Table 2. Morphological, biological and phylogenetical distinguishing features of *Heterorhabditis* and *Steinernema* (Forst and Clarke, 2002)

Phenotypic trait	*Heterorhabditis*	*Steinernema*
First generation adults	Hermaphroditic	Males and females
Bacterial location	Last 2/3 of intestine	Within specialized intestinal vesicle
Phenotypic relationships[a]	Rhabditida (rhabditidae) and Stongylida	Rhabditida (Strongyloididae) and Rhabditida (Panagrolaimidae)
Retetion of secondary form of bacteria	No	Yes
Infective juveniles	— With cuticular tooth — Excretory pore below nevre ring — Lateral field with 2 lines	— Without cuticular tooth — Excretory pore above nevre ring — Lateral field with 6–8 lines
First generation males	— With bursa — 9 pairs of bursal rays (genital papillae) or a reduction of this number	— Without bursa — 10 or 11 pairs and 1 single genital papillae

[a] After Blaxter *et al.* (1998).

9.3.1. Morphological Characterization of Steinernematidae (Chitwood and Chitwood, 1937) (Adams and Nyguyen, 2002)

Males: Male morphological characters are the most important in the taxonomy of Steinernematidae. Head of males is swollen or not, four cephalic papillae often larger than six labial papillae. Anterior end with or without perioral disc. Posterior region of the males, the number of genital papillae is important. Spicule head is either longer than wide or as long as wide, or wider than long; blade almost straight or curved; velum and rostrum present or absent. Gubernaculum, in ventral view it tapers anteriorly abruptly or gradually; the cuneus is V-shaped, Y-shaped or arrowhead-shaped (Figure 8 (right)). *Females:* Some female characters are important in the taxonomy of Steinernematidae. Anterior end is similar for all species. The appearance, and the present or absent of epiptygma is important. Tail tip round, with mucron, bluntly pointed, can be used as a diagnostic character. *Infective juveniles:* Head with four cephalic papillae and six labial papillae or with only four cephalic papillae without labial papillae. The lateral field pattern and the arrangement of ridges in the filed from head to tail is an important taxonomic character.

9.3.2. Morphological Characterization of Heterorhabditidae (Poinar, 1976) (Adams and Nyguyen, 2002)

Males: Bursa with nine papillae. The bursa structure is similar in all species. Spicules and gubernaculums in ventral view can be used as diagnostic characters (Figure 8. (left)). *Females:* Head region with six forward-directed papillae and is similar for almost all species, except the labial papillae is curved outward in *Heterorhabditis hawaiiensis*. The pattern of the vulva is a good diagnostic character. *Infective juveniles:* Morphological characters of infective juveniles have been poorly investigated and are unreliable for species differentiation. Identification is based primarily on morphometrics and molecular data. In addition to these morphological characters, groups of EPNs are described in direction to body length of infective juveniles. These features are summarized in Table 3.

Table 3. Characterization of *Heterorhabditis* and *Steinernema* species with average body length of IJs (Stock and Hunt, 2005)

Heterorhabditis	*Steinernema*
indica-group (IJs < 550 μm)	carpocapsae-group (IJs < 600 μm)
H. poinari H. indica H. taysaerae	S. asiaticum -S. anatoliense S. siamkayai -S. thermophium S. ritteri -S. carpocapsae S. rarum -S. scapterisci S. tami -S. websteri S. abbasi -S. kushidai
bacteriophora-group (IJs=550-700 μm)	intermedium-group (IJs = 600-800 μm)
H. bacteriophora H. baujardi H. zealandica H. marelata	S. riobrave -S. monticolum S. intermedium -S. sangi S. pakistanense -S. bicornutum S. affine -S. ceratophorum
Heterorhabditis	*Steinernema*
megidis-group (IJs > 550 μm)	feltiae-group (IJs = 800-1000 μm)
H. megidis H. downesii	S. feltiae -S. karii S. thanhi -S. kraussei S. neocurtillae -S. origonense S. scarabaei -S. loci
	glaseri-group (IJs > 1000 μm)
	S. longicornum -S. puertoricense S. caudatum -S. cubanum S. glaseri

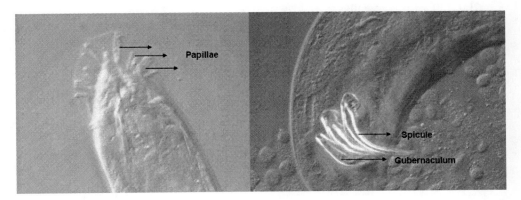

Figure 8. First generation male of tail of *Heterorhabditis* sp. with papillae (left) and that of *Steinernema kraussei* without papillae (right) (Photo: A. Susurluk, Raisdorf, Germany).

9.3.3. Molecular Identification

9.3.3.1. DNA Extraction

Phenol/Chloroform method[*]:

- Add 1 ml extraction buffer to nematode pellet (0.1 M Tris, pH 8.0; 0.2 M NaCl; 5 mM EDTA; 1% SDS; and 2 mg/ml Proteinase K). Incubate at 37 °C for a min. of 4 hours with occasional gentle mixing (N.B. overnight incubation will increase yield of DNA extracted).
- Add an equal volume of phenol equilibrated with 1x TE buffer (10 mM Tris, pH 8.0; 1 mM EDTA). Gently invert tube for 10 minutes and centrifuge at 10.000 g for 5 minutes. Transfer the upper aqueous phase to a fresh 2.0 ml tube and repeat two further times.
- Add an equal volume of chloroform/isoamyl alcohol (24:1 v/v) to the aqueous phase from the step above. Gently invert tube for 10 minutes and centrifuge at 10.000 g for 5 minutes. Transfer the upper aqueous phase to a fresh 2.0 ml tube.
- Add RNAseA to a final concentration of 2 mg/ml and incubate at 37 °C for 1 hour. Add an equal volume of chloroform/isoamyl alcohol (24:1 v/v). Gently invert tube for 10 minutes and centrifuge at 10.000 g for 5 minutes. Transfer the upper aqueous phase to a fresh 10 ml tube.
- Add 100 µl of 3 M Sodium acetate, pH 5.5 and 3 ml of 96% ethanol. Mix contents gently to allow DNA to precipitate. Wash precipitate twice with 70% ethanol (the DNA can be "hooked out" of the ethanol using a heat-sealed slightly bent Pasteur pipette). Briefly air dry the DNA precipitate for approximately 30 seconds and resuspend in 100 to 200 µl of sterile distilled water. Store DNA samples at either 4 °C or – 20 °C.
- The concentration of the DNA can either be determined by UV spectrophotometer or by running a small amount on an agarose gel against known standards. Use 25-50 ng per 50 µl PCR reaction.

DNeasy tissue kit-Qiagen method:

- 200-400 µl infective juvenile (dead or alive 2500-3000 IJs) into 2 ml cap, then centrifuge 1 minute at 14000 rpm. Remove liquid from the cap.
- Add 180 µl Buffer ATL to the cap then vortex for 30 seconds.
- Add 20 µl Proteinase K then vortex for 30 seconds; incubate at 3-5 hours in waterbath at 55 °C then vortex for 15 seconds.
- Add 200 µl Buffer AL then vortex for 1 minute; incubate for 10 minutes at 70 °C in the waterbath.
- Add 200 µl ethanol (96%) then vortex for 1 minute (in this step, the sample must be gently homogenized).
- Pour all samples from this cap to Bind DNA cap, then centrifuge for 1 minute at 8000 rpm. Remove the outher part of the Bind DNA cap. Put on new cap, then add 500 µl Buffer AW1, then centrifuge for 1 minute at 8000 rpm.

---

[*] from Alex Reid, Entomopathogenic Nematode Molecular Identification Course Text

- Put on again new cap, then replace 500 µl Buffer AW2 (highly toxic and wear gloves when handling), then centrifuge for 3 minutes at 14000 rpm.
- Put on another cap, then add 100 µl Buffer AE, wait for 1 minute at room temperature, then centrifuge for 1 minute at 8000 rpm (this step must be done two times).
- DNA of the sample gathered in the cap (beneath the Bind).
- Extracted DNA must be stored at -27 °C until using.

Electrophoresis of DNA:

- Five µl of a DNA product is mixed with 3 µl loading suspension (30 % glycerol, 50 mM EDTA, 0.001 % bromophenol blue and 0.001 % xylencyanol) and 2 µl aqua dest.
- Transferred into the gel pockets.
- One pocket is loaded with 5 µl of a marker ranging from 100 to 1.000 base pairs, which had been treated like the DNA sample.
- The fragments are photographed under UV light (Figure 9).

9.3.3.2. PCR Conditions

PCR amplification of the ITS region is carried out in a reaction volume of 100 µl for each strain, containing 75.6 µl of H_2O, 2 µl of dNTPs (2mM), 10 µl of 10 x PCR-Buffer, 1 µl of Primer Forward (200 µM), 1 µl of Primer Reverse (200µM), 0.4 µl of Taq polymerase (5U/µl) and 10 µl of purified DNA. The primers 18S (5`-TTGATTACGTCCCTGCCCTTT-3`) and 26S (5`-TTTCACTCGCCGTTACTAAGG-3`) are used as forward and reverse primers, respectively (Vrain *et al.,* 1992).

The amplification is carried out using a DNA thermal cycler (Perkin Elmer Cetus, Emeryville, CA, USA). The samples are placed in the thermal cycler, which was preheated to 95 °C and incubated at 94 °C for 3 min.

Amplification is started at annealing temperatures for 1 min at 55 °C followed by an extension period at 72 °C for 1 min 30 s and a denaturalization period at 94 °C for 30 s´. After 40 of these cycles, a final step of 1 min at 94 °C, 3 min annealing temperatures at 55 °C and 5 min at 72 °C of extension to ensure that all of the final amplification products are full length (Reid and Hominick, 1998). At the end of the 40 cycles, the samples are stored at 4 °C.

Electrophoresis of PCR products: Extracted or amplified DNA is subjected to 180 V for 35 minutes. The fragments are photographed under UV light (Reid and Hominick, 1998).

9.3.3.3. Restriction Fragment Length Polymorphism (RFLP) and Electrophoresis of Digested PCR Products

Nine different enzymes were used for the digestion. The digestion enzymes Alu I, Hae III, Hind III were supplied from company Eurogentec, and Dde I, Hha I, Hinf I, Hpa II, Rsa I and Sau 3Al.

The resulting fragments were separated with 2 % agarose gel containing ethidium bromide (0.2 %) in 0.5xTBE (54.0 g Tris base+27.5 g Boric acid+20 ml 0.5 M EDTA (pH 8.0) make up to 1 L with distilled water at 100 volt for 1.5 hours.

The fragments were photographed under UV light. RFLP patterns are compared with published information on described species (Figure 10) (Reid and Hominick, 1998).

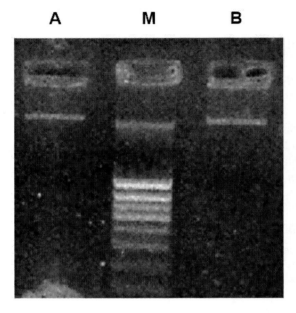

Figure 9. Lane A and B show total genomic DNA of *Steinernema feltiae* and *Heterorhabditis bacteriophora* respectively. Lane M is the molecular weight marker and band size are shown in base pairs (1000, 800, 700, 600, 500, 400, 300, 200, 100).

CONCLUSIONS

The many isolates and species of EPNs found in the world provide a lot of opportunities for biological control strategies. Thus, it is still needed that so many studies on EPNs and apply their results to in/out door experiments. In these studies; isolation, preservation, characterization and identification of EPNs contain very important main methods. Firstly, these initial methods are done, and then additional experiments can be performed. Before study with unknown EPN species or survey study for EPN determination, the following steps must be correctly done; 1. An EPN species is isolated from soil or insects, 2. The isolated species must be preserved at cool temperature and then 3. Examining its characters and to identify the species (characterization and identification). If these steps are completed, then other studies can be performed. The field of entomopathogenic nematology has experienced exponential growth over the past decade. By means of intensive study on EPNs, a great deal knowledge is learned on their methodology. However, day by day, new and useful methods take place in methodological references of EPNs, except some embraced methods. The significance of the relationship between correct methodology and success of an experiment has been recognized for many years. Moreover, to do reliable study, two or more methods can be integrated for getting valuable and certain results. For example; in taxonomic study, besides morphometric results of a species, molecular identification methods should also be used. Addition to definitive methods, some scientist can be improved some useful methods for their study. In this point, it is important to be approved this method by authorities. Moreover, cheapness of a using method is very important factor for its preferring in an experiment. All useful and common methods and their materials were expressed with a particular order in this chapter.

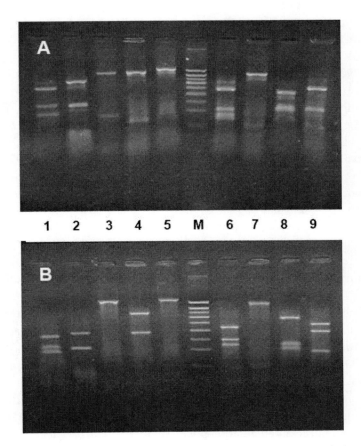

Figure 10. PCR amplified products from the ITS region digested with 9 restriction enzymes. EPNs species are: A. *Steinernema feltiae*; B. *Heterorhabditis bacteriophora*. Lanes 1-9 indicate the following enyzmes: 1. Alu I; 2. Dde I; 3. Hae III; 4. Hha I; 5. Hind III; M. Molecular weight markers (band sizes 1000, 800, 700, 600, 500, 400, 300, 200, 100 base pairs) 6. Hinf I; 7. Hpa II; 8. Rsa I (Afa I); 9. Sau 3 AI.

REFERENCES

Adams, B.J. and Nguyen, K.B. 2002. Taxonomy and Systematic. In: Gaugler, R. (Ed.): *Entomopathogenic Nematology*. CABI Publishing, Oxon, UK, pp. 1-28.

Akhurst, R.J. 1986. Controlling insects in soil with entomopathogenic nematodes. *Fundamental and applied aspects of invertebrate pathology*. In. Samson, R.A, Vlak, J.M and Peters, D. (Eds). *Proceeding of 4th International Colloquium of Invertebrate Pathology*: 265-267.

Barbercheck, M.E. and Kaya, H.K. 1991. Effect of host condition and soil texture on host finding by the entomogenous nematodes *Heterorhabditis bacteriophora* (Rhabditida: Heterorhabditidae) and *Steinernema carpocapsae* (Rhabditida: Steinernematidae). *Environmental Entomology* 20: 582-589.

Bedding, R.A. and Akhurst, R.J. 1975. A simple technique for the detection of insect parasitic nematodes in soil. *Nematologica* 21, 109-110.

Bednarek, A. 1998. The agricultural system, as a complex factor, effects the population of entomopathogenic nematodes (Rhabditida: Steinernematidae) in the soil, *IOBC Bulletin* 21, 155–216.

Begley, J.W. 1990. Efficacy against insects in habitats other than soil.. In: Randy, G. and. Harry, K.K. (Eds), *Entomopathogenic nematodes in biological control-* Boca Raton, CRC Press Inc. Florida, USA, pp. 215-231.

Blaxter, M.L., De Ley, P., Garey, J., Liu, L.X., Scheldeman, P., Vierstraete, A., Vanfleteren, J., Mackey, L.Y., Dorris, M., Frisse, L.M., Vida, J.T. and Thomas, W.K. 1998. A molecular evolutionary framework for the phylum Nematoda. *Nature* 392, 71–75.

Boemare, N.E., Laumond, C. and Mauleon, H. 1996. The entomopathogenic nematode-bacterium complex: Biology, life cycle and vertebrate safety. *Biocontrol Science and Technology* 6: 333-346.

Chitwood, B.G. and Chitwood, M.B. 1937. *An Introduction to Nematology*.Monumental Printing Company, Baltimore, Maryland, 213 pp.

Curran, J. and Heng, J. 1992. Comparison of three methods for estimating the number of entomopathogenic nematodes present in soil samples. *Journal of Nematology* 24(1): 170-176.

Ehlers, R.-U. 1996. Current and future use of nematodes in biocontrol: Practice and commercial aspects in regard to regulatory policies. *Biocontrol Science and Technology* 6(3): 303-316.

Ehlers, R.-U. 2001. Mass production of entomopathogenic nematodes for plant protection. *Applied Microbiological Biotechnology* 56: 623-633.

Ehlers, R.-U. and Hokkanen, H.M.T 1996. Insect biocontrol with non-endemic entomopathogenic nematodes (*Steinernema* and *Heterorhabditis* sp.): OECD and COST workshop on scientific and regulation policy issue. *Biocontrol Science and Technology* 6: 295-302.

Ehlers, R.-U. and Peters, A. 1995. Entomopathogenic nematodes in biological control: Feasibility, perspectives and possible risks. In: Hokkanen H.M.T and. Lynch J.M. (Eds), *Biological Control: Benefits and risks*. Cambridge University Press, Cambridge, UK, pp. 119-136.

Forst, S. and Clarke, D. 2002. Bacteria-nematode symbiosis. In: Gaugler, R. (Ed.): *Entomopathogenic Nematology*. CABI Publishing, Oxon, UK, p. 61.

Forst, S. and Nealson, K. 1996. Molecular biology of the symbiotic pathogenic bacteria *Xenorhabdus* and *Photorhabdus* spp. *Microbiology Review*, 60, 21–43.

Fuchs, 1915. Die Naturgeschichte der Nematoden und einiger Parasitien. *Zoologie Jahrbücher Abteilung System* 38.

Gaugler, R. 1988. Ecological considerations in the biological control of soil-inhabiting insects with entomopathogenic nematodes. *Agricultural Ecosystems Environment* 24: 351-360.

Georgis, R. 1992. Present and future prospects for entomopathogenic nematode products. *Biocontrol Science and Technology* 2, 83–99.

Georgis, R. and Kaya, H.K. 1998. Formulation of entomopathogenic nematodes. In: Burges, H.D. (Ed), *Formulation of microbial biopesticides: Beneficial microorganisms, nematodes and seed treatments*. Dordrecht, Kluwer Academic Publishers, pp. 289-308.

Georgis, R. and Mamweiler, S.A. 1994. Entomopathogenic nematodes: A developing biological control technology. *Agricultural Zoology Review* 6, 63-94.

Glazer, I. 2002. Survival biology. In: Gaugler, R. (Ed.), *Entomopathogenic Nematology*, CABI Publishing, Oxon, UK, pp. 169-187.

Griffin, C.T., Boemare, N.E. and Lewis, E.E. 2005. Biology and Behaviour. In: Grewal, P.S., Ehlers, R.-U. and Shapiro-Ilan, D.I. (Eds), *Nematodes as Biocontrol Agents*. CABI Publishing, Oxfordshire, UK, pp. 47-64.

Hass, B., Griffin, C.T. and Downes, M.J. 1999. Persistence of Heterorhabditis infective juveniles in soil: comparison of extraction and infectivity measurements. *Journal of Nematology* 31(4): 508-516.

Hominick, W.M., Reid, A.P., Bohan, D.A. and Briscoe, B.R. 1996. Entomopathogenic nematodes: biodiversity, geographical distribution and the Convention on Biological Diversity. *Biocontrol Science and Technology* 6, 317–331.

Hui, E. and Webster, J.M. 2000. Influence of insect larvae and seedling roots on the host-finding ability of *Steinernema feltiae* (Nematoda: Steinernematidae). *Journal of Invertebrate Pathology* 75: 152-162.

Johnigk, S.-A. and Ehlers, R.-U. 1999. Juvenile development and life cycle of *Heterorhabditis bacteriophora* and *H. indica* (Nematoda: Heterorhabditidae). *Nematology* 1(3): 251-260.

Kaya, H.K. and Stock, P. 1997. Techniques insect nematology. In: Lacey, L. (Ed.), *Manual of techniques in insect pathology*, Academic Press San Diego, USA, pp. 282-324.

Koppenhöfer, A.M., Campbell, J.F., Kaya, H.K. and Gaugler, R. 1998. Estimation of entomopathogenic nematode population density in soil by correlation between bait insect mortality and nematode penetration. *Fundamental Applied Nematology*. 21(1): 95-102.

Kung, S.P., Gaugler, R. and Kaya, H.K. 1990. Soil type and entomopathogenic nematode persistence. *Journal of Invertebrate Pathology* 55: 401-406.

Poinar, G.O. 1976. Description and biology of a new insect parasitic rhabditoid, *Heterorhabditis bacteriophora* n.gen.n.sp.(Rhabditida: Heterorhabditidae n.fam.). *Nematologica* 21, 463–470.

Reid, A.P. and Hominick, W.M. 1993. Cloning of the rDNA repeat unit from a British entomopathogenic nematode (Steinernematidae) and its potential for species identification. *Parasitology* 107, 529–536.

Reid, A.P. and Hominick, W.M. 1998. Molecular taxonomy of *Steinernema* by RFLP analysis of the ITS region of the ribosomal DNA repeat unit. In: Abad, P., Burnell, A., Laumond, C., Boemare, N. and Coudert, F. (Eds), *COST 819 Entomopathogenic nematodes -Genetic and molecular biology of entomopathogenic nematodes*. Brussels, Luxembourg, 18261: 87-93.

Saunders, J.E. and All, J.N. 1982. Laboratory extraction methods and field detection of entomophilic rhabditoid nematodes from soil. *Environmental Entomology* 7: 605-607.

Stock, S.P. and Hunt, D.J. 2005. Morphology and Systematics of Nematodes Used in Biocontrol. In: Grewal, P.S., Ehlers, R.-U. and Shapiro-Ilan, D.I.(Eds), *Nematodes as Biocontrol Agents.* CABI Publishing, Oxfordshire, UK, pp. 3-40.

Sturhan, D. 1995. Untersuchungen über sympatrisches Vorkommen entomopathogener Nematoden. *Nachrichtenbülten Deutsche Pflanzenschutz,* 47: 54.

Sturhan, D. and Mracek, Z. 2000. Comparison of the Galleria baiting technique and a direct extraction method for recovering *Steinernema* (Nematoda: Rhabditida) infective-stage juveniles from soil. *Folia Parasitologica* 47(4): 315-318.

Susurluk, A. and Ehlers, R.-U. 2008. Sustainable control of black vine weevil larvae, *Otiorhynchus sulcatus* (Coleoptera: Curculionidae) with *Heterorhabditis bacteriophora* in strawberry. *Biocontrol Science and Technology,* 18 (6), 635-640.

Susurluk, A., Dix, I., Stackebrandt, E., Strauch, O., Wyss, U. and Ehlers, R.-U. 2001. Identification and ecological characterisation of three entomopathogenic nematode-bacterium complexes from Turkey. *Nematology* 3 (8): 833-841.

Susurluk, I.A. 2009. Seasonal and vertical distribution of the entomopathogenic nematodes, *Heterorhabditis bacteriophora* (TUR-H2) and *Steinernema feltiae* (TUR-S3), in turf and fallow areas. *Nematology,* 11 (3): 321–327.

Wiesner, A. 1993. Die Induktion der Immunabwehr eines Insekts (*Galleria mellonella,* Lepidoptera) durch synthetische Materialien und arteigene Haemolymphfaktoren. Berlin University, Berlin, Germany.

Vrain, T.C., Wakarchuk, D.A., Levesque, A.C. and Hamilton, R.I. 1992. Intraspecific rDNA restriction fragment length polymorphism in the *Xiphinema americanum* group. *Fundamental Applied Nematology* 15(6): 536-573.

In: Microbial Insecticides: Principles and Applications   ISBN: 978-1-61209-223-2
Editors: J. Francis Borgio, K. Sahayaraj, et al.   © 2011 Nova Science Publishers, Inc

Chapter 10

MODE OF ACTION AND FIELD EFFICACY
OF ENTOMOPATHOGENIC NEMATODES

*Atwa, A. Atwa**

King Abdul Aziz University (KAU), Jeddah, King of Saudi Arabia (KSA)
Plant Protection Research Institute (PPRI), Agricultural Research Center(ARC), Ministry
of Agriculture and Land Reclamation, Dokkii, Giza, Egypt

ABSTRACT

Entomopathogenic nematodes (EPNs) in the genera *Heterorhabditis* and *Steinernema* (Rhabditida) are obligate parasites of insects. Infective juveniles (IJs), the only stage of the nematodes found in the soil, enter hosts through natural openings such as the mouth, anus or spiracles, but IJs of some species can also enter through the cuticle. The two nematode families *Steinernematidae* (61 species) and *Heterorhabditidae* (14 species), contain the insect parasitic nematode species. The most commonly used beneficial nematodes are *Steinernema carpocapsae, S. feltiae, S. glaseri, S. scarabaei, Heterorhabditis bacteriophora* and *H bacteriophora*. Nematodes that are endoparasites of insects attack a wide variety of agricultural pests. Different EPNs species and strains exhibit differences in searching behavior which make them more or less suitable for insect pest infectivity, e.g., *S. glaseri* dispersed up to 90 cm in a sandy soil, while some species of *Heterorhabditis* migrate very actively through the soil, others such as *S. carpocapsae* migrate less and many nictate on solid surface when relative humidities are high, but become inactive in soil in the absence of hosts. Generally, *Heterorhabditid* IJs more migratory than those of *Steinernematids*. *Steinernema carpocapsae, S. scapterisci, S. glaseri*, and *H. bacteriophora* responded with behavioral changes to host feces. Head thrusting was presumed to be a penetration behavior that could be related to successful parasitism. EPNs can penetrate to the hemocoel of a host through the cuticle, through the wall of the gut via the anus/mouth or through the tracheal cuticle via the spiracles. Thus, putative penetration stimuli may be located at these locations. Since penetration into non-suitable hosts is a dead end for that specific EPNs genotype, it can be expected

---

* E-mail: atwaradwan@hotmail.com, (00966-54)2108625 and (00966-2) 6951945

that EPNs recognize suitable hosts and that the response to host-associated stimuli and the ultimate success in establishing and propagating in a host is correlated. Influence of EPNs factors on the decision; both extrinsic (presence of other nematodes), and intrinsic (such as sex and age) to the deciding EPN will be discuss in this chapter. For a *Steinernematid* to reproduce, there must be at least one member of the opposite sex present in the cadaver. The *Steinernema* life cycle includes two developmental pathways: juveniles develop directly to adults within a host cadaver as long as conditions are favorable, but in less favorable conditions (generally believed to be due to crowding and resource depletion, developmentally arrested IJs are formed which leave the cadaver to search for another host. Field application showed that the EPNs of the genera *Steinernema* and *Heterorhabditis* are effective biological control agents against a wide variety of soil insect pests and for various cropping systems, such as the black vine weevil, *Otiorhynchus sulcatus* (F.), diaprepres root weevil, *Diaprepes abbreviates* (L.), fungus gnats (Diptera; Sciaridae), various white grubs (Coleoptera; Scarabaeidae) and some lepidopterous insects "the leopard moth, *Zeuzera pyrina*, the Egyptian cotton leafworm, *Spodoptera littoralis,*and the cabbage looper, *Pieris brassica*. Nematode attributes like infectivity, persistence, availability, and cost are important factors growers must consider when deciding to utilize these biological alternatives. The ability to maintain an active nematode population in the soil for an extended period of time is an important factor in long-term management of this pest and could be a major economic factor in the adoption of nematodes by the growers. Finally, this chapter will discus and explain the mode of action and field efficacy of entomopathogenic nematodes.

10.1. INTRODUCTION

Nematodes generally, are simple, colorless, unsegmented, roundworms, lacking appendages. Nematodes may be free-living, predaceous, or parasitic, and many of the parasitic species cause important diseases of plants, animals, and humans. The only insect parasitic nematodes or entomopathogenic nematodes (EPNs) possessing an optimal balance of biological control attributes are entomopathogenic (also referred to as "beneficial" or "insecticidal") nematodes in the genera *Steinernema* and *Heterorhabditis*. EPNs are extraordinarily lethal to many important soil insect pests, yet are safe for plants and animals. Most biologicals require days or weeks to kill, yet nematodes, working with their symbiotic bacteria, kill insects in 24-48 hr. Dozens of different insect pests are susceptible to infection, yet no adverse effects have been shown against non-targets in field studies.

Entomopathogenic nematodes (EPNs) in the genera *Heterorhabditis* and *Steinernema* (Rhabditida) are obligate parasites of insects (Poinar, 1990). These nematodes have a symbiotic relationship with bacteria in the genera *Photorhabdus* and *Xenorhabdus*, respectively (Forst and Clarke, 2002). Infective juveniles (IJs), the only stage of the nematodes found in the soil, enter hosts through natural openings such as the mouth, anus or spiracles, but IJs of some species can also enter through the cuticle. After penetrating into the host's hemocoel, the nematodes release their symbiotic bacteria, which usually kill the host within 24 to 48h. The bacteria are also responsible for antibiotic production and for providing nutrition for the nematodes (Dowds and Peters, 2002). The nematodes feed, develop, mate, and often complete 2–3 generations within the host cadaver. When resources within the cadaver are depleted, a new generation of IJs is produced and leaves the cadaver to search for new hosts (Kaya and Gaugler, 1993).

In fact, EPNs possess many attributes of an ideal microbial insecticidal or biological control agents: They have a wide host spectrum, are environmentally safe, can be produced in large-scale bioreactors, are easily applied, are compatable with most chemical pesticides, are applied in deverse climatic condition and, are capable of finding hosts in soil (Garcia Del Pino and Palomo , 1996). In addition, the use of naturally occurring nematodes in a particular area as biological control agents may also reduce the risk to nontarget organisms when compared with exotic isolates (Blackshaw, 1988). We can conclude that; EPNs are beneficial for six reasons. First, they have such a wide host range that they can be used successfully on numerous insect pests. The nematodes' nonspecific development, which does not rely on specific host nutrients, allows them to infect a large number of insect species. Second, nematodes kill their insect hosts within 48 hours. As mentioned earlier, this is due to enzymes produced by the *Xenorhabdus* and *Photorhabdus* bacteria. Third, nematodes can be grown on artificial media. This allows for commercial production which makes them a more available product. The nematodes can stay viable for months when stored at the proper temperature. Usually three months at a room temperature of 15 to 25 $\pm$ 2 $^{\circ}$C and six months when refrigerated at 5 to 10°C. They can also tolerate being mixed with various insecticides, herbicides and fertilizers. Also, the infective juveniles can live for some time without nourishment as they search for a host. Fifth, there is no evidence of natural or acquired resistance to the *Xenorhabdus* and *Photorhabdus* bacteria. Though there is no insect immunity to the bacteria, some insects, particularly beneficial insects, are possibly less parasitized because nematodes are less likely to encounter beneficials which are often very active and escape nematode penetration by quickly moving away.

Finally, there is no evidence that EPNs or their symbiotic bacteria can develop in vertebrates. This makes nematode use for insect pest control safe and environmentally friendly. The United States Environmental Protection Agency (EPA) has ruled that nematodes are exempt from registration because they occur naturally and require no genetic modification by man. In this chapter we will discuss the review into two sections that describe the mode of action and field efficacy of EPNs. The first section describes the penetration mechanism or how EPNs recognize hosts and how they gain entrance into the host hemocoel, nematodes-bacteria complex interaction and host-finding behaviors and strategies. In the second section, we will submit some models of EPNs field application and factors affecting on nematodes efficacy.

10.2. MODE OF ACTION OF ENTOMOPATHOGENIC NEMATODES

Nematodes are morphologically, genetically and ecologically diverse organisms occupying more varied habitats than any other animal group except arthropods. These naturally occurring organisms are microscopic, unsegmented round worms that live in the soil and, depending on the species, infect plants and animals. The two nematode families *Steinernematidae* (61 species) and *Heterorhabditidae* (14 species), contain the insect parasitic nematode species. The most commonly used beneficial nematodes are *Steinernema carpocapsae, S. feltiae, S. glaseri, S. scarabaei, Heterorhabditis bacteriophora* and

H bacteriophora. Nematodes that are endoparasites of insects attack a wide variety of agricultural pests.

10.2.1. Host-Finding Ability

Different EPNs species and strains exhibit differences in searching behavior which make them more or less suitable for insect pest infectivity, e.g., *S. glaseri* dispersed up to 90 cm in a sandy soil (Kaya, 1990), while some species of *Heterorhabditis* migrate very actively through the soil (Smits *et al.,* 1991; Susurluk, 2009), others such as *S. carpocapsae* migrate less and many nictate on solid surface when relative humidities are high, but become inactive in soil in the absence of hosts (Ishibashi *et al.,* 1994). Generally, heterorhabditid IJs more migratory than those of steinernematids (Downes and Griffin, 1996). In seeking new hosts, EPNs that search by moving throughout their environment to find hosts are termed "cruisers" while those that wait for hosts to come to them are termed "ambushers". *S. glaseri* is cruisers that actively move in the soil, respond strongly to host cues and are adapted to infect sedentary hosts (Campbell and Gaugler, 1993). In contrast, *S. carpocapsae* is ambusher that stay near the soil surface and does not disperse into the soil, is unresponsive to host cues (Lewis *et al.,* 1992), and is adapted to infect mobile hosts on the soil surface (Campbell and Gaugler, 1993). However, cruiser and ambusher behaviors reflect different balances of advantage for the species that display them. Movement increases the probability of encounter with a stationary host, but also with the nematode's natural enemies. Furthermore, an active nematode undoubtedly uses up its limited reserves more quickly. Regarding the attraction of nematodes to insect hosts, EPNs have been shown to respond positively to chemical gradient around the host, carbon dioxide and thermal gradients, materials from hosts or their feces. Further, they can be activated by thermal or mechanical shock, and by certain chemicals (Gaugler and Campbell, 1991).

When a nematode has reached the surrounding area of an insect host, it must change its behavior to gain entrance into the hemocoel of that host. These behaviors could be energetically expensive and penetration into an unsuitable host could potentially be fatal to a nematode, due to either defense against infection (a lethal composition of gut fluids or a strong immune response against the nematode) or some other condition that would kill the infecting nematode. Ultimately, nematodes choosing an unsuitable host will not produce offspring and those choosing the suitable ones will. A certain capability of recognizing a suitable host before trying to enter can therefore be expected in EPN populations. In contrast to "host finding," the term "host recognition" describes the reaction towards host stimuli which ultimately result in penetration into the host hemocoel (Lewis *et al.,* 2006).

The mechanism which EPNs to reach the host, most likely that EPNs react to chemical stimuli or that they sense the physical structure of the insect's integument. The stimuli might be associated with the host directly or with their by-products, like feces or volatile by-products of the insect's metabolism. Whether signals from the plants on which insect hosts feed are involved in host recognition has not been investigated (Lewis *et al.,* 2006). The penetration rate, which is generally determined by counting the number of nematodes that enter an insect host, could serve as an indicator of how "stimulated" a population of EPNs was by a particular host. However, there are too many other factors affecting penetration, like the physiological status of the insect, the impact of wounds to the insect's cuticle which might stimulate penetration of further nematodes, or the establishment of the symbiotic bacteria in the host which might trigger secondary invasion (Hay and Fenlon, 1995) or deter nematodes from entering (Glazer, 1997). A reliable screening for penetration stimuli must rely on specific events in the penetration cascade, not just how many EPNs end up inside a host. Recording the electrical activity of neurons is widely used to screen for stimuli in insects. Dolan *et al.* (2002) have used nucleic acid binding SYTO dyes to detect early events in dauer juvenile recovery of *Heterorhabditis bacteriophora* Poinar. By analyzing the regions where transcriptional activity is evoked by host stimuli, the physiological changes during host recognition can be elucidated. For instance, if enzymes are involved in the penetration process, transcriptional activity in the salivary glands of the dauer juveniles would be expected (Lewis *et al.,* 2006).

Steinernema carpocapsae, S. scapterisci, S. glaseri, and H. bacteriophora responded with behavioral changes to host feces (Grewal *et al.,* 1993b). Head thrusting was presumed to be a penetration behavior that could be related to successful parasitism. EPNs can penetrate to the hemocoel of a host through the cuticle, through the wall of the gut via the anus/mouth or through the tracheal cuticle via the spiracles. Thus, putative penetration stimuli may be located at these locations. Since penetration into non-suitable hosts is a dead end for that specific EPNs genotype, it can be expected that EPNs recognize suitable hosts and that the response to host-associated stimuli and the ultimate success in establishing and propagating in a host is correlated (Lewis *et al.,* 2006). This was partially found for *S. glaseri, S. carpocapsae, S. scapterisci,* and *H. bacteriophora* with the hosts *Spodoptera exigua* (Hübner), *Popillia japonica* Newman, *Blatella germanica* (L.), and *Achaeta domestica* (L.) (Grewal *et al.,* 1993a). The reaction of *S. carpocapsae* to contact with the integument of nine different insect species was studied by Lewis *et al.* (1996). Excitement was indicated by significantly faster movement towards volatile host cues after contact with the cuticle of a putative host. The nematodes responded differently to different host species and even to different stages of the same host species [larvae versus pupae of *Agrotis ipsilon* (Hufnagel). The levels of response were positively correlated with the susceptibility of the insect hosts tested and with reproductive success of the nematodes within the host (Lewis *et al.,* 2006). A notable exception to this rule was larvae of *P. japonica* which were not susceptible although *S. carpocapsae* responded positively to cuticle contact with this insect. It is likely that mechanical barriers (sieve plates that are located on the entrance to the spiracles in this species) exclude *S. carpocapsae* from the hemocoel. Hence, for *P. japonica*, host specificity acts at the time of penetration rather than recognition. The ambusher *S. carpocapsae* resides at the soil surface and would therefore rarely encounter root-feeding scarab grubs. Due to this spatial separation, avoiding the attempted penetration into *P. japonica* was probably not under strong selection pressure in *S. carpocapsae*. The nature of the materials in the insect's hemocoel that stimulate penetration remains unclear (Lewis *et al.,* 2006). Comparing three lepidopteran hosts, Khlibsuwan *et al.* (1992) reported a positive correlation between larval susceptibility and the attractiveness of their cell-free hemolymph to *S. carpocapsae*. Aqueous surface washings from *T. oleracea* triggered penetration behavior in *S. feltiae*. Possibly, hydrophilic components of the hemolymph which may be small enough to diffuse through the insect's integument trigger penetration behavior (Lewis *et al.,* 2006). Those substances should also be present in the tracheae and the intestine. Higher concentrations can be expected to diffuse at the intersegmental membranes which are known to be preferred penetration sites (Bedding and Molyneux, 1982; Blossey and Ehlers, 1991).

Many of publications illustrated another site of penetration by entomopathogenic nematodes is the integument or the intersegmental membranes of an insect. Penetration through the integument was shown to be the main route of entry for *S. feltiae* into leatherjackets (Peters and Ehlers, 1994). Another port of entry to adult arthropods is the gonad openings. This is the main entry port for nematodes into ticks (Samish and Glazer, 1992). After successfully entering the tracheal system or the intestine, the IJ still must pass through the tracheole or the gut wall, respectively. The fragile tracheole wall might be penetrated just by the mechanical pressure of the pointed nematode head. The gut wall, however, is in part protected by the peritrophic membrane, which can be a serious obstacle for nematodes (Forschler and Gardner, 1991). *Steinernema glaseri* IJs take 4–6 h to penetrate from the midgut to the hemocoel of *P. japonica* (Cui *et al.,* 1993); the nematodes preferably penetrate the midgut region and the gastric caecae. Infective juvenile *H. bacteriophora* were observed using their proximal tooth to rupture the insect body wall (Bedding and Molyneux, 1982) and it was long believed that only *Heterorhabditis* could penetrate tissues like the insect's integument, since IJs of *Steinernema* spp. lack the apical tooth. Reports of superior penetration of *S.glaseri* compared to *H. bacteriophora* through gut tissues of grubs (Wang and Gaugler, 1998) and of the penetration of *S. feltiae* through the integument of leatherjackets (Peters and Ehlers, 1994) challenged this perspective.

There is evidence that nematode secretions are involved in penetration, at least in *Steinernema* species. Protease inhibitors decreased penetration of *S. glaseri* through the gut wall of *P. japonica* (AbuHatab *et al.,* 1995) and the midgut epithelium cells of *G. mellonella* showed a marked histolysis in response to secretions of axenic *S. carpocapsae* (Simoes, 1998). Penetration by *S. feltiae* through the cuticle of leatherjackets might be attributed to the absence of an epicuticular wax layer, which may block the activity of histolytic enzymes (Dowds and Peters, 2002). When penetrating into the insect's hemocoel, the IJ encounters the non-self response by the immune system of the host. Nematodes can be trapped into cellular or noncellular capsules. The non-cellular capsules, consisting of melanin, are formed rapidly. In *T. oleracea*, IJs of *S. feltiae* have been found stuck in the cuticle and completely melanized (Gouge, 1994; Peters, 1994). Encapsulation of entomopathogenic nematodes has been reported in Orthoptera, Coleoptera, Diptera, and Lepidoptera (Dowds and Peters, 2002). Whether or not encapsulation occurs depends on the particular nematode species–insect species combination. In *Acheta domesticus*, the nematodes *S. carpocapsae* and *H. bacteriophora* are encapsulated, whereas *S. scapterisci* is not (Wang *et al.,* 1994). Interestingly, *S. scapterisci* has been found naturally associated with the orthopteran *Scapteriscus vicinus* Scudder. Similarly, *S. glaseri*, which is often found associated with scarab larvae, is not encapsulated in *P. japonica*, in contrast to *S. carpocapsae* and *H. bacteriophora*, both of which elicit a strong encapsulation response. These Wndings suggest that nematodes are not encapsulated in hosts similar to those with which they are naturally associated. Nematodes may resist encapsulation in insects by either avoidance of being recognized (evasion), by overwhelming the immune system by multiple infections and disrupting encapsulation (tolerance), or by actively suppressing the encapsulation response (suppression) (Dowds and Peters, 2002). The presence of the symbiotic bacteria increased encapsulation of *S. feltiae* in leatherjackets (Peters and Ehlers, 1997). At the same time, however, the bacteria suppress the immune response since they adhere to and kill the hemocytes (Dunphy and Webster, 1988). The period from nematode invasion to bacterial release is hence crucial for counteracting the encapsulation response. The physiological and behavioral changes following host recognition are likely to increase the energy consumption of the nematodes. Increased energy consumption will subsequently shorten the life of the non-feeding IJ. Hence, unsuccessful penetration attempts must be costly. It is not exactly known, however, at what step in the cascade of events the IJ has irreversibly switched to a parasitic stage (Lewis *et al.,* 2006).

10.2.2. Host and EPNs Action Mechanism

EPNs not use a host insect represents not only a source of food but also a mating rendezvous, and so the decision to infect may be shaped by the need both to find suitable partners and resources and to avoid competition. The decision to infect (as well as the outcome of the attempt to infect subsequent to that decision) will obviously be influenced by host species and stage (Lewis *et al.,* 2006). Influence of EPNs factors on the decision; both extrinsic (presence of other nematodes), and intrinsic (such as sex and age) to the deciding EPN will be discuss in this chapter. For a steinernematid to reproduce, there must be at least one member of the opposite sex present in the cadaver. The Steinernema life cycle includes two developmental pathways: juveniles develop directly to adults within a host cadaver as long as conditions are favorable, but in less favorable conditions (generally believed to be due to crowding and resource depletion (San-Blas *et al.,* 2008), developmentally arrested IJs are formed which leave the cadaver to search for another host. IJs may develop from eggs laid by generations under poor conditions such as a crowded cadaver, but may also develop in each generation from eggs which hatch within the mother i.e., in response to locally crowded conditions (Baliadi *et al.,* 2004). The short life cycle, ready availability of native hosts and diversity of species make Steinernema an attractive model for addressing many biological questions (Stock, 2005). Even for *heterorhabditids* species in which most of the IJs develop into self-fertile adults "hermaphrodites" (Stock *et al.,* 2004), while the presence of conspecifics may be advantageous, facilitating outcrossing in subsequent generations in both *heterorhabditids* and *Steinernema* species. There may in addition be a requirement for "mass attack" or invasion by sufficient number of IJs to overcome the host's defenses (Peters and Ehlers, 1997). Above this minimum number required to provide mating partners and co-attackers, every additional invading nematode is also a potential competitor. As crowding increases, the reproductive output per invading nematode is reduced (Lewis *et al.,* 2006) and at very high densities no IJs at all are produced from the cadaver. However, in the laboratory, both steinernematids and heterorhabditids continue to invade crowded hosts, reaching numbers well in excess of the host's carrying capacity. Moreover, it was reported that no change in the proportion of nematodes invading over a range of exposure concentrations (Ryder and Griffin, 2002). Other studies, particularly those including a wider range of concentrations, have noted a decline in the proportion of nematodes invading with increasing concentration (Booth *et al.,* 2000) suggesting that under these conditions at least some nematodes detect and avoid overcrowded hosts. Such experiments, in which insects are simultaneously exposed to large numbers of IJs, probably do not reflect conditions in the Weld, where encounters would be spread out over longer periods, and so may fail to detect mechanisms for avoiding or deterring invasion into occupied hosts (Lewis *et al.,* 2006).

Moreover it is possible that the nematodes natural hosts emit signals in response to crowding that are not produced by the "unnatural" wax moth host used in lab studies.

EPNs male of steinernematids are the colonizing sex; that they disperse, locate and establish in distant hosts before females, and that parasitism by males renders the hosts more attractive to females (Grewal *et al.,* 1993c). Evidence that males are more highly represented amongst the dispersing fraction than in the base population was presented for four of the five *Steinernema* species tested, including *S. glaseri.* The male colonization hypothesis remains controversial in relation to *Steinernema. S. feltiae,* the only species not showing male-biased dispersal in Grewal *et al.,* (1993c) assays, may actually have a female-first strategy: Bohan and Hominick (1997) reported a markedly female-biased sex ratio during the initial stages of the infection of *G. mellonella,* but with later invasion the sex ratio became balanced, as was the underlying population sex ratio. May be the small size of *S. feltiae's* dipteran hosts accounted for it not having a male-colonizer strategy (Grewal *et al.,* 1993c).

Infectivity of EPNs was studied by Hominick and Reid (1990) whom proposed that on emergence from a host, some individuals are immediately infective, while others become dormant for a time. In this way, IJs produced from the same mother avoid competition with each other in crowded hosts near the source (the infected host from which they emerged). There are currently two concepts of "phased infectivity" in EPNs. The first model, is based on a switch between non-infectious and infectious nematodes, while the second, is based on graded levels of infectiousness (Lewis *et al.,* 2006). According to the more popular switch model, individuals in a population are either infectious or non-infectious, and the proportion of non-infectious individuals in the population may change over time. However, subsequently it was shown that over the period during which there is an increase in infectivity of *H. megidis,* there is a decrease in locomotion and responsiveness to host volatiles, but an increase in "head-thrusting" in the absence of a host, which may reflect an increase in penetration motivation (Dempsey and Griffin, 2002). An increase in infectivity was also reported for *H. bacteriophora* stored at 22°C (Wojcik *et al.,* 1986); this may represent (at least in part) a switch of some individuals from a non-infectious to infectious state (phased infectivity). Campbell *et al.* (1999) found that, evidence for the existence of a non-infectious proportion in *H. bacteriophora.*

Steinernema and *Heterorhabditis* nematodes have similar life histories. The non-feeding IJs seek out insect hosts, especially in the soil environment. They search out susceptible hosts, primarily insect larvae, by detecting excretory products, carbon dioxide and temperature changes. When a host has been located, the nematodes penetrate into the insect body, usually through natural body openings (mouth, anus, spiracles) or areas of thin cuticle. Heterorhabditid nematodes can also pierce through the insect's body wall using the dorsal tooth. The juvenile form of the nematode carries *Xenorhabdus* sp. bacteria (for steinernematids) and *Photorhabdus* sp. (for Heterorhabditids) in their pharynx and intestine. Once the bacteria are introduced into the insect host, death of the host usually occurs in 24 to 48 hours. When the bacteria enzymatically breaks down the internal structure of the insect, the nematodes feed upon the bacteria and the insect tissue that has been broken down by the bacteria (liquefying insect), and mature into adults. The life cycle of *Photorhabdus* and *Xenorhabdus* begins and ends with the colonization of the intestinal tract of a soil-dwelling and non-feeding stage of the nematode known as the IJ (Figure 1). Once their development has reached the third juvenile stage, the nematodes exit the remains of the insect body. Thus, EPNs are a nematode-bacterium complex.

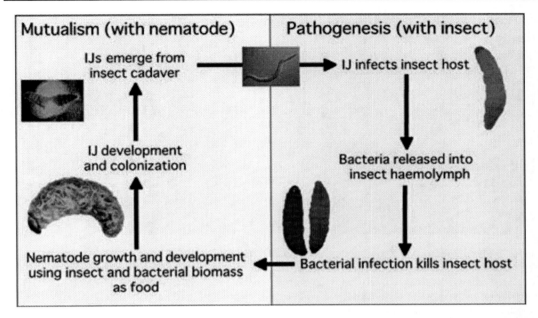

Figure 1. The life cycle of Photorhabdus and Xenorhabdus. The bacteria are normally found colonizing the gut of infective juvenile (IJ) stage nematodes from the families Heterorhabditidae and Steinernematidae respectively. During their life cycle both Photorhabdus and Xenorhabdus have pathogenic and mutualistic interactions with their different invertebrate hosts, i.e. the insect and nematode. After Goodrich-Blair and Clarke 2007.

The nematode may appear as little more than a biological syringe for its bacterial partner, yet the relationship between these organisms is one of classic mutualism. Nematode growth and reproduction depend upon conditions established in the host cadaver by the bacterium. In turn, the bacterium contributes anti-immune proteins to assist the nematode in overcoming host defenses, and anti-microbials that suppress colonization of the cadaver by competing secondary invaders. Steinernematid infective juveniles may become males or females, whereas heterorhabditids develop into self-fertilizing hermaphrodites although subsequent generations within a host produce males and females as well. The life cycle is completed in a few weeks, and hundreds of thousands of new infective juveniles emerge in search of fresh insect hosts.

10.2.3. Nematode Colonization

A key stage in the nematode–bacteria association is bacterial recolonization of IJ, as it is the IJ that vectors the symbiont to a new insect host (Figure 2). This relationship is specific and each species of *Heterorhabditis* or *Steinernema* nematode isolated from nature is colonized by a specific bacterial phylotype and nematodes grown in the laboratory on non-cognate bacteria typically produce uncolonized IJs. The level of specificity observed in the bacteria–nematode interaction is undoubtedly dictated by events occurring at the molecular and cellular interface between the host and the bacteria during the colonization process (Goodrich-Blair and Clarke 2007). *Photorhabdus* colonize the proximal region of the gut of the IJ and intimately contact the gut epithelia (Figure 2).

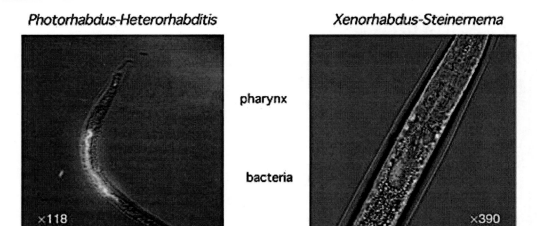

Figure 2. Photorhabdus and Xenorhabdus colonize different sites in the IJ stage of their nematode hosts. Nematodes were grown on GFP-labelled bacteria, the IJs were collected and the bacteria were visualized using fluorescence microscopy. In both photomicrographs the bacteria are located just posterior to the pharynx of the nematode. However, it is clear that Photorhabdus bacteria occupy a substantial fraction of the lumen of the nematode gut while Xenorhabdus appear to be localized within a vesicle. The numbers indicate the level of magnification used to capture the image. After Goodrich-Blair and Clarke 2007.

Until recently it was assumed that *Photorhabdus* colonizing bacteria were derived from the insect cadaver. However, careful microscopic observations revealed that they are instead maternally transmitted. *Heterorhabditis* IJ formation requires endotokia matricida, a process whereby juvenile nematodes develop within and consume the body cavity of the mother. Prior to endotokia matricida the mother's rectal glands become infected with Photorhabdus cells, which subsequently serve as the source of inoculum for the progeny IJs (Goodrich-Blair and Clarke 2007).

A low number of Photorhabdus cells (probably one to two) colonize each IJ and these bacteria reproduce within the lumen of the nematode gut to give a mature population of between 50 and 150 cfu IJ-1. Therefore, *Heterorhabditis* IJ colonization is defined by three stages: (i) colonization of the rectal gland of the mother, (ii) colonization of the IJ gut and (iii) outgrowth. *Photorhabdus* mutants defective in the first two stages have been identified and are being characterized.

However, at this time, the only published report of a *Photorhabdus* locus required for IJ colonization is the pbgPE operon. The pbgPE operon is also required for virulence and, as discussed earlier, this operon is predicted to be involved in the adaptive response of *Photorhabdus* to CAMP production by the insect (Bennett and Clarke, 2005). Interestingly, nematodes have been shown to have an innate immune response that includes the production of CAMP-like proteins.

Therefore, it is exciting to consider the possibility that the colonization defect of the *Photorhabdus* pbgPE mutant may result from an inability of this mutant to resist nematode antimicrobial activity encountered during infection of the maternal rectal glands. Such antimicrobial activity may serve as a positive selection for symbiotic bacteria. Indeed, the specificity of the *Heterorhabidtis–Photorhabdus* relationship may be dictated in part by selective processes occurring during maternal rectal gland colonization (Goodrich-Blair and Clarke 2007).

10.3. FIELD EFFICACY OF EPNS

Sustainable agriculture in the 21st century will rely increasingly on alternative interventions for pest management that are environmentally friendly and reduce the amount of human contact with chemical pesticides. The role of microbial pesticides in the integrated management of insect pests has been recently reviewed for agriculture forestry and public health. EPNs from the families Steinernematidae and Heterorhabditidae have proven to be the most effective as biological control agents. They are soil-inhabiting agents and can be used effectively to control soilborne insect pests and insect in cryptic habitats (Jansson, *et al.*, 1991). As a group of nearly 75 species of EPNs (61 specis of *Stinernama* and 14 species of *Heterorhabdities*), each with its own suite of preferred hosts, beneficial nematodes can be used to control a wide range of insect pests including a variety of caterpillars, cutworms, grubs, and weevils. Beneficial nematodes have been released extensively in the field with negligible effects on nontarget insects and are regarded as exceptionally safe to the environment (Bedding, 1976).

Field application showed that the EPNs of the genera *Steinernema* and *Heterorhabditis* are effective biological control agents against a wide variety of soil insect pests and for various cropping systems, such as the black vine weevil, *Otiorhynchus sulcatus* (F.), diaprepres root weevil, *Diaprepes abbreviates* (L.), fungus gnats (Diptera; Sciaridae), various white grubs (Coleoptera; Scarabaeidae) (Atwa, 2003) and some lepidopterous insects "the leopard moth, *Zeuzera pyrina*, the Egyptian cotton leafworm*, Spodoptera littoralis,*and the cabbage looper, *Pieris brassica.* (Atwa, 1999, Atwa and Shamseldean, 2008). The inoculate release of nematode-based biological control agents are thought to succeed when: 1) the pest is present throughout most of the year, 2) the pest has a high economic threshold, and 3) soil conditions are favourable to nematodes survival (Atwa, 2009). All these criteria can be met in turf system, in which the univoltine scarabs have larvae present in the soil for most of the year and the turf is irrigated during dray conditions favourable to nematodes (Atwa, 2009). In this part of the chapter we will focus on some models of EPNs used under field conditions.

10.3.1. EPNs as Biological Control Agents

EPNs have received increasing attention because of their potentiality as biological control agents of soil insect pests easily found in soil. Poinar (1975) reported that *Heterorhabditis bacteriophora* appeared to be an important pathogen of *Heliothis punctigera* in alfalfa fields in South Australia. Cabanillas and Raulston (1994) stated that *S. riobravis* appears to be endemic in Texas, where it was found parasitizing prepupae and pupae of both corn earworm, *Helicoverpa zea,* and fall armyworm, *Spodoptera frugiperda.* Actuality, EPNs possess many attributes of an ideal biological control agent: they have a wide host spectrum, are environmentally safe, can be produced in large-scale bioreactors, are easily applied, are compactable with most chemical pesticides, are applied in diverse climatic condition and, are capable of finding hosts in soil. In addition, the use of naturally occurring nematodes in a particular area as biological control agents may also reduce the risk to nontarget organisms when compared with exotic isolates (Blackshaw, 1988).

Nematode attributes like infectivity, persistence, availability, and cost are important factors growers must consider when deciding to utilize these biological alternatives. The ability to maintain an active nematode population in the soil for an extended period of time is an important factor in long-term management of this pest and could be a major economic factor in the adoption of nematodes by the growers (Williams *et al.* 2010). With respect to the survival of IJs in soil, both biotic and on a biotic factors are involved, whereby unfavorable conditions can drive IJ population to the point of extinction in a particular place. Hence, to establish the ecological factors that are more directly associated to EPN natural population dynamics in Mediterranean areas, a long-term study taking into account EPN occurrence, the host insect population dynamics and EPN vertical migration in relation to climatic conditions should be carried out (Campos-Herrera *et al.*2010). Seasonal approach conducted on the occurrence of native EPNs in the Mediterranean area, as well as on their relationship to agricultural management and soil characteristics. Although the data recorded in this area suggest a seasonal trend for EPN populations, no significant differences were found for this factor. More long-term surveys under different environmental and management conditions would provide more accurate data on this subject. Agricultural practices such as tillage, fertilization and irrigation induced changes in soil properties which affect soil microbiota, including EPNs, although the extent of such an effect needs to be linked to the climatic features (Campos-Herrera *et al.*2010).

10.3.2. Virulence of EPNs

Selection of appropriate EPNs as biological control agents includes bioassays in the laboratory to identify virulent strains and evaluating efficacy under simulated field conditions (Jansson *et al.*, 1993). Gray and Webster (1986) demonstrated that differences in virulence among nematode strains were influenced by temperature. It affects their motility, infectivity, pathogenicity, survival and reproduction (Glazer, *et al.*, 1996). For example, Grewal *et al.* (1993[d]) stated that *Heterorhabditis bacteriophora* adapted to cold or warm temperature by improving reproduction, but not virulence, whereas *Steinernema anomali* improved virulence but not reproduction. Additionally, coinhabiting nematode species may reduce competition in their niche by having different thermal optima (Freckman and Caswell, 1985). Because of the above-mentioned reasons, temperature may be one of the most important factors limiting the success of *Heterorhabditis* and *Steinernema* spp. in biological control of insect pests.

10.3.3. Release Strategy of EPNs in Field Trials

The field release strategy of EPNs may be risky and may result in poor efficacy if environmental conditions are not favorable and/or nematode quality is poor (Jansson *et al.*, 1991).

Similarly, possible reasons for poor control or rapid resurgence of insect populations by insecticide applications may be due to the increase in insect–tolerance and/or the suppression of natural enemies. Accordingly, reducing the use of insecticides in pest management by mixing low compatible concentrations with virulent EPNs may maximize the pest management and minimize environmental hazards (Sammour and Saleh, 1996). However, different factors play different roles in co-applications of nematodes, which reflect on field efficacy against noxious insect pests. For example temperature, humidity and UV, for these reasons application techniques, using of evaporation retardant agents and proper time of application may be reduce the environment hazards on EPNs under field conditions.

10.3.4. Application Techniques:

Regarding the application of EPN studies, Ishibashi and Takii (1993) indicated that *S. carpocapsae* applied to soil may survive relatively longer than when foliarly applied. In the meantime, soil applications should include insecticide acephate or permethrin to maintain nematode activity for a long time without having a detrimental effect on these nematodes. Nevertheless, good results for *S. carpocapsae* stimulation in field trials for control of cabbage worms by spraying mixed applications of nematodes and dichlorovos. Some reports indicated that compatible combinations were sharply effective against *Z. pyrina* larvae infesting apple trees. Othman (1994) reduced significantly the larval population of this pest by injecting the galleries with phenthoate, diazinon, vydate and inserting the suspension of *N. carpocapsae* into the interance holes of galleries by using a soaking cotton–wool. According to him the larval galleries can be left open after treatment with the chemical insecticides but in case of the nematode suspensions, plugging the galleries is preferable. More recently, Sammoure and Saleh (1996) obtained some differences due to the application techniques in controlling *Z. pyrina* larvae in infested apple trees. The injection technique achieved better control than spray technique in separate applications of either *Heterorhabditis sp.* and *Steinernema sp.* (1000 nematodes/ml) or the chemical insecticides cidial 50% E.C. and basudin 60 % E.C. (3000 ppm). The best results were obtained by injecting basudin 750ppm with *Heterorhabditis sp.* 500 Ijs/ml (64.74% mortality) or by injecting 1500ppm with *Heterorhabditis sp.* 500 Ijs/ml (63.89% mortality). Atwa and Shamseldean, 2008 obtained that, the superiority of *Steinernema* sp. (EGB20) strain over *Heterorhabditis bacteriophora* (EGB13) and *H. indica* (EBN16) when applied for control *Zeuzera pyrina* with 1000 IJs/ml of EPNs. Injection of the tested nematode suspension into the insect galleries of *Z. pyrina* was more effective than the spray technique. The addition of an evaporation retardant and sticker agent was associated with efficient insect control. Moreover, *Steinernema glaseri* (NJ strains) was tested in the field against *Temnorhynchus baal* infestation on strawberry plants with a percentage of population reduction varying from 89.2% to 96.8% after four field applications. The overall percentage of population reduction after eight field applications was 96.3% to 99.1% (Atwa, 2009). The results also showed that; both EGB13 and *Steinernema* spp. (EGB20) nematode isolates were more effective in reducing larval population of *S. littoralis* and *P. brassica* on cabbage plants than the isolate EBN16 (Atwa and Shamseldean, 2008).

10.3.5. Time of Application:

Almost nothing is known about the controlling of insect pests if co-applications of EPNs was properly timed. Additionally, the proper time determined for separate application of EPNs was limited to few records in literature. With reference to the EPNs as biological control agents against the leopard moth, field applications are recommended during autumn and spring to control different developmental stages of this insect pest (Deseö *et al.*, 1984). Also, Sikora *et al.* (1978) mentioned that the *S. littoralis* in Egypt moves to Egyptian clover producing generations at the end of October, and the beginning of November, the end of February and the beginning of March. Therefore, EPNs applications would have to occur on clover at these times, because of the negative effect of low temperatures on EPNs in December and January and adverse effects of high temperatures after March. Furthermore, the formation and maintenance of a continuous blanket of dew on the foliage at night would be of extreme importance in the level of insect mortality by nematodes. Also, Atwa (2009) demonstrated that, the nematodes *S. glaseri* used for controlling the larvae of white grubs (*T. baal*) was highly effective when applied on the soil surface at a concentration of $20000IJ/m^2$ more than *H. bacteriophora* with the same concentration using drip irrigation system before dusk (sunset). These former illustrative examples gave clues to the relationship between the proper time of EPNs applications and the maximum resultant insect moralities.

10.3.6. Field Release of EPNs

The requirements of EPNs for survival depend upon the expectations of the user. If short term (within-generation) control in the goal, survival of individual infective juvenils in the soil requirements, and it is necessary only to assure that sufficient nematodes live long enough to find and infect enough hosts. Alternatively, when long-term control is desired host suitability (the ability of host to enable the nematodes to recycle), host distribution, and the duration of host availability must be considered for EPNs population maintenance (Atwa, 2009). Whatever the goal of the control, the first step toward conservation biological control must be to monitor the fat of EPNs in the habitats where they are used.

The nearly ubiquitous distribution of EPNs in so many diverse habitats suggests that conservation of natural populations should be considered for pest control, especially in habitats that are perennial such as forests, orchids, and turf. The development of EPNs as commercially viable biological control insecticides has relegated ecological consideration, formulation to the background until recently. Most of the research was emphasis for the mass production, formulation, establishing laboratory host ranges, and conducting field efficacy trials. However, the unexpectedly poor performance of some nematode species in field trials has in many cases been explained by ecological constraints that were not considered previous to their release. Recent emphasis on entomopathogenic nematodes ecology, especially in terms of host affiliation, has lead to conservation strategies which didn't previously receive serious consideration, especially in the long-term (Atwa, 2003). For example, as a group, EPNs show astounding diversity in habitat requirements and host affiliations yet within species these requirements and relationships tend to be much more specific. Therefore, only the most general requirements can be considered for all species; most must be employed on a more species-or even strain-specific basis (Campbell, *et. al.*, 1997).

Abiotic soil condition to favor EPNs survival includes adequate moisture and temperatures warm enough to allow infection. Biotic conditions necessary for infective juvenile's survival are less well understood and offer many opportunities for new research direction. Survival strategies employed by nematodes in the wild are poorly understood, including those that enable nematodes to endure extremely hostile environs. Competition for hosts between natural populations and applied nematodes is rarely, if ever, considered. The influence of nematode antagonists (i.e., bacteria, fungi, predaceous nematodes and mites) on application of EPNs has been addressed in a few studies but no conclusions have been reached (Kaya and Koppenhöfer, 1996).

Long-term population level survival of EPNs in the soil is even more difficult to address. The records of long-term survival of applied nematodes indicate that recycling through hosts must have occurred. Records of epizootics suggest that under certain conditions dense populations of EPNs occur, presumably in response to host abundance. However, the presence of dense populations of acceptable hosts doesn't seem to be the sole requirement for EPN epizootics. For example, outbreaks of scarab grubs in turf in the northeastern U.S.A. do not always give rise to dense nematodes populations (Akhurst, et. al. 1992). Studies of EPNs population dynamics reveal only that they generally lack seasonality. To establish more complete guidelines for conservation of natural populations we need first to understand the requirements and structure of natural populations. Conservation practices for EPNs can potentially decrease costs of control and increase the efficacy and predictability of control for both inundative and inoculative release (Atwa, 2009). Long-term predictability of EPNS influence on host populations legs far behind short-term predictability (Gaugler, et. al., 1992).

Steinernematid nematodes, along with their associated bacteria (*Xenorhabdus* spp.), possess many of the attributes of an ideal biological control agent, including broad host rang, high virulence, host – seeking capability, and ease of mass production. Applications have been made against foliage, aquatic, and cryptic insects with varying degrees of success. Failures, particular for foliage–feeding insects, have most often been linked to poor nematode persistence. The soil, by comparison, is the natural reservoir of steinernematid nematodes, provides shelter from environmental extremes, and offers the greatest potential for nematode persistence, establishment, and recycling. Obvious advantages of using steinernematid nematodes in habitats to which they are ecological adapted have resulted in a gradual but nearly completed research shift to the soil as a target habitat (Kung, et. al., 1991). In contrast, if Steinernematid nematodes are to be used effectively against soil insects, information is needed on identifying the relationships between a biotic factors and nematode persistence, i.e., survival and pathogenicity. Soil type, pH, and oxygen effects on steinernematid nematodes persistence are best in sandy soils, and decreases with decreasing oxygen concentrations. Relative humidity (RH), a critical nematode persistence factor in controlled environment is also important in the soil environment and influences nematode survival.

10.3.7. Field Release of EPNs for Soil Insects

EPNs (*Steinernematidae* and *Heterorhabditidae*) can provide good control of white grubs for examples, the larvae of scarab beetles that are serious pests of turf grass in the United States and strawberries in Egypt. These EPNs offer an environmentally safe alternative to chemical insecticides that are still the prevailing method for white grub control, (Koppenhöfer, et. al., 1999). Understanding the biology and ecology of insecticidal EPNs is the first step of many in using them for inundative or inoculative release (Atwa, 2009). Clearly, insecticial EPNs being living animals cannot be treated like inert, chemical compounds. Although they can be applied like a chemical insecticide, knowing there positive attributes and their limitations will enhance their successful use against target pests. They cannot tolerate temperature extremes, rapid desiccation, or ultraviolet light, and therefore, applications should be made early in the morning, late in the afternoon, or on cloudy days. Yet, many other issues such as foraging strategies and matching nematode species with the target insect, biology and ecology of the pest species, the cropping system, formulation, quality control, and application techniques require your attention to optimize pest control with the insecticidal EPNs.

White grubs in the coleopterous family Scarabaeidae, has a worldwide impact on agriculture as pests of various crops and is the most serious pests of turf grass, cranberry in the United States and strawberry in Egypt. The prevailing methods for white grubs control involves the application of chemical insecticides, but the development of pesticides tolerance, pesticides degradation by soil microorganisms and concerns about environmental contamination and safety have increased the interest in alternative strategies. EPNs such as *Steinernema glaseri* (Steiner) or *Heterorhabditis bacteriophora* (Poinar) can provide good control of scarab larvae in the field, but often the results are not consistent (Atwa, 2009; Georgis and Gaugler, 1991). Although control by chemical pesticides can also be inconsistent, applications of living organisms such as EPNs demand more understanding than the application of chemical pesticides, i. e., knowledge about nematode biology and interaction with the environment. In addition, scarabaeid larvae have coevolved with EPNs and developed some defense mechanisms against EPN infection. However, new EPNs isolates and species may prove to be more effective for control of white grubs, (Koppenhöfer, and Kaya, 1997).

Biopesticides based on microbial control agents such as EPNs offer a biorational alternative to conventional insecticides. Initial investigations demonstrated that *Steinernema carpocapsae* Weiser suppressed black vine weevil, *Otiorhynchus sulcatus* in cranberry, but high costs, short storage life, and limited availability of commercial product slowed implementation. In recent years, strains of EPNs of superior searching ability (*Heterrhabditis bacteriophora* Poinar), and lower temperature thresholds of activity (*H. marelatus*) have become less expensive and more accessible, (Booth *et. al.,* 2002).

White grubs, the root − feeding larvae of scarab beetles, cause significant damage to many agriculture and horticulture plants. Their subterranean habit makes them the most difficult turf grass pests to control in the United State. In contrast, white grubs outbreaks are difficult to predict nature and the difficulty of sampling for white grubs in general and their eggs and first instar in particular. Despite the need for stronger implementation of Integrated Pest Management (IPM) because of the proximity of turf to humane and pests, and growing public concerns about safety, chemical pesticides are still the first choice of turf grass managers for the control of white grubs and other turf grass pests. Nevertheless, the presence of natural enemies, such as predatory and parasitic insects and insect pathogens, and the availability of biological control agents coupled with the perennial nature of turfgrass should be conductive to the development of more sustainable turfgrass systems with minimal reliance on chemical pesticides. Biological control is one of the cornerstones of IPM, but its implementation in turf is constrained by the limited availability and reliability of commercially available natural enemies, for example, entomopathogenic nematodes (*Steinernematidae* and *Heterorhabditidae*) are relatively expensive, have limited field persistence, and do not always provide consistent control levels. Inconsistent results may, in part, be attributed to inappropriate handling of these living organisms, but they are also sensitive to many a biotic and biotic factors that affect their infectivity and persistence, (Koppenhöfer, *et. al.,* 2000). In contrast, Gaugler *et. al.,* (2002) reported that, renewed regulatory pressure on soil insecticides has generated unparalleled opportunities for biological alternative. There are few biological alternatives to entomopathogenic nematodes for use against diverse insect pests in high value markets. Nematode production to meet these markets is based upon in vivo or in vitro culture technologies.

10.3.8. Root Feeders:

The black vine weevil (*Otiorhynchus sulcatus* Fabricius, Coleoptera: *Curculionidae*) as well as several other root weevil, (*O. ovatus* Linnaeus, *Sciopithes obscurus* Horn, and *Nemocestes incomptus* Horn) are key subterranean pests of all small fruit crops in most northern and some southern temperature growing regions. Scrubs are also fairly ubiquitous.

10.3.8.1. Strawberry Growing and Pathogen Interaction

Strawberry is a perennial plant but it is often cultured as an annual crop in worm locales, and bi or tri-annuals in cooler regions. Most varieties bloom briefly in the spring, but ever-bearing varieties has an extended bloom. Dormant crowns are usually planted in fumigated fields during spring in cool regions or during fall at wormer sites. During spring, foliage and inflorescence tissues develop quickly and plants are quite large by early summer. Simultaneously, axially buds on the crown developed stools from vegetative tendrils 'runners' that sprout new 'daughter' plants, (Booth *et. al.,* 2000). Such a simple method of vegetative reproduction is advantageous for fast and efficient plant propagation, but can also result in a reduction of nutrients away from developing fruit. Accordingly, multi–year plantings are usually pruned post–harvest to promote crown growth before winter causing a thick under story of foliage and duff to develop by mulching straw berry hills with straw or black plastic. These ground layer habitats are conductive to the survival and development of microbials, deleterious or otherwise. Fields are usually fumigated between plantings to suppress harmful pathogens; this practice limits persistence of EPNs.

Field plots are often easier to establish in strawberry than in other small fruit crops, as the size and location of most informations can be more easily discerned. Plots may be partial rows, but buffer rows should be included for most pest and pathogens. Strawberries grow in a variety of soil types, which can strongly affect of EPNs (Georgis and Poinar, 1983), so this should be considered when planning EPN trials.

In order to obtain the best field efficacy of entomopathogenic nematodes (EPNs) against particular pest, one must use the most infective nematode and optimize the application regime. In figure 3. Steinernema glaseri and Heterorhabdities bacteriophora were applied with untraditional application methods (inoculative and inundative release) via drip irrigation system to control scarab beetle with strawberries fields in Egypt. Further studies, under natural conditions, are needed to optimize application efficiency and evaluate the commercial utilization of the EPNs. On the other wise field efficacy showed the EPNs in the genera Steinernema and Heterorhabditis can be effective biological control agents against a wide variety of soil insect pests and for various cropping systems, such as black vine weevil (Otiorhynchus sulcatus F.) in cranberry bogs and strawberry fields, citrus root weevils (*Diaprepes abbreviatus* L. and *Pachnaeus litus* Germar) in citrus groves, or the alfalfa fields (Fife, *et al.,* 2003). The inoculate or inundate release of nematode-based biological control agents are thought to succeed when; 1) the pest is present throughout most of the year or at the time of application, 2) the pest has a high economic threshold, and 3) soil conditions are favorable to nematodes survival. All these criteria can be met in crops irrigate with drip irrigation system, in which the scarabs have larvae present in soil for the most of the year and the crops is irrigated during dry conditions unfavorable to nematodes, these conditions are closed to conditions of strawberry field in Egypt (Atwa, 2009).

The field experiment (table I) explain the difference and significantly effect of both nematodes application strategy and untreated control whereas, there are a significant differentiation of S. glaseri inoculative and inundative release. The inoculative release was effective (F value, 3.47, mean square "M.S.", 1.65; P<0.05 after three inoculate release of nematodes) more than the inundative release (F value, 2.61, M.S., 1.26; P<0.05 after the first of nematodes inundate release).

Furthermore, the data after six application and second application for the inoculate and inundate release respectively showed the same trend with *S. glaseri* (table I). Data analysis with inoculative and inundative release of Heterorhabdities bacteriophora for comparing the differentiation of host mortality required to different application strategy mentioned a highly significant variation of both strategy (F value, 3.71, "M.S.", 1.41; P<0.05 after three inoculate release of nematodes) On the other wise, (F value, 11.15, M.S., 2.49; P<0.05 after the first of nematodes inundate release). While, there is no significant variation between the two methods of application with H. bacteriophora after inocultive and inundative release (Atwa, 2009). The nematodes work extremely well with the presence of large numbers of white grubs in the soil (high larval population), because, the more dense of the grubs population, the greater the chance of EPNs finding their host, and the more chance of a second wave of parasitism (where thousands of infective juveniles produced in the cadavers of white grubs killed by the initial application move back into the soil. Atwa (2009), mentioned that the long-term population level survival of EPNs in the soil is even more difficult to addressed. The records of long-term survival of applied nematodes indicate that recycling through hosts must have occurred. Records of epizootics suggest that under certain conditions dense populations of EPNs occur, presumably in response to host abundance.

However, the presence of dense populations of acceptable hosts doesn't seem to be the sole requirement for EPNs epizootics. Finally Atwa (2009) demonstrated that, the nematodes S. glaseri was highly effective when applied on the soil surface at a concentration of 20000IJ/m2 more than H. bacteriophora with the same concentration in both method of application inundative and inoculative release. This nematodes species is most likely adapted to the climatic conditions in these regions and will be suitable for application in the strawberries field for controlling scarab beetle.

10.3.9. The Effect of Natural Enemies and Other Antagonists on EPNs

The infective juveniles of EPNs occur in the soil rhizosphere in which they encounter a wide array of antagonistic microbial successful as exemplified by their wide distribution in various soil habitats throughout the world. The infective juveniles can evade antagonists by their foraging behavior or have morphological structures (e. g., second stage cuticle) and physiological factors (e. g., lack of receptor sites on the cuticle) that protect them from nematophagus fungi. The production of antibiotic by the mutualistic bacteria within a host creates a favorable environment for nematode development.

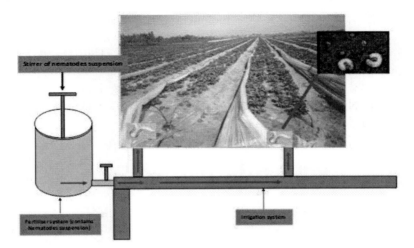

Figure 3. Field application of EPNs via drip irrigation system in Egypt for controlling white grubs in strawberries field (Sfter Atwa, 2003).

Table 1. Comparative population reduction of Temnorhynchus ball estimated before and after nematode applications of *Steinernema glaseri* and *Heterorhabdities bacteriophora* in strawberry field plots in both inoculative and inundative release in Egypt

Treated plots with different nematodes application		Mean numbers before application	After the 3rd application		after the 6th application	
			Mean numbers after application*	% population reduction	Mean numbers after application*	% population reduction
Control		2.95	4.2 d	--	6.7 d	--
Steinernema glaseri	1 #	2.65	0.3 a	92.1	0.2 a	96.7
	2 #	2.75	0.25 a	93.6	0.15 a	97.6
	3 #	2.6	0.15 a	96.0	0.25 a	95.8
	Average	2.67	0.23 a	94.0	0.2 a	96.7
	1 **	3.05	0.7 b	83.9	0.45 b	93.5
	2 **	3.0	0.75 b	82.4	0.6 b	91.2
	3 **	2.6	0.65 b	82.4	0.5 b	91.5
	Average	2.88	0.7 b	82.9	0.52 b	92.1
Heterorhabdities	1 #	3.1	1.05 c	76.21	1.2 c	83.0
	2 #	2.75	1.05 c	73.2	1.05 c	83.2
	3 #	2.85	1.1 c	72.9	1.0 c	84.6
	Average	2.9	1.07 c	74.1	1.15 c	82.5
	1 **	3.1	0.7 b	84.1	0.8 b	88.6
	2 **	3.1	0.85 b	80.7	0.75 b	89.4
	3 **	3.2	0.75 b	83.5	0.8 b	89.0
	Average	3.13	0.77 b	82.7	0.78 b	89.0

* Values followed by the same letter within rows or columns are not significantly different (LSD test, $P < 0.05$).

\# Field plots treated with inoculative release.

** Field plots treated with inundative release.

Although there is no evidence that bioluminescence associated with *Photorhabdus* protects the cadaver from scavengers, there is a strong evidence that a factor (s) deters scavengers from consuming heterorhabditid-killed insects. The faster life cycle of steinernematids allows infective juveniles to exit the host quicker and many minimize the detrimental effects of scavenging activity. By understanding how natural enemies and antagonists affect EPNs, use of these EPNs in inundative or augmentative biological control programs can be enhanced, (Kaya, 2002).

CONCLUSIONS

Entomopathogenic nematodes (EPNs) of the families Steinernematidae and Heterorhabditidae possess impressive attributes for biological control. They have a worldwide distribution as they have been isolated from every inhabited continent and many islands. They have been isolated from different soil types, from sea level to high altitudes, and from natural habitats of disturbed agro ecosystems. Because the entomopathogenic nematodes are obligate parasites in nature, they need to recycle in their hosts to maintain their presence in the environment. In this respect, EPNs belong to the families *Heterorhabditidae* and *Steinernematidae* have already been successfully used throughout the world for the control of important agricultural insect pests. Qualities that make EPNs excellent biocontrol agents are the broad host range, the ability to search actively for their hosts, and to kill them relatively quickly, the economic mass production, being non injurious to vertebrates, easily applied, compatible with most chemical insecticides and environmentally safe. Actuality, EPNs possess many attributes of an ideal biological control agents: they have a wide host spectrum, are environmentally safe, can be produced in large-scale bioreactors, are easily applied, are compatible with most chemical pesticides, are applied in diverse climatic condition and, are capable of finding hosts in soil. In addition, the use of naturally occurring nematodes in a particular area as biological control agents may also reduce the risk to non-target organisms when compared with exotic isolates. Although results from many laboratory tests with EPNs have been promising in controlling insect pests, field evaluation results have often been highly variable particularly against well hidden insects of cryptic habitats such as soil (scarabs) and tunnel living (leopard moth and red palm weevil) insects. They are well protected from chemical insecticides with a high rate of survival. Thus, these insect hosts are capable of producing large populations and new generations that subsequently disperse or migrate or both to more susceptible plant hosts where more control measures are required. Therefore, field trials were conducted to validate laboratory findings. In addition, the efficacy of multiple inoculative releases of EPNs against target insect pests at long-term were also monitored. The field release strategy of nematodes may be risky and may result in poor efficacy if environmental conditions are not favorable and/or nematode quality is poor. Similarly, possible reasons for poor control or rapid resurgence of insect populations by insecticide applications may be due to the increase in insect– tolerance and/or the suppression of natural enemies. Accordingly, reducing the use of insecticides in pest management by mixing low compatible concentrations with virulent EPNs may maximize the pest management and minimize environmental hazards. However, different factors play different roles in co-applications of nematodes and chemical insecticides, which reflects on field efficacy against noxious insect pests. Some of these factors are application techniques, using of evaporation retardant agents and proper time of application.

Long-term population level survival of EPNs in the soil is even more difficult to addressed. The records of long-term survival of applied nematodes indicate that recycling through hosts must have occurred. Records of epizootics suggest that under certain conditions dense populations of EPNs occur, presumably in response to host abundance. However, the presence of dense populations of acceptable hosts doesn't seem to be the sole requirement for EPN epizootics. Finally ENPs can be used exclusively to control insect pests which live in cryptic habitats such as tree borers and soil insects e. g. the red palm weevil and white grubs of scarab pests of strawberries.

REFERENCES

AbuHatab, M., S. Selvan and R. Gaugler. (1995). Role of proteases in penetration of insect gut by the entomopathogenic nematode *Steinernema glaseri* (Nematoda: Steinernematidae). *J. Invertebr. Pathol.* 66, 125–130.

Akhutst, R. J., R. A. Bedding, R. M. Bull, and K. R. J. Smith. (1992). An epizootic of *Heterorhabdities* spp. (Heterorhabditidae: Nematoda) in sugar cane scarabaeids (Coleoptera). *Fund. Appl. Nematol.* 15, 71-73.

Atwa, A, A. (1999). Interaction of certain insecticides and entomopathogenic nematodes in controlling some insect pests on fruit and vegetable crops. M.Sc. thesis, Faculty of Agriculture, University of Ain Shams at Shobra El-Khaima, Cairo, Egypt. 161 pp. (Egypt)

Atwa, A. A. (2003). Identification, mass culture, and utilization of entomopathogeninc nematodes against insect pests. Ph. D. thesis, Faculty of Agriculture, University of Cairo at Giza, Egypt: l72pp.

Atwa, A. A. (2009). comparison between inoculative and inundative release for controlling scarab beetles in strawberry using entomopathogenic nematodes under field conditions. Bull.Fac, Agric.,Cairo Univ.,60 (2009):197- 205.

Atwa, A. Atwa and M. M. Shamseldean. (2008). Entomopathogenic nematodes as a model of field application against some important insect pest in Egypt. *Alex. J. Agric. Res. 53 (2) 41 – 47.*

Baliadi, Y., T. Yoshiga and E. Kondo. (2004). Infectivity and post-infection development of infective juveniles originating via Endotokia matricida in entomopathogenic nematodes. *Appl. Entomol. Zool.* 39, 61–69.

Bedding R.A. (1976). New methods increase the feasibility of using *Neoaplectana* spp. (Nematoda) for the control of insect pests. Division of Entomology, CSIRO, Hobart, Australia: 250-254.

Bedding, R.A. and A.S. Molyneux. (1982). Penetration of insect cuticle by infective juveniles of *Heterorhabditis* spp. (Heterorhabditidae: Nematoda). *Nematologica* 28, 354–359.

Bennett, H.P.J., and Clarke, D.J. (2005) The pbgPE operon in Photorhabdus luminescens is required for pathogenicity and symbiosis. *J. Bacteriol.* 187: 77–84.

Blackshaw, R. P. (1988). A survey of insect parasitic nematodes in Northern Ireland. *Ann. Appl. Biol.* 113: 561-565.

Blossey, B. and R.-U.Ehlers. (1991). Entomopathogenic nematodes (*Heterorhabditis* spp. and *Steinernema anomali*) as potential antagonists of the biological weed control agent *Hylobius transversovittatus* (Coleoptera: Curculionidae). *J. Invertebr. Pathol.* 58, 453–454.

Bohan, D.A. and W. M. Hominick. (1996). Investigations on the presence of an infectious proportion amongst populations of *Steinernema feltiae* (Site 76 strain) infective stages. *Parasitology*. 112, 113–118.

Booth, S. R., F. A. Drummound, and E. Groden. (2000). Evaluation of entomopathogens in specific system, vegetable row crops, small fruits. In: Lacey, L. A., and H. K. Kaya. Field manual of Techniques in Invertebrate Pathology. Kluwer Academic publishersm p. o.17, 3300 AH Dorecht, the Netherlands, pp. 597 – 616.

Booth, S. R., L. K. Tanigoshi, and C. H. Shanks, JR. (2002). Evaluation of entomopathogenic nematodes to many root weevil larvae in Washington State cranberry, strawberry, and red raspberry. *Environ. Entomol.* 31 (5): 859 – 902.

Cabanillas, H. E. and J. R. Raulston. (1994). Pathogenicity of *Steinernema riobravis* against corn earworm, *Helicoverpa zea* (Boddie). *Fundam. appl. Nematol.* 17: 219-223.

Campbell, J.F., A. M. Koppenhöfer, H. K. Kaya and B. Chinnasri. (1999). Are there temporarily non-infectious dauer stages in entomopathogenic nematode populations: a test of the phased infectivity hypothesis. *Parasitology* 118, 499–508.

Campbell, J. F., E. E. Lewis., F. Yoder, and R. Gaugler. (1997). Entomopathogenic nematode (Heterorhabditidae and Steinernematidae) spatial distribution in turfgrass. *Parasitology*. *113*, 473 – 482.

Campbell, J.F., and R. Gaugler. 1993. Nictation behavior and its ecological implications in the host search strategies of entomopathogenic nematodes (Heterorhabditidae and Steinernematidae). *Behavior* 126: 155-169.

Campos-Herrera, R.; A. Piedra-Buena,M. Escuer, B. Montalbán and C. Gutiérrez. (2010). Effect of seasonality and agricultural practices on occurrence of entomopathogenic nematodes and soil characteristics in La Rioja (Northern Spain). *Pedobiologia*, 53, 253–258.

Cui, L., R. Gaugler and Y. Wang. (1993). Penetration of steinernematid nematodes (Nematoda: Steinernematidae) into Japanese beetle larvae, *Popillia japonica* (Coleoptera: Scarabaeidae). *J. Invertebr. Pathol.* 62, 73–78.

Dempsey, C.M. and C. T. Griffin. (2002). Phased activity in *Heterorhabditis megidis*. Parasitology 124, 605–613.

Deseö, K.V.; S. Grassi; F. Foschi and L. Rovesti. 1984. A system of biological control against the leopard moth (*Zeuzera pyrina* L.; Lepidoptera, Cossidae). *Atti-Giorante-Phytopathologishe*, 22: 403-414.

Dolan, K.M., J.T. Jones and A. M. Burnell. (2002). Detection of changes occurring during recovery from the dauer stage in *Heterorhabditis bacteriophora*. *Parasitology* 125, 71–81.

Dowds, B.C.A., and A. Peters. (2002). Virulence mechanisms. In: Gaugler, R. (Ed.), *Entomopathogenic Nematol.* CABI Publishing, New York, NY, pp. 79–98.

Downes ,M. J. And C. T. Griffin. 1996. Dispersal behavior and transmission strategies of the entomopathogenic nematodes of Heterorhabditis and steinernema . *Biocontrol Science and Technology*, 6, 447-356.

Dunphy, G.B. and J. M. Webster. 1988. Virulence mechanisms of *Heterorhabditis heliothidis* and its bacterial associate, *Xenorhabdus luminescens*, in non-immune larvae of the greater wax moth, *Galleria mellonella*. *Int. J. Parasitol.* 18, 729–737.

Fife, J.P., R.C. Derksen, H. E. Ozkan, and P. S. Grewal. (2003). Effect of pressure differentials on viability and infectivity of entomopathogenic nematodes. *Biological Control*, 27, 65-72.

Forschler, B.T. and W. A. Gardner. (1991). Parasitism of *Phyllophaga hirticula* (Coleoptera: Scarabaeidae) by *Heterorhabditis heliothidis* and *Steinernema carpocapsae*. *J. Invertebr. Pathol.* 58, 396–407.

Forst, S. and D. Clarke. (2002). Bacteria-nematode symbiosis. In: Gaugler, R. (Ed.), *Entomopathogenic Nematol.* CABI Publishing, New York, NY, pp. 57–77.

Freckman, D. W. and E. P. Caswell. 1985. The ecology of nematodes in agroecosystems. *Annual Review of phytopathology* 23, 275 – 296.

Garcia Del Pino, F. and A. Palomo. 1996. Natural occurrence of entomopathogenic nematodes (Rhabditida: Steinernematidae and Heterorhabditidae) in Spanish soils. *Journal of invertebrate Pathology* 68, 84-90.

Gaugler, R., I. Brown, D. Shapiro-Ilan, and A. A. Atwa. (2002). Automated technology for *in vivo* mass production of entomopathogenic nematodes. *Biological control,* 24, 199-206.

Gaugler, R. and J. Campbell. (1991). Behavioural response of the entomopathogenic nematodes Steinernema carpocapsae and Heterorhabditis bacteriophora to oxamyl. *Annals of Applied Biology* 119, 131-138.

Gaugler, R., J. F. Campbell, S. Selvan, and E. E. Lewis. (1992). Large – scale, inoculative release of the entomopathogen: *Steinernema glaseri*, assessment 50 years later. *Biol. Control. 2, 181-187.*

Georgis, R., and G. O. poinar. (1983). Effect of soil texture on the distribution and infectivity of *Neoaplectana carpocapsae* (Nematoda: Steinernematidae). *J. Nematol.* 15, 308-311.

Georgis, R., and R. Gaugler. (1991). Predictability in biological control using entomopathogenic nematodes. *J. Econ. Entomol.* 84, 713-720.

Glazer, I. (1997). EVects of infected insects on secondary invasion of steinernematid entomopathogenic nematodes. *Parasitology* 114, 597–604.

Glazer, I., E. Kozodoi., G. Hashmi, and R. Gaugler. (1996). Biological characteristics of the entomopathogenic nematode *Heterorhabditis* sp. IS-5: A heat tolerant isolate from Israel . *Nematologica ,* 24, 481 – 492.

Goodrich-Blair, H. and D. J. Clarke. (2007). Mutualism and pathogenesis in *Xenorhabdus* and *Photorhabdus*: two roads to the same destination. *Molecular Microbiology.* 64(2), 260–268.

Gouge, D.H. (1994). Biological control of sciard Xies (Diptera: Sciaridae) with entomopathogenic nematodes (Nematoda: Rhabditida), including reference to other Diptera. PhD Thesis, Department of Agriculture, University of Reading, UK, pp. 251.

Gray, B.D., and J.M. Webster, (1986). Temperature effects on the growth and virulence of *Steinernema feltiae* strains and *Heterorhabditis heliothidis*. *Journal of Nemaotology* 18(2): 270-272.

Grewal, P.S., R. Gaugler and E. E. Lewis. (1993[a]). Host recognition by entomopathogenic nematodes: behavioral response to contact with host feces. *J. Chem. Ecol.* 19, 1219–1231.

Grewal, P.S., R. Gaugler and E. E. Lewis. (1993[b]). Host recognition behavior by entomopathogenic nematodes during contact with insect gut contents. *J. Parasitol.* 79, 495–503.

Grewal, P.S., R. Gaugler and E. E. Lewis. (1993[c]). Male insect-parasitic nematodes: a colonizing sex. *Experientia* 49, 605–608.

Grewal, P.S.; R. Gaugler, H. K. Kaya and M. Wusaty. (1993[d]). Infectivity of the entomopathogenic nematode *Steinernema scapterisci* (Nematoda: Steinernematidae). *J. Invertebr. Pathol.*62, 22-28.

Hay, D.B., and J. S. Fenlon. (1995). A modiWed binomial model that describes the infection dynamics of the entomopathogenic nematode *Steinernema feltiae* (Steinernematidae: Nematoda). *Parasitology.* 111, 627–633.

Hominick, W.M., and A. P. Reid. (1990). Perspectives on entomopathogenic nematology. In: Gaugler, R., Kaya, H. (Eds.), *Entomopathogenic nematodes in biological control.* CRC Press, Boca Raton, FL, pp. 327–345.

Ishibashi, N. And S. Takii. (1993). Effectes of insecticides on movement, nictation, and infectivity of *Steinernema carpocapsae. J. Nematology* 25(2) 204-213.

Ishibashi, N.; N. Takii, S. and E. Kondo. (1994). Infectivity of nictating juveniles of Steinernema carpocapsae (Rhabditida: Steinernematidae). *Japanese Journal of Nematology* 24, 20-29.

Jansson, R. K., H. L. Scott and R. Gaugler. (1991). Comparison of single and multiple releases of *Heterorhabiditis bacteriophora* poinar (Nematoda: Heterorhabditidae) for control of *Cylas formicarius* (Fabricius) (Coleoptera: Apionidae). *Biological control* 1, 320-328.

Jansson, R. K., H. L. Scott and R. Gaugler. (1993). Field efficacy and persistence of entomopathogenic nematodes (Rhabditida: Steinernematidae, Heterorhabditidae) for control of sweetpotato weevil (Coleoptera: Apionidae) in southern Florida. . *J. of Economic Entomol.* 86 (1), 1055 – 1063.

Kaya, H. K. (1990). Soil ecology, in *Entomopathogenic Nematodes in Biological Control* (Gaugler, R. and Kaya, H.K., Eds) CRC Press, Boca Raton, FL, pp. 93-115.

Kaya, H. K. (2002). Natural enemies and other antagonists. In: Gauglr, R. Eentomopathogenic Nematology. CABI publishing, CAB International, Wallingford, Ox 10 8DE, UK, pp. 189-203.

Kaya, H. K., and A. M. Koppenhöfer. (1996). Effect of microbial and other antagonistic organisms and competition on entomopathogenic nematodes. *Biocont. Sci. Technolo.* 6, 357-371.

Kaya, H.K. and R. Gaugler. 1993. Entomopathogenic nematodes. *Annu. Rev. Entomol.* 38, 181–206.

Khlibsuwan, W., N. Ishibashi and E. Kondo. (1992). Response of *Steinernema carpocapsae* infective juveniles to the plasma of three insect species. *J. Nematol.* 24, 156–159.

Koppenhöfer, A. M., and H. K. Kaya. (1997). Additive and synergistic interaction between entomopathogenic nematodes and *Bacillus thuringiesis* for scarab grub control. *Biologiacl control,* 8, 131 – 137.

Koppenhöfer, A. M., H. Y. Choo, H. K. Kaya, O. W. Lee, and W. D. Gelernter. (1999). Increased field and greenhouse efficacy against scarab grubs with combination of an entomopathogenic nematodes and *Bacillus thuringiesis. Biologiacl control, 14,* 37 – 44.

Koppenhöfer, A. M., I. M. Brown, R. Gaugler, P. S. Grewal, H. K. Kaya, and M. G. Klein. (2000). Synergism of entomopathogenic nematodes and Imidacloprid against white grubs: greenhouses and field evaluation. *Biological control,* 19, 245 – 251.

Kung, S. P., R. Gaugler, and H. K. Kaya. (1991). Effect of soil temperature, moisture, and relative humidity on entomopathogenic nematode persistence. *J. Invertebr. Pathol.,* 57, 242-249.

Lewis, E. E., J. Campbell, C. Griffin, H. Kaya and A. Peters. (2006). Behavioral ecology of entomopathogenic nematodes. *Biological Control* 38, 66–79.

Lewis, E.E., M. Ricci and R. Gaugler. (1996). Host recognition behavior predicts host suitability in the entomopathogenic nematode *Steinernema carpocapsae.* (Rhabditida: Steinernematidae). *Parasitology* 113, 573–579.

Lewis, E.E., R. Gaugler and R. Harrison. (1992). Entomopathogenic nematode host finding: response to host contact cues by cruise and ambush forgers. *Parasitology* 105, 103-107.

Othman, K.S.A. (1994). Injection of various pest control agents against *Zeuzera pyrina* L. in apple trees. *J. Agric. Sci. Mansoura Univ.,* 19(5): 1867-1875.

Peters, A. (1994). Interaktionen zwischen den Pathogenitɔtsmechanismen entonopathogener Nematoden und den Abwehrmechanismen von Schnakenlarven (*Tipula* spp.) sowie Mglichkeiten zur Virulenzsteigerung der Nematoden durch Selektion. PhD Thesis, University Kiel, Germany.

Peters, A. and R.-U. Ehlers. (1994). Susceptibility of leatherjackets (*Tipula paludosa* and *Tipula oleracea*; Tipulidae; Nematocera) to the entomopathogenic nematode *Steinernema feltiae. J. Invertebr. Pathol.* 63, 163–171.

Peters, A. and R.-U. Ehlers. (1997). Encapsulation of the entomopathogenic nematode *Steinernema feltiae* in *Tipula oleracea. J. Invertebr. Pathol.* 69, 218–222.

Poinar, G.O.Jr. (1975). *Entomogenous nematodes. A manual and host list of insects nematodes associations.* Leiden, Netherlands, E.J. Brill.

Poinar Jr., G.O. (1990). Taxonomy and biology of Steinernematidae and Heterorhabitidae. In: Gaugler, R., Kaya, H.K. (Eds.), *Entomopathogenic Nematodes in Biological Control.* CRC Press, Boca Raton, FL, pp. 23–62.

Ryder, J.J. and C.T. Griffin. (2002). Density dependent fecundity and infective juvenile production in the entomopathogenic nematode, *Heterorhabditis megidis. Parasitology* 125, 83–92.

Samish, M. and I. Glazer. (1992). Infectivity of entomopathogenic nematodes (Steinernematidae and Heterorhabditidae) to female ticks of *Boophilus annulatus* (Arachnida: Ixodidae). *J. Med. Entomol.* 29, 614–618.

Sammour, E. A. and M. E. Saleh. (1996). Combination of entomopathogenic nematodes and insecticides for controlling the apple borer, *Zeuzera pyrina* L. (Lepid., cossidae). *J. Union. Arab. Biol., 5(A) Zoology,* 369-380.

San-Blas, E., Gowen, S.R., Pembroke, B. (2008). Steinernema feltiae: ammonia triggers the emergence of their infective juveniles. *Exp. Parasitol.* 119, 180–185.

Sikora, R.A., I.E.M. Salem and F. Klingauf. (1978). Susceptibility of *Spodoptera littoralis* to *Neoaplectana*, and observations on the effect of environmental factors on insect mortality levels. Proc. 4th Conf. Pest Control, NRC, II: 940.

Simoes, N., (1998). Pathogenicity of the complex *Steinernema carpocapsae– Xenorhabdus nematophilus*: molecular aspects related with the virulence. In: Simoes, N., Boemare, N., Ehlers, R.-U. (Eds.), *Pathogenicity of Entomopathogenic Nematodes Versus Insect Defence Mechanisms: Impact on Selection of Virulent Strains.* European Commission, Brussels, pp. 73–83.

Smits ,P. H., J. T. M. Groenen, and G. De Raay. (1991). Characterization of Heterorhabditis isolates using DNA restriction fragment length polymorphisms. Revue de Nematologie 14: 445-453.

Stock, S.P. (2005). Insect-parasitic nematodes: From lab curiosities to model organisms. *J. Invertebr. Pathol.* 89, 57–66.

Stock, S.P., C.T. Griffin and R. Chaerani. (2004). Morphological and molecular characterisation of *Steinernema hermaphroditum* n. sp. (Nematoda: Steinernematidae), an entomopathogenic nematode from Indonesia, and its phylogenetic relationships with other members of the genus. *Nematology* 6, 401–412.

Susurluk, I. A. (2009). Seasonal and vertical distribution of the entomopathogenic nematodes, *Heterorhabditis bacteriophora* (TUR-H2) and *Steinernema feltiae* (TUR-S3), in turf and fallow areas. *Nematology*, Vol. 11 (2): 309–315.

TKaya, H.K. and R. Gaugler. (1993). Entomopathogenic nematodes. *Annu. Rev. Entomol.* 38, 181–206.

Wang, Y. and R. Gaugler. (1998). Host and penetration site location by entomopathogenic nematodes against Japanese beetle larvae. *J. Invertebr. Pathol.* 72, 313– 318.

Wang, Y., R. Gaugler and L. Cui. (1994). Variations in immune response of *Popillia japonica* and *Acheta domesticus* to *Heterorhabditis bacteriophora* and *Steinernema* species. *J. Nematol.* 26, 11–18.

Williams R.N., D.S. Fickle, P.S. Grewal and J. Dutcher. (2010). Field efficacy against the grape root borer *Vitacea polistiformis* (Lepidoptera: *Sesiidae*) and persistence of *Heterorhabditis zealandica* and *H. bacteriophora* (Nematoda: *Heterorhabditidae*) in vineyards. *Biological Control.* 53, 86–91.

Wojcik, W., I. Popiel and D.Grove. (1986). Is the pathogenicity of *Heterorhabditis heliothidis* dependent on prior history of temperature? In: 4th International Colloquium of Invertebrate Pathology, Veldhoven, The Netherlands, p. 319.

In: Microbial Insecticides: Principles and Applications
Editors: J. Francis Borgio, K. Sahayaraj, et al.

ISBN: 978-1-61209-223-2
© 2011 Nova Science Publishers, Inc

Chapter 11

GENETICS OF ENTOMOPATHOGENIC NEMATODES

*You-Jin Hao[1, 2], Mitzi Flores-Ponce[3] and Rafael Montiel[3, *]*

[1] CIRN, Departamento de Biologia, Universidade dos Acores, Ponta Delgada, 9501-80, Azores, Portugal
[2] Department of Physics, the University of Chicago, Chicago, Illinois, 60637, USA
[3] Laboratorio Nacional de Genómica para la Biodiversidad, CINVESTAV-IPN, 36821, Irapuato, Mexico.

ABSTRACT

Entomopathogenic Nematodes (EPNs) are classified into Steinernematidae and Heterorhabditidae families and have great potential as biological control agents against insect pests. These nematodes are soil organisms, which symbiotically associate with the bacterium *Xenorhabdus* and *Photorhabdus*, respectively. A recent genetic-based classification places the *Steinernematidae* within the suborder *Tylenchina*, which also includes insect parasitic allantonematids and neotylenchids. The *Heterorhabditidae* were positioned within the suborder *Rhabditina* which includes, among others, free-living *Rhabditidae* and animal parasitic *Strongylidae*. Despite the unknown systematic status for the majority of nematode and bacterial taxa, nematode-bacterium entomopathogens are some of the best-studied members of these highly diverse groups of organisms. Genome sizes are estimated at 2.3×10^8 bp in *S. carpocapsae* and 3.9×10^7 bp in *H. bacteriophora*. Repetitive DNA content represents 39% and 51% of these respective genomes. Compared with animal and plant nematodes, EPN genomics is running slowly. In June 2005, it was announced the targeting of *H. bacteriophora* for high quality coverage, but until now the sequencing project is still in process. *S. carpocapsae* genome sequencing project was recently put up by Rafael Montiel and Nelson Simoes, and financed by a Mexican agency. However, functional gene libraries construction and annotation is a helpful platform to provide a basis for genomic studies, accelerating research towards a better understanding of the events that occur in the parasitic process of EPNs, including those mechanisms involved in symbiosis with bacteria. cDNA and SSH libraries have been done for both *S. carpocapsae* and *H. bacteriophora*, and used for gene cloning and screens to isolate new gene activities. Serine proteases, chitinases,

---

* Tel. + (52) 462 1663016; Fax: (52) 462 6245846; E-mail: montiel@ira.cinvestav.mx

metalloproteases, protease inhibitors and other molecules have been described. All this information is enhancing the use of ENPs as model organisms, as part of a tripartite model, to study both symbiosis and parasitism.

11.1. INTRODUCTION

Entomopathogenic nematodes (EPNs), which are classified into Steinernematidae and Heterorhabditidae families, have great potential as biological control agents against a large number of insect pests with great economical impact (Kaya *et al.*, 2006; Ehlers, 2001). Their activity against different insect pests is already well studied (Kaya and Gaugler, 1993). Entomopathogenic nematodes are soil organisms, which symbiotically associate with the bacterium *Xenorhabdus* and *Photorhabdus*, respectively. Infective juvenile-stage nematodes (IJs) live in soil and infect suitable insect hosts primarily by gaining entry through natural body openings (Koppenhofer *et al.*, 2007) or thin areas of the host's cuticle (Burnell and Stock, 2000). They then undergo a recovery process, become parasitic and release their symbiotic bacteria (Sicard *et al.*, 2004). The nematodes penetrate digestive tract tissues and invade the hemocoel of hosts. Both pathogens kill the host within 2–3 days and multiply within the insect cadaver. When food reserves are depleted, nematode reproduction ceases and offspring develop into IJs, which disperse from the dead host and are able to survive in the environment to seek new hosts (You-Jin *et al.*, 2008).

11.2. NEMATODES BIODIVERSITY

Nematodes are the most abundant animals on earth (Adams *et al.*, 2006) and are important since the parasitic nematodes threaten the health of plants, animals and humans on a global scale (Blaxter *et al.*, 1998). A lack of clearly homologous characters and the absence of an informative fossil record have hampered the development of a consistent evolutionary framework for the phylum (Blaxter *et al.*, 1998). And even though they seem to have uniform body plans, nematodes seem to be more diverse at the molecular level than was previously recognized (Adams *et al.*, 2006; Bird *et al.*, 2005). The phylogenetic position of the Nematoda relative to other metazoans, and among metazoans in general, is currently controversial and hotly contested (Adams *et al.*, 2006). As a result, nematodes have been hypothesized to share a most recent common ancestor with arthropods, kinorhynchs, nematomorphs, onychophorans, priapulids, and tardigrade (Adams *et al.*, 2006).

In 1998, Blaxter *et al.*, (Blaxter *et al.*, 1998) presented a phylogenetic analysis, using 53 small subunit ribosomal DNA sequences from a wide range of nematodes. This study made possible for the fist time the comparison of all taxa using the same defined measurement. In this study, five major clades were identified within the phylum, all of which include parasitic species, suggesting that animal parasitism arose independently at least four times, and plant parasitism three times (Blaxter *et al.*, 1998). Moreover, each of the Secernentean vertebrate-parasitic clades is associated with arthropod-parasitic or -pathogenic taxa (*Heterorhabditis* is associated with *Strongylida*; *Steinernema* is associated with *Strongyloides*; and *Rhigonematida* is associated with *Ascaridida*, *Spirurida* and *Oxyurida*) (Blaxter *et al.*, 1998).

There is a more recent classification of Nematoda by De Ley and Blaxter (De Ley *et al.,* 2002). This new classification scheme places the *Steinernematidae* within the suborder *Tylenchina*, which also includes insect parasitic allantonematids and neotylenchids. The *Heterorhabditidae* were positioned within the suborder *Rhabditina* which includes, among others, free-living *Rhabditidae* and animal parasitic *Strongylidae* (Adams *et al.,* 2006). Despite the unknown systematic status for the majority of nematode and bacterial taxa, nematode-bacterium entomopathogens are some of the best-studied members of these highly diverse groups of organisms (Adams *et al.,* 2006).

It has been speculated that in the mid-Paleozoic (approximately 350 million years ago) ancestors of the *Heterorhabditidae* and *Steinernematidae* began to independently explore mutualistic relationships with Gram-negative enteric bacteria (Poinar, 2003). But even that the genera *Steinernema* and *Heterorhabditis* share the strategy of entomopathogenesis they do not share an exclusive common ancestry (Blaxter *et al.,* 1998). And is not a surprise, given that these nematodes last shared a common ancestor more than 300 million years ago, the degree of synteny is not extensive across the genome (Ghedin *et al.,* 2004) but exhibits some local conservation (Guiliano *et al.,* 2002).

11.3. PHYLOGENETIC RECONSTRUCTION AND THE EMERGENCE OF PHYLOGENOMIC

Phylogenetic is the reconstruction of evolutionary history. This reconstruction relies on the use of mathematical methods to infer the past from features of contemporary species. It involves the identification of homologous characters shared between species and the inference of phylogenetic trees, with the use of the fossil record (Delsuc *et al.,* 2005). Indeed, the comparative anatomy of fossils and extant species has proved to be powerful in some respects and with some species (Delsuc *et al.,* 2005). In nematodes, the reconstruction of an evolutionary history based on the analysis of morphological and ultrastructural characters has been difficult, due to the lack of nematode fossil records (Adams *et al.,* 2006).

After the introduction of the use of molecular data, a few genes became reference markers across organisms due to its considerable degree of conservation. For example, Blaxter *et al.* (Blaxter *et al.,* 1998) developed a molecular evolutionary framework for the phylum Nematoda from small subunit ribosomal DNA sequences from a wide range of nematodes. However the information provided from a single gene is not enough to obtain substantial information for phylogenetic trees and phylogenomics has arisen, which uses the phylogenetic principles to the genomic data (Delsuc *et al.,* 2005). In fact, during the past decades the develop of genomic approaches is offering a considerable amount of information, which with functional genomics, transcriptomics and proteomics complements the development of a more complete reconstruction in many taxa, including nematodes (Parkinson *et al.,* 2004; Mitreva *et al.,* 2007).

There are two general approximations to phylogenomics. One is based on sequence data after performing multiple-sequence alignments and the other one is based on whole-genome characteristics, such as gene order and gene content, or other aspects of genome structure, like rare genomic changes (RGCs) which include insertions, deletions, intron positions, retroposon integrations, and gene fusions and fission events (Delsuc *et al.,* 2005).

There are many examples that show the achievements of phylogenomics. By 2008, 5321 complete genomes had been sequenced (1896 eukaryotes, 1129 bacteria, 73 archaeans and 2223 viruses, virods and plasmids), in contrast of the 260 complete genomes sequenced by 2005 (http://www.ncbi.nlm.nih.gov/sites/entrez?db=genome).

11.3.1. Nematodes Phylogenomics

With all the information that phylogenomics generates now the challenge is to understand the evolutionary history of organisms and their genomes, the functions of their genes, and how this relates to their interactions with the environment (Delsuc *et al.*, 2005).

In the case of the nematode phylum it is especially important to understand what sorts of new nematode genes are evolving, especially for parasites where specific and unique selection pressures are present (Bird *et al.*, 2005). That is why completed genomes will serve both as a source of information on the presence and absence of genes.

The whole genome information from broadly selected nematode taxa will help to resolve the deepest branches in the phylum. Such efforts are concordant with the Nematode Tree-of-Life Project (NemATOL: http://nematol.unh.edu/) and, conversely, each nematode to be sequenced will have to be placed in a phylogenetic context for effective use of its data. In addition to better understanding evolutionary relationships within the phylum, nematode genome data will contribute to understanding the relationship of Nematode to other metazoan phyla (Bird *et al.*, 2005).

Addressing the fundamental question of where nematodes fit in animal evolution will greatly benefit from the inclusion of multiple, diverse nematode genome sequences.

Furthermore, having these genome sequences from diverse nematode species will support future studies into the biology and evolution of this phylum and the ecological association that nematodes might have with other organisms, in particular, parasitic associations (Bird *et al.*, 2005).

11.4. GENOME SIZE, COMPLEXITY AND STATUS

The genome size and the repetitive DNA diversity are fundamental information for molecular genetics investigations. Genome sizes are estimated at 2.3×10^8 bp in *S. carpocapsae* and 3.9×10^7 bp in *H. bacteriophora*. Repetitive DNA content represents 39% and 51% of these respective genomes (Grenier *et al.*, 1997). The highly repetitive components are similar in proportion for both entomopathogenic nematodes. Sequences such as satellite DNA have been described in several *Heterorhabditis* and *Steinernema* species (Grenier *et al.*, 1996; Grenier *et al.*, 1995).

The satellite DNA has been found to represent between 5% and 10% of their genomes, respectively. These results can account for the genome differences, both in size and complexity, between *H. bacteriophora* and *S. carpocapsae*. The ability of entomopathogenic nematodes to produce enormous populations in a short period is likely reinforced by the small size of their genome, which also harbour an additional genome- a mitochondrial genome.

The complete mitochondrial genome is 13,925 bp (Montiel *et al.,* 2006) of *S. carpocapsae* and 18,128 bp of *H. bacteriophora* (http://www.ncbi.nlm.nih.gov/entrez/ viewer.fcgi?db= nucleotideandval =EF043402).

Compared with animal and plant nematodes, entomopathogenic nematode genomics is much sluggish. In June 2005, NHGRI (*National Human Genome Research Institute*) announced that it was targeting *H. bacteriophora* for a high quality (6×) coverage, but until now the sequencing project is still in process. Genomic DNA from the *H. bacteriophora* TTO1 inbred for 13 generations by self-fertilizing individual IJs from axenic IJs had been undergoing heterozygosity testing in Washington University Genome Sciences Center and approximately 53616 sequence traces generated for heterozygosity testing have been deposited in the NCBI Trace Archive.

Fortunately, *S. carpocapsae* genome sequencing project was recently put up by Rafael Montiel and Nelson Simoes, and financed by a Mexican agency (Fondo Mixto CONACYT-Estado de Hidalgo, Mexico).

Currently there are insufficient genome data for understanding of the molecular genetics of ENPs, functional gene libraries [cDNA library and suppression subtractive hybridization (SSH) library] construction and annotation is a helpful platform to provide a basis for genomic studies, specifically for accelerating research towards a better understanding of the events that occur in the parasitic process of entomopathogenic nematodes. These events include infection of mid-gut lumen, invasion of hemocoelium, adaptations to insect innate immunity and stress responses, and toxicity. The identification of key genes in the parasitic processes must provide applicable tools for the improvement of EPNs as a biological agent through recombinant DNA.

Furthermore, it can be used to study specific insights in the larval development and differentiation processes, and to understand the mechanisms involved in symbiosis with bacteria.

The advantage of cDNA library is that it contains only the coding region of a genome. Information in cDNA libraries is a powerful and useful tool since gene products are easily identified, the libraries lack information about enhancers, introns, and other regulatory elements found in a genomic DNA library. cDNA libraries are most useful in reverse genetics where the additional genomic information is of less use.

Suppression subtractive hybridization (Diatchenko *et al.,* 1999; Diatchenko *et al.,* 1996) is an efficient and widely used PCR-based method to obtain subtracted libraries and identify differentially expressed genes under two biological conditions. And modified SSH methods had been widely employed in functional genomic studies (Hepworth *et al.,* 2007;Altincicek and Vilcinskas, 2007;Salvador *et al.,* 2005). SSH includes a normalization step, which makes this approach preferable for cloning low abundance transcripts (Ji *et al.,* 2002).

cDNA libraries and SSH library have been done for both *S. carpocapsae* and *H. bacteriophora*, and used for gene cloning and for gain of function screens to isolate new gene activities (7; Sandhu *et al.,* 2006; Tyson *et al.,* 2007) . The increased availability of good cDNA libraries has allowed for a significant increase in the cDNA end-sequences.

Community contributions of cDNA library combined with the efforts of nematode sequencing centre has lead to the generation of more entomopathogenic nematodes ESTs. These sequences are useful for gene discovery and the corresponding cDNA clones can be used for generate probes for expression pattern analysis and gain of function studies.

11.5. PARASITISM-RELATED PROTEINS IDENTIFICATION BY COMPARATIVE FUNCTIONAL GENOMICS ANALYSIS

11.5.1. Surface-Coated Proteins

Insects defence themselves against bacteria or parasite infections with cellular and humoral immune responses. The immediately response against nematode is encapsulation. After penetration of the IJs into the hemolymph, insect's non-self response system initially deals with only nematodes. Nematodes may resist encapsulation by immuno-evasion strategies which often involved the parasite body surface (i.e. cuticle, and antigen) that seems to play a key role in the interactions with the host immune reactions (Brivio *et al.,* 2000;Maizels *et al.,* 1993;Blaxter *et al.,* 1992). EST transcripts isolated from *H. bateriophora* is identified to be significantly similar to surface-associated antigen (SAA1) in *A. caninum*, which is proved to be an immunodominant molecule. Stage-specific surface differences are known to occur in a variety of nematode species, including the free-living species *C. elegans* (Philipp and Rumjaneck, 1984;Politz *et al.,* 1987;Politz and Philipp, 1992). The potential importance of a stage-specific surface composition to parasitism is suggested by *Trichinella spiralis* infections of mammals. Later stages that express different surface antigens escape the immune attack directed against the stages present early in infection (Wakelin and Denhem, 1983). Surface composition can also change within a single stage, e.g., in response to a new host or host tissue, and surface molecules can be shed in response to binding of immune effectors or antibodies (Blaxter *et al.,* 1992; Philipp and Rumjaneck, 1984; Politz and Philipp, 1992). Therefore, surface-associated antigen gene in *H. bacteriophora* nematode would help nematode evade insect immune attacks.

11.5.2. Protease

Mortality is attributed mainly to virulence factors produced by the associated bacteria, however the same lethality time was observed for insect exposed to axenic *S. carpocapsae* (Laumond *et al.,* 1989), suggesting that parasitic nematode is able to produce insecticidal factors. Lethal factors have been detected in insect parasitized with axenic *S. carpocapsae* (Boemare *et al.,* 1982), as well as in the culture medium of axenic nematodes (Burman, 1982). Furthermore, analysis of excreted products from the parasitic stage of *S. carpocapsae* showed the presence of a large number of different proteins and distinct activities, namely cytotoxicity, proteolysis, immunosupression, and apoptosis. Proteolytic activity has been shown to be higher in excreted products from a virulent strain than in that from a less virulent strain (Simoes *et al.,* 2000).

Proteases encompass a board of hydrolytic enzymes that play essential roles in cellular, developmental and digestive process, blood coagulation, inflammation, wound healing and hormone processing. Parasite proteases, some of which are in the excretory-secretory (ES) products, facilitate the invasion of host tissues, aid in the digestion of host proteins, help parasites evade the host immune response and mediate molting in parasitic nematodes. Serine proteases are among the most representative compounds in nematode excreted products, which also includes cysteine, aspartic- and metallo- proteases, and others.

11.5.3. Serine Protease

Several notable classes of predicted proteins with potential roles in pathogenesis of parasitic nematode are identified in *S. carpocapsae* and *H. bacteriophora* parasitic stage ESTs. Despite secretion of proteases being a common feature of both free-living and parasitic organisms, the use of proteases to degrade host extracellular matrix appears to be obligatory in parasitic organisms (Lackey *et al.*, 1989). Also specific release of digestive enzymes after infection of a host serves an integral function in the transition of a larva to parasitism (Hawdon *et al.*, 1995; Gamble and Mansfield, 1996). In parasites, in addition to facilitating invasion of host tissues by digestion, proteases help parasites evade the host immune response, prevent blood coagulation (McKerrow, 1989) and have potential effects on growth (Phares, 1996).

Ten ESTs are identified with homology to diverse serine proteases including trypsine-like, elastase, and serine carboxypeptidase in *S. carpocapsae* parasitic stage. Among them five genes (one chymotrypsin-like protease, two trypsin-like protease, one elastase and one serine carboxypeptidase) products are identified in the excreted product from parasitic stage by 2D LTQ MS/MS (Simoes, in prep).

In addition, three chymotrypsin-like serine proteases were purified from the excreted product of *S.carpocapsae*. One was proved to be cytotoxic to insect cells by apoptosis mechanism (Toubarro *et al.*, 2009), one revealed anticapsulation activities (Balasubramanian *et al.*, 2009).

The gene encoding elastase was cloned and comparative analyzed. Its expression was upregulated in the initial parasitic stage and localized in the esophageal cells. Taken together, elastase activity found in secretions of parasitic nematodes may play a role in the degradation of intestinal tissues and in facilitating either the penetration of the parasites into host or the release of nutrients (Todorova and Stoyanov, 2000).

11.5.4. Chitinase

Chitinases belong to the large family of O-glycosyl hydrolases and catalyze the hydrolysis of b-1, 4-Nacetyl-D-glucosamine linkages in chitin polymers. Chitinase activity has been detected in many parasitic nematodes such as *Brugia malayi* (Fuhrman *et al.*, 1992), *Onchocera volvulus* (Yang *et al.*, 2001)and *Acanthocheilonema viteae* (Ralf *et al.*, 1996), and may play a key role in hatching, molting and transmission. Nematode chitinases are extracellular proteins, and while presumably they have a biological role in egg hatching (Arnold *et al.*, 1993), but the existence of multiple genes and stage-specific expression indicates chitinases may have other functional roles in the nematode life cycle (Yang *et al.*, 2001).

Filarial chitinases have been proposed to have a role in facilitating the migration through host tissues (Ralf *et al.*, 1996). Genes coding chitinase are also identified and expressed in the glandular oesophagus of third-stage larvae (L3) of *Acanthocheilonema viteae* and *O. volvulus*, and thought to be has potential function in the infection (Yang *et al.*, 2001; Ralf *et al.*, 1996). ESTs coding chitinase is first identified in entomopathogenic nematode *H. bacteriophora* (You-Jin Hao, in prep).

Chitin is a major component of the insect cuticle, also an integral part of insect peritrophic matrices, which function as a permeability barrier between the food bolus and the midgut epithelium, enhance digestive processes and protect the brush border from mechanical disruption as well as from attack by toxins and pathogens (Tellam, 1996). In nematode infections, nematode must penetrate insect gut and entry into hemocoel. Chitinase gene expression is up-regulated in the early parasitic stage *H. bacteriophora* recovered *in vitro* by insect hemolymph, which suggests that chitinase would play a role in the penetration of host gut.

11.5.5. Metalloprotease

One of several candidate proteases involved in the tissue invasion of parasitic nematodes are zinc metalloproteases (Tort *et al.,* 1999). Generally, extracellular metalloproteases have been found to be involved in a variety of proteolytic processes including cell migration, organogenesis and wound healing.

In nematode infections, metalloproteases have been suggested to be operative in invasion of infective larvae of various parasites, and degradation of host's tissues (Tort *et al.,* 1999). They are also shown to be strong immunodominant antigens which can stimulate allergy type and protective immune responses. In parasitic nematodes, astacin proteases of the met-zincin superfamily seem to play crucial parts in tissue invasion. Astacins have been identified in parasitic nematodes, *Trichinella* spiralis (Lun *et al.,* 2003), *Strongyloides stercoralis* (Gomez *et al.,* 2005), *Ancylostoma caninum* (Zhan *et al.,* 2002; Angela *et al.,* 2006), and *Ostertagia ostertagi* (Lun *et al.,* 2003; Maere *et al.,* 2005).

Furthermore, they have been found in the free-living nematode *C. elegans* (Mohrlen *et al.,* 2003). In parasitic nematodes, infective larvae were demonstrated to release various molecules that presumably aid in infection and establishment of parasitism. Interference with their function may be a potentially successful strategy for prevention of disease. An astacin from the dog hookworm *Ancylostoma caninum* is considered as a vaccine candidate by the human hookworm vaccine initiative (Hotez *et al.,* 2003).

In entomopathogenic nematode, metalloprotease activities are first found in the excretory-secretory product of *H. bacteriophora* (Duarte Toubarro, personal communication). A gene encoding metalloprotease (named *as Hb-AST-MET*) is cloned from *H. bacteriophra*, and some ESTs encoding metalloprotease are also identified in parasitic stage *S. carpocpasae*. Hb-AST-MET exhibits significant similar to Zinc-dependent metalloprotease from *C. briggsae*, metalloprotease dpy-31 from *C. elegans*, and metalloprotease mp1from *O. volvulus*. Expression analysis by realtime RT-PCR shows that *Hb-AST-MET* is up-regulated in the early parasitic stage, which suggests that it may be crucial for parasite migration (You-Jin Hao, in prep).

11.5.6. Protease Inhibitors

Parasite-derived protease inhibitors have been shown to play a variety of roles in the survival of the parasite by the inhibition of exogenous host proteases (Peanasky *et al.,* 1984; Hartmann and Lucius, 2003) or endogenous origin.

11.5.6. Serpin

Serine protease inhibitors (serpin) genes comprise a large gene family (Potempa *et al.,* 1994), and their protein products regulate a wide variety of protease-dependent physiological functions, such as complement activation, fibrinolysis (Collen and Lijnen, 1991), coagulation (Carrell *et al.,* 1991) and inflammation in different parasitic nematodes (Potempa *et al.,* 1994). Ten clusters encoding proteins with similarity to serine protease inhibitors were identified from the cDNA library of *S. carpocapsae* parasitic stage in vitro induced by insect homogenate (Hao *et al.,* unpublished data). Two of them were identified by 2D- MS/MS to appear in the excreted products of parasitic stage *S.carpocapsae* induced *in vitro* by insect homogenate (Nelson Simoes, unpublished data).

11.5.7. Cystatin

Cystatins are a widely-distributed family of cysteine protease inhibitors which play essential roles in a spectrum of physiological processes (Barrett, 1986). The known cystatins can be grouped into three families on the basis of amino acid sequences. Type I cystatins (Stefins) are generally unglycosylated proteins (×100 amino acids) that lack disulfide bridges. Type II
cystatins are about 120 amino acids and have two intrachain disulfide bonds. Type III cystatins (Kininogens) are single-chain glycoproteins that contain three cystatin-like domains. Both type II and III are considered to be evolutionarily more advanced than type I cystatin.

Cystatins have been found in free-living nematode and in parasitic nematodes *Onchocerca volvulus* (Lustigman *et al.,* 1992), *Nippostrongylus brasiliensis* (Dainichi *et al.,* 2001), *Haemonchus contortus* (Newlands *et al.,* 2001) and *Brugia malayi* (Murray *et al.,* 2005). Entomopathogenic nematode cystatin was first reported in *S. carpocpasae*, and though to be involved in host-parasite interactions (7). The expression of *Sc-CYS* is upregulated in *S. carpocapsae* parasitic stage *in vitro* induced by insect hemolymph.

11.5.8. Other Molecules

Four ESTs encoding acetylcholinesterase were identified in *S. carpocapsae* cDNA library. Acetylcholinesterase secreted by nematodes was thought to modulate the immune system of the host (Pritchard *et al.,* 1994). Acetylcholinesterase secreted by *H. contortus* and *Ostertagia circumcincta* that inhabit the stomach or abomasums of small ruminants, seems to reduce inflammation and local ulceration by hydrolyzing acetylcholine, which stimulates gastric acid secretion (Konigova *et al.,* 2008; Sutherland and Lee, 1993). Acetylcholine has also been recorded to have numerous effects on leukocytes, including stimulation of chemotaxis and lysosomal enzyme secretion by neutrophils, inflammatory mediators, histamine and leukotriene release by mast cells, and augmentation of lymphocyte-mediated cytotoxicity (Lee, 1996). Therefore, acetylcholinesterase activity might help to prevent stimulation of cellular and humoral response to parasite infection.

One EST had 49% similarity to a fatty acid retinoid binding protein (FAR) in *O. ostertagi*. FARs were described in *S. carpocase*, which has been reported in animal parasitic nematodes including *A. caninum* (Basavaraju *et al.*, 2003), *B. malayi* (Kennedy *et al.*, 1995) and *Globodera pallida* (Prior *et al.*, 2001), and are thought to be involved in complex host–parasite interactions.

Another interesting discovery of *S.carpocpasae* cDNA library was that two EST (Hao *et al.* in preparation) shared homology to saposin-like protein in *E. invadens* and *B. malayi*. The saposin-like protein family comprises pore-forming peptides, which have been identified in a variety of organisms including the secreted products of blood-feeding nematodes *H. contortus* (Fetterer and Rhoads, 1997) and *A. caninum* (Don *et al.*, 2007). In *C. elegans*, a family comprising 29 genes of saposin-like protein/saposin-like domain containing protein has been identified. Among them, spp1 (Gene ID: T07C4.4) had been expressed as a recombinant in *E. coli* (Banyai and Patthy,1998) and its antibacterial activity assayed. spp7 (Gene ID: ZK616.9) was also reported as a candidate antimicrobial gene in *C. elegans* (Scott *et al.*, 2007). We hypothesize that the saposin-like proteins identified in this study participate in the maintenance of the monoxenic symbiosis established by this nematode with the bacteria *X. nematophila*.

11.6. GENETIC IMPROVEMENT

These nematodes parasitize and kill a large number of insects (in laboratory tests, *S. carpocapasae* alone infect more than 250 species of insects from over 75 families in 11 orders) (Poinar, 1975) and have been established as an important biological control agent among the commercially available products. However, like other biological control agents, nematodes are constrained by being living organisms that require specific conditions to be effective. Thus, desiccation or ultraviolet light rapidly inactivates insecticidal nematodes. Similarly, nematodes are effective within a narrower temperature range than chemicals, and are more impacted by suboptimal soil type, thatch depth, and irrigation frequency (Georgis and Gaugler, 1991). Moreover the rate of parasitism and mortality that a nematode strain can cause is insect- and developmental stage-specific, thus suggesting a differential nematode susceptibility to stress caused by each insect. Therefore the markets for entomopathogenic nematodes are relatively small (Gaugler *et al.*, 1993).

Then, improvements in traits including IJ longevity, bacterial retention, tolerance to extreme environments (particularly heat, ultra violet radiation, and desiccation), resistance to encapsulation in the hemocoel encountered in some key insects, and trait stability are required. Molecular biology has provided fundamental knowledge and technical protocols for the improvement of entomopathogenic nematodes by genetic engineering (Gaugler and Hashmi, 1996). Genetic improvement of the nematodes has been based on selective breeding, mutation, or genetic engineering (Gaugler *et al.*, 1989). Molecular markers to study gene expression have been found. Several genes that encode for useful traits have been identified, and efficient transformation methods have been developed. Although many countries have regulations on the release of genetically engineered organisms, the strategy to receiving approval of the release may include inducing a commercial rather than ecological gene into the nematodes and over-expressing an existing gene rather than introducing a foreign one. To better understand nematode genetic, the genomic information can rapidly promote scientific progress in the field of genetic improvement of entomopathogenic nematodes.

11.7. ENTOMOPATHOGENIC NEMATODES AS MODEL ORGANISMS

In general, parasitic taxa are difficult to culture and analyze independently of their hosts, and this makes difficult their use as genetic models; even though the suggestion that free-living sister taxa may act as good models for parasitic groups (Blaxter *et al.,* 1998). Notwithstanding, entomopathogenic nematodes offer good model for parasitism, initially by taking advantage of molecular phylogenetic inference to directly aid in the investigation of the biology of parasitism, and its dynamics of morphological and developmental evolution (Blaxter *et al.,* 1998). It is well known that the entomopathogenic nematodes of the genera *Heterorhabditis* and *Steinernema* provide effective biological control that infect and kill insect larvae (Hallem *et al.,* 2007; Stock *et al.,* 2008). Therefore the interest in these nematodes has grown in the past years, as biocontrol agents for insect pests and disease vectors and as potential models for human parasitic nematodes (Hallem *et al.,* 2007).

Both nematode families offer useful and favorable characteristics as biocontrol agents, like their small size, their broad host range, the formation of a durable infective stage, the specificity to insects which make them safe to organisms including humans, other vertebrates and plants, and do not pollute the environment (Hallem *et al.,* 2007; Stock *et al.,* 2008). In fact, entompathogenic nematodes are being used as model organisms as part of a tripartite model nematodes-bacteria-insect, like *Heterorhabditis-Photorhabdus*-Insect (Hallem *et al.,* 2007) and *Steinernema-Xenorhabdus*-Insect (Cowles and Goodrich-Blair, 2008), in which both models offer a way to study the symbiotic associations of mutualism and parasitism.

An example in this model is the association between *Heterorhabditis–Photorhabdus-Drosophila,* where the infection caused by the nematode with the bacteria elicits a dynamic immune response in *Drosophila* (Hallem *et al.,* 2007). One of the interesting and best features of the model is that each part of it can be studied separately or with different combinations, enabling pathogenesis and mutualism to be studied individually or together (Hallem *et al.,* 2007; Goodrich-Blair, 2007). An example is how the axenic infective juveniles (IJs) of nematodes can kill insect larvae as well as the symbiotic IJs. The difference is the efficiency of the killing that becomes low without the bacteria (Hallem *et al.,* 2007; Han and Ehlers, 2000).

Even thought in the past few years there has been a big increase in the understanding of the molecular model nematode-bacteria symbiosis, there is still is a lot to be unveiled (Goodrich-Blair, 2007). Continued exploration of the bacteria-nematode symbiosis will could help the understanding on how the bacteria and/or nematode symbiotic associations with mammals and plants work. It will also reveal similarities with other systems, contributing to defining the basic principles underlying long-term mutually and parasitic relationships (Goodrich-Blair, 2007; Goodrich-Blair and Clarke, 2007). Also symbiotic associations could give some insights about some human diseases, since there are reports of strains of *Photorhabdus* that attack humans (Tounsi *et al.,* 2006). In order to get more information about entomopathogenic nematodes there is a need to obtain nematodes from different parts of the world, to evaluate their differences caused by their adaptation and distribution to a diverse climates and habitats. Some of the differences include soil type, availability of suitable hosts, and physiological and behavioral adaptations (Adams *et al.,* 2006; Stock *et al.,* 2008). In the present there are different ways and sources to get genetic information about entomopathogenic nematodes. These tools will give advances in the concepts of nematode and bacterial species and the molecular genetics and phenotypic data will contribute to an increasingly stable systematic framework (Adams *et al.,* 2006; Bird *et al.,* 2005).

Regarding the use of entomopathogenic nematodes as models of human pathogens, it is interesting to note that *Steinernema* is very close to *Strongyloides*, a nematode that represent a sever health problem in tropical countries. Currently, we have a genome project to sequence *Steinernema carpocapsae*, as described in other sections of this chapter.

REFERENCE

Adams, BJ; Fodor, A; Koppenhöfer, HS; Stackebrandt, E; Stock, SP; Klein, MG. (2006). Biodiversity and systematics of nematode–bacterium entomopathogens. *Biological Control*, 37, 32–49.

Altincicek, B; Vilcinskas, A. (2007). Analysis of the immune-inducible transcriptome from microbial stress resistant, rat-tailed maggots of the drone fly *Eristalis tenax*. *BMC Genomics*, 8, 326-337.

Angela, LW; Sara, L; Yelena, O; Vehid, D; Jordan, P; Susana, M; Bin, Z; Maria, EB; Peter, JH; Alex, L. (2006). *Ancylostoma caninum* MTP-1, an astacin-like metalloprotease secreted by infective hookworm larvae, is involved in tissue migration. *Infection and Immunity*, 74(2), 961-967.

Arnold, K; Brydon, LJ; Chappell, LH, Gooday, GW. (1993). Chitinolytic activities in *Heligmosomoides polygyrus* and their role in egg hatching. *Molecular and Biochemical Parasitology*, 58(2), 317-323.

Balasubramanian, N; Hao, YJ; Toubarro, D; Nascimento, G; Simões, N. (2009). Purification, biochemical and molecular analysis of a chymotrypsin protease with prophenoloxidase suppression activity from the entomopathogenic nematode *Steinernema carpocapsae*. *International Journal of Parasitology*, 39, 975-984.

Banyai, L; Patthy, L. (1998). Amoebapore homologs of *Caenorhabditis elegans*. *Biochimica et Biophysica Acta*, 1429(1), 259-264.

Barrett, AJ. (1986). The cystatins: a diverse superfamily of cysteine peptidase inhibitors. *Biomedica Biochimica Acta*, 45(11-12), 1363-1374.

Basavaraju, S; Bin, Z; Kennedy, MW; Yue, YL; Hawdon, J; Hotez, PJ. (2003). Ac-FAR-1, a 20 kDa fatty acid- and retinol-binding protein secreted by adult *Ancylostoma caninum* hookworms: gene transcription pattern, ligand binding properties and structural characterisation. *Molecular and Biochemical Parasitology*, 126(1), 63-71.

Bird, DM; Blaxter, ML; McCarter, JP; Mitreva, M; Sternberg, PW; Thomas, WK. (2005). A white paper on nematode comparative genomics. *Journal of Nematology*, 37(4), 408–416.

Blaxter, ML; De Ley, P; Garey, JR; Liu, LX; Scheldeman, P; Vierstraete, A; Vanfleteren, JR; Mackey, LY; Dorris, M; Frisse, LM; Vida, JT; Thomas, WK. (1998). A molecular evolutionary framework for the phylum Nematoda. *Nature*, 392, 71-75.

Blaxter, ML; Page, AP; Rudin, W; Maizels, RM. (1992). Nematode surface coats: actively evading immunity. *Parasitology Today*, 8(7), 243-247.

Boemare, NL; Laumond, C; Luciani, J. (1982). Mise en evidence d'une toxicogenese provoquee par le nematode axenique entomophague *Neoaplectana carpocapsae* Weiser chez 1-insect axenique. *Galleria mellonella* L. Conte Rendues de *l'Academie de les Sciences, Serie* D, 543–546.

Brivio, MF, Egulleor, MD; Grimaldi, A; Vigetti, D; Valvassori, R; Lanzavecchia, G. (2000). Structural and biochemical analysis of the parasite *Gordius villoti* (Nematomorpha, Gordiacea) cuticle. *Tissue Cell*, 32(5), 366-376. *Gordius villoti*.

Burman, AM. (1982). *Neoaplectana carpocapsae*: Toxin production by axenic insect parasitic nematodes. *Nematologica*, 28, 62-70.

Burnell, AM; Stock, SP. (2000). Heterorhabditis, Steinernema and their bacterial symbionts, lethal pathogens of insects. Nematology, 2, 31-42.

Carrell, RW; Evans, DL; Stein, PE. (1991). Mobile reactive centre of serpins and the control of thrombosis. *Nature*, 353, 576-578.

Collen, D; Lijnen, HR. (1991). Basic and clinical aspects of fibrinolysis and thrombolysis. *Blood*, 78, 3114-3124.

Cowles, CE; Goodrich-Blair, H. (2008). The *Xenorhabdus nematophila* nilABC genes confer the ability of *Xenorhabdus* spp. to colonize *Steinernema carpocapsae* nematodes. *Journal of Bacteriology*, 190(12), 4121-4128.

Dainichi, T; Maekawa, Y; Ishii, K; Himeno, K. (2001). Molecular cloning of a cystatin from parasitic intestinal nematode, *Nippostrongylus brasiliensis*. *The Journal of Medical Investigation,* 48(1-2), 81-87.

De Ley, P; Blaxter, ML. Systematic position and phylogeny. In: The Biology of Nematodes, Lee, D.L.(Eds.), London: Taylor and Francis. 2002, 1–30.

Delsuc, F; Brinkmann, H; Philippe, H. (2005). Phylogenomics and the reconstruction of the tree of life., 6, 361-375.

Diatchenko, L; Lau,YF; Campbell, AP; Chenchik, A; Moqadam, F; Huang, B; Lukyanov, S; Lukyanov, K; Gurskaya, N; Sverdlov, ED; Siebert, PD. (1996). Suppression subtractive hybridization: a method for generating differentially regulated or tissue-specific cDNA probes and libraries. the Proceeding of National Academy of Science, 93, 6025-6030.

Diatchenko, L; Lukyanov, S; Lau, YF; Siebert, PD. (1999). Suppression subtractive hybridization: a versatile method for identifying differentially expressed genes. *Methods in Enzymology*, 303, 349-380.

Don, TA; Oksov, Y; Lustigman, S; Loukas, A. (2007). Saposin-like proteins from the intestine of the blood-feeding hookworm, *Ancylostoma caninum*. *Parasitology*, 134(3), 427-436.

Ehlers, RU. (2001). Mass production of entomopathogenic nematodes for plant protection. *Applied Microbiology and Biotechnology*, 56(5-6), 623-633.

Fetterer, RH; Rhoads, ML. (1997). Characterization of haemolytic activity from adult *Haemonchus contortus*. *International Journal of Parasitology*, 27(9),1037-1040.

Gamble, HR; Mansfield, LS. (1996). Characterization of excretory-secretory products from larval stages of *Haemonchus contortus* cultured in vitro. *Veterinary Parasitology*, 62(3-4), 291-305.

Gaugler R; McGuire T; Campbell J. (1989). Genetic variability among strains of the entomopathogenic nematode *Steinernema feltiae*. *Journal of Nematology*, 21, 247-253.

Gaugler, R; Glazer, I; Campbell, JF; Liran, N. (1993). Laboratory and field evaluation of an entomopathogenic nematode genetically selected for improved host-finding. *Journal of Invertebrate Pathology*, 63, 68-73.

Gaugler, R; Hashmi, S. (1996). Genetic engineering of an insect parasite. *Genetic Engineering*, 135-155.

Georgis, R; Gaugler, R. (1991). Predictability in biological control using entomopathogenic nematodes. *Journal of Economic Entomology*, 84, 713-720.

Ghedin, E; Wang, S; Foster, JM; Slatko, BE. (2004). First sequenced genome of a parasitic nematode. *Trends in Parasitology*, 20(4), 151-153.

Gomez, GS; Loukas, A; Slade, RW; Neva, FA; Varatharajalu, R; Nutman, TB; Brindley, PJ. (2005). Identification of an astacin-like metallo-proteinase transcript from the infective larvae of *Strongyloides stercoralis*. *Parasitology International*, 54(2),123-133.

Goodrich-Blair, H. (2007). They've got a ticket to ride: *Xenorhabdus nematophila–Steinernema carpocapsae* symbiosis. *Current Opinion in Microbiology*, 10: 225–230.

Goodrich-Blair, H; Clarke, DJ. (2007). Mutualism and pathogenesis in *Xenorhabdus* and *Photorhabdus*: two roads to the same destination. *Molecular Microbiology*, 64(2),260-268.

Grenier, E; Bonifassi, E; Abrad, P; Laumond, C. (1996). Use of species-specific satellite DNAs as diagnostic probes in the identification of Steinernematidae and Heterorhabditidae entomopathogenic nematodes. *Parasitology*, 113(5), 483-489.

Grenier, E; Catzeflis, FM; Abad, P. (1997). Genome sizes of the entomopathogenic nematodes *Steinernema carpocapsae* and *Heterorhabditis bacteriophora* (Nematoda:Rhabditida). *Parasitology*, 114 (5), 497-501.

Grenier, E; Laumond, C; Abad, P. (1995). Characterization of a species-specific satellite DNA from the entomopathogenic nematode *Steinernema carpocapsae*. *Molecular and Biochemical Parasitology*, 69(1), 93-100.

Guiliano, DB; Hall, N; Jones, SMJ; Clark, LN; Corton, CH; Barrell, BG; Blaxter, ML. (2002). Conservation of long-range synteny and microsynteny between the genomes of two distantly related nematodes. *Genome Biology*, 3(10), Research0057.

Hallem, E.A., Rengarajan, M., Ciche, T.A. and Sternberg P.W. (2007). Nematodes, Bacteria, and Flies: A Tripartite Model for Nematode Parasitism. *Current Biology*, 17, 898–904.

Han, R; Ehlers, RU. (2000). Pathogenicity, development, and reproduction of *Heterorhabditis bacteriophora* and *Steinernema carpocapsae* under axenic in vivo conditions. *Journal of Invertebrate Pathology*, 75(1), 55-58.

Hartmann, S; Lucius, R. (2003). Modulation of host immune responses by nematode cystatins. *International Journal of Parasitology*, 33(11), 1291-1302.

Hawdon, JM; Jones, BF; Perregaux, MA; Hotes, PJ. (1995) *Ancylostoma caninum*: metalloprotease release coincides with activation of infective larvae in vitro. *Experimental Parasitology*, 80(2), 205-211.

Hepworth, PJ; Leatherbarrow, H; Hart, CA; Winstanley, C. (2007). Use of suppression subtractive hybridisation to extend our knowledge of genome diversity in *Campylobacter jejuni*. *BMC Genomics*, 8, 110-120.

Hotez, PJ; Ashcom, J; Zhan, B; Bethony, J; Loukas, A; Hawdon, J; Wang, Y; Jin, Q; Jones, KC; Dobardzic, A; Dobardzic, R; Bolden, J, Essiet, I; Brandt, W; Russell, PK; Zokk, BC; Howard, B; Chacon, M. (2003). Effect of vaccination with a recombinant fusion protein encoding an astacinlike metalloprotease (MTP-1) secreted by host-stimulated

Ancylostoma caninum third-stage infective larvae. the *Journal of Parasitology*, 89, 853-855.

Ji, W; Wright, MB; Cai, L; Flament, A; Lindpaintner, K. (2002). Efficacy of SSH PCR in isolating differentially expressed genes. *BMC Genomics*, 3,12-18.

Kaya, HK; An, M; Alumai, A; Choo, HY; Torre, M; Fodor, A; Ganguly, S; Hazir, S; Lakatos, T; Pye, A; Wilson, M; Yamanaka, S; Yang, H; Ehlers, RU. (2006). Status of entomopathogenic nematodes and their symbiotic bacteria from selected countries or regions of the world. *Biological Control*, 38, 134-1552.

Kaya, HK; Gaugler, R. (1993). Entomopathogenic nematodes. *Annual Review of Entomology*, 38, 181-206.

Kennedy, MW; Allen, JE; Wright, AS; McCruden, AB; Cooper, A. (1995). The gp15/400 polyprotein antigen of *Brugia malayi* binds fatty acids and retinoids. *Molecular and Biochemical Parasitology*, 71(1), 41-50.

Konigova, A; Hrchova, G; Velebny, S; Corba, J; Varady, M. (2008). Experimental infection of *Haemonchus contortus* strains resistant and susceptible to benzimidazoles and the effect on mast cells distribution in the stomach of *Mongolian gerbils (Meriones unguiculatus)*. *Parasitology Research*, 102(4), 587-595.

Koppenhofer, AM; Grewal, PS; Fuzy, EM. (2007). Differences in penetration routes and establishment rates of four entomopathogenic nematode species into four white grub species. *Journal of Invertebrate Pathology*, 94(3),184-195.

Lackey, A; James, ER; Sakanari, JA; Resnick, SD; Brown, M; Blanco, AE; Mckerrow, JH. (1989). Extracellular proteases of *Onchocerca*. *Experimental Parasitology,* 68(2),176-185.

Laumond C; Simoes, N; Boemare, N. (1989). Toxines des nematodes entomoparasites. Pathogenicite de *Steinernema carpocapsae*: Perspectives dapplications en genie genetique. *Comptes Rendus de l'Academie d'Agriculture de France*, 75, 135–138.

Lee, DL. (1996). Why do some nematode parasites of the alimentary tract secrete acetylcholinesterase? *International Journal of Parasitology*, 26(5), 499-508.

Lun, HM; Mak, CH; Ko, RC. (2003). Characterization and cloning of metallo-proteinase in the excretory/secretory products of the infective-stage larva of *Trichinella spiralis*. *Parasitology Research*, 90(1), 27-37.

Lustigman, S; Brotman, B; Huima, T; Prince, AM; McKerrow, JH. (1992). Molecular cloning and characterization of onchocystatin, a cysteine proteinase inhibitor of *Onchocerca volvulus*. the *Journal of Biological Chemistry*, 267(24),17339-17346.

Maere, VD; Vercauteren, I; Gelghof, P; Gevaert, J; Vercruysse, J; Claerebout, E. (2005). Molecular analysis of astacin-like metalloproteases of Ostertagia ostertagi. *Parasitology*, 130(1), 89-98.

Maizels, RM; Blaxter, ML; Selkirk, ME. (1993). Forms and functions of nematode surfaces. *Experimental Parasitology*, 77(3), 380-384.

McKerrow, JH. (1989). Parasite proteases. *Experimental Parasitology*, 68(1), 111-115.

Mitreva, M; Zarlenga, DS; McCarter, JP; Jasmer, DP. (2007). Parasitic nematodes - from genomes to control. *Veterinary Parasitology*, 148: 31-42.

Mohrlen, F; Hutter, H; Zwilling, R. (2003). The astacin protein family in *Caenorhabditis elegans*. European *Journal of Biochemistry*, 270(24), 4909-4920.

Montiel, R; Lucena, MA; Medeiros, J; Simoes, N. (2006). The complete mitochondrial genome of the entomopathogenic nematode *Steinernema carpocapsae*: insights into nematode mitochondrial DNA evolution and phylogeny. *Journal of Molecular Evolution*, 62(2), 211-225.

Murray, J; Manoury, B; Balic, A; Watts, C; Maizeis, RM. (2005). Bm-CPI-2, a cystatin from Brugia malayi nematode parasites, differs from *Caenorhabditis elegans* cystatins in a specific site mediating inhibition of the antigen-processing enzyme AEP. *Molecular and Biochemical Parasitology*, 139(2), 197-203.

Newlands, GFJ; Skuce, PJ; Knox, DP; Smith, WD. (2001). Cloning and expression of cystatin, a potent cysteine protease inhibitor from the gut of *Haemonchus contortus*. *Parasitology*, 122(3), 371-378.

Parkinson, J; Mitreva, M; Whitton, C; Thomson, M; Daub, J; Martin, J; Schmid, R; Hall, N; Barrell, B; Waterston, RH; McCarter, JP; Blaxter, ML. (2004). A transcriptomic analysis of the phylum nematoda. *Nature Genetics*. 36(12), 1259-1267.

Peanasky RJ; Bentz,Y; Paulson, B; Graham, DL; Babin, DR. (1984). The isoinhibitors of chymotrypsin/elastase from *Ascaris lumbricoides*: isolation by affinity chromatography and association with the enzymes. *Archives of Biochemistry and Biophysics,* 232(1), 127-134.

Phares, K. (1996). An unusual host-parasite relationship: the growth hormone-like factor from plerocercoids of spirometrid tapeworms. *International Journal of Parasitology*, 26(6), 575-588.

Philipp, M; Rumjaneck, FD. (1984). Antigenic and dynamic properties of helminth surface structures. *Molecular Biochemical Parasitology*, 10(3), 245-268.

Poinar, G. (2003). Trends in the evolution of insect parasitism by nematodes as inferred from fossil evidence. *Journal of Nematology*. 35(2),129-132.

Poinar, GO. (1975). *Entomogenous nematodes*: A manual and host list of insect-nematode associations. Leiden, E.J. Brill.

Politz, SM; Chin, KJ; Herman, DL. (1987). Genetic analysis of adult-specific surface antigenic differences between varieties of the nematode *Caenorhabditis elegans*. *Genetics*, 117(3), 467-76.

Politz, SM; Philipp, M. (1992). *Caenorhabditis elegans* as a model for parasitic nematodes: a focus on the cuticle. *Parasitology Today*, 8(1), 6-12.

Potempa, J; Korzus, E; Travis, J. (1994). The serpin superfamily of proteinase inhibitors: Structure, function, and regulation. the *Journal of Biological Chemistry*, 269, 15957-15960.

Potempa, J; Korzus, E; Travis, J. (1994). The serpin superfamily of proteinase inhibitors: structure, function, and regulation. *Journal of Biological Chemistry*, 269(23), 15957-15960.

Prior, A; Jones, JT; Blok, VC; Beauchamp, J; McDermontt, L; Cooper, A; Kennedy, MW. (2001). A surface-associated retinol- and fatty acid-binding protein (Gp-FAR-1) from the potato cyst nematode *Globodera pallida*: lipid binding activities, structural analysis and expression pattern. *Biochemical Journal*, 356(2), 387-394.

Pritchard, DI; Brown, A; Toutant, JP. (1994). The molecular forms of acetylcholinesterase from *Necator americanus* (Nematoda), a hookworm parasite of the human intestine. *European Journal of Biochemistry*, 219(1-2), 317-323.

Ralf, A; Brigtte, K; Werner, R; Thomas, F; Thomas, M; Richard, L. (1996). Identification of chitinase as the immunodominant filarial antigen recognized by sera of vaccinated rodents. *Journal of Biological Chemistry*, 271(3),1441-1447.

Salvador, H; Tsanko, G; Petra, LB; William, JM; Ruud, AM. (2005). *Bacillus thuringiensis* Cry1Ca-resistant *Spodoptera exigua* lacks expression of one of four Aminopeptidase N genes. *BMC Genomics*, 6, 96-105.

Sandhu, SK; Jagdale, GB; Hogenhout, SA; Grewal, PS. (2006). Comparative analysis of the expressed genome of the infective juvenile entomopathogenic nematode, *Heterorhabditis bacteriophora*. *Molecular and Biochemical Parasitology*, 145(2), 239-244.

Scott, A; Sandra, JM; Brad, L; Jonathan, HF; David, AS. (2007). Specificity and complexity of the Caenorhabditis elegans innate immune *response. Molecular and Cellular Biology*, 2007, 27(15), 5544-5553.

Sicard, M; Ferdy. JB; Page, S; Le, BN; Godelle, B; Boemare, N; Moulia, C. (2004). When mutualists are pathogens: an experimental study of the symbioses between *Steinernema* (entomopathogenic nematodes) and *Xenorhabdus* (bacteria). *Journal of Evolutionary Biology*, 17(5), 985-993.

Simoes, N; Caldas, C; Rosa, JS; Bonifassi, E; Laumond, C. (2000). Pathogenicity caused by high virulent and low virulent strains of *Steinernema carpocapsae* to Galleria mellonella. *Journal of Invertebrate Pathology*, 75(1), 47-54.

Stock, S.P., Banna L.A., Darwish, R. and Katbeh, A. (2008). Diversity and distribution of entomopathogenic nematodes (Nematoda: Steinernematidae, Heterorhabditidae) and their bacterial symbionts (c-Proteobacteria: Enterobacteriaceae) in Jordan. *Journal of Invertebrate Pathology*, 98, 228–234.

Sutherland, IA; Lee, DL. (1993). Acetylcholinesterase in infective-stage larvae of *Haemonchus contortus, Ostertagia circumcincta* and *Trichostrongylus colubriformis* resistant and susceptible to benzimidazole anthelmintics. *Parasitology*, 107(5), 553-557.

Tellam, RL. (1996). The peritrophic matrix. In Editor Lehane, M. J., and Billingsley, P. F. Editor (Eds). Biology of the Insect Midgut (86-114). Cambridge: Chapman and Hall.

Todorova, VK; Stoyanov, DI. (2000). Partial characterization of serine proteinases secreted by adult *Trichinella spiralis. Parasitology Research*, 86(8), 684-687.

Tort, J; Brindley, PJ; Knox, D; Wolfe, KH; Dalton, JP. (1999). Proteinases and associated genes of parasitic helminths. *Advances in Parasitology*, 43,161-266.

Toubarro, D; Lucena-Robles, M; Nascimento, G; Costa, G; Montiel, R; Coelho, AV; Simões, N. (2009). An apoptosis-inducing serine protease secreted by the entomopathogenic nematode *Steinernema carpocapsae*. *International Journal of Parasitolology*. [Epub ahead of print].

Tounsi, S; Blight, M; Jaoua, S; de Lima Pimenta, A. (2006). From insects to human hosts: Identification of major genomic differences between entomopathogenic strains of *Photorhabdus* and the emerging human pathogen *Photorhabdus asymbiotica*. *International Journal of Medical Microbiology*, 296, 521–530.

Fuhrman, JA; Lane, WS; Piessens, WF; Perler, FB. (1992). Transmission-blocking antibodies recognize microfilarial chitinase in brugian lymphatic filariasis. the *Proceeding of National Academy of Science*, 89(5),1548-1552.

Tyson, T; Reardon, W; Browne, JA; Burnell, AM. (2007). Gene induction by desiccation stress in the entomopathogenic nematode *Steinernema carpocapsae* reveals parallels with drought tolerance mechanisms in plants. *International Journal Parasitology*, 37(7), 763-776.

Wakelin, D., Denhem, D. A. (1983). The immune responses. In: Campbell WC. (Eds.), *Trichinella* and trichinosis (p.265–308). New York: Plenum Press.

Yang, W; Gillian, E; Anthony, PU; Shohei, S; Albert, B. (2001). Expression and secretion of a larval-specific chitinase (family 18 glycosyl hydrolase) by the infective stages of the parasitic nematode, *Onchocerca volvulus*. *Journal of Biological Chemistry*, 276(45), 42557-42564.

You-Jin, H; Rafael, M; Gisela, N; Duarte, T; Nelson, S. (2008). Identification, characterization of functional candidate genes for host-parasite interactions in entomopathogenetic nematode *Steinernema carpocapsae* by suppressive subtractive hybridization. *Parasitology Research*,103(3), 671-683.

Zhan, B; Hotez, PJ; Wang, Y; Hawdon, JM. (2002). A developmentally regulated metalloprotease secreted by host-stimulated *Ancylostoma caninum* third-stage infective larvae is a member of the astacin family of proteases. *Molecular and Biochemical Parasitology*, 120(2), 291-296.

PART III:
ENTOMOPATHOGENIC VIRUSES

In: Microbial Insecticides: Principles and Applications
Editors: J. Francis Borgio, K. Sahayaraj, et al.

ISBN: 978-1-61209-223-2
© 2011 Nova Science Publishers, Inc

Chapter 12

BIOLOGICAL CONTROL POTENTIAL OF ENTOMOPATHOGENIC VIRUSES

Neiva Knaak and Lidia Mariana Fiuza*

Microbiology Laboratory, PPG-Biology, Universidade do Vale do Rio dos Sinos; Av. Unisinos, 950. 93001-970, São Leopoldo, RS, Brasil

ABSTRACT

Viruses are a potential source of bio-pesticides for control of pests that attack food-crop plantations. Entomopathogenic viruses are found in many kinds of insects but those of the Lepidoptera and Hymenoptera.

The Baculovirus (granulose and nuclear polyhedrosis virus) are very specific in action and in general utilize the arthropods as hosts - this is a very important characteristic from a safety standpoint when they are used for selective control of pests that attack economically important cultures. VPNs and VGs have been used all over the world for many years against insect pests - especially insects of the Lepidoptera order, as *Anticarsia gemmatalis* Nuclear Polyhedrosis Virus (AgMNPV) and *Spodoptera frugiperda* Nuclear Polyhedrosis Virus (SfMNPV). More recently, they have also been used as vectors of expression and are widely utilized in medicine as therapeutic agents, prophylactics (vaccines) and in diagnosis. In addition, the VPNs and VGs have contributed to the production of genetically modified viral insecticides, thus improving the pathological characteristics of the Baculovirus as agents of biological control and widening their host spectrum.

12.1. ENTOMOPATHOGENIC VIRUSES

Viruses are a potential source of bio-pesticides for control of pests that attack food-crop plantations. Entomopathogenic viruses are found in many kinds of insects but those of the Lepidoptera and Hymenoptera orders have received considerable attention because they include many important pests (Table 1). Baculovirus, reovirus, poxvirus, picornavirus,

---
* Phone (+5551) 35908477, E-mail: neivaknaak@gmail.com

iridovirus and parvovirus all infect insects, but amongst these, only the Baculovirus have been isolated in hosts of the arthropod class and are the only group of viruses that have undergone extensive safety tests and been shown to be inoffensive to microorganisms, other non-insect invertebrates and plants (Payne, 1986; Groner, 1989; Kitajima, 1994).

The Baculovirus (granulose and nuclear polyhedrosis virus) are very specific in action and in general utilize the arthropods as hosts - this is a very important characteristic from a safety standpoint when they are used for selective control of pests that attack economically important cultures.

Although there are many entomopathogenic viruses, those of the Baculoviridae family are of greatest interest in applied entomology because they include many viruses already widely used and others that appear to have great potential as biological insecticides (Ribeiro *et al.,* 1998; Moscardi, 1999; Castro *et al.,* 1999; Cory and Myers, 2003).

Table 1. Classification of entomopathogenic virus and their target pests

Family	Genera	Virus	Targets
Reoviridae	Cypovirus	CPV	*Bombyx mori*
Poxviridae	Entomopoxvirus	EPVs	Coleoptera (*Melolontha melolontha*), Lepidoptera (*Amsacta moorei*), Orthoptera Diptera (*Chironomus luridus*)
Iridoviridae	Iridovirus, Chloriridovirus	IVs	Lepidoptera, Diptera, Coleoptera, Hemiptera, Hymenoptera
Parvoviridae	Densovirus		Lepidoptera (*Bombyx mori*)
Picornaviridae	Not yet established		Diptera (*Drosophila* sp.
Rhabdoviridae	Lyssavirus		Diptera (*Drosophila melanogaster*)
Polydnaviridae	Ichnovirus Bracovirus	GsV GmV	*Compoletis sonorensis* *Cotesia melanoscela*
Birnaviridae	Birnavirus	DXV	Diptera (*Drosophila* sp.)
Togaviridae	Alphavirus		Virus sindbis
Flaviviridae	Flavivirus		Diptera
Buyanviridae	Bunyavirus, Phlebovirus, Nairovirus, Hantavirus, Tospovirus		Artropodes, vertebrates
Tetraviridae	Not yet established		Lepidopteros (*Nudaurelia beta*)
Nodaviridae	Nodavirus		Diptera, Coleoptera, Lepidoptera
Baculoviridae	Nuclear polyhedrosis virus	VPN	Lepidoptera (*Autographa californica, Bombyx mori, Mamestra brassicae, Orgya pseudotsugata, Heliothis zea, Trichoplusia ni*);
	Granulosis virus	VG	Lepidoptera (*Plodia interpunctella, Heliothis zea*)

VPNs and VGs have been used all over the world for many years against insect pests - especially insects of the Lepidoptera order.

More recently, they have also been used as vectors of expression and are widely utilized in medicine as therapeutic agents, prophylactics (vaccines) and in diagnosis (O'Reilly, *et al.*, 1992; Richardson, 1995; Bonning and Hammock, 1996).

In addition, the VPNs and VGs have contributed to the production of genetically modified viral insecticides (Wood and Granados, 1991; Maeda, 1995; Bonning and Hammock, 1996), thus improving the pathological characteristics of the Baculovirus as agents of biological control (Cory *et al.*, 1994; Miller, 1995) and widening their host spectrum (Kamita and Maeda, 1993; Maeda *et al.*, 1993; Chen *et al.*, 1998).

12.2. BACULOVIRUSES

One of the first attempts to control insects with a virus was made in 1892 with the introduction of a VPN into *Lymantria monacha* populations found in German forests (Hubner, 1986).

In the United States of America, the first biological tests were made on *Lymantria dispar* (Lepidoptera: Lymantriidae) in 1913 (Cunningham, 1990). The first successful tests with VPN produced in laboratories were performed by Steinhaus and Thompson (1949) in California (USA) to control *Colias eurytheme* (Lepidoptera: Pieridae).

From the sixties onwards considerable effort was expended to develop quality controlled and standardized virus-based commercial products that could legally registered. The first success in this work was obtained in 1975 with the registry in the USA of a viral insecticide against *Helicoverpa zea* (Lepidoptera: Noctuidae) (Ignoffo and Couch, 1981; Hubner, 1986). This registration had a significant influence on the later development and use of other Baculovirus (Moscardi and Sósa-Gomez, 1996).

12.2.1. Commercial Products of the Baculoviruses

Majority of the viral insecticides inhabits only a small number of hosts and therefore can be commercially successful only against large or extensive infestations. Some insecticides have been registered for a wide range of hosts such as, for example, Mamestrin MNPV for the lepidopteros species: *Mamestra brassicae*, *Helicoverpa armigera*, *Plutella xylostella*, *Phthorimaea opereulella* and *Lobesia botrana*.

However, the real commercial potential of these products is much smaller. At the moment, only a relatively small number of entomopathogenic viruses have been registered and these are exclusively of the Baculoviridae family (Table 2).

The first commercial products registered in Brazil were the *A. gemmatalis* (AgMNPV) nuclear polyhedrosis virus, and the first in Argentina was the *Cydia pomonella* (CpGV) granular virus (Sosa-Gómez *et al.*, 2008).

Table 2. Commercial products of the Baculovirus to control pest-insects

Target-pests	Crop	Virus	Product name	Producer	Country
Adoxophyes orana	Apple	GV	Capex 2	Andermatt Biocontrol	Switzerland
Adoxophyes sp.	Tea	GV			Japan
Agrotis segetum		GV			China
Anagrapha falcifera	Cotton, vegetables	NPV			Brazil
Anticarsia gemmatalis	Soybean	NPV	Polygen, Multigen, Baculoviron, Baculovirus Nitral, Coopervirus, Protege VPN	Agroggen Embrapa Agricola El Sol	Brazil Brazil Guatemala
Autographa californica	Cabbage, Cotton, vegetables	NPV	VPN-80 Gusano	Agrícola El Sol Thermo Trilogy	Guatemala USA
Buzura suppressaria	Tea, tung oil tree	NPV	-	-	China
Cryptophlebia leucotreta	Citrus	GV	Cryptex	Biocontrol	Switzerland
Cydia pomonella	Apples, walnuts, pears	GV	Granupom Madex Virin-GyAp Carpovirusine Carpovirus Plus Cyd-x Granusal	AgrEvo A. Biocontrol NPO Vector NPP Calliope Agro Roca T. Trilogy BBA	Germany, Spain Russia France Argentina USA Germany
Erynnys ello	Cassava	GV	-	-	Brazil
Helicoverpa armigera	Cotton Bollworm, Corn Earworm, Tobacco	NPV	Helicovex Virin-HS	Biocontrol	Switzerland Russia
Helicoverpa zea	Cotton, vegetables	NPV	Elcar Gemstar	Novartis Thermo Trilogy	USA
Heliothis virescens	Cotton	NPV	Elcar Gemstar	Novartis Thermo Trilogy	USA
Homona magnanima	Tea	NPV			Japan
Hypantrea cunea	Forest, mulberry	NPV	Virin-ABB	NPO Vector	Russia
Lymantria dispar	Deciduous forests	NPV	Gypcheck Disparvirus Virin-ENSH	U.S. F.Service NPO Vector -	USA Canada Russia

Target-pests	Crop	Virus	Product name	Producer	Country
Mamestra brassicae	Cabbage	NPV	Virin-EKS Mamestrin	NPO Vector NPP Calliope	Russia France
Neodipriom sertifer	Pine forests	NPV	Leconteivirus Monisarmiovir us Virox	Canadian Kemira Oxford	Canada Finland UK
Orygia pseudotsugata	Douglas-fir forests	NPV	TM Biocontrol-1	U.S. Forest Service	USA Canada
Phthorimaea operculella	Field and stored potatoes	GV	PTM baculovirus		Peru, Tunisia, Egypt
Spodoptera exigua	Vegetables flowers	NPV	Spod-X Spodopterin	Thermo Trilogy Bio NPP	USA Switzerland France
Spodoptera frugiperda	Maize	NPV			Brazil
Spodoptera Littoralis	Cotton	NPV	Spodopterin Littovir	NPP Biocontrol	France Switzerland
Spodoptera sunia	Vegetables	NPV	VPN 82		Guatemala

GV=granulovirus, NPV= nuclear polyedrosis virus.
Based: Rechcigl and Rechcigl (1999), Moscardi (1999), Pimentel (2002).

The group of VPNs has been isolated in the following organisms: Neuroptera, Trichoptera, Lepidoptera, Diptera, Hymenoptera and Coleoptera, as in the Crustacea and Arachnida classes (Tinsley and Kelly, 1985; Bilimoria, 1986, 1991; Martignoni and Iwai, 1986; Adams and McClintock, 1991; O'Reilly *et al.,* 1992), as well as in Thysanura and Homoptera (Federici, 1986). In the USA, even the VPNs registered by the Environmental Protection Agency - EPA are not commercially available. The reasons for this state of affairs are complex, but basically involve the continued availability of effective wide-spectrum synthetic chemical products for pest control, limited markets (and hence low profit potential for the manufacturers) and questionable efficiency. In countries where the registration process is less complex and expensive and labor costs are lower, VPNs are frequently utilized for controlling butterfly pests in green vegetables and other cultures. Examples of this are the use of *Anticarsia gemmatalis* VPNs in soya (*Glycine max*) in Brazil; *Spodoptera exigua* VPN in kitchen gardens in South East Asia, especially Thailand; *Spodoptera* VPNs in various cultures in Central America and in the USA Mid West and *Heliothis armigera* VPNs (Hübner) in cotton (*Gossypium hirsutum*) in China (Federici, 1993). However, information on these is frequently out of date and not always available in the literature.

12.3. ENTOMOPATHOGENIC VIRUSES AS BIOLOGICAL CONTROL AGENTS

Worldwide nowadays there are innumerable groups dedicated to basic and applied research on the entomopathogenic viruses all seeking biological control of insects that are important in agriculture and are disease vectors in the public health areas as shown in Table 3.

Table 3. Some studies with entomopathogenic virus applied in biological control

Target-insect	Type virus	Reference
Agrotis ipsilon	NPV	Alves, 1986, Lucchini *et al.*, 1984, Schmitt, 1985, Chagas *et al.*, 1986, Moscardi, 1987, Valicente and Cruz, 1991, Kitajima, 1986.
Alabama argillacea	NPV	Alves, 1986, Lucchini *et al.*, 1984, Schmitt, 1985, Chagas *et al.*, 1986, Moscardi, 1987, Valicente and Cruz, 1991, Kitajima, 1986.
Anagasta Kuehniella	NPV	Alves, 1986, Lucchini *et al.*, 1984, Schmitt, 1985, Chagas *et al.*, 1986, Moscardi, 1987, Valicente and Cruz, 1991, Kitajima, 1986.
Anthonomus grandis	NPV	Alves, 1986, Lucchini *et al.*, 1984, Schmitt, 1985, Chagas *et al.*, 1986, Moscardi, 1987, Valicente and Cruz, 1991, Kitajima, 1986.
Anticarsia gemmatalis	NPV AgMNPV-2D	Alves, 1986, Lucchini *et al.*, 1984, Schmitt, 1985, Chagas *et al.*, 1986, Moscardi, 1987, Moscardi, 1989, Valicente and Cruz, 1991, Kitajima, 1986. Almeida *et al.*, 2009
Autographa californica	AcMNPV	Hodgson et al, 2009 Guo *et al.*, 2005, Wang *et al.*, 2009
Automeris memusae	CPV	Alves, 1986, Lucchini *et al.*, 1984, Schmitt, 1985, Chagas *et al.*, 1986, Moscardi, 1987, Valicente and Cruz, 1991, Kitajima, 1986.
Bombyx mori	BmMNPV	Torquato *et al.*, 2006, Brancalhão *et al.*, 2009; Pereira *et al.*, 2008, Kato *et al.*, 2009, Ribeiro *et al.*, 2009, Yun *et al.*, 2009.
Cadra cautella	NPV	Alves, 1986, Lucchini *et al.*, 1984, Schmitt, 1985, Chagas *et al.*, 1986, Moscardi, 1987, Valicente and Cruz, 1991, Kitajima, 1986.
Colias lesbia pyrhotea	NPV	Alves, 1986, Lucchini *et al.*, 1984, Schmitt, 1985, Chagas *et al.*, 1986, Moscardi, 1987, Valicente and Cruz, 1991, Kitajima, 1986.
Condylorrhiza vestigialis	CvMNPV	Almeida *et al.*, 2008, Castro *et al.*, 2008
Corcyra cephalonica	NPV	Alves, 1986, Lucchini *et al.*, 1984, Schmitt, 1985, Chagas *et al.*, 1986, Moscardi, 1987, Valicente and Cruz, 1991, Kitajima, 1986.
Cydia pomonella	CpGV	Eberle *et al.*, 2009
Diatraea flavipenella	GV	Alves, 1986, Lucchini *et al.*, 1984, Schmitt, 1985, Chagas *et al.*, 1986, Moscardi, 1987, Valicente and Cruz, 1991, Kitajima, 1986.
Diatraea saccharalis	GV	Alves, 1986, Lucchini *et al.*, 1984, Schmitt, 1985, Chagas *et al.*, 1986, Moscardi, 1987, Valicente and Cruz, 1991, Kitajima, 1986.
Dione juno juno	NPV	Alves, 1986, Lucchini *et al.*, 1984, Schmitt, 1985, Chagas *et al.*, 1986, Moscardi, 1987, Valicente and Cruz, 1991, Kitajima, 1986.
Aquatic mosquitos	CuNI-NPV	Bernardino *et al.*, 2009

Target-insect	Type virus	Reference
Eacles imperialis magnifica	CPV	Alves, 1986, Lucchini *et al.*, 1984, Schmitt, 1985, Chagas *et al.*, 1986, Moscardi, 1987, Valicente and Cruz, 1991, Kitajima, 1986.
Ectropis obliqua	EcobSNPV	Ma *et al.*, 2006
Elasmopalpus lignosellus	GV	Alves, 1986, Lucchini *et al.*, 1984, Schmitt, 1985, Chagas *et al.*, 1986, Moscardi, 1987, Valicente and Cruz, 1991, Kitajima, 1986.
Epinotia aporema	GV	Alves, 1986, Lucchini *et al.*, 1984, Schmitt, 1985, Chagas *et al.*, 1986, Moscardi, 1987, Valicente and Cruz, 1991, Kitajima, 1986.
Erinnyis ello	GV	Alves, 1986, Lucchini *et al.*, 1984, Schmitt, 1985, Chagas *et al.*, 1986, Moscardi, 1987, Valicente and Cruz, 1991, Kitajima, 1986.
Eupseudoma involuta	NPV	Alves, 1986, Lucchini *et al.*, 1984, Schmitt, 1985, Chagas *et al.*, 1986, Moscardi, 1987, Valicente and Cruz, 1991, Kitajima, 1986.
Eupseudosom a aberrans	NPV	Alves, 1986, Lucchini *et al.*, 1984, Schmitt, 1985, Chagas *et al.*, 1986, Moscardi, 1987, Valicente and Cruz, 1991, Kitajima, 1986.
Euselasia sp.	NPV	Alves, 1986, Lucchini *et al.*, 1984, Schmitt, 1985, Chagas *et al.*, 1986, Moscardi, 1987, Valicente and Cruz, 1991, Kitajima, 1986.
Galleria mellonella	NPV	Alves, 1986, Lucchini *et al.*, 1984, Schmitt, 1985, Chagas *et al.*, 1986, Moscardi, 1987, Valicente and Cruz, 1991, Kitajima, 1986.
Glena sp.	CPV	Alves, 1986, Lucchini *et al.*, 1984, Schmitt, 1985, Chagas *et al.*, 1986, Moscardi, 1987, Valicente and Cruz, 1991, Kitajima, 1986.
Grapholita molesta	GV	Alves, 1986, Lucchini *et al.*, 1984, Schmitt, 1985, Chagas *et al.*, 1986, Moscardi, 1987, Valicente and Cruz, 1991, Kitajima, 1986.
Helicoverpa armigera	HearNPV	Cherry *et al.*, 2000. Gypta *et al.*, 2007.
Heliothis virescens	NPV and CPV	Alves, 1986, Lucchini *et al.*, 1984, Schmitt, 1985, Chagas *et al.*, 1986, Moscardi, 1987, Valicente and Cruz, 1991, Kitajima, 1986.
Heliothis zea	NPV, CPV and GV	Alves, 1986, Lucchini *et al.*, 1984, Schmitt, 1985, Chagas *et al.*, 1986, Moscardi, 1987, Valicente and Cruz, 1991, Kitajima, 1986.
Iragoidae fasciata	IrfaNPV	Yang *et al.*, 2009
Manduca sexta	NPV and GV	Alves, 1986, Lucchini *et al.*, 1984, Schmitt, 1985, Chagas *et al.*, 1986, Moscardi, 1987, Valicente and Cruz, 1991, Kitajima, 1986.
Nystalea niseus	IR	Alves, 1986, Lucchini *et al.*, 1984, Schmitt, 1985, Chagas *et al.*, 1986, Moscardi, 1987, Valicente and Cruz, 1991, Kitajima, 1986.
Oiketicus kirbyi	CPV	Alves, 1986, Lucchini *et al.*, 1984, Schmitt, 1985, Chagas *et al.*, 1986, Moscardi, 1987, Valicente and Cruz, 1991, Kitajima, 1986.

Table 3. (Continued)

Target-insect	Type virus	Reference
Pectinophora gossypiella	NPV and CPV	Alves, 1986, Lucchini *et al.*, 1984, Schmitt, 1985, Chagas *et al.*, 1986, Moscardi, 1987, Valicente and Cruz, 1991, Kitajima, 1986.
Phthorimaea operculella	NPV and CPV	Alves, 1986, Lucchini *et al.*, 1984, Schmitt, 1985, Chagas *et al.*, 1986, Moscardi, 1987, Valicente and Cruz, 1991, Kitajima, 1986.
	PhopGV	Mascarin *et al.*, 2009
Plodia interpuctella	NPV	Alves, 1986, Lucchini *et al.*, 1984, Schmitt, 1985, Chagas *et al.*, 1986, Moscardi, 1987, Valicente and Cruz, 1991, Kitajima, 1986.
Plutella xylostella	NPV and GV	Alves, 1986, Lucchini *et al.*, 1984, Schmitt, 1985, Chagas *et al.*, 1986, Moscardi, 1987, Valicente and Cruz, 1991, Kitajima, 1986.
Pseudoplusia includens	NPV	Alves, 1986, Lucchini *et al.*, 1984, Schmitt, 1985, Chagas *et al.*, 1986, Moscardi, 1987, Valicente and Cruz, 1991, Kitajima, 1986, Zanardo *et al.*, 2009.
Rachiplusia nu	NPV	Alves, 1986, Lucchini *et al.*, 1984, Schmitt, 1985, Chagas *et al.*, 1986, Moscardi, 1987, Valicente and Cruz, 1991, Kitajima, 1986.
Sabulodes caberata	NPV and GV	Alves, 1986, Lucchini *et al.*, 1984, Schmitt, 1985, Chagas *et al.*, 1986, Moscardi, 1987, Valicente and Cruz, 1991, Kitajima, 1986.
Sarcina violacens	NPV	Alves, 1986, Lucchini *et al.*, 1984, Schmitt, 1985, Chagas *et al.*, 1986, Moscardi, 1987, Valicente and Cruz, 1991, Kitajima, 1986.
Sibine sp.	Densovirus	Alves, 1986, Lucchini *et al.*, 1984, Schmitt, 1985, Chagas *et al.*, 1986, Moscardi, 1987, Valicente and Cruz, 1991, Kitajima, 1986.
Spdoptera exigua	SeMNPV	Cheng and Lynn, 2009
Spodoptera frugiperda	SfGV SfMNPV	Oliveira and Bolonheiz, 2009 Alves, 1986, Cheng *et al.*, 2000, Chagas *et al.*, 1986, Lucchini *et al.*, 1984, Moscardi, 1987, Kitajima, 1986. Paiva *et al.*, 2008, Schmitt, 1985, Tuelher *et al.*, 2009, Valicente and Cruz, 1991, Valicente *et al.*, 2009.
Spodoptera latifascia	NPV	Alves, 1986, Lucchini *et al.*, 1984, Schmitt, 1985, Chagas *et al.*, 1986, Moscardi, 1987, Valicente and Cruz, 1991, Kitajima, 1986.
Thyrinteina arnobia	NPV and GV	Alves, 1986, Lucchini *et al.*, 1984, Schmitt, 1985, Chagas *et al.*, 1986, Moscardi, 1987, Valicente and Cruz, 1991, Kitajima, 1986.
Thysanoplusia orichalcea	ThorNPV ThorMNPV	Cheng and Carner, 2000 Cheng and Lynn, 2009
Tuta absoluta	PhopGV	Mascarin *et al.*, 2009b
Urbanus proteus	NPV	Alves, 1986, Lucchini *et al.*, 1984, Schmitt, 1985, Chagas *et al.*, 1986, Moscardi, 1987, Valicente and Cruz, 1991, Kitajima, 1986.

12.4. PRECAUTIONS TO INCREASE THE EFFICACY OF PRODUCTS

In formulations for viral bio-pesticides it is recommended that agricultural technicians take the following precautions to increase the efficacy of these products (McNeil, 2009):

a) Insect viruses are highly specific so that it is important to ensure that the target insect is correctly identified.

b) Apply the virus when the target insects are still young - they feed most actively at that stage in their development.

c) Be sure to apply the virus all over the effected area. Young plants can even be dipped in the viral particle solution so as to completely cover the leafy areas.

d) Apply in the early morning, at night or on cloudy days to reduce as far as possible degradation by the solar radiation.

e) Avoid applications on rainy days - the rain may wash the virus particles off the leafy surfaces and increase the longevity.

12.5. APPLICATION OF THE VIRUS FOR CONTROL OF PEST-INSECTS IN AGRICULTURE

Viruses may be utilized in the field in various ways depending on the objective of the application program, the target insect involved and the plant species under attack. Successful use of any control strategy depends on biotic and extra-biotic factors that affect the virus-host-plant system and which must be thoroughly understood before any entomopathogens are applied to open-air fields (Moscardi, 1998).

12.6. *ANTICARSIA GEMMATALIS* NUCLEAR POLYHEDROSIS VIRUS (AGMNPV)

One of the principle pests that preys on soya cultivations is the soya caterpillar, *Anticarsia gemmatalis* (AgMNPV) (Hübner, 1818), of the Noctuidae family, which attacks the aerial portion of the plants and causes the leaves to fall away, and in severe cases, so reduces photosynthesis that development of the culture is retarded and the plants produces less grain (Morales *et al.,* 1995). The soya worm is a key pest and is the main cause of defoliation of the soya plants from the North of Argentina to the USA Southeast.

An occurrence of NPV in this species was reported for the first time in insects collected in alfalfa grasses in Peru (Steinhaus and Marsh, 1962). Pioneering research undertaken in Brazil in 1977 by CNPSO (The National Center for Soya Research), an institution belonging to EMBRAPA (The Brazilian Company for Agricultural Research) located in the city of Londrina in the Brazilian State of Parana, resulted in the large-scale production and use in commercial soya plantations of a bioinsecticide, denominated "Baculovirus anticarsia" (AgMNPV). For two successive harvests (1980/81 and 1981/82), the VPNAg was tested on various farms in the adjacent States of Paraná and Rio Grande do Sul and was found to be efficient in reducing the populations of the pest and maintaining favorable levels of potential production in the soya fields when compared to similar areas under chemical control or without the application of control measures or witnesses (Moscardi, 1986). Baculovirus use increased progressively in subsequent harvests, reaching one million hectares of treated plants in 1989/90. Moscardi and Santos (2004) states that the areas planted with soya treated with this virus increased rapidly in Brazil - from 2000 hectares in 1982/83 to more than 2 million hectares in the 2002/03 harvests which demonstrates the success of this biological control method (applied in about 12% of the entire area of soya cultivation in Brazil). Some 40% of this increase is concentrated in the State of Paraná. However, the use of this biological insecticide is also increasing rapidly in the center-south of Brazil. In the central state of Mato Grosso, the Baculovirus was commercialized for use on 350 mille hectares in the 2001/02 harvest and the total area treated with this biological insecticide reached 1.6 million hectares in the 2001/02 harvest. In the 2004/05 crop, the AgMNPV was produced and commercialized for treatment of 2 million hectares, or about 10% of the area cultivated with soya in Brazil (Sosa-Gómez et al., 2008). In vast soya plantations in the Center and South of Brazil, AgMNPV has shown itself efficient and safe for the control of the targeted pest. Larval mortality has been above 80% and defoliation has been held below the liminar for economic damage when the bioinsecticide was applied in the recommended manner (Moscardi, 1983, 1989; Moscardi and Correa-Ferreira, 1985; Moscardi and Sosa-Gómez, 1996). Furthermore, laboratory and field studies with various periods of utilization of AgMNPV, demonstrated that its virulence was sustained after more than 15 years of use as a biological insecticide (Berino, 1995; Batista, 1997). Neither was the development of any resistance observed in the natural insect populations of the regions submitted to many applications of the virus when compared to such populations in other areas where the virus had never been used (Abot et al., 1995). According to Fuxa et al. (1993), this may be because the agents have not been used in quantity or frequency sufficient to foster the selection of an insect population resistant to AgMNPV. Therefore, monitorization of the natural populations for their susceptibility to the virus, and evaluation of the selection process pressures in laboratories is extremely important when we consider that the AgMNPV, not only occurs naturally, but has been used extensively as a pesticide in Brazil. Montor (2003), states that the program being developed in Brazil for the control of the soya caterpiller has already produced direct economies of more than 100 million dollars for the country in the purchase of chemical agro-toxicants, in addition to the environmental benefits obtained from the elimination of the need to apply about 11 million liters of such products. Not only Brazil, but other Latin American countries also, Argentina, Paraguay and Bolivia, now use AgMNPV as an agent of microbial control. The Mexican program for production and use of AgMNPV started in 1998 and the virus is now produced on site by the INIFAP (The Mexican National Institution for Forest, Agriculture and Animal Husbandry Research) which sells the product to the Mexican soya producers. In the 2003 harvest approximately 3500 hectares (Ávila-Valdez and Rodriguez del Bosque, 2003) was applied and had reached 20 mille hectares by 2005 - this then represented 20% of the entire area of soya plantations in Mexico (Sosa-Gómez et al., 2008). Paraguay commenced using AgMNPV in 1986 (Dickel, 1990) and by the 1990/91 harvest the use of the virus had reached 60 mille hectares - it is now estimated to be used on 150 mille hectares annually. The product is imported from Brazil by the Paraguayan company Agro Fertil which buys the product from Coodetec in Brazil for resale to the Paraguayan soya producers (Sosa-Gómez et al., 2008). In Uruguay, the AgMNPV was produced locally for the 1990/91 harvest and applied in about 600 hectares (Chiaravalle, 1996). In Argentina the control of the soya caterpilar with the

AgMNPV virus commenced in 1985 at the Cirpon Institute (*The Center for research into the Regulation of Noxious Organisms*) using inoculate donated by Embrapa and imported from Brazil in 1984. Cirpon produced sufficient virus to treat 15 mille hectares (Sosa-Gómez and Moscardi, 1991), but the program was discontinued in 1989.

12.6.1. *Spodoptera Frugiperda* NUCLEAR POLYHEDROSIS VIRUS (SfMNPV)

The army fallworm, *Spodoptera frugiperda*, is a polyphagos insect with wide distribution worldwide and is considered the most dangerous pest preying on corn cultivations in Brazil. It attacks at all stages of the growth of the corn plant and can cause production losses of up to 38.7 % (Cruz *et al.*, 1996). The program for the control of the fall armyworm is coordinated in Brazil by Embrapa (*The national Corn and Sorghum Research center*) in the town of Sete Lagoas in the Brazilian State of Minas Gerais. From an army fallworm infected in the field, a Nuclear Polyhedrosis Virus was identified as SfMNPV and purified (Valicente *et al.*, 1989). Garcia (1979) studied the efficiency of the biotic and abiotic factors in the regulation of the *S. frugiperda* population, and observed that the VNP was a promising agent for control of the pest. Field application of VNP in the irrigation water caused a mortality greater that 50% in *S. frugiperda* by making it possible for parasitoids to infect the sampled armyworms (Valicente and Cruz, 1991; Valicente and Costa, 1995). The use of SfMNNPV in Brazil grew rapidly and as much as 5000 hectares of corn were treated with this virus annually, but problems in producing the virus (such as cannibalism) made the final product too expensive and Embrapa has temporarily suspended the program (Sosa-Gómez *et al.*, 2008). Observations of the resistance of the insects to the SfMNPV detected low rates of resistance and a rapid reversion of this resistance to susceptibility, probably due to selection effects on the resistant gene or genes in the absence of NPV (Fuxa *et al.*, 1988, 1993; Fuxa and Richter, 1989, 1991). The application of Nuclear Polyhedrosis Virus was also applied via the irrigation water with some success in the USA against the fall armyworm (Hamm and Hare, 1982).

12.7. OTHER BACULOVIRUSES

12.7.1. *Erinnyis Ello* Granular Virus

An on-going Brazilian program for the control of the *Erinnyis ello* (Linnaeus, 1758) (Lepidoptera: Sphingidae), main pest of the manioc plant, utilizes a granular virus (EeGV) (Ribeiro *et al.*, 1998). The extreme virulence of the pathogen has been demonstrated clearly - 90% mortality after nine days of infection. At present, this virus is produced by the EPAGRI (*Company for research into agro and related products of the State of Santa Catarina*) and by IAPAR (*Agronômico Institute of the State of Paraná*). The product is produced in the form of a mince of infected worms. The bio-pesticide is applied in the States of Santa Catarina and Parana and in some locations in the Brazilian North-East (Schmitt, 1983, 1984). This GV was also tested in Columbia (Belotti *et al.*, 1992) and is being used successfully in extensive plantations of manioc in Venezuela - reductions in the use of chemical insecticides have been almost 100% as a result (Smith and Bellotti, 1996). Since then, pathogens obtained from insects collected in the field have been applied for control of this pest.

12.7.2. CvMNPV

Another program under development in Brazil involves the use of the NPV of the Poplar Tree caterpillar, *Condylorrhiza vestigialis* (CvMNPV). The Poplar (*Populus* spp.) is planted in wetlands in Brazil to produce light lumber for the match-making industry. The poplar caterpillar can provoke extensive defoliation of the trees and retard their growth. At the Paraná State University a pilot production line was established in the laboratory to produce the insect on an artificial diet, with a view to inoculation of the 3rd. instars for the production of the virus. In this initial phase, sufficient virus is being obtained to treat one hectare per day (365 hectares per year) but the target is to produce enough to treat approximately 2 mille hectares per year - or a considerable portion of the 5.5 mille hectares of Alamo planted annually in Brazil. The virus has been applied *in situ* and resulted in very efficient control (about 93%) of the pest (Sosa-Gómez *et al.,* 2008).

Also in an initial phase is a program to control the *Opsiphanes cassiae*, which is the principle pest afflicting the *dendê* plant and which causes high mortality in annual occurrences by means of a still unidentified virus. In plantations belonging to the Agropalma Company in Brazil, larvae killed by the virus are collected and frozen for later application in an aqueous solution and this results in of control of the insects by inducing epizooties in the caterpillar population (Sosa-Gómez *et al.,* 2008). *Dendê* pests such as *Sibine fusca*, may also be damaging the cultivation in Columbia. This pest can be efficiently controlled by an adenovirus aerosol (micronair) (Genty and Mariau, 1975). Orlato (2002), in studies realized with the caterpillar *Thyrinteina arnobia,* (one of the main pests that attack the eucalyptus tree) (Stoll, 1782) (Lepidoptera: Geometridae) indicated a granular virus as a potential biological agent for the control of this pest.

12.7.3. *S. Exigua* VPN

NPVs have also been developed and used against *S. litura* (China, India and Taiwan); *S. littoralis* (Egypt), *S. exigua* (USA, Guatemala, Tyland), and *S. sunia* (Guatemala). In Thailand, an *S. exigua* VPN is produced in laboratories for distribution to agricultural workers for production of the field virus (Deseo-Kovaks and Rovesti, 1992). In China, an *S. litura* VPN was applied successfully to control this insect in cotton, rice and ground-nut plantations (Yi and Li, 1989).

12.7.4. PrGV

The cabbage GV, *Pieris rapae,* has been mass-produced in China as PrGV since 1978 and applied in many regions to a total of 100 mille hectares (Yi and Li, 1989). *Plutella xylostella* and *Agrotis segetum* GVs have been used in extensive areas of the Chinese center and northwest respectively (Yi and Li, 1989).

12.7.5. *Pseudaletia* NPV

In wheat fields of the south Brazil, one nuclear polyhedrosis virus (NPV) was isolated from the caterpillar of the genus *Pseudaletia*. In laboratory assays, this NPV was highly pathogenic for small and medium larvae of *P. sequax*, providing more than 80% mortality, while for larger larvae, mortality of 60% (Fiuza, 1991).

12.7.6. CPGV

In many experiments conducted in a number of countries, GVs isolated from *C. pomonella* (CpGV) - a worldwide pest of apples, pears and nuts - were highly virulent against the attacking insects and protected the fruit from economic damage (Falcon, 1985; Falcon and Berlowitz, 1986; Hubner, 1986; Jem *et al.,* 1997). The potential for use of this GV is very great because of the importance of *C. pomonella* all over the world, but so far it has been little used.

12.7.7. PHOPGV

In Guatemala, commercial products based on the Nuclear Polyhedrosis Virus of *Autographa californica* and *Spodoptera albula*, are readily available and are recommended for various types of lepidopteron (http://www.agricolaelsol.com). The Baculovirus of the potato worm *Phthorimaea operculella*, has attained considerable relevance in Peru because the worm is a serious threat to the potato in the Andes regions and both a Granular Virus and a Nuclear Polyhedrosis Virus are found associated to its populations. The virus most frequently used is the PhopGV, producing mortality between 70 and 100%, and this virus can persist in the field for two months or more (Raman and Alcázar, 1992). In Bolivia, the Institution of the Papal Program of Investigations (PROINPA) initiated a program of control of *P. operculella* in 1993/94, and now produces a granular virus known commercially as *Matapol Plus*. During the 1995/96 and 1996/97 harvests, the production of *Matapol Plus* reached 2 tons of the formulated products per year (Sosa-Gómez *et al.,* 2008). Other viruses, also potentially of use, have been isolated from the pests of cultures such as sugar cane, cotton, wheat, rice, fruit bearers, green vegetables, pastures and forests (Moscardi and Correa-Ferreira, 1985; Kitajima, 1986; Valicente and Cruz, 1991; Ribeiro *et al.,* 1997), all of which show promise for development for microbial control of insects.

12.8. ADVANTAGES OF UTILIZATION OF THE VIRUS

The Baculovirus have become models for animal virology, because the insects have short lifespan which permits mass multiplication in laboratories. In nature, these viruses occur in insect populations and persist in the environment where they may cause epizooties causing mortality of a large number of larvae (Benz, 1986; Evans, 1986).

According to Pavan (1988), the Baculovirus present various degrees of specificity and are highly virulent to their original hosts. This characteristic permits control of the population of pest-insects in the cultures, both natural and artificial, without causing damage to beneficial arthropods. The author also states that the careful utilization of analyses of the isolates is fundamental from the point of view of their use in biological pest control, because the complexity, plasticity and genetic variability found in the viruses demonstrate that permanent Quality Control should accompany all application programs - also that the implementation of this control should be focused both on the efficiency and on the safety of pathogen utilization.

A system of production *in vivo* requires the maintenance of a population of the host species in the laboratory from which the larvae can be obtained for the successive phases of inoculation, incubation, extraction, purification and storage of the virus. The larvae from the natural population, in addition to their seasonal cycle and heterogeneity, are normally affected by parasites or pathogens, which may influence the quantity and quality of the virus produced. It is therefore convenient to use populations created in laboratories and follow basic standards to eliminate or reduce possible contaminants (Maracajá *et al.,* 1999).

Alves (1998) comments that the viruses that attack insects in nature are consider specific and safe for man and other vertebrates that have acidic digestive tracts. This safety can be affected by the mass production of these agents. The genetic variations of the pathogen and the contaminations by potentially dangerous bacteria (such as *Salmonella* and others) cause problems in mass production. There is also the theoretical risk that the Baculovirus DNAcould act on mammal genomes. Therefore, the author suggests that rigorous safety tests and extensive quality control should always be in place when manipulating the "new" viruses indicated for pest control.

The VPN of *Pseudaletia* has been used successfully to control this insect which is an importance pest in agriculture (Neelgund, 1975). On the other hand, before large scale use, its possible effect on beneficial insects should be investigated. Dhaduti and Mathad (1979) realized tests with *Bombyx mori*, in which they observed that *Pseudaletia* VPN is safe for silk-worms. The effect of *Pseudaletia* VPN was investigated in colonies of *Apis cerana indica*, after both topic and oral administration, without significant damage to the colonies (Dhaduti and Mathad, 1980).

It is of vital importance to stress that the use of microbial insecticides is fundamentally different to the utilization of chemical insecticides. The Baculovirus are safe for man, other non-targeted organisms and for the environment. They require time for incubation in the host, and therefore take longer to eliminate a pest and also demand specific application techniques that have been elaborated in such a way as to avoid damage to the culture in question. Because of their capacity to multiply themselves and disseminate in the host population, the application of biological insecticides based on Baculovirus may result in adequate control of the targeted insects with a lower number of applications than would be required if the chemical insecticides were used instead (Moscardi and Souza, 2002).

The possibility of economical production on a commercial scale has been possible so far only *in vivo,* both for insects raised on an artificial diet in laboratories and those grown under field conditions (for example: Baculovirus of the soya caterpillar, *Anticarsia gemmatalis* – VPN*Ag*). However, in the first case the production of different Baculovirus is onerous, because of the cost of the diet, equipments, labor etc. In the field, while the production costs may be low, the quantity produced each year can vary considerably as a function of the density of the host population and the influence of biotic and climatic factors (occurrence of other natural enemies etc.) For these reasons, to be able to mass produce a Baculovirus in the laboratory at a competitive cost by more efficient methods is ever more necessary when we consider that the commercial production of these agent *in vitro* in bioreactors utilizing insect cells is still meeting technical and economic problems, notwithstanding the recent advances achieved (Moscardi, 2002).

12.9. DISADVANTAGES OF VIRUS USE

The majority of viruses take several days to kill the target insect and during this period the insect is still inflicting damage on the crops. The death of the insect is dependent on the dose and not infrequently very large doses are necessary to control a population of insects. Also, as they grow older, the insects become less susceptible to infection by the virus. (McNeil, 2009).

Another important aspect is the fact that the Polyhedrosis viruses are very sensitive to solar radiation, especially to the ultraviolet frequencies, and for this reason the virus may become inactive in two to five days after application. This problem can be resolved by adding solar protection to the Baculovirus formulas, but this will certainly increase the final cost of the product (Moscardi and Souza, 2002). The contribution of other factors to the inactivation of insect viroses under field conditions do not appear to be so significant when compared to the solar radiation. The air humidity has very little effect on the virus. However, humidity resulting from mist forming on the leaf surface can reduce the stability of the microorganism. In addition to all this, some agricultural practices can reduce the permanence of the microorganisms from season to season - for example direct planting which buries the viral particles in the ground.

12.10. PERSPECTIVES

Sustainable agriculture cannot dispense with the ever increasing utilization of biological control in general and especially of the entomopathogenic viruses. The Baculovirus have been developed as biological insecticides for use in various countries and against many pests - especially those of the Lepidoptera order - but the practical application has been restricted. The most successful cases are the *Anticarsia gemmatalis*, *Helicoverpa/Heliothis*, *Lymantria dispar* VPNs and the *Cydia pomonella* GVs which demonstrate clearly that utilization of these agents is a viable alternative to the use of chemical insecticides.

However, the use of the Baculovirus will depend on research so that they can used as expression vectors, thus providing scope for widespread utilization in medicine as therapeutics, prophylactics (vaccines) and for diagnosis of diseases. In addition, they could be used for the production of genetically modified viral insecticides which would make it possible to improve the pathogenic characteristics of Baculovirus as agents of biological control and the amplification of their host spectrums. On the other hand, the methods to be used to increase the efficacy of the Baculovirus in microbial pest control - such as the incorporation of protective substances and viral activity potentialization agents in formulations - need to be evaluated under field conditions bearing in mind that the use of the

Baculovirus will certainly increase in the medium term. Research should also be directed to determining the possibilities that the insect populations could develop resistance to the viral insecticides. Finally, efforts should be intensified to produce these agents *in vitro* for large-scale commercial production and utilization with final user costs at least competitive to those of the chemical.

CONCLUSION

Virus intended to be used as microbial insecticides must be highly virulent to host, in order to maintain its population below the economic injury level (Fuxa, 1987). However, if a baculovirus in addition to being effective in reducing the population of the insect host also shows good recyclability (production and dissemination), this attribute is important because it should result in a lower number of applications against the target insect compared with chemical insecticides (Moscardi, 1999). Despite the development of several baculoviruses as biological insecticides for use in different countries, its practical application has been restricted, the greatest accomplishment is the NPV of *A. gemmatalis* in Brazil, which is the largest program of using baculovirus in the world, showing that the use of these agents is feasible. However the expansion of the use of baculovirus depend on further research in the areas of recombinant baculovirus and *in vitro* production of these agents, due to its high specificity and relatively slow action on the host

REFERENCES

Abot, A.R., Moscardi, F., Fuxa, J.R., Sosa-Gómez, D.R. and Richter, A.R. 1995. Susceptibility of populations of *Anticarsia gemmatalis* from Brazil and the United States to a nuclear polyhedrosis virus. *Journal of Entomology Science*, 30:62 – 69.

Adams, J.R. and Mcclintock, T.J. 1991. Baculoviridae: nuclear polyhedrosis virus. In: Adams, J.R. and Bonani, J.R. (Eds.). *Atlas of invertebrate vir*uses. Boca Raton: CRC. p.87-204.

Almeida, A.F., Macedo, G.R., Souza, M.L. and Pedrini, M.R.S. 2009. Capacidade de produção de corpos de oclusão de isolados do baculovírus AgMNPV utilizando células SF21 em suspensão. In.: *X Simpósio de Controle Biológico*, 1 a 5 junho de 2009, Bento Gonçalves, RS.

Almeida, G.F., Paula, D.P., Souza, M.L. and Castro, M.E.B. 2008. Clonagem e sequenciamento de um gene de virulência (p74) de um vírus isolado da lagarta-do-álamo *condylorrhiza vestigiali*s. In: *XXII Congresso Brasileiro de Entomologia*, 24 a 29 agosto 2008, Uberlândia – MG.

Alves S.B. 1998. *Controle Microbiano de Insetos*. Piracicaba, Brazil: Fealq. 2nd ed.

Alves, S.B. 1986. Vírus entomopatogênicos. In: Alves, S.B., *Controle Microbiano de Insetos*. Ed. Manole, São Paulo, p. 171-187.

Ávila-Valdez, J., and Rodriguez Del Bosque, L.A. 2003. Uso del nulcleopoliedrovirus de *Anticarsia gemmatalis* como principal estrategia del MIP en soya de la región sur de

Taumalipas. In: *XXVI Congresso Nacional de Control Biológico*. Memorias. México, Guadalajara, p.327-330.

Batista, T.F.C. 1997. Fatores que limitam a eficiência de *Baculovirus anticarsia* sobre *Anticarsia gemmatalis* Hübner, 1818. MSc. thesis. Univ. Fed. Pelotas, Brasil.

Belotti, A.C., Arias, V.B. and Guzman, O.L. 1992. Biological control of the cassava hornworm *Erinnyis ello* (Lepidoptera: Sphingidae). *Fla. Entomol.* 75:506 – 15.

Benz, G.A. 1986. Introduction: historical perspectives. In: Granados, R.R., Federici, B.A. (Eds.). *The biology of baculoviruses*. Boca Raton: CRC,. v.1, p.1-35.

Berino, E.C.S. 1995. Determinação da atividade biológica de isolados geográficos e temporais de VPN de *Anticarsia gemmatalis* Hübner, 1818 (Lepidoptera: Noctuidae). MSc. thesis. Univ. Fed. Rio Grande do Sul, Porto Alegre, Brasil.

Bernardino, T.C., Tuan, R., Cunha, A.B.P., Araújo-Coutinho, V. and Cunha, C.J.P. 2009. Levantamento e caracterização de vírus entomopatogênicos em larvas de Culicídeos no Estado de São Paulo. In.: *X Simpósio de Controle Biológico*, 1 a 5 junho de 2009, Bento Gonçalves, RS.

Bilimoria, S.L. 1986. Taxonomy and identification of baculoviruses. In: Granados, R.R. and Federici, B.A. (Eds.). *The biology of baculoviruses*. Boca Raton: CRC. v.1, p.37-59.

Bilimoria, S.L. 1991. The biology of nuclear polyhedrosis viruses. In: Kurstak, E. (Ed.). *Viruses of invertebrates*. New York: Marcel Dekker. p.1-72.

Bonning, B.C. and Hammock, B.D. 1996. Development of recombinant baculovirus for insect control. *Annual Review of Entomology,* 41:191 – 210.

Brancalhão, R.M, Torquato, E.F. and Fernandez, M.A. 2009. Cytopathology of *Bombyx mori* (Lepidoptera: Bombycidae) silk gland caused by multiple nucleopolyhedrovirus. *Genet Mol Res.*, 8(1):162 - 72.

Castro, M.E.B. , Souza, M.L., Sihler,W., Rodrigues,J.C.M. and Ribeiro,B.M. 1999. *Biologia molecular de baculovírus e seu uso no controle biológico de pragas no Brasil*. Pesq.agropec.bras., Brasília, v.34, n.10, p. 1733-1761.

Castro, M.E.B., Almeida, G.F., Paula, D., Ribeiro, Z.M. and Souza, M.L. 2008. Identificação e caracterização de um novo vírus patogênico à mariposa-do-álamo, *condylorrhiza vestigialis*. In: *XXII Congresso Brasileiro de Entomologia*, 24 a 29 agosto 2008, Uberlândia – MG.

Chagas, M.C.M., Kitajima, E.W. and Costa, O.G. 1986. Utilização de vírus no controle da lagarta verde do cajueiro, *Eacles imperiales magnífica*. Resumo *X Congresso Brasileiro de Entomologia*, RJ. p. 212.

Chen, C., Quentin, M.E., Brennan, L.A., Kukel, C. and Thiem, S.M. 1998. *Lymantria dispar* nucleopolyhedrovirus *hrf-1* expands the larval host range of *Autographa californica* nucleopolyhedrovirus. *Journal of Virology*, 72:2526 – 2531.

Cheng, X.W. and Carner, G.R. 2000. Characterization of a single-nucleocapsid nucleopolyhedrovirus of *Thysanoplusia orichalcea* L. (Lepidoptera:Noctuidae) from Indonesia. *Journal of Invertebrate Pathology*, 75; 279 - 287.

Cheng, X.W. and Lynn, D.E. 2009. Chapter 5 baculovirus interaction: *In vitro* and *in vivo*. *Advances in Applied Microbiology*, 68:217 - 239.

Cheng, X.W., Carner, G.R. and Arif, B.M. 2000. A new ascovirus from *Spodoptera exigua* and its relatedness to the isolate from *Spodoptera frugiperda*. *Journal of General Virology*, 81:3083 - 3092.

Cherry, A.J., Rabindra, R.J., Parnell, M.A., Geetha, N., Kennedy, J.S. and Grazywacz, D. 2000. Field evalution of *Helicoverpa armigera* nucleopolyhedrovirus formulation for control of the chickpea pod borer, *H. armigera* (Hubn.) on chickpea (*Cicer arietinum* var. Shoba) in Southern India. *Crop Prot.,* 19:51 - 60.

Chiaravalle, W.R. 1996. Informe sobre el avance Del control biológico em el Uruguay. In: Zapater, M.C. (Ed.). *El control biológico em América Latina.* Buenos Aires:SRNT-IOBC. P.93-98.

Cory, J.S. and Myers, J. 2003. The ecology and evolution of insect baculovírus. *Annu. Rev. Ecol. Syst.,* 34:239 -272.

Cory, J.S., Hist, M.L., Williams, T., Hails, R.S., Goulson, D., Green, B.M., Carty, T.M., Possee, R.D., Caylay, P.J. and Bishop, D.H.L. 1994. Field trial of a genetically improved baculovirus insecticide. *Nature*, 370:138 – 140.

Cruz, I., Oliveira, L.J., Oliveira A. C. and Vasconcelos, C.A. 1996. Efeito do nível de saturação de alumínio em solo ácido sobre os danos de *Spodoptera frugipereda* (J. E. Smith) em milho. *An. Soc. Entomol. Brasil*, 25:293.

Cunningham, J.C. 1990. Use of microbials for control of defoliating pests of conifers. In *Proc. Int. Coll. Invertebr. Pathol and Microbial Control*, 5th, Adelaide, pp. 164–68. Adelaide, Aust.: Soc. Invertebr. Pathol.

Deseo-Kovaks, K.V. and Rovesti L. 1992. *Lotta Microbiologica Contro I Fitofagi*: Teoria a Pratica. Bologna, Italy: Edagricole.

Dhaduti, S.G. and Mathad, S.B. 1979. Effect of NPV of the armyworm *Mythima* (*Pseudaletia*) *separata* on the silkworm *Bombyx mori. Experientia*, 35:81.

Dhaduti, S.G. and Mathad, S.B. 1980. Effect of NPV of the armyworm *Mythima* (*Pseudaletia*) *separata* on colonies of *Apis cerana indica*. Karnataka University, India. *Journal of Apicultural research*, 19(1): 77 - 78.

Dickel, C. 1990. Control biológico con *Baculovirus anticarsia*. Cooperativa Colonias Unidas, Obligado, Paraguay:Boletín de divulgación, n 2B, 13p.

Eberle, K.E., Sayed. S., Rezapanah, M., Shojai-Estabragh S. and Jehle, J.A. 2009 Diversity and evolution of the *Cydia pomonella* granulovirus. *J. Gen. Virol.,*90(3):662 – 71.

Evans, H.F. 1986. Ecology and epizootiology of baculoviruses. In: Granados, R.R. and Federici, B.A. (Eds.). *The biology of baculoviruses.* Boca Raton: CRC, v.2, p.89-132.

Falcon, L.A. 1985. Development and use of microbial insecticides. In: *Biological Control in AgriculturalIPMSystems*, ed. MA Hoy, DC Herzog, pp. 229–42. Orlando, FL: Academia

Falcon, L.A. and Berlowitz, A. 1986. Experiences in field-testing codling moth granulosis virus in the Pacific rim countries. In *Fundamental and Applied Aspects of Invertebrate Pathology*, Int. Coll. Invertebr. Pathol., 4th, Veldhoven, ed. RA.

Federici, B. A. 1993. Viral pathobiology in relation to insect control, p. 81-101. In:Beckage, N.E., Thompson, S.N. and Federici, B.A. (Eds.), *Parasites and Pathogens of Insects*, vol.2. San Diego, CA, Academic Press, 294p.

Federici, B.A. 1986. Ultrastructure of baculoviruses, p. 61-88. In: Granados, R.R. and B.A. Federici (Eds.). *The biology of Baculoviruses.* vol. 1. Boca Raton, CRC Press, 275p.

Fiuza, L.M. 1991. Avaliação do vírus de poliedrose nuclear (VPN) em lagartas de *Pseudaletia sequax* Franclemont, 1951 (Lepidoptera: Noctuidae). Dissertação de Mestrado em Agronomia, Faculdade de Agronomia, Universidade Federal do Rio Grande do Sul, Porto Alegre, 82p.

Fuxa, J.R. 1987. *Spodoptera frugiperda* susceptibility to nuclear polyhedrosis virus isolates with reference to insect migration. *Environmental Entomology*, 16 (1):218 – 223.

Fuxa, J.R. and Richter, A.R. 1989. Reversion of resistance by *Spodoptera frugiperda* to nuclear polyhedrosis virus. *Journal of Invertebrate Pathology*, 53:52 – 56.

Fuxa, J.R. and Richter, A.R. 1991. Selection for an increased rate of vertical transmission of *Spodoptera frugiperda* (Lepidoptera: Noctuidae) nuclear polyhedrosis virus. *Environmental Entomology*, 20:603 – 609.

Fuxa, J.R., Abot, A.R., Moscardi, F., Sosa-Gómez, D.R. and Richter, A.R. 1993. Selection for *Anticarsia gemmatalis* resistance to NPV, and susceptibility of field populations to the virus. *Resistant Pest Management*, 5(1):39 - 41.

Fuxa, J.R., Mitchell, F.L. and Richter, A.R. 1988. Resistance of *Spodoptera frugiperda* [Lep.: Noctuidae] to a nuclear polyhedrosis virus in the field and laboratory. Entomophaga, 33:55 - 63.

Garcia, M.A. 1979. Potencialidade de alguns fatores bióticos e abióticos na regulação populacional de *Spodoptera frugiperda* (Abbot e Smith, 1797) (Lep. Noctuidae). Campinas: UNICAMP, 96p. Tese de Mestrado.

Genty, P. and Mariau, D. 1975.Utilización de un germen entompopatógeno en la lucha contra *Sibine fusca*. *Oléagineux*, 30:349-354.

Groner, A. 1989. Safety to nontarget in invertebrates of baculoviruses. In: Lacey, L. and Davison, E.W. (Eds). *Safety of Microbial Insecticides*. CRC Press. p.135-147.

Guo, T., Wang S., Guo, X. and Lu, C. 2005. Productive infection of *Autographa californica* nucleopolyhedrovirus in silkworm *Bombyx mori* strain Haoyue due to the absence of a host antiviral factor. *Virology*, 341:231 -237.

Gypta, R.K., Raina, J.C. and Bali, A.K. 2007. Selection and field effectiveness of nucleopolyhedrovirus isolates against *Helicoverpa armigera* (Hubner). *International Journal of virology*, 3(2):45 - 59.

Hamm, J.J. and Hare, W.W. 1982. Application of entomopathogens in irriwater for control of fall armyworm and corn earworms (Lepidoptera:Noctuidae) on corn. *Journal of Economic Entomology*, 75(6):1074 - 1079.

Hodgson, J., Arif, B.M. and Krell, P.J. 2009. *Autographa californica* multiple nucleopolyhedrovirus and *Choristoneura fumiferana* multiple nucleopolyhedrovirus v-cath genes are expressed as pre-proenzymes. *J. Gen. Virol.*, 90:995 – 1000.

Hubner J. 1986. Use of baculovirus in pest management programs. *See Ref.*, 62:181 – 202.

Ignoffo, C.M. and Couch, T.L. 1981. The nucleopolyhedrosis virus of *Heliothis* species as a microbial insecticide. In: *Microbial Control of Pests and Plant Diseases*, ed. HD Burges, pp. 329–62. London: Academic.

Jem, K.J., Gong, T., Mullen, J. and Georgis, R. 1997. Development of an industrial insect cell culture process for large scale production of baculovirus biopesticides. *See Ref.*, 96:173 – 80.

Kamita, S.G. and Maeda, S. 1993. Inhibition of *Bombyx mori* nuclear polyhedrosis virus (NPV) replication by the DNA helicase gene of *Autographa californica* NPV. *Journal of Virology*, 67:6239 – 6245.

Kato, T., Manohar, S.L., Tanaka, S. and Park, E.Y. 2009. High-titer preparation on of *Bombyx mori* nuclepolyhedrovirus (BmNPV) displaying recombinant protein in silkworm larvae by size exclusion chromatography and its characterization. *BMC Biotechbology*, 9:55.

Kitajima, E.W. 1986. Perspectivas de controle biológico de insetos por vírus no Brasil. *Revista Brasileira de Fruticultura*, 8:65 – 71.

Kitajima, E.W. 1994. Taxonomia, caracterização bioquímica e molecular de vírus de insetos. In: Controle Microbiano de Insetos. Cenargem/Embrapa, Brasília, Brasil. 18p.

Lucchini, F., Morin, J.P. and Rocha Souza, J.C. 1984. Perspectivas do uso de entomovirus para o controle de *Sibine* sp. Desfolhante do dendê. Resumo *IX Congresso Bras. Entomologia*, p.153.

Ma, X.C., Xu, H.J., Tang, M.J., Xiao, Q., Hong, J. and Zhang C.X. 2006. Morphological, Phylogenetic and Biological Characteristics of *Ectropis obliqua* Single-Nucleocapsid Nucleopolyhedrovirus. *The Journal of Microbiology*, 44 (1):77 - 82.

Maeda, S. 1995. Further development of recombinant baculovirus insecticides. *Current Opinion in Biotechnology*, 6:313 – 319.

Maeda, S., Kamita, S.G. and Kanodo, A. 1993. Host range expansion of *Autographa californica* nuclear polyhedrosis virus (NPV) following recombination of a 0.6-kilobase-pair DNA fragment originating from *Bombyx mori* NPV. *Journal of Virology*, 67:6234 – 6238.

Maracajá, P.B., Osuna, E.V. and Álvarez, C.S. 1999. Histopatologia do vírus da poliedrose nuclear de *Heliothis armigera*(VPNHa) em larvas dos hospedeiros primários e alternativo *Spodoptera exigua*. *Caatinga*, 12(1/2):35 - 40.

Martignoni, M.E. and Iwai, P.J. 1986. *A Catalog of viral diseases of insects, mites, and ticks*. 4.ed. Portland, OR: USDA-Forest Service. 51p. (USDA. PNW-195).

Mascarin, G.M., Alves, S.B. and Delalibera-Júnior, I. 2009. Virulência de um vírus de granulose (PhopGV) para lagartas de *Phthorimaea operculella* (Lepidoptera, Gelechiidae) sob diferentes temperaturas e dietas. In.: *X Simpósio de Controle Biológico*, 1 a 5 junho de 2009, Bento Gonçalves, RS.

Mascarin, G.M., Rampelotti-Ferreira, F.T., Alves, S.B., Delalibera Júnior, I. and Vendramim, J.D. 2009b. Concentration mortality of granulovirus PhopGV to control of *Tuta absoluta* (Lepidóptera:Gelechiidae). In.: *X Simpósio de Controle Biológico*, 1 a 5 junho de 2009, Bento Gonçalves, RS.

McNeil, J. 2009. *Vantagens e desvantagens Baculovirus*. Available: http://www.extension. org/article/18927. Visited on 20 may 2009.

Miller, L.K. 1995. Genetically engineered insect virus pesticides: present and future. *Journal of Invertebrate Pathology*, 65:211 - 216.

Montor, W.R. 2003. Insetos como biofábricas de proteínas humanas. São Paulo. In: *Ciência Hoje*, 33(196):17 - 23.

Morales, L., Moscardi, F., Kastelic, J.G., Sosa-Gómez, D.R., Paro, F.R. and Soldorio, I.L. 1995. Suscetibilidade de *Anticarsia gemmatalis* Hübner e *Chrysodeixis includens* Walker (Lepidoptera:Noctuidae), a *Bacillus thuringiensis* Berliner. *An. Soc. Entomol. Brasil*, 24: 593 - 598.

Moscardi, F. 1983. Utilização de *Baculovirus anticarsia* para o controle da lagarta-da-soja, *Anticarsia gemmatalis*. Londrina: Embrapa-CNPSo, 13p. (Embrapa-CNPSo. Comunicado técnico, 23).

Moscardi, F. 1986. Utilização de vírus para o controle da lagarta da soja, pp. 188-202. In: S.B.Alves (Ed.), *Controle Microbiano de Insetos*, primeira ed. São Paulo: Ed. Manole.

Moscardi, F. 1987. *Uso de virus no controle de praga*. Anais do 1° Encontro Sul Brasileiro de Controle Biológico de Pragas, Passo Fundo, RS. P. 191-262.

Moscardi, F. 1989. The use of viruses for pest control in Brazil: the case of the nuclear polyhedrosis virus of the soybean caterpillar *Anticarsia gemmatalis*. *Memórias do Instituto Oswaldo Cruz,* 84:51 – 56.

Moscardi, F. 1998. Utilização de vírus entomopatogênico em campo. In: Alves,S.B. (Ed.). *Controle microbiano de insetos.*2. ed. Piracicaba: Fealq, p.509-539.

Moscardi, F. 1999. Assessment of application of baculoviruses for control of Lepidoptera. *Annual Review of Entomology,* 44:257 – 289.

Moscardi, F. 2002. Inseto como Inseticida? *Cultivar Grandes Culturas*, v. 40.

Moscardi, F. and Correa-Ferreira, B.S. 1985. Biological control of soybean caterpillars. In: *World soybean research conference,* Ames. Proceedings. Boulder: Westview, p.703-711.

Moscardi, F. and Santos, B. 2004. Further developments in the commercial laboratory production of the nucleopolyhedrovirus of *Anticarsia gemmatalis* in Brazil.. In: XXXVII Annual meeting of the society for invertebrate pathology, 2004, Helsinki. book of abstracts - XXXVII annual meeting of the society for invertebrate pathology. Helsinki : *Society for Invertebrate Pathology*, p. 97-97.

Moscardi, F. and Sosa-Gómez, D.R. 1996. Utilizacion de virus a campo. In: *Microorganismos Pat´ogenos Empleados en el Control Microbiano de Insectos Plaga*, ed. nRE Lecuona, pp. 261–76. Buenos Aires: Mass.

Moscardi, F. and Souza, M.L. 2002. Baculovirus para o controle de pragas: Panacéia ou realidade. *Biotecnologia, Ciência e Desenvolvimento*, 24: 22 - 29.

Neelgund, Y.F. 1975. Studies on nuclear polyhedrosis of the armyworm *Mythima (Pseudaletia) separata* Walker. Karnataka University, Dharwad: Ph.D. Thesis.

Oliveira, N.C. and Bolonheiz, H. 2009. Eficiência de *Baculovirus spodoptera* em comparação ao lufenuron no controle de diferentes instares e densidades populacionais da lagarta-do-cartucho em milho. In.: *X Simpósio de Controle Biológico*, 1 a 5 junho de 2009, Bento Gonçalves, RS.

O'Reilly, D.R., Miller, L.K. and Luckow, V. A. 1992. *Baculovirus expression vectors: a laboratory manual*. Salt Lake City, UT: W.H. Freeman, 347p.

Orlato, C. 2002. Avaliação do TaV (*Thyrinteina anobia* virus) no controle de *Thyrinteina anobia* (Lepidoptera:Geometridae). Botucatu – SP. Dissertação (Mestrado em Agronomia) - Curso de Pós-Graduação em Ciências Agronômicas, área de concentração em Proteção de Plantas, da Universidade Estadual Paulista, Campus de Botucatu.

Paiva, C.E.C., Tuelher, E.S. and Valicente, F.H. 2008. Avaliação da mortalidade de *Spodoptera frugiperda* por dois isolados de *Baculovirus spodoptera* e por parasitismo, em Janaúba e em Sete Lagoas /MG. In: *XXII Congresso Brasileiro de Entomologia*, 24 a 29 agosto 2008, Uberlândia – MG.

Pavan, O.H.O. 1988. Baculovirus e Controle Biológico. In: *Controle biológico dos insetos.* Anais 2, Soc. Entomol. do Brasil. Campinas, SP. Fundação Cargill.

Payne, C.C. 1986. Insect pathogenic viruses as pest control agents. *Fortschritteder Zoologie*, 32:183 - 200.

Pereira, E.P., Conte, H., Ribeiro, L.F.C., Zanatta, D.B., Bravo, J.P., Fernandez, M.A. and Brancalhão, R.M.C. 2008. Cytopathological process by multiple Nucleopolyhedrovirus in the testis of Bombyx mori L., 1758 (Lepidoptera: Bombycidae). *Journal of Invertebrate Pathology*, 99:1 - 7.

Pimentel, D. 2002. *Encyclopedia of pest management*. CRC Press. 929 pg.

Raman, K.V. and Alcázar, J. 1992. Biological control of Potato tuber moth using Phthorimaea Baculovirus. CIP *Training Bulletin*, 2. 27p.

Rechcigl, J.E. and Rechcigl, N.A. 1999. *Biological and biotechnological control of insect pests*. CRC Press. 374pg.

Ribeiro, B.M., Souza, M.L. and Kitajima, E.W. 1998. Taxonomia, caracterização molecular e bioquímica de vírus de insetos. In: Alves, S.B. (Coord). *Controle microbiano de insetos*. 2 ed. Piracicaba : FEALQp.481-504.

Ribeiro, B.M., Zanotto, P.M.A., McDowell, S., Souza, M.L. and Kitajima, E.W. 1997. Characterization of a baculovirus infecting the passion fruit caterpillar *Dione juno juno*. *Biocell*, 21:71 – 82.

Ribeiro, L.F.C., Torquato, E.F.B. and Brancalhão, R.M.C. 2009. Infecção das células da válvula estomodeal de *Bombyx mori* ao Nucleopolyhedrovirus, subgrupo múltiplo, BmMNPV. In.: *X Simpósio de Controle Biológico*, 1 a 5 junho de 2009, Bento Gonçalves, RS.

Richardson, C.D. 1995. *Baculovirus expression protocols*. Totowa: Humana, 418p.

Schmitt, A.T. 1983. Ocorrência de inimigos naturais de *Erinnyis ello* (L.) no Estado de Santa Catarina. *Revista Brasileira de Mandioca*, 2:59 – 62.

Schmitt, A.T. 1984. Inimigos naturais do *Erinnyis ello* da mandioca. In: *Encontro Nacional de Fitossanitaristas*, Florianópolis. Anais. Brasília: Ministério da Agricultura, p.201-208.

Schmitt, A.T. 1985. Eficiência da aplicação de *Baculovirus erinnyis* no controle do mandarová da mandioca. *Comunicado técnico* 88, Empasc.

Smith, L. and Bellotti, A.C. 1996. Successful biocontrol projects with emphasis on the Neotropics. Cornell Com. Conf. Biol.Control, Ithaca. http://www.nysaes.cornell.edu/ent/bcconf/talks/bellotti.html, 12 pp.

Sosa-Gómez, D.R. and Moscardi, F. 1991. Microbial control and insect pathology in Argentina. *Ciência e cultura*, 43:375 - 379.

Sosa-Gómez, D.R., Moscardi, F., Santos, B., Alves, L.F.A. and Alves, S.B. 2008. Produção e uso de vírus para o controle de pragas na América Latina. In: Alves, J.B. and Lopes, R.B. (Eds.). *Controle Microbiano de Pragas na América Latina*. FEALQ, Brasil, p. 49-68.

Steinhaus, E.A. and Marsh, G.A. 1962. Report of diagnosis of diseased insects. *Hilgardia*, 33:349 - 90.

Steinhaus, E.A. and Thompson, C.G. 1949. Granulosis disease in the buckeye caterpillar, *Unonia caenia* Hubner. *Science*, 110:276 - 278.

Tinsley, T.W. and Kelly, D.C. 1985. Taxonomy and nomenclature of insect pathogenic viruses, p.3-26. In: Maramorosh, K. and Sherman, K.E. (Eds.). *Viral insecticides for biological control*. London, Academic Press, 809p.

Torquato, E.F., Neto, M.H. and Brancalhão, R.M.C. 2006. Nucleopolyhedrovirus infected central nervous system cells of *Bombyx mori* (L.) (Lepidoptera: Bombycidae). *Neotrop Entomol.*, 35(1):70 – 4.

Tuelher, E.S., Paiva, C.E., Sans, A.C. and Valicente, F. H. 2009. Concentração letal e produção viral de nucleopoliedrovirus de *Spodoptera frugiperda*. In.: *X Simpósio de Controle Biológico*, 1 a 5 junho de 2009, Bento Gonçalves, RS.

Valicente, F.H. and Cruz, I. 1991. Controle biológico da lagarta-do-cartucho, *Spodoptera frugiperda*, com o baculovírus. Sete Lagoas: Embrapa-CNPMS, 23p. (Embrapa-CNPMS. Circular técnica, 15).

Valicente, F.H. and Costa, E.F. 1995. Control of the fall armyworm, *Spodoptera frugiperda* (J.E. Smith) with *Baculovirus spodoptera* through irrigation water. *An. Soc. Entomol. Bras.* 24: 61-67.

Valicente, F.H., Peixoto, M.J.V.D., Paiva, E. and Kitajima, E.W. 1989. Identificação e purificação de um vírus de poliedrose nuclear da lagarta-do-cartucho *Spodoptera frugiperda*. *Anais da Sociedade Entomológica Brasileira*, 18:71 – 82.

Valicente, F.H., Tuelher, E.S., Paiva, C.E.C. and Sans, A.C. 2009. Inoculação de nucleopoliedrovírus de *Spodoptera frugiperda* em dieta artificial para produção de bioinseticida. In.: *X Simpósio de Controle Biológico*, 1 a 5 junho de 2009, Bento Gonçalves, RS.

Wang, L., Salem, T.Z., Campbell D.J., Turney, C.M., Kumar, C.M. and Cheng, X.W. 2009. Characterization of a virion occlusion defectiver *Autographa californica* multicapsid nucleopolyhedrovirus (AcMNPV) mutant lacking the p26, p10 and p74 genes. *J. Gen. Virol.*

Wood, A.W. and Granados, R.R. 1991. Genetically engineered baculovirus as agents for pest control. *Annual Review of Microbiology*, 45:69 – 87.

Yang, L.R., Qiang, X., Zhang, B.Q., Tang, M.J. and Zhang, C.X. 2009. Characterization of a baculovirus newly isolated from the tea slug moth, *Iragoidae fasciata. The journal of microbiology*,47(2):208 – 213.

Yi, P. and Li, Z. 1989. Microbial pest control in China: viruses and fungi. Symp. Sino-*American Biocontrol Res.*, San Antonio.

Yun, E.Y., Lee, J.K., Kwon, O.Y., Hwang,J.S., Kim, I. and Kang, S.W. 2009. *Bombyx mori* transferrin: Genomic structure, expression and antimicrobial activity of recombinant protein. Article in press.

Zanardo, A.B.R., Ávila, C.J., Duarte, M.M. and Silva, J.A. 2009. Efeito de um vírus de poliedrose nuclear no consumo foliar de soja e na mortalidade de três tamanhos de lagartas de *Pseudoplusia includens* (Walker, 1857) (Lepidoptera:Noctuidae). In.: *X Simpósio de Controle Biológico*, 1 a 5 junho de 2009, Bento Gonçalves, RS.

In: Microbial Insecticides: Principles and Applications
Editors: J. Francis Borgio, K. Sahayaraj, et al.

ISBN: 978-1-61209-223-2
© 2011 Nova Science Publishers, Inc

Chapter 13

MASS PRODUCTION OF ENTOMOPATHOGENIC VIRUSES

Ali Mehrvar[*]

Department of Plant Protection, Faculty of Agriculture, University of Maragheh – 55181-83111, Golshahr, Maragheh, East-Azerbaijan, Iran

ABSTRACT

Formulation of any entomopathogenic agent such as viruses at an economically reasonable cost and effects of production processes on the viability of the virus up to application point should be entirely considered in production levels of a bioinsecticide. Many factors are involved in production of viral insecticides which influence all the processing stages of the mass production. Several studies however are focused on domestication of microorganisms by overcoming the related processing parameters as they are living organisms in nature. Type of host insects, their biology and behaviour; age, stage and sex of the larvae used for virus production; the rearing environment such as temperature, humidity, photoperiod and nutritional quality of the insect diet are listed by several workers as the factors affecting mass production of entomopathogenic viruses. However, any factor that influencing the larval growth in post-inoculation period should be considered as a virus production element which can come under the process considerations. So, standardization of mass production, purification and enumeration of the virus particles as well as formulation of the virus products are of economically important views in any IPM disciplines.

13.1. INTRODUCTION

Insect viruses as microbial pesticides and their application have been reviewed sufficiently by several workers (Ignoffo, 1968; Stairs, 1971, 1972; Benz, 1981; Yearian and Young, 1982; Entwistle and Evans, 1985; Jacques, 1985; Jayaraj, 1985, 1998; Podgwaite, 1985; Shieh, 1989; Jayaraj *et al.,* 1989, 1992, 1994; Rabindra and Jayaraj, 1990; King *et al.,*

[*] E-mail: mehrvar@mhec.ac.ir, Fax: +(98)421-2276060, Phone: +(98)421-2276068

1994; Easwaramoorthy, 1998; Hunter-Fujita *et al.,* 1998; Moscardi, 1999; Young, 2000; Rabindra *et al.,* 2003). An insect pathogenic virus product is supposed to supply in a reasonable market cost in advance. In a small-scale laboratory production of a virus, related costs might be managed under an economic program, whereas, in the case of large-scale crop protection regimes by viral biopesticides, administration of virus production process and all its viable aspects, could brings the product under an industrially and commercially status. Hence, standardization of mass production of a virus deals with some factors to produce a marketable product. On the other hand, virus bioefficacy as well as its ease of mass production are important to select a virus for pest management programs. Some viral insecticides used in insect pest management are indicated in Table 1.

More than 1270 insect-virus associations had been recognized by 1981 (Martignoni and Iwai, 1981), in which more than 70 per cent involving Lepidopterans as host (van Driesche and Bellows, 1996). Viruses of two families, viz., Baculoviridae and Tetraviridae are known to be multiplied only in invertebrates especially in arthropods, whereas, most other families contain viruses associated with different hosts involving vertebrates and other non-arthropods and thus are of less interest as potential insect control agents, because of their presumptive effects on mammals.

Table 1. List of some commercially products of viral insecticides

Virus Name	Trade/Manufacturer Name	Pests Controlled
Granulosis virus	Capex	leafrollers
Cydia pomonella GV	Cyd-XT	Codling moth
Autographa californica NPV	VFN80	Alfalfa looper (*Autographa californica*)
Helicoverpa zea NPV	Gemstar LCT	American bollworm, tobacco budworm, corn earworm, tomato fruit borer
Helicoverpa armigera NPV	Helicide® PCI(India)	American bollworm, tobacco budworm, corn earworm, tomato fruit borer
NPV *Helicoverpa armigera* 2% AS (min. OB 1×10^9/ml)	Bio-Pest Management Bangalore, India	American bollworm (*Helicoverpa armigera*)
NPV *Helicoverpa armigera* (min. OB 1×10^9/gm)	Bio Tech. Idus. Ltd., New Delhi, India	American bollworm (*Helicoverpa armigera*)
Spodoptera exigua NPV	Spod-X LC	Beet armyworm, lesser armyworm, pig weed caterpillar, small mottled willow moth
Spodoptera litura NPV	Spodo-Cide™ (0.50% AS), PCI(India)-Bangalore	Tobacco caterpillar (*Spodoptera litura*)
NPV *Spodoptera litura* 0.5% AS (min. OB 1×10^9/gm)	Pest Control India Mumbai	*Spodoptera litura*

Virus Mass Production

Optimal production of the entomopathogenic viruses is one that results in the greatest yield of biologically active virus and that confirms to the quality control standards (Shapiro, 1986; Shuler *et al.,* 1995). Hygiene and sanitation of virus production and product along with some other factors such as availability of required materials and their costs, quality control, and skilled labors engaged in the production program are of major factors which administrate *in vivo* production of viral biopesticides. Shapiro (1986) and Shapiro *et al.* (1981) listed several factors noticeably influencing the mass production of viruses like type of host insects; its biology and behaviour; age, stage and sex of the larvae used for virus production; the rearing environment which includes factors such as temperature, humidity, photoperiod and nutritional quality of the insect diet. As a matter of fact, any factor that influencing the larval growth in post-inoculation period would be considered as a virus production element which can come under the process considerations (Hunter-Fujita *et al.,* 1998; Burges, 1998; Bell *et al.,* 1980a, b; Shieh and Bohmfalk, 1980; Shapiro, 1986; Ebora *et al.,* 1990; Moscardi *et al.,* 1997; Hunter-Fujita *et al.,* 1998; Subramanian *et al.,* 2006).

Baculoviruses are the most common and well studied group which has been of greatest interest as potential microbial control agents. The family Baculoviridae divides into three subgroups (genera) including the nuclear polyhedrosis viruses (NPV), the granulosis viruses (GV), and, formerly, the non-occluded viruses (NOV). Francki *et al.* (1991) placed the first two genera in the subfamily "Eubaculovirinae" and third subgroup in the subfamily "Nudibaculovirinae" (Tanada and Kaya, 1993). The genus NPV is of occluded viruses within polyhedrin proteins containing randomly-occluded virions. Poinar and Thomas (1984) are classifying this genus into two subgenera: The single nucleocapsid NPVs (SNPV) and the multi-nucleocapsid NPVs (MNPV). More than 520 NPVs have been identified in insects by 1986 (Martignoni and Iwai, 1986).

13.2. MASS PRODUCTION OF BACULOVIRUSES

13.2.1. *In Vivo* Production

In vivo production of baculoviruses has been reviewed by Shapiro in 1982. In summary, susceptible insects are reared to an optimum stage and then infected with virus. The insects are then reared for a further period and harvested just prior to, or after, death. Research on optimizing production yields has concentrated on determining the appropriate virus dose administered, age/instar/weight of host insect, and length and temperature of incubation, as well as automating some steps (e.g. dispensing diet, egg placement).

Normally, the amount of virus produced per insect is positively correlated to larval weight. Thus conditions are optimized so that larvae reach maximum weight before dying from viral infection. Alternatively, virus can be obtained by field collection of virus-killed insects or production 'in the field' by farmers. Harvested larvae contain a mixture of virus, insect debris and contaminant microorganisms (bacteria, fungi, protozoa, etc.). Insect debris and other contaminants can alter the results of a bioassay either by affecting the virus directly, resulting in partial or total inactivation, antagonism or synergism, or by affecting the test insect/cell, resulting in death, or interfering in virus infection/replication or reducing insect feeding and hence virus uptake. These effects are often unpredictable and variable, particularly as the amount and type of contaminants can vary.

Many production techniques have been designed to minimize contamination by microorganisms, for example harvesting of infected insects whilst still alive results in reduced numbers of spore-forming bacteria in comparison with harvesting after death. Production within a closed automated system also minimizes contamination. It is generally important to minimize potential sources of contamination through proper preparation of insect diets, promotion of a high standard of operator hygiene and selection and maintenance of healthy insect colonies. These are also essential to ensure predictable and even growth of insects, which is essential for optimum production and accurate bioassays.

13.2.2. *In Vitro* Production

The large-scale production of virus *in vitro* is too expensive at present, although rapid advances have been made in recent years. *In vitro* production involves the production and infection of a susceptible insect cell line in a bioreactor. Ignoffo and Hink summarized the requirements of successful *in vitro* production as: (i) the development of robust, prolific insect cell lines that yield high pathogen titres; (ii) the availability of simple and cheap culture media; and (iii) development of plant-scale equipment and efficient, routine production procedures. Numerous cell lines are now available, along with suitable simple and serum-free media, as well as improved bioreactors and procedures.

In many systems, however, there are still problems of production reverting to mutants with only budded virus. Also, further bioreactor improvements are required to achieve oxygen levels required in vessels larger than 250 litres. Cell culture of NPV is the best established, with a number of cell lines capable of supporting the replication of *Spodoptera* and *Heliothis* NPVs. Cell culture of GV is less well advanced, with only a few cell lines available that are capable of supporting virus replication. A necessary feature of cell culture systems is high levels of sterility, so the contamination problems encountered with virus produced in vivo do not occur, but of course this requires the availability of facilities that allow sterile handling of equipment.

13.3. NPV MASS PRODUCTION

13.3.1. Standardization of NPV Mass Production

Under large-scale production of nuclear polyhedrosis viruses there must be a large-scale culture of their insect hosts. Maintenance of such a host culture requires availability of natural plant materials, which constraints the production of viral biopesticides, due to their season dependent availability.

To overcome these difficulties artificial (semi-synthetic) diets are used without any dependency to the host and season in production process, resulting optimization of the virus product under more controlled environment.

13.3.1.1. Optimization of Larval Artificial Diet

Production of nuclear polyhedrosis virus of *Helicoverpa* spp. which began in 1961, progresses through various research and developmental phases and attained technical realization as the first commercial viral pesticide (Ignoffo, 1973). A boarder and more complete account of some aspects on virus production and role of virus in insect pest control have been given by Burges (1981). To maintain an acceptable quality of host insect culture, continuous availability of larvae of uniform age and size should be considered. The components of artificial diet and their quality are greatly influence the host production which directly reflex in virus propagation. By the way, Ignoffo (1966) demonstrated the feasibility of mass production of *Helicoverpa armigera* NPV by utilizing the technique of rearing larvae individually in multicavity plastic trays. Subsequent development for mass production process, improvements in viral recovery procedures and formulation of the viral product made it possible for commercialization of *H. armigera* nucleopolyhedovirus (Shieh, 1978). Further, Ignoffo and Couch (1981) improved the method of mass production of *Baculovirus heliothis* from laboratory reared Helicoverpa larvae through which about seven to nine times more active and two to five times more OB/larvae were obtained from virosed larvae. Altering the concentration of vitamins in diet of *Lymantrya dispar* larvae, by 30 percent increased the yield of *Ld*NPV (Shapiro *et al.*, 1981). Likewise, Kelly *et al.* (1989) developed a high wheat germ diet for the production of *Euproctis chrysorrhoea* NPV. Several works focused on those alterations in larval diet which cause any increase in larval growth and virus yield productivity. In a study conducted by Mehrvar *et al.* (2007a) it was suggested that, sunflower oil at 6000 ppm in larval diet showed the highest yield per larva with a productivity ratio of 1.646×10^4 OB/larva which was significantly on par with that of soybean oil. Also, French bean flour was identified better than chickpea flour in enhancing the virus yield of HearNPV in *H. armigera* artificial diet (Srinivasan *et al.*, 1994).

13.3.1.2. Effect of Larval Age and Weight on Virus Yield

The time required for multiplication of the virus in the body of host is mostly in related to the age of larvae inoculated. In case, young larvae of lower instars and weights inoculated with a lethal dosage of virus used for mass production, the virus yield harvested per larva would not be in an optimum and economically accepted level. Because, death of the infected larvae will occur as soon as the larval body could enough grow to produce optimum number of OBs. Moreover, inoculation of later instars of the larvae will result the same virus yield as the previous case mentioned. So, larval age and weight at the time of acquisition of inoculum is important in obtaining appreciable number of OBs at death. Considerable studies were previously confirmed the same status on different insects (Whitlock, 1977; Evans, 1981; Shapiro *et al.*, 1981; Rabindra and Jayaraj, 1986; Smith and Vlak, 1988; Teakle and Byrne, 1989; Rabindra, 1993; Cherry *et al.*, 1997; Moscardi *et al.*, 1997; Kalia *et al.*, 2001; Mehrvar *et al.*, 2007b). Teakle and Byrne (1989) could successfully attain a 100-fold increase in the yield of *H. armigera* NPV in six-day treated larvae compared to one-day old ones. Similarly, Evans (1981) achieved 170-fold increase in the yield of virus from first to fifth instars of *Mamestra brassicae* larvae. Mehrvar *et al.* (2007b) found a significant influence of larval age and weight on the percent larval mortality, yield/larva, and productivity ratio of *H. armigera* NPV. Of the different age groups studied (mid fourth, late fourth, early fifth and mid fifth instars), early fifth instar with a weight range of 65.46-68.13 mg recorded significantly the highest yield of 6.73×10^9 OB/larva (Figure 1).

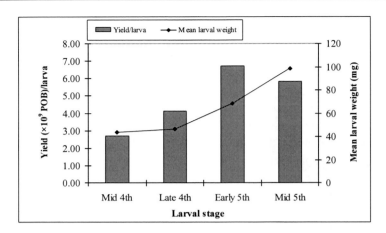

Figure 1. Effect of larval stage and weight on the yield of HearNPV at an inoculation dose of 5×10^5 OB/larva and an incubation temperature of $25 \pm 1°C$ (From Mehrvar *et al.*, 2007b).

Mid fourth, late fourth, and mid fifth instars larvae registered lower OBs production in all the virus isolates evaluated.

Entwistle and Evans (1985) suggested a positive correlation between larval weight and number of occlusion bodies recovered. A yield of 9.2×10^6 to 4.3×10^7 OB/mg of larval body weight in the order Lepidoptera was achieved (Entwistle and Evans, 1985).

13.3.1.3. *Effect of Virus Inoculation Dose on Virus Yield*

The objective is identification of minimal dose which is responsible for obtaining maximum virus yield. Ignoffo and Couch (1981) stated that the optimal dose varies with the virulent virus and age of the host. Bell (1991) could attain 1.35 fold increase in occlusion bodies yield of *H. armigera* NPV in *Heliothis virescens* larvae with a lower inoculation dose of 54 OB/mm^2 compared to the highest dose of 2708 OB/mm^2. Similar studies were so far conducted by several authors on various insects (Shieh, 1978, 1989; Santharam, 1986; Kelly and Entwistle, 1988; Tuan *et al.*, 1989; Jones *et al.*, 1994; Moscardi *et al.*, 1997; Subramanian, 1998; Subramanian *et al.*, 2001; Kumar and Rabindra, 2003). Mehrvar *et al.* (2007b) could achieve the highest yield of *H. armigera* NPV (6.87×10^9 OB/early fifth instar larva) at an inoculation dose of 1965.87 OB/mm^2 compared to the other virus dosages evaluated.

13.3.1.4. *Effect of Inoculation Temperature on Virus Yield*

Several workers demonstrated that temperature regulates the potential of virus replication (Morris, 1971; Stairs, 1978; Boucias *et al.*, 1980; Shapiro *et al.*, 1981; Kelly and Entwistle, 1988; Ribeiro and Pavan, 1994; Subramanian *et al.*, 2006).

Standardization of incubation temperature could significantly optimize the virus yield production. Lower and higher temperature adversely affect on the on mass multiplication of the virus as it can influence the larval regular growth as well as virus activity. Work done by Morris (1971) indicated that LT$_{50}$ values of *Lambdina fiscellaria lugubrosa* NPV progressively decreased as incubation temperature increased.

However, higher temperatures (above 40°C) directly inactivate virus infectivity and replication (Stairs, 1978). Subramanian *et al.* (2006) showed that among the different incubation temperature treatments, mortality of the *Spodoptera liture* larvae was significantly higher in the case of 25°C and 30°C than at 35°C and room temperature. Similar findings were also demonstrated by Mehrvar *et al.* (2007b).

The data on different incubation temperatures of *H. armigera* larvae indicated maximum yield/larva was obtained at 25°C followed by room temperature and 30°C. In *H. armigera* larvae, the virus multiplies at a slow pace at 25°C allowing the fat bodies to proliferate simultaneously.

At higher temperatures the virus multiplies faster destroying the fat body before it could grow to provide greater substrate volume. Therefore, a good mass production facility should possess a temperature-controlled incubation chamber to provide a constant temperature of 25±1°C.

13.3.1.5. Virus Yield in Relation to the Period of Harvest

The actual time to larval harvest has a critical effect on the commercial NPV-production units. However, the information on the optimal time to harvest the infected larvae is very few.

Characteristics of baculovirus preparations processed from living and dead larvae of *H. zea* have been investigated by Ignoffo and Shapiro (1978). Results showed that the *Hz*NPV processed from living larvae was 7-9 times less potent that those from dead hosts. Shapiro and Bell (1981) studied the *in vivo* mass production of Gypsy moth nuclear polyhedrosis virus (*Ld*NPV). The number of occlusion bodies harvested from cadavers was up to 7 times more than from infested living larvae (2×10^9 OB/larva).

Similarly, Ignoffo and Shapiro (1978) showed that the number of NPV occlusion bodies attained from dead larvae of *Heliothis* spp. was significantly more that that of the occluded bodies processed from living larvae. Also, the biological activity (LC$_{50}$ = 85 OB/ml) of those viruses harvested from cadavers was more than the viruses obtained from living larvae (LC$_{50}$ = 599 OB/ml). Mehrvar *et al.* (2007b) stated that, NPV yield is high when virosed larvae of *H. armigera* harvested as cadavers than when harvested at different days after inoculation (Figure II).

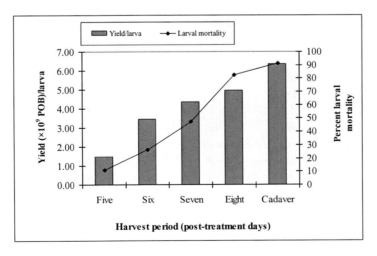

Figure 2. Effect of harvest period on the yield of HearNPV and larval mortality achieved from early fifth instar larvae inoculated with 5×10^5 OB/larva at 25±1°C (From Mehrvar *et al.,* 2007b).

13.4. PURIFICATION OF VIRUSES

Ideally, virus suspensions should be purified so that other material or microorganisms present do not interfere with actual viral count. Numbers of purification techniques are available which have different efficiencies; some of these methods themselves may also affect virus activity.

13.4.1. Acetone Co-Precipitation

This was first developed for viruses. Aqueous lactose and acetone solution is slowly added to a virus suspension. This causes the virus to precipitate from the suspension. The suspension is then filtered and washed with sterile water. This technique removes, for instance, insect protein, as well as killing some vegetative bacteria.

13.4.2. Density-Gradient Centrifugation

This is the most often-used method for producing highly purified suspensions (Box 1).

Box 1
Density-gradient centrifugation
Infected larvae are macerated in 0.1% (w/v) sodium dodecyl sulphate filtered through a double layer of muslin
centrifuged at 100 g for 30 s to remove gross debris
supernatant is then centrifuged at 2500 g for 10 min for NPV
pellet is resuspended in 0.1% SDS
layered on a 45–60% (w/v) sucrose gradient
centrifuged at 50,000 g for 1 h for NPV
purified virus forms an opaque band at a sucrose density of 54–56%
band is removed with a syringe or pipette

13.4.3. Semi-Purification

Semi-purification of virus, which removes large insect debris and some contaminant microorganisms, can be used if the equipment is not available for density-gradient centrifugation. The following methodology for semi-purification of NPV has been successfully used on a laboratory scale by F.R. Hunter-Fujita and K.A. Jones in Thailand and results in a suspension that can be quantified using a haemocytometer (Box II).

13.5. ENUMERATION OF VIRUSES

For occluded viruses, such as NPVs, GVs, Cytoplasmic polyhedrosis viruses (CPVs) and entomopox viruses (EPVs), quantification is generally measured in terms of the number of

OB. The OB of NPVs, CPVs and EPVs are large enough to be easily seen under a light microscope and can be counted using haemocytometer and phase-contrast optics (×400). Due to the considerably smaller size of GVs, accurate enumeration using this method is very difficult and dark-field optics should be used. Alternatively, electron microscopy can be used.

Box II
Semi-purification
1. Macerate larvae and filter through a double layer of muslin.
2. Dilute 1:4 with 0.1% SDS in distilled water.
3. Centrifuge at 600 rpm in a bench-top centrifuge with a fixed-angle rotor to pellet insect debris. Discard the pellet.
4. Centrifuge the supernatant at 3150 rpm for 10 min and discard the supernatant. The virus will be pelleted in two layers, a darker layer on the bottom and a lighter layer on top.
5. Collect the darker layer and resuspend in 0.1% SDS. Store in a refrigerator.
6. Resuspend the lighter layer in 0.1% SDS and centrifuge at 4050 rpm for 12 min. Discard the supernatant. Again the virus pellet will be in two layers, a darker lower layer and a lighter upper layer.
7. Collect the darker layer and add to the suspension collected earlier in step 5.
8. Resuspend the lighter layer in sterile distilled water and centrifuge at 4050 rpm. Discard supernatant and resuspend the pellet in sterile distilled water and repeat this step. This should be repeated at least twice. The virus should now be stored in the refrigerator or freezer.
9. Take the 'dark' suspension and centrifuge at 4050 rpm for 10 min and discard the supernatant. Remove any virus in a light layer and treat as in step
10. For the dark layer repeat from step 5. This can be repeated as necessary. Finally, wash the remaining dark layer in distilled water as described in step8 and store separately, or discard.

13.5.1. Direct Visual Enumeration

These techniques are often used as they are simple and do not require expensive equipment.

13.5.1.1. Haemocytometer

Improved Neubauer ruling consists of two grids of 25 squares, which are further subdivided into 16 smaller squares, each with an area of 0.0025 mm^2. When a coverslip is firmly placed on the slide, such that Newton's rings are visible, a gap of 0.1 mm is left between the top surface of the slide and the undersurface of the coverslip. The small square thus marks the area above which is a volume of 0.1 × 0.0025 mm^3 or 0.00025 ml. As a maximum, 20 large squares (320 small squares, 160 from each of the two grids drawn on the haemocytometer) are viewed in a predefined pattern. The concentration of the virus suspension is then calculated as:

$$\text{OB ml}^{-1} = \frac{D \times X}{N \times K}$$

Where D = dilution of suspension dispensed into the haemocytometer, X = number of OB counted, N = number of small squares counted, and K = volume above a small square (in ml) (for a 0.1-mm-deep improved Neubauer haemocytometer this is 2.5×10^{-7} ml).

13.5.1.2. Dry Films

Known volumes of a virus suspension are evenly spread in circles of known size on a glass microscope slide and, after drying, stained with Giemsa. Using a light microscope and an oil-immersion objective ($\times 1000$), the number of OBs are counted at specific points along the radius of the stained sample.

13.5.1.3. Proportional Count

Virus suspension is mixed with a known concentration of polystyrene beadsand smeared on to a microscope slide. The slide is then stained and counted. Normally several counts (up to 40) are made per slide. The concentration of the virus suspension is calculated as:

$$R \times S \times \frac{V_1}{V_2}$$

Where R = mean ratio of spores or beads to virus, i.e.

$$\frac{\sum p \text{ (individual counts of virus)}}{\sum b \text{ (individual counts of beads/spores)}}$$

S = no. of spores/beads per ml, V1= volume of bead/spore suspension, V2= volume of virus suspension.

13.5.1.4. Impression-Film Technique

The technique uses clear, double-sided adhesive tape to remove the virus, which is then stained with Buffalo Black for 5 min at 40–45°C and viewed with a light microscope fitted with an eyepiece graticule, under oil immersion ($\times 1000$). OBs appear as round, black objects, being distinguished from other such particles by their size. Indirect methods of enumeration.

13.5.2. Indirect Methods of Enumeration

Virus suspensions can also be enumerated using non-visual methods.

13.5.2.1. Protein Assay

A calibration curve of protein concentration is obtained for a standard solution of bovine plasma albumin by measuring the absorbance of solutions of known concentration in a spectrophotometer (at 540 nm). Similarly, the absorbance of the test virus suspensions is determined and the protein concentration determined from the calibration curve.

13.5.2.2. Enzyme-Linked Immunosorbent Assay

Serological techniques provide a method of virus detection and quantification which is both sensitive and specific and enzyme-linked immunosorbent assay (ELISA) has been used to detect and, in some cases, quantify insect viruses. ELISA uses an antibody to detect the virus and a subsequent enzyme-mediated reaction to quantify it. The end result is a colour reaction, the intensity of which is proportional to the amount of virus present. ELISA can be used to detect virus particles, rather than occlusion body proteins. Hence it is suitable for non-occluded viruses.

13.5.2.3. Coulter Counter

Highly purified suspensions of occluded virus can be enumerated automatically using a Coulter counter.

13.5.2.4. Larval Equivalents

Virus can also be quantified in terms of 'larval equivalents', which is taken as the amount of virus produced by a virus-killed insect. This is regularly considered as 6×10^9 OB/ml as one larval equivalent (LE) of *Helicoverpa armigera* NPV by several workers and academic institutes.

13.6. FORMULATION OF NPV

Infestation by pests in agriculture, forestry, and public health sectors has been traditionally controlled by chemical pesticides, some of which have been currently replaced by ecofriendly biopesticides. The development of biopesticides formulation closely paralleled that of the pesticides (Angus and Luthy, 1971). Formulation comprises aids to preserving organisms, to delivering them to their targets and - once there - to improving their activities (Jones and Burges, 1998).

Burges (1998) issued a complete reference of all factors influencing the formulation of microbial control agents. The major functions of the products or formulation which have been mentioned by most of the workers are listed out as stabilization, handling and application, environmental persistence, and improvement of action (Couch and Ignoffo, 1981; Entwistle and Evans, 1985; Dhandapani *et al.,* 1987; Rabindra and Jayaraj, 1988, Shapiro and Robertson, 1990; Rabindra and Jayaraj, 1992; Dhandapani *et al.,* 1993; Rabindra and Jayaraj, 1995; Muthuswami *et al.,* 1994; Burges, 1998; Tamez-Guerra *et al.,* 2000; Young, 2000; Shapiro, 2001; Shapiro and Argauer, 2001; Behle *et al.,* 2003; Farrar *et al.,* 2003; Cook *et al.,* 2003; Murillo *et al.,* 2003; Martinez, *et al.,* 2004; Arthures *et al.,* 2005; Dougherty *et al.,* 2006; Mehrvar *et al.,* 2008a, b).

13.6.1. Types of Formulations

A technical concentrate of an organism that has been formulated is termed as a formulation, or a product, which may be stored and put on sale commercially. A product often does not fully serve all the requirements of use on all crops. More additives may be needed to achieve optimum application on some crops (Jones and Burges, 1998). Formulation of biopesticides includes two major groups, *viz.*, basic (shelf) and tank mix formulations (Couch and Ignoffo, 1981; Rhodes, 1993; Jones and Burges, 1997; Burges, 1998).

There are a wide variety of formulation types, both liquid and solid. The main types currently used for organisms have been classified by Rhodes (1993) into dry products (dusts, granules and briquettes) and suspensions (oil- or water-based and emulsions).

13.6.1.1. Basic Formulations

Seaman (1990) and later Burges (1998) reviewed trends in the development of basic formulations. Development of suspension concentrates; oil-in-water emulsions and water dispersible granules have been highlighted in the commercial phase.

13.6.1.1.1. Dry Products

These products can be classified based on their particle or aggregate size as dusts, granules and briquettes. Also wettable powders are included in this group (Jones and Burges, 1998).

For preparation of dry products, powders of baculoviruses are prepared by spray and freeze drying techniques and diluted with inert carriers. Minerals such as clays are often the first choice of diluents, but silica minerals are also used (Burges, 1998). Diluents with high surface acidity or alkalinity are usually avoided as they tend to form an unstable product (Polon, 1973). Inert fillers used with insect pathogens are listed by some workers (Angus and Luthy, 1971; Couch and Ignoffo, 1981; Polon, 1973; Becher, 1973).

Dusts typically contain less than 10% of an organism by weight, whereas this amount is 5-20% and 50-80% for granules and wettable powders, respectively (Burges, 1998). There are three granules, *viz.*, (i) the organisms are attached to the outer surface of a granular carrier in a rotating drum by a sticker (Lisansky *et al.,* 1993), (ii) the organisms are sprayed onto a rotating granular carrier without a sticker (Sjogren, 1996) and (iii) the organisms are incorporated into a carrier paste or powder which sets as a matrix (Connick *et al.,* 1991).

Wettable powders which predominated among early commercial products contain 50-80% technical powder, 15-45% filler, 1-10% dispersant and 3-5% surfactant by weight (McKinely *et al.,* 1989). These products are formulated as dry powder, designed to be added to a liquid carrier, normally water, just before application (Jones and Burges, 1998).

With dry products, deterioration accelerates if the product moisture content is allowed to increase progressively above 5% (Couch and Ignoffo, 1981). Once exposed freely to ambient air, the moisture content of a material always approaches equilibrium with the air humidity. Equilibrium moisture content increases as the air humidity increases and may be excessive for hygroscopic materials (Jones and Burges, 1997).

Freeze-drying, although expensive, has been the commonest method of drying and stabilizing virus (Martignoni, 1978; Shapiro, 1982; Young and Yearian, 1986; Jones and Burges, 1998). It gave the most active and most stable technical concentrate out of a number of methods compared by Ignoffo *et al.* (1976) for *H. virescens* nuclear polyhedrosis virus. McKinely *et al.* (1989) described a method for the development of freeze-dried *S. littoralis* NPV. Clumping during freeze-drying may create storage and tank-mixing difficulties, but these can usually be prevented by suitable additives (McKinely *et al.,* 1989).

Spray-drying technique can improve the shelf-life and persistence of virus under field situations (Huber, 1986; Jones and Burges, 1997). However, spray-drying destroyed most of the activity of *Choristoneura fumiferana* NPV (Young, 1989), and reduced the shelf-life of *Cydia pomonella* GV (Huber, 1986) and Sandoz-406 (Young and Yearian, 1986).

Air-drying technique has also considered for producing NPV-dry products (*e.g.* powders of *Anticarsia gemmatalis* NPV) especially in some developing countries. Such powders are made by pouring a thin layer of virus suspension onto clean surfaces (like large tables' surface) and leaving it to dry at room temperature for several hours, then scraping it off. In this method bacteria noticeably multiply during drying phase, but speed of drying can be increased by fans and warm air (Moscardi, 1989).

13.6.1.1.2. Liquid Suspensions

In these formulations the organism is carried in a liquid, normally oil (Cherry *et al.,* 1994; Thennarasan, 1997) or water (Ignoffo *et al.,* 1972). However, the viruses are reported as stable in water normally (Ignoffo and Garcia, 1966; Andrews and Sikorowski, 1973, Narayanan, 1979). The commonest products are suspension concentrates and emulsions.

In suspensions the particles account for 10-40%, suspender ingredient 1-3%, dispersant 1-5%, surfactant 3-8% and carrier liquid (oil or water) 3-5% by weight. Viscosity should roughly equal the settling rate of the particles. This is achieved by the use of colloidal clays, polysaccharide gums, cellulose or synthetic polymers (Theng, 1979).

Emulsification reduces sedimentation of particles during storage and in the spray tank, because buoyancy of the oil counteracts high, relative densities of the particles (Burges, 1998). Both oil in water (normal emulsions) and water in oil (invert emulsions) have been applied in preparation of microbial pesticides. Invert emulsions showed high viscosity, so are less likely to separate. They produce larger drops with most sprayers and, because the external phase is oil, evaporation from drops is minimal, both factors combining to present less drift hazard. Invert emulsions can be made by mixing two phases at the spray nozzle, or are pre-mixed with the addition of fatty amine salts. The stability and possible phytotoxicity of these formulations need to be carefully assessed (Jones and Burges, 1998).

Efforts have been continued to minimize the number of microbial contaminants in suspensions. Bacterial contamination arising from the gut of dead larvae must be less than 1×10^7 colony-forming units (c.f.u.)/g (Podgwaite *et al.,* 1983; Grzywacz *et al.,* 1997) in the final products. Lipa (1998) prepared water-glycerin based (25-50%) suspensions for control of crop pests. The bacterial titer in the preparation was 1×10^7/g and the pH of the suspension was between 7.0-7.2. In India, Subramanian (1998) incorporated glycerin in a suspension containing *S. litura* NPV. He found that after six months storage, the formulated NPV had 1.42-1.46 times lesser bacterial load than the unformulated virus and the pH remained at neutral condition.

Stability of baculoviruses in different types of oils has been evaluated by some workers (Cherry *et al.,* 1994; Thennarasan, 1997; Lipa, 1998; Subramanian, 1998). Thennarasan (1997) reported decrease of virulence in the oil based formulations of *Ha*NPV due to storage and development of rancidity. Similarly, Cherry *et al.* (1994) found that storage of *S. littoralis* NPV at 4°C for 18 months in groundnut oil-based formulation decreased 7 per cent of viral activity.

Behle *et al.* (2003) studied the storage stability of *Anagrapha falcifera* MNPV in spray-dried lignin-based formulations. They demonstrated that the lignin-based *Af*MNPV formulation had a shelf life of up to 3 months at 30°C and up to 30 months at 4°C and with longer residual insecticidal activity in the field compared with unformulated or a glycerin-based formulation.

13.6.1.1.3. Microencapsulations

Sher (1977) listed 22 commonly used encapsulating materials, including gelatin, starch, cellulose and several types of polymers (Jones and Burges, 1998). A thin film of the mentioned materials is deposited around solid core to produce microcapsules. The microcapsules give good protection from environmental factors, such as sunlight and leaf-surface chemicals. Dyes can be incorporated into capsule walls to increase UV protection. Also stickers and wetters can be adsorbed to the capsule surfaces to improve retention on the target (Burges, 1998).

Helicoverpa NPV has been developed as microencapsulated product using ethyl cellulose gelatin polymer (Ignoffo and Batzer, 1971); styrene maleic anhydride half ester (SMA-2625A) (Bull *et al.*1976) and starch (Dunkle and Shasha, 1988; Ignoffo *et al.*, 1991; Moguire, 1992). The size of the capsules in these preparations ranged from 10 μm to 106 μm.

13.6.2. Tank Mix Formulations

These formulations normally added just before field application by the end user and the final formulation applied is termed a tank mix. Wetting agents, stickers, thickeners, humectants, botanical additives, protectants and enzymes are certain additives for application (Muthuswami, 2001). The first *Helicoverpa* NPV formulation was prepared by Montoya *et al.* (1966) and Allen and Pate (1966) by combining water extracts of corn and corn silks respectively with NPV and found it effective in laboratory and field tests. Similar results were achieved with tank mix formulations of *Ha*NPV incorporated with different kinds of adjuvants (Rabindra and Jayaraj, 1992; Muthuswami *et al.*, 1994).

13.7. SECONDARY CONTAMINATIONS AND QUALITY CONTROL

The bacterial content arising from dead larvae directly affects in NPV mass production and considered as major problem of production by the producers. This contamination must be in a determined range deposited by registration requirements. In some countries, present of 10^7 colony-forming units (c.f.u.)/g is considered as the limitation range in case of no primary human pathogenic species. The insect debris and secondary microbial contaminants can be separated from the viral suspension by techniques such as differential gradient centrifugation (Harrap *et al.*, 1977; Rabindra and Jayaraj, 1986); however, this increases costs by 4 times (McKinely *et al.*, 1989). Control of the quality and improving NPV production should mainly encompass the suppression of spore-forming bacteria. Use of antibiotics in larval food, good hygiene and stabilization of larval cadavers as soon as possible after maximum virus production or death of larvae, for example by deep-freezing are the techniques considered. Bacterial invasion can be minimized by harvesting virus-infected larvae just before death (McKinely *et al.*, 1989). Most workers have relied on the collection of cadavers and subsequent freezing to delay the bacterial build up (Chauthani and Claussen, 1968; Bell *et al.,* 1980a). Bacteria might be curbed by low-temperature incubation of large larvae, so that virus replication and cell lysis continue with much less bacterial growth (McKinely *et al.,* 1989). Vegetative bacteria can be killed by manipulation of subsequent spray-drying (Huber, 1986).

However, use of purified inoculants and harvesting of live infected larvae suggested as the possible methods to keep the bacterial load within acceptable limits (McKinely *et al.,* 1989). Purification is another area that can provide unlimited storage life to the product. Viruses purified by density gradient centrifugation (Harrap *et al.,* 1977; Dobrata and Hinton, 1992) can be maintained under room temperature for several years without degradation. Nevertheless, use of the method is restricted by the cost factor (Burges and Jones, 1997).

The pH of a product must be stabilized within certain ranges. Very high and low pH conditions will normally inactivate agents (Griffiths, 1982; Salama and Morris, 1993). A buffer may therefore be required, and certain additives with extreme pH values must be avoided. A useful method of minimizing growth of contaminants in a product is to maintain the pH of the suspension at a value outside the optima for growth of contaminants, but not inhibitory of infection, growth or replication of the biocontrol organism after application. To improve the quality of the product, it may also be necessary to include an additive that restricts the growth of contaminants (Jones and Burges, 1997).

13.8. NPV MASS PRODUCTION UNDER A SIMPLE AND ROUTINE PRACTICE

To produce an *in vivo* viral product, presence of a healthy host insect culture preferably in a facility away from virus production units required. So, the following trend shall stepwise follow to propagate a HearNPV product.

A. Provide the virus preparation as the preliminary viral stock.
B. Prepare a hygiene semi-synthetic diet based on a standard protocol and put in 5 ml glass vials (or in multicavity trays if available) up to the half of their heights.
C. Inoculate the diet surface with 10 µl of a viral suspension of 1×10^7 OB/ml using a micropipette.
D. Spread the inoculation liquid uniformly by the blunt end of a 6 mm glass rod, and then allow the surface to air-dry well.
E. Transfer early fifth instar larvae of *H. armigera* individually and then plug the vials with cotton swabs.
F. Incubate the vials at $25 \pm 1°C$ in incubator or an air-conditioned room.
G. Collect the cadavers after 5 days of inoculation as the death occurred.
H. Homogenize the collected dead larvae using sterilled pestle and mortar (or by an electric grinder if available) and dilute the concentrate with 9 parts of distilled water.
I. Filter the homogenate through a sterile, double-layered muslin cloth, and then repeat the filtration to receive to recover the maximum extraction of the polyhedra (crude extract of NPV).

J. Spin the obtained NPV slurry using a centrifuge 3 minutes at 600 rpm to remove larval debris and large particles.

K. Centrifuge the supernatant for 20 minutes at 5000 rpm to pellet viral polyhedra.

L. Discard supernatant and resuspend the pellet using least volume of distilled water and wash twice for obtaining semi-purified inoculum.

M. Enumerate OBs of the inoculum using a double-ruled improved Neubauer haemocytometer.

N. Maintain the suspension at a refrigerator (3°C and preferably -20°C).

CONCLUSION

Microorganisms' formulations used against insect pests such as entomopathogenic viruses are based on living organism and are particulate in nature. The viability of these organisms will have to be maintained at acceptance levels during formulation processes as well as storage conditions and finally application by the end users. An insect pathogenic virus product is supposed to supply in a reasonable market cost in advance. Optimal production of the entomopathogenic viruses is one that results in the greatest yield of biologically active virus and that confirms to the quality control standards. Under large-scale production of nuclear polyhedrosis viruses there must be a large-scale culture of their insect hosts. Maintenance of such a host culture requires availability of natural plant materials, which constraints the production of viral biopesticides, due to their season dependent availability. To overcome these difficulties artificial/semi-synthetic diets are used without any dependency to the host and season in production process, resulting optimization of the virus product under more controlled environment. Hygiene and sanitation of virus production and product along with some other factors such as availability of required materials and their costs, quality control, and skilled labors engaged in the production program are of major factors which administrate production of viral biopesticides. As a matter of fact, any factor that influencing the larval growth in post-inoculation period would be considered as a virus production element which can come under the process considerations. Quite a lot of major parameters, such as larval age and weight, inoculation dose, incubation temperature, and time of harvesting the larvae, on the production of nucleopolyhedrovirus (NPV) were evaluated by several authors. Many other factors are also engaged in virus production such as virus purification, enumeration or virus quantity and quality evaluations, and finally production of virus formulations. Formulation comprises aids to preserving organisms, to delivering them to their targets and to improving their activities. There are a wide variety of formulation types, both liquid and solid. The main types currently used for organisms have been classified into dry products (dusts, granules and briquettes) and suspensions (oil-based or water-based and emulsions). The major functions of the formulations are listed out as stabilization, handling and application, environmental persistence, and improvement of action.

REFERENCES

Allen, G.E. and Pate, T.L. 1966. The potential role of a feeding stimulus in connection with the nuclear polyhedrosis virus of *Heliothis*. *Journal of Invertebrate Pathology*, 8: 129-131.

Andrews, G.L. and Sikorowski, P.P. 1973. Effects of cotton leaf surface on the nuclear polyhedrosis virus of *Heliothis zea* and *Heliothis virescens* (Lep., Noctuidae). *Journal of Invertebrate Pathology*, 22: 290-291.

Angus, T.A. and Luthy, P. 1971. Formulation of microbial insecticides. In: Burges, H.D. and Hussey, N.W. (Eds.), *Microbial control of insects and mites*. Academic Press, London, pp. 623-638.

Arthures, S.P., Lacey, L.A. and Fritts, R. 2005. Optimizing use of codling moth granulovirus: effects of application rate and spraying frequency on control of codling moth larvae in Pacific Northwest apple orchards. *Journal of Economic Entomology*, 98: 1459-1468.

Becher, P. 1973. The emulsifier. In: van Walkenburg, W. (Ed.). *Pesticides formulations*. Marcel Dekker, New York, pp. 65-92.

Behle, R.W., Tamez-Guerra, P. and McGurie, M.R. 2003. Field activity and storage stability of *Anagrapha falcifera* NPV in spray-dried lignin-based formulations. *Journal of Economic Entomology*, 96: 1066-1075.

Bell, M.R. 1991. *In vivo* production of a nuclear polyhedrosis virus utilizing a tobacco budworm multicellular larval rearing container. *Journal of Entomological Sciences*, 26: 69-75.

Bell, R.A., Owens, C.D., Shapiro, M. and Tardiff, T.R. 1980a. Effectiveness of nuclear polyhedrosis virus. *Journal of Economic Entomology,* 71: 350-352.

Bell, R.A., Owens, C.D., Shapiro, M. and Tardiff, T.R. 1980b. Effectiveness of the nuclear polyhedrosis virus. In: Doane, C.C. and McManus, M.L. (Eds.), *The gypsy moth research, research towards integrated pest management*. Forest Service, USDA, Washington DC., pp. 599-655.

Benz, G. 1981. Use of viruses for insect suppression. In: Papavizas, G.C. (Ed.), *Biological Control in Crop Production*. Maryland, United States, pp. 259-272.

Boucias, D.G., Johnson, D. and Allen, G.E. 1980. Effect of host age, virus dosage and temperature on the infectivity of a nuclear polyhedrosis virus against velvet bean caterpillar, *Anticarcia gemmatalis* larvae. *Environmental Entomology,* 9: 59-61.

Bull, D.L., Ridgway, R.L., House, V.S. and Pryor, N.D. 1976. Improved formulations of the *Heliothis* nuclear polyhedrosis virus. *Journal of Economic Entomology,* 69: 731-736.

Burges, H.D. 1981. Progress in the microbial control of pests. In: Burges, H.D. (Ed.), *Microbial control of pests and plant diseases* (1970-1980). Academic Press, London, pp. 1-6.

Burges, H.D. 1998. *Formulation of microbial biopesticides*. Kluwer Academic Publishers, the Netherlands, 412 pp.

Burges, H.D. and Jones, K.A. 1997. Formulation of bacteria, viruses and protozoa to control insects. In: Burges, H.D. (Ed.), *Formulation of Microbial biopesticides, Beneficial Microorganisms and Nematodes*, Chapman and Hall, London, 818 pp.

Chauthani, A.R. and Claussen, D. 1968. Rearing Douglas-fir tussock moth larvae on semi-synthetic media for the production of nuclear polyhedrosis virus. *Journal of Economic Entomology,* 61: 101-103.

Cherry, A.J., Parnell, M.A., Grywacz, D. and Jones, K.A. 1997. The optimization of *in vivo* nuclear polyhedrosis virus production of *Spodoptera exempta* and *S. exigua. Journal of Invertebrate Pathology*, 70: 50-58.

Cherry, A.J., Parnell, M.A., Smith, D. and Jones, K.A. 1994. Oil formulation of insect viruses. *IOBC Bulletin*, 17: 254-257.

Connick, W.J., Boyette, C.D. and McAlpine, J.R. 1991. Formulation of mycoherbicides using a pasta-like process. *Biological Control*, 1: 281-287.

Cook, S.P., Webb, R.E., Podgwaite, J.D. and Reardon, R.C. 2003. Increased mortality of gypsy moth *Lymantria dispar* (L.) (Lepidoptera: Lymantriidae) exposed to gypsy moth nuclear polyhedrosis virus in combination with the phenolic glycoside salicin. *Journal of Economic Entomology*, 96: 1662-1667.

Couch, T.L. and Ignoffo, C.M. 1981. Formulation of insect pathogens. In: Burges, H.D. (Ed.), *Microbial control of pests and plant diseases*. Academic Press, London, pp. 621-634.

Dhandapani, N., Jayaraj, S. and Rabindra, R.J. 1987. Efficacy of ULV application of nuclear polyhedrosis virus with certain adjuvants for the control of *Heliothis armigera* on cotton. *Journal of Biological Control,* 1: 111-117.

Dhandapani, N., Jayaraj, S. and Rabindra, R.J. 1993. Cannibalism on nuclear polyhedrosis virus infected larvae by *Heliothis armigera* (Hbn.) and its effect on viral infection. *Insect Science and Application*, 14: 427-430.

Dobrata, M. and Hinton, R. 1992. Conditions for density gradient separations. In: Richwood, A. (Ed.), *Preparative centrifugation, a practical approach*. IRL Press, New York, pp. 77-142.

Dougherty, E.M., Narang, N., Leob, M., Lynn, D.E. and Shapiro, M. 2006. Fluorescent brightener inhibits apoptosis in baculovirus-infected gypsy moth larval midgut cells in vitro. *Biocontrol Science and Technology*, 16: 157-168.

Dunkle, R.L. and Shasha, B.S. 1988. Starch encapsulated *Bacillus thuringiensis*, a potential new method for increasing the environmental stability of entomopathogens. *Environmental Entomology*, 17: 120-126.

Easwaramoorthy, S. 1998. World survey, Indian subcontinent. In: Hunter-Fujita, F.R., Entwistle, P.F., Evans, H.F. and Crook, N.E. (Eds.), *Insect Viruses and Pest Management*. John Wiley, New York, pp. 232-243.

Ebora, R.V., Shepard, B.M. and Gadapan, E.P. 1990. Mass propagation and factors affecting virulence of nuclear polyhedrosis virus of *Spodoptera litura. Philippine Journal of Biotechnology*, 1: 138-148.

Entwistle, P.E. and Evans, H.F. 1985. Viral control. In: Kerkut, G.A. and Gilbert, L.I. (Eds.), *Comprehensive insect physiology, biochemistry and pharmacology*, 12: 347-412.

Evans, H.F. 1981. Quantitative assessment of the relationship between dosage and response of the nuclear polyhedrosis virus of *Mamestra brassicae. Journal of Invertebrate Pathology,* 37: 101-109.

Farrar, R.R., Shapiro, M. and Javaid, I. 2003. Photostabilized titanium dioxide and a fluorescent brightener as adjuvants for a nucleopolyhedrovirus. *Biocontrol*, 48: 543-560.

Griffiths, I.P. 1982. A new approach to the problem of identifying baculoviruses. In: Kurstak, E. (Ed.), *Microbial and viral pesticides*, Marcel-Dekker, New York, pp. 527-583.

Grzywacz, D., Makinley, D.J., Jones, K.A. and Moawas, G. 1997. Microbial contamination in *Spodoptera littoralis* nuclear polyhedrosis virus produced in insects in Egypt. *Journal of Invertebrate Pathology,* 69: 151-156.

Harrap, K.A., Payne, C.C. and Robertson, J.S. 1977. The properties of three baculoviruses from closely related hosts. *Virology*, 79: 14-31.

Huber, J. 1986. *In vivo* production and standardization. In: Samson, A., Vlak, J.M. and Peters, D. (Eds.), *Fundamental and applied aspects of invertebrate pathology*, IV *International Colloquium on Invertebrate Pathology*, Wageningen, Society for Invertebrate Pathology, pp. 87-89.

Hunter-Fujita, F.R., Entwistle, P.F., Evans, H.F. and Crook, N.E. 1998. Insect viruses and pest management. John Wiley and Sons, New York. 620 pp.

Ignoffo, C.M. 1966. Standardization of products containing viruses. *Journal of Invertebrate Pathology*, 8: 547-548.

Ignoffo, C.M. 1968. Viruses, living insecticides. In: Maramrosh, K. (Ed.), *Current Topics in Micro biology and Immunology*, Springer Verlag, Berlin, 42: 129-167.

Ignoffo, C.M. 1973. Development of a viral insecticide, concept to commercialization. *Experimental Parasitology*, 33: 380-406.

Ignoffo, C.M. and Batzer, O.F. 1971. Microencapsulation and ultraviolet protectants to increase sunlight stability of an insect virus. *Journal of Economic Entomology*, 64: 850-853.

Ignoffo, C.M., Bradley, J.R., Gilliland, F.R., Harris, F.A., Falcon, L.A., Larson, L.V., McGarr, R.L., Sikorowski, P.P., Watson, T.F. and Yearian, W.C. 1972. Field studies on stability of *Heliothis* nuclear polyhedrosis virus at various site throughout the cotton belt. *Environmental Entomol*ogy, 2: 388-390.

Ignoffo, C.M. and Couch, C.L. 1981. The nuclear polyhedrosis virus of *Heliothis* spp. In: Burges, H.D. and Hussey, N.W. (Eds.), *Microbial Control of Pests and Plant Diseases*. Academic Press, New York, pp. 329-362.

Ignoffo, C.M. and Garcia, C. 1966. The relation of pH to the activity of inclusion bodies of a *Heliothis* nuclear polyhedrosis, *Journal of Invertebrate Pathology*, 8: 426-427.

Ignoffo, C.M., Hostetter, D.L., and Smith, D.B. 1976. Gustatory stimulant, sunlight protectant, evaporation retardant, three characteristics of a microbial insecticidal adjuvant. *Journal of Economic Entomology*, 69: 207-210.

Ignoffo, C.M. and Shapiro, M. 1978. Characteristics of baculovirus preparations processed from living and dead larvae. *Journal of Economic Entomology*, 71: 186-188.

Ignoffo, C.M., Shasha, B.S. and Shapiro, M. 1991. Sunlight ultraviolet protection of the *Heliothis* nuclear polyhedrosis virus through starch-encapsulation technology. *Journal of Invertebrate Pathology*, 57: 134-136.

Jacques, R.P. 1985. Stability of insect viruses in the environment. In: Maramorosch, K. and Sherman, K.E. (Eds.), *Viral Insecticides for Biological Control*. Academic Press, Orlando, Florida, pp. 285-360.

Jayaraj, S. 1985. Biological suppression of insect pests through the use of pathogens. In: Jayaraj, S. (Ed.), *Integrated Pest and Disease Management*. Tamil Nadu Agricultural University, Coimbatore, India, pp. 319-336.

Jayaraj, S. 1998. Virus biopesticides and integrated pest management for sustainable agricultural development. *Proceedings of National Symposium on Development of Microbial Pesticides and Insect Pest Management*, Pune, India. pp. 3-4.

Jayaraj, S., Anathakrishnan, T.N. and Veeresh, G.K. 1994. *Biological control in India, progress and perspectives*. Project No.2, Rajiv Gandhi Institute for Contemporarily Studies, New Delhi, India, 101 pp.

Jayaraj, S., Rabindra, R.J. and Narayanan, K. 1989. Development and use of microbial agents for control of *Heliothis* spp. (Lepidoptera: Noctuidae). In: King, K.G. and Jackson, V. (Eds.), *Proceedings of International Workshop on Biological Control of Heliothis, increasing the effectiveness of natural enemies*. New Delhi, India. Pp. 483-504.

Jayaraj, S., Sathiah, N. and Sundara Babu, P.C. 1992. Biopesticides in India, present and future status. In: Vasantharaj David, B. (Ed.), *Pest Management and Pesticides, Indian Scenario*. Namrutha publications, Madras, India, pp. 144-156.

Jones, K.A. and Burges, H.D. 1997. Product stability from experimental preparation to commercial reality. In: Evans, H.F. (Ed.), *Microbial insecticides, novelty or necessity?* Symposium Proceedings No. 68, British Crop Protection Council, Farnham. pp. 163-171.

Jones, K.A. and Burges, H.D. 1998. Technology of formulation and application. In: Burges, H.D. (Ed.), *Formulation of microbial biopesticides*. Kluwer Academic Publishers, the Netherlands, pp. 7-30.

Jones, K.A., Ketunuti, U. and Grzywacz, D. 1994. Production and use of NPV to control *Heliothis armigera* and *Spodoptera litura* in Thailand (Abstract). *Proceedings of Sixth International Colloquium on Invertebrate Pathology and Microbial Control*, Montpellier, France. pp. 177.

Kalia, V., Chaudhari, S. and Gujar, G.T. 2001. Optimization of production of nuclear polyhedrosis virus of *Helicoverpa armigera* through larval stages. *Phytoparasitica*, 29: 1-6.

Kelly, P.M. and Entwistle, P.F. 1988. *In vivo* mass production in cabbage moth (*Mamestra brassica*) of a heterologous (*Panolis*) and a homologous (*Mamestra*) nuclear polyhedrosis virus. *Journal of Virological Methods*, 25: 93-100.

Kelly, P.M., Speight, M.R. and Entwistle, P.F. 1989. Mass production a purification of *Euproctis chrysorrhoea* (L.) nuclear polyhedrosis virus. *Journal of Virological Methods*, 19: 249-256.

King, L.A., Posse, R.D., Hugues, D.S., Atkinson, A.E., Plamer, C.P., Marlong, S., Pickering, J.M., Joyce, K.A., Lawrie, A.M., Miller, D.P. and Beadle, D.J. 1994. Advances in insect virology. *Advances in Insect Physiol*ogy, 25: 1-73.

Kumar, C.M. and Rabindra, R.J. 2003. Influence of dietary vegetable oils on the tobacco cutworm, *Spodoptera litura* (Fabricius) and its nuclear polyhedrosis virus production. *Journal of Biological Control*, 17: 57-61.

Lipa, J.J. 1998. World Survey, Eastern Europe and former Soviet Union. In: Hunter-Fujita, F.R., Entwistle, P.F., Evans, H.F. and Crook, N.E. (Eds.), *Insect Viruses and Pest Management*. John Wiley, New York, pp. 216-231.

Lisansky, S.G., Quinlan, R.J. and Tassoni, G. 1993. *The Bacillus thuringiensis production handbook*. CPL Press, Newbury, pp. 18-35.

Martignoni, M.E. 1978. Viruses in biological control, production, activity and safety. In: Brooks, M.J.H., Stark, R.W. and Campbell, R.W. (Eds.), *The Douglas-fir Tussock Moth, a Synthesis*. USDA, Washington, DC., pp. 140-147.

Martignoni, M.E. and Iwai, P.J. 1981. A catalogue of viral diseases of insects, mites and ticks. In: Burges, H.D. (Ed.), *Microbial control of pests and plant diseases*. Academic Press, London, pp. 897-911.

Martignoni, M.E. and Iwai, P.J. 1986. *A catalogue of viral diseases of insects, mites, and ticks*. USDA For. Serv. Pac. Northwest Res. Stn. Gen. Tech. Rep. PNW-195.

Martinez, A.M., Caballero, P., Villanueva, M., Miralles, N., Martin, I.S., Lopez, E. and Williams, T. 2004. Formulation with an optical brightener does not increase probability of development resistance to *Spodoptera frugiperda* nucleopolyhedrovirus in the laboratory. *Journal of Economic Entomology*, 97: 1202-1208.

McKinely, D.J., Moawad, G., Jones, K., Grzywacz, D. and Turner, T. 1989. The development of nuclear polyhedrosis for the control of *Spodoptera littoralis* (Boisd.) in cotton. In: Green, M.B. and Lyon, D.J. (Eds.), *Pest Management in Cotton*. Chichester, Ellis Horwood, pp. 93-100.

Mehrvar, A., Rabindra, R.J., Veenakumari, K. and Narabenchi, G.B. 2007a. Optimization of yield productivity in seven geographic isolates of nucleopolyhedrovirus of *Helicoverpa armigera* (Hübner) (Lepidoptera: Noctuidae) using plant origin oils. *National conference on organic waste utilization and eco-friendly technologies for crop protection*, Plant Protection Association of India, March 15-17, Hyderabad, India. pp. 205-207.

Mehrvar, A., Rabindra, R.J., Veenakumari, K. and Narabenchi, G.B. 2007b. Standardization of mass production of three isolates of nucleopolyhedrovirus of *Helicoverpa armigera* (Hübner). *Pakistan Journal of Biological Sciences*, 10(22): 3992-3999.

Mehrvar, A., Rabindra, R.J., Veenakumari, K. and Narabenchi, G.B. 2008a. Evaluation of adjuvants for increased efficacy of HearNPV against *Helicoverpa armigera* (Hübner) using suntest machine. *Journal of Biological Sciences*, 8 (3): 534-541.

Mehrvar, A., Rabindra, R.J., Veenakumari, K. and Narabenchi, G.B. 2008b. Molecular and biological characteristics of some geographic isolates of nucleopolyhedrovirus of *Helicoverpa armigera* (Hübner) (Lepidoptera: Noctuidae). *Journal of Entomological Society of Iran*, 28 (1): 39-60.

Moguire, M.R. 1992. Starch encapsulation of microbial pesticides (Abstract). *Twenty fifth Annual Meeting of Society of Invertebrate Pathology*, Heidelberg, Germany, pp 263.

Montoya, E.L., Ignoffo, C.M. and McGarr, R.L. 1966. A feeding stimulant to increase effectiveness of and a field test with a nuclear polyhedrosis virus of *Heliothis*. *Journal of Invertebrate Pathology*, 8: 320-324.

Morris, O.N. 1971. The effect of sunlight, ultraviolet and gamma radiations and temperature on the infectivity of nuclear polyhedrosis virus. *Journal of Invertebrate Pathology*, 18: 292-294.

Moscardi, F. 1989. Use of viruses for pest control in Brazil: the case of the nuclear polyhedrosis virus of the soybean caterpillar, *Anticarsia gemmatalis*. Mem. Inst. Oswaldo Cruz, Rio de Janeiro. 84: 51-56.

Moscardi, F. 1999. Assessment of the application of baculovirses for control of lepidoptera. *Annual Review of Entomology*, 44: 257-289.

Moscardi, F., Leite, L.G. and Zamataro, C.E. 1997. Production of nuclear polyhedrosis virus of *Anticarsia gemmatalis* Hübner (Lepidoptera: Noctuidae): effect of virus dosage, host density and age. *Anais da Sociedade Entomolgica do Brasil*, 26: 121-132.

Murillo, R., Lasa, R., Goulson, D., Williams, T., Munoz, D. and Caballero, P. 2003. Effects of Tinopal LPW on the insecticidal properties and genetic stability of the nucleopolyhedrovirus of *Spodoptera exigua* (Lepidoptera: Noctuidae). *Journal of Economic Entomology*, 96: 1668-1674.

Muthuswami, M. 2001. Adjuvants for nuclear polyhedrosis viruses. In: Rabindra, R.J., Kennedy, J.S., Sathiah, N., Rajasekaran, B. and Srinivasan, M.R. (Eds.), *Microbial control of crop pests*, Tamil Nadu Agricultural University, Coimbatore, India, pp. 275-282.

Muthuswami, M., Rabindra, R.J. and Jayaraj, S. 1994. Evaluation of certain adjuvants as phagostimulants and UV protectants of nuclear polyhedrosis virus of *Helicoverpa armigera* (Hbn.). *Journal of Biological Control*, 8: 27-33.

Narayanan, K. 1979. Studies on the nuclear polyhedrosis virus of the gram pod borer, *Heliothis armigera* (Hübner) (Noctuidae: Lepidoptera). Tamil Nadu Agricultural University, Coimbatore, India.

Podgwaite, J.D. 1985. Strategies for field use of baculoviruses. In: Maramorosch, K. and Sherman, K.E. (Eds.), *Viral Insecticides for Biological Control*. Academic Press, New York, pp. 775-797.

Podgwaite, M.D., Bruen, R.B. and Shapiro, M. 1983. Microorganisms associated with production lots of the nuclear polyhedrosis virus of the gypsy moth, *Lymantria dispar*. *Entomophaga*, 28: 9-16.

Poinar, G.O. and Thomas, G.M. 1984. *Laboratory guides to insect pathogens and parasites*. Plenum Press, 392 pp.

Polon, J.A. 1973. Formulation of pesticide dusts, wettable powders and granules. In: van Walkenburg, W. (Ed.), *Pesticides formulations*. Marcel Dekker, New York, pp. 143-234.

Rabindra, R.J. 1993. Virus harvest in *Heliothis armigera*. *National Training on Mass multiplication of Biocontrol agents*, Tamil Nadu Agricultural University, Coimbatore, India, pp. 75.

Rabindra, R.J., Geetha, N., Grzywacz, D. and Brown, M. 2003. Comparative efficacy of two isolates of nuclear polyhedrosis virus against *Helicoverpa armigera* (Hbn.). In: Rananavare, H.D., Naik, S.R. and Dongre, T.K. (Eds.), *Microbial pesticides and insect pest management*, pp. 127-131.

Rabindra, R.J. and Jayaraj, S. 1986. Multiplication and use of nuclear polyhedrosis viruses for the control of important crop pests. In: Jayaraj, S. (Ed.), *Pests and Disease Management in Oilseeds, Pulses, Millets and Cotton*. Tamil Nadu Agricultural University, Coimbatore, India, pp. 51-61.

Rabindra, R.J. and Jayaraj, S. 1988. Efficacy of nuclear polyhedrosis virus with adjuvants as high volume and ultra low volume applications against *Heliothis armigera* (Hbn.) on chickpea. *Tropical Pest Management*, 34: 441-444.

Rabindra, R.J. and Jayaraj, S. 1990. Microbial control of *Heliothis armigera*. In: Jayaraj, S., Uthamasamy, S. and Gopalan, M. (Eds.), *Heliothis Management. Proceedings of National Workshop*, Tamil Nadu Agricultural University, Coimbatore, India. pp. 154-164.

Rabindra, R.J. and Jayaraj, S. 1992. Efficacy of extracts of certain host plants as adjuvants for NPV of *Helicoverpa armigera* (Hbn.) and its dust formulation. *Journal of Biological Control*, 6: 80-83.

Rabindra, R.J. and Jayaraj, S. 1995. Management of *Helicoverpa armigera* with nuclear polyhedrosis virus on cotton using different spray equipment and adjuvants. *Journal of Biological Control*, 9: 34-36.

Rhodes, D.J. 1993. Formulation of biological control agents. In: Jones, D.G. (Ed.), *Exploitation of microorganisms*. Chapman and Hall, London, pp. 411-439.

Ribeiro, H.C.T. and Pavan, O.H.O. 1994. Effect of temperature on the development of baculoviruses. *Journal of Applied Entomology*, 118: 316-320.

Salama, H.S. and Morris, O.N. 1993. The use of *Bacillus thuringiensis* in developing countries. In; Entwistle, P.F., Cory, J.S., Bailey, M.J. and Higgs, S. (Eds.), *Bacillus thuringiensis, an environmental biopesticide, theory and practice*. John Wiley, Chichester, pp. 237-253.

Santharam, G. 1986. Studies on the nuclear polyhedrosis virus of tobacco cut worm *Spodoptera litura* (Fabricius) (Lepidoptera: Noctuidae). Tamil Nadu Agricultural University, Coimbatore, India.

Seaman, D. 1990. Trends in the formulation of pesticides, an overview. *Pesticides Science*, 29: 437-449.

Shapiro, M. 1982. *In vivo* mass production of insect viruses. In: Kurstak, E. (Ed.), *Microbial and viral Pesticides*. Marcel Dekker, New York, pp. 463-492.

Shapiro, M. 1986. *In vivo* production of baculoviruses. In: Granados, R.R. and Federici, B.F. (Eds.), *The Biology of Baculoviruses,* Vol. II, Boca Raton, CRC Press, F.L., Lewis, pp. 11-61.

Shapiro, M. 2001. The effects of cations on the activity of the gypsy moth (Lep.: Lymantriidae) nuclear polyhedrosis virus. *Journal of Economic Entomology*, 94: 1-6.

Shapiro, M. and Argauer, R. 2001. Relative effectiveness of selected stilbene optical brighteners as enhancers of the beet armyworm (Lepidoptera: Noctuidae) nuclear polyhedrosis virus. *Journal of Economic Entomology*, 94: 339-343.

Shapiro, M. and Bell, R.A. 1981. Biological activity of *Lymantria dispar* nucleopolyhedrosis virus from living and virus killed larvae. *Annals of Entomological Society of America*, 74: 27-28.

Shapiro, M. and Robertson, J.L. 1990. Laboratory evaluation of dyes as ultraviolet screens for the gypsy moth (Lepidoptera: Lymantridae) nuclear polyhedrosis virus. *Journal of Economic Entomology*, 83: 168-172.

Shapiro, M., Bell, R.A. and Owens, C.D. 1981. *In vivo* mass production of gypsy moth nuclear polyhedrosis virus. In: Doane, C.C. and McManus, M.L. (Eds.), *The Gypsy Moth Research, Research Towards Integrated Pest Management*. Forest Service, USDA, Washington DC., pp. 633-655.

Sher, H.B. 1977. Microcapsulated pesticides. In: Sher, H.B. (Ed.), *Controlled release pesticides*, American Chemical Society, Washington, D.C., pp. 126-144.

Shieh, T.R. 1978. Characteristics of a viral pesticide Elcar®. *Proceedings of International Colloquium on Invertebrate Pathology*. pp. 191-194.

Shieh, T.R. 1989. Industrial production of viral pesticides. *Advances in Virus Research*, 36: 315-343.

Shieh, T.R. and Bohmfalk, G.T. 1980. Production and efficacy of baculoviruses. *Biotechnology and Bioengineering*, 22: 1357-1375.

Shuler, M.L., Granados, R.R., Hammer, D.A. and Wood, H.A. 1995. Overview of baculovirus, insect cell system. In: Shuler, M.L., Granados, R.R., Hammer, D.A. and Wood, H.A., (Eds.), *Baculovirus expression systems and biopesticides*. John Wiley, New York, pp. 1-11.

Sjogren, R.D. 1996. *Insecticidal composite timed released particle*. US Patent 5484600.

Smith, P.H. and Vlak, J.M. 1988. Quantitative and qualitative aspects of the production of a nuclear polyhedrosis virus in *Spodoptera exigua* larvae. *Annals of Applied Biology*, 112: 249-257.

Srinivasan, G., Sundara Babu, P.C., Sathiah, N. and Balasubramanian, G. 1994. Standardization of diet for mass culturing of *Helicoverpa armigera* (Abstract). In: Verma, A.K., Bhardwaj, S.P., Sheiker, C. and Sood, A. (Eds.), *Proceedings of National Symposium on Emerging Trends in Pest Management*, Solan, India. pp. 11.

Stairs, G.R. 1971. Use of viruses for microbial control of insects. In: Burges, H.D. and Hussey, N.W. (Eds.), *Microbial Control of Insects and Mites*. Academic Press, London, pp. 97-124.

Stairs, G.R. 1972. Pathogenic microorganisms for the regulation of forest insect populations. *Journal of Invertebrate Pathology,* 17: 355-372.

Stairs, G.R. 1978. Effect of wide range of temperature on the development of *Galleria mellonella* to its specific baculovirus. *Environmental Entomology,* 7: 297-299.

Subramanian, S. 1998. Studies on the nuclear polyhedrosis virus of *Spodoptera litura* (Fabricius). Tamil Nadu Agricultural University, Coimbatore, India.

Subramanian, S., Santharam, G. and Rabindra, R.J. 2001. Optimization of stage and dose of virus inoculum for maximizing the *Spodoptera litura* (Fab.) NPV yield. In: Singh, D., Dilawari, V.K., Mahal, M.S., BRAR, K.S., Sohi, A.S. and Singh S.P. (Eds.), *Proceedings for quality crop protection in the current millennium*. Punjab Agricultural University, Ludhiana, India. pp. 79-80.

Subramanian, S., Santharam, G., Sathiah, N. and Rabindra, R.J. 2006. Influence of incubation temperature on productivity and quality of *Spodoptera litura* nucleopolyhedrovirus. *Biological Control*, 37: 367-374.

Tamez-Guerra, P., McGuire, M.R., Behle, R.W., Sumner, H.R., and Shush, B.S. 2000. Sunlight persistence and rainfastness of spray-dried formulations of baculovirus isolated from *Anagrapha falcifera* (Lepidoptera: Noctuidae). *Journal of Economic Entomology*, 93: 210-218.

Tanada, Y. and Kaya, H. 1993. *Insect pathology*. Academic Press, 666 pp.

Teakle, R.E. and Byrne, V.S. 1989. Nuclear polyhedrosis virus production in *Heliothis armigera* infected at different larval ages. *Journal of Invertebrate Pathology*, 53: 21-24.

Theng, B.K.G. 1979. *Formulations and properties of clay polymer complexes*. Elsevier, New York.

Thennarasan, M. 1997. Studies on the development of oil formulations of nuclear polyhedrosis virus of *Helicoverpa armigera* (Hbn.). Tamil Nadu Agricultural University, Coimbatore, India.

Tuan, S.J., Tang, L.C. and Hou, R.F. 1989. Factors affecting the pathogenicity of NPV preparations to the corn earworm, *Heliothis armigera*. *Entomophaga*, 34: 541-549.

van Driesche, R.G. and Bellows, T.S. 1996. Biological control. Chapman and Hall. 539 pp.

Whitlock, V.H. 1977. Effect of larval maturation and mortality induced by nuclear polyhedrosis and granulosis infections in *Heliothis armigera*. *Journal of Invertebrate Pathology*, 32: 386-387.

Yearian, W.C. and Young, S.Y. 1982. Control of insect pests of agricultural importance by viral insecticides. In: Kurstak, E. (Ed.), *Microbial control and Viral Pesticides*, Marcel and Dekker, New York, pp. 387-723.

Young, S.Y. 1989. Problems associated with the production and use of viral pesticides. Mem. Inst. Oswaldo Cruz, Rio de Janeiro, 84: 67-73.

Young, S.Y. 2000. *Persistence of viruses in the environment*. Online www.agctr.lsu.edu/s265/young.htm.

Young, S.Y. and Yearian, W.C. 1986. Movement of a nuclear polyhedrosis virus from soil to soybean and transmission in *Anticarsia gemmatalis* (Hübner) (Lep., Noctuidae) populations on soybean. *Environmental Entomology*, 15: 573-580.

PART IV:
ENTOMOPATHOGENIC BACTERIA

In: Microbial Insecticides: Principles and Applications   ISBN: 978-1-61209-223-2
Editors: J. Francis Borgio, K. Sahayaraj, et al.   © 2011 Nova Science Publishers, Inc

Chapter 14

BIOASSAY PROCEDURES AND MASS PRODUCTION TECHNOLOGY OF ENTOMOPATHOGENIC BACTERIA

Ninfa María Rosas-García[*]

Laboratorio de Biotecnología Ambiental. Centro de Biotecnología Genómica- IPN. Blvd. del Maestro s/n Reynosa, Tamp. México CP 88710

ABSTRACT

This chapter comprises in a descriptive way the basic procedures needed to evaluate and produce the entomopathogenic bacterium *Bacillus thuringiensis* as it is the most popular biological control agent. The production of the entomopathogenic bacterium involves isolation, bioassays procedures, massive production and formulation process for finally being applied to the field. Bioassays are important to determine toxic activity and for choosing the best entomopathogenic strains toward different insect pests. The culture media, inoculums, and other conditions such as pH and temperature, are essential factors for mass production technology. Different bioassay procedures such as diet-based bioassays, disk bioassays, green house bioassays and field tests are included with their main characteristics and applicability. In the same way, the mass production in a fermentation process is described as well as formulation development for the field application.

14.1. INTRODUCTION

Insect pests are susceptible to a great variety of microorganisms that includes virus, bacteria, fungi and protozoans. Nowadays more than 100,000 microbial species are known, among them 750 fungal species, 700 virus, 300 protozoans and 100 bacterial species possess entomopathogenic properties (Galán-Wong *et al.,* 1996). Although some microorganisms are used to control agriculturally, forestry, and public health importance insect pests, the majority

---

[*] Email: nrosas@ipn.mx, ninfarosasg@yahoo.com.mx, Tel: 52-899-924-36267 Ext. 7721; Fax: 52-899-925-2889

of them are unknown. However a number of varieties of the bacterium *Bacillus thuringiensis* have been studied with more detail and its potential has been widely developed and exploited.

The discovery of *B. thuringiensis* as entomopathogenic bacterium led to the production of biological insecticides to control a great variety of insect pests, mainly from economically important crops. The production of these kinds of bioinsecticides requires standardized processes carefully managed, as living organisms should be mass produced, and their management require high quality control standards. Studies related to the search of new *B. thuringiensis* strains are associated to toxic activity measurement through well planned bioassays. The production of entomopathogenic bacteria has included the study and development of management and maintenance methods, a wide variety of culture media as well as scaling process for massive production. Up to date, fermentation technologies continue offering the best alternative for production of entomopathogenic microorganisms, since it is possible to produce a number of them in vitro at laboratory scale, and in this way to determine in most cases the nutritional requirements and physicochemical parameters necessary for large scale production. At any production level, maximum quality and highest yield are always seek, this means an excellent management and control of a series of fermentation parameters technically and economically important, that allow to define the most relevant unit operations in the whole process. These concepts involve complicated actions related to productivity, yielding, and costs frequently observed in fermentation processes mainly in scale up processes. To initiate a massive production process for entomopathogenic bacteria it is important to carry out a suitable selection of bacterial strains which should be produced. On a regular basis, bacterial strains are obtained from soil, and sometimes from dead or diseased insect larvae. Toxic activity should be tested through bioassays and once selected the best strains other parameters will become of fundamental concern such as to determine bacteria nutritional requirements, to develop suitable culture media and to establish optimum developing conditions to favor their propagation. The main point for an entomopathogenic bacterium to be massively produced is its toxic activity, and laboratory bioassays become essential in this determination. Although bioassays ultimately define the toxic capacity of a bacterial strain, the arrival of molecular techniques has help to predict toxic activity, reducing time-consuming bioassays. The search and finding of genes associated to toxicity allow in an easy way to select those strains with potential toxic activity.

14.2. INSECT BIOASSAYS

Bioassays are easy techniques to determine host range, relative activity, killing speed and some other factors that are important in the selection of strains and formulation of ingredients (McGuire *et al.*, 1997). Bioassays with B. *thuringiensis* differ from those with synthetic insecticides because the spore-crystal complex produced by this bacterium (Figure. 1) has to be ingested to be toxic to sensitive target species. In 1971 representatives of different industries agreed to adopt the methodology established by H.T. Dulmage as a standard bioassay to ensure consistency in tested products (Galán-Wong *et al.*, 1996, McGuire *et al.*, 1997). Bioassays should be planned according to desired activities to obtain broad information, for this reason is necessary to define exactly what is needed to be evaluated. Among the most important studies is evaluation of toxic activity of new isolates, determination of role of spores or toxins or their interactions, as well as testing of different ingredients to produce toxic formulations. It is important to define the criteria that will be considered to evaluate such activity, among them can be considered mortality, morbidity, growth reduction (McGuire *et al.,* 1997), and effects of environmental conditions.

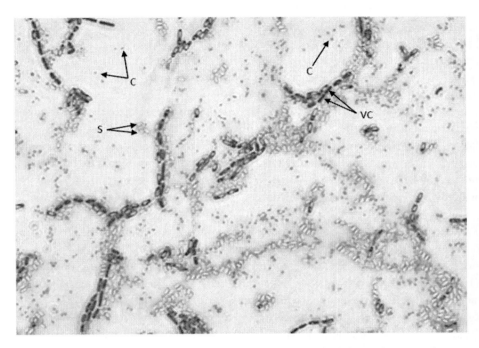

Figure 1. A native strain of Bacillus thuringiensis cultivated during 72 h in nutrient agar shows vegetative cells (VC), spores (S), and crystals (C). Provided by the author.

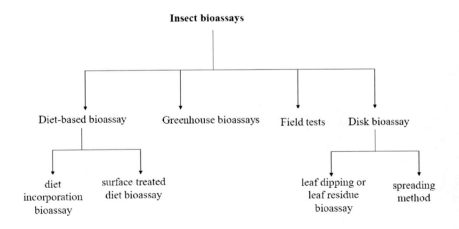

14.2.1. Diet-Based Bioassay

There are several types of bioassays, the diet-based bioassay can be divided in diet incorporation bioassay and surface treated diet bioassay. The choice will depend on the insect behavior. For example, an insect such as *Trichoplusia ni*, that feeds on diet surface, the surface treated diet bioassay will be preferable, otherwise if the insect bores the diet such as *Diatraea saccharalis*, a diet incorporation bioassay will be the most suitable. To conduct these bioassays an artificial diet should be prepared.

Frequently diet incorporation bioassays have been used to evaluate the toxicity of spore-crystal complexes against many lepidopterans such as *Heliothis virescens*, *Spodoptera exigua*, and *T. ni* (Tamez-Guerra *et al.*, 1998), and to test the biological activity of novel commercial preparations of *B. thuringiensis* in laboratory using colonies of *Plutella xylostella* (Asano *et al.*, 1993), as well as to study behavioral responses and sublethal effects on other lepidopterans (Johnson *et al.*, 1991) (Figure. 2).

The surface treated diet bioassay has been helpful to distinguish between a susceptible and a resistant field population of the diamondback moth (Ferré *et al.*, 1991); however Perez *et al.* (1997) recommend this kind of bioassay to evaluate resistance to *B. thuringiensis* in Indian meal moth and almond moth. On the other hand, this bioassay has been use to determine natural variation in *Choristoneura occidentalis* as an important fact that should be previously determine to avoid erroneous data to detect resistance and other biological changes (Robertson *et al.*, 1995).

For these bioassays, containers should contain a small amount of solidified diet, one larva should be placed per container, and a plastic cap or a cardboard lid must be use to avoid larvae escape and moisture retention. For lepidopterans an easy to prepare diet is shown in Table 1, this could be use for bioassay or rearing, some modifications on ingredients could be made according to insect nutritional requirements.

For example, in the case of *D. saccharalis* diet is prepared adding more sucrose because this lepidopteran feeds on sugarcane, and more agar is added to harden the diet for larvae easily bore it (Rosas-García, 2002).For a complete bioassay, ordinarily three replicates are done per dilution to be tested. 75 cups with diet and one larva per cup are prepared with the corresponding dilution. 25 cups are placed inside paper bags and incubated at 25-30°C and 60-70% RH. Dissolve agar in boiling water, and mix with all solid ingredients in a blender for 10 min, immediately after add liquid ingredients and mix for 5 min. Finally vitamin mix is added and blended for 2 min. Diet should be put in recipients immediately.

Figure 2. Diet incorporation bioassays. A) Small cups with plastic lid for larvae that bore or may escape. B) Small cups with cardboard lid for less active larvae or to avoid moisture excess. Provided by the author.

Table 1. Ingredients to prepare artificial diet for bioassays of lepidopterans

Ingredients	Amount (g/l)
Soy flour	71.1
Wheat germ	31.7
Wesson salt	10.6
Sucrose	13.0
Sorbic acid	1.0
Methylparaben	1.6
Agar	14.0
Ascorbic acid	4.3
Acetic acid (25%)	12.0
Formaline (10%)*	4.4
Choline chloride[+]	7.3
Vanderzant vitamin mix	3.5

*Prepare 37% formaldehyde and dilute the necessary volume to prepare a 10% solution.
[+] Dilute to a total volume of 1 liter by adding 150 ml of choline chloride

14.2.2. Disk Bioassay

The disk bioassays include two different methods; one of them is the spreading method in which an area of the leaf is marked with a permanent marker and then a carefully measured amount of solution is dropped on the leaf surface. The drop is spread across the leaf surface with a glass rod and allowed to dry. This method effectively controls the concentration of the active ingredient and can be used to LC_{50} determinations (McGuire *et al.,* 1997).

The other method is leaf dipping or leaf residue bioassay. In this method the complete leaf is immersed in a suspension of the spore-crystal complex and then allowed to dry. In this technique all surfaces of the leaf are covered, but the amount of tested material could be quite different in the leaf surface. This is a common procedure for assessing natural variation in *Leptinotarsa decemlineata* and *P. xylostella* (Robertson *et al.,* 1995) and resistance in *P. xilostella* to commercial formulations of *B. thuringiensis* (Tabashnik *et al.,* 1990; Shelton *et al.,* 1993), and spores and crystals (Tang *et al.,* 1996). The same procedure was used to determine efficacy of *B. thuringiensis* strain 4D21 against 7 day old castor semilooper larvae. *B. thuringiensis* spray solution was sprayed at 0.1 and 0.2% concentrations on castor leaves whose stalks were dipped in wet cotton. After the solution dried up, the larvae were released on the sprayed leaves. Larval mortality was recorded every day up to 5 days after spraying (Devi *et al.,* 2005).

LC_{50} determination requires experimentation with high and low concentrations of the spore-crystal complex. These initial range-finding concentrations are usually 50 and 500 µg/ml of the spore-crystal complex to determine the relative activity of each strain. Several concentrations based on preliminary results are established, and a range of concentrations from 7 to 10 serial dilutions of test material should be prepared. A control should always be included for a correct data interpretation. In 1972 a reference standard HD-1-S-1971 with a potency of 18000 IU/mg with a higher potency than the international standard E61 was established. Due to the scarcity HD-1-S-1971, ten years later a new standard H-1-S-1980 was adopted with 16000 IU (Galán-Wong *et al.,* 1996).

Particularly for *B. thuringiensis* var. *israelensis*, the international standard for all preparations is IPS-82. In this case the potency is measured in international toxic units (ITU/mg) against fourth instar larvae of *A. aegypti* and it is obtained by using the following equation:

$$\text{Product potency} = \frac{\text{LC}_{50}(\text{IPS-82}) \times 15{,}000}{\text{LC}_{50} \text{ (test product)}}$$

Where LC_{50} = the necessary lethal concentration to cause 50% larval mortality after 24 h of exposure.

The potency of these strains may be assigned by comparison with the IPS82 international sample standard of *B. thuringiensis* H14 provided by the Pasteur Institute, Paris, France (Sarrafzadeh *et al.,* 2007). *Bacillus sphearicus* based products are comparatively assayed against a standard power called SPH88, although bioassay is conducted in a similar way to that of *B. thuringiensis* var. *israelensis*, the LC_{50} is recorded after 48 h of exposure. The potency of this standard has been determined by comparison with the first standard RB80 with a potency of 1000 ITU/mg (Maldonado-Blanco 2005). Values of 50% lethal concentration (LC_{50}) are determined by Probit analysis (Finney, 1962).

Several bioassays are conducted in order to test experimental formulations based on the spore-crystal complex. Generally experimental preparations are formulated as granules or as wettable powders. Bioassays to test these preparations are conducted in a different way. For example toxic activity from granules is tested placing the granule in a Petri dish and allowing larvae to feed for 24 h, survivor larvae are subsequently transfer to fresh artificial diet to continue feeding and observations of mortality are recorded during 4 days. This bioassay could suffer some variations as that mentioned by Tamez-Guerra *et al.,* (1998), where granules were blended with the artificial diet. Wettable powders are usually applied with sprayers, and these formulations could be applied on the surface of the diet or by mixing with diet.

14.2.3. Greenhouse Bioassays

Bioassays mentioned before are very useful to indicate the toxic activity of *B. thuringiensis* against some kind of pest, however they do not indicate if formulations will be acceptable by larvae, this means, if formulations will be easily eaten by the larvae in the environment where they are exposed to the plant which is their natural feeding substrate. The well known anti-feeding effect caused by *B. thuringiensis* hampers consumption of a lethal dose of this pathogen (Heimpel, 1967; Yendol *et al.,* 1975; Gillespie *et al.,* 1994). For this reason a palatable formulation in granular form should be develop for this purpose and feeding preference bioassays must be conducted. Several types of formulations are prepared using different ingredients to be tested. These are *B. thuringiensis*-free formulations in order to determine the best ingredients that will be accepted by the insect. The bioassays are performed by confronting pairs of granules (Bartelt *et al.,* 1990) (Figure. 3).

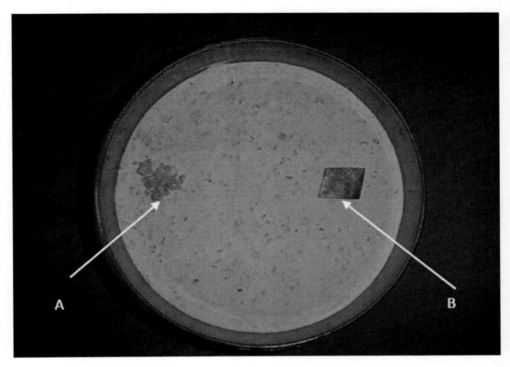

Figure 3. Feeding preference bioassay. Petri dish (5 cm diam.) with a bottom layer of plaster of Paris and activated charcoal (15:1). Granules to be tested are confronted by pairs and larvae are placed on the center and allow feed during 16 h. After that, Petri dishes are frozen at -20°C for 2-3 days and larvae are counted in each pile. A) granular formulation, B) fresh maize leaf (control). Provided by the author.

Once the best ingredients are selected a bioinsecticide is formulated, including *B. thuringiensis* at different concentrations. This is a very important step since only in green house bioassays the most effective *B. thuringiensis* concentration could be determined. So both parameters such as acceptance of the bioinsecticide and *B. thuringiensis* concentration are determined by this assay. It is a challenge to evaluate the effectiveness of the bioinsecticide since pest is infesting the plant, and bioinsecticide, in some way, should be more palatable than the plant or at least as palatable as the plant.

14.2.4. Field Tests

Finally the work done at laboratory and greenhouse should be transferred to the field, where multiple factors and environmental conditions cannot be controlled. Among them sunlight plays an important role since easily inactivates *B. thuringiensis* crystals, and rainfall also washes *B. thuringiensis* deposits off foliar surfaces (McGuire *et al.,* 1996), then the real efficacy of the bioinsecticide will be truly demonstrated in this last test. However field tests involve larger areas so as the amounts of bioinsecticide to be applied are increased, that it is why this process mainly requires of the massive production of *B. thuringiensis* and the recovery of the spore-crystal complex (Figure. 4) in an easy and cost-effective way. The following content deals with the massive production of *B. thuringiensis* as it is the most important biological control agent.

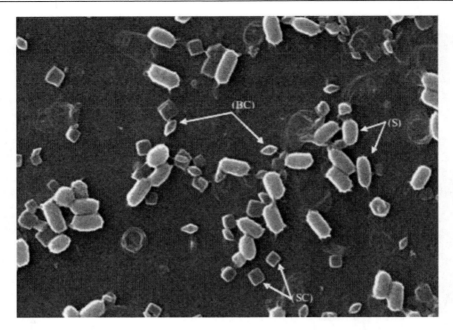

Figure 4. Microphotograph of a strain of *Bacillus thuringiensis* at scanning electron microscope (5000X) showing spores (S), bypiramidal crystals (BC), and square crystals (SC). Provided by the author.

14.3. MASSIVE PRODUCTION OF *BACILLUS THURINGIENSIS*

14.3.1. Strain Isolation and Stock Cultures Maintenance

Bacillus thuringiensis could be easily isolated from soil, which is its main habitat, but it could also be isolated from insect cadavers, water, leaves, etc. Generally soil samples for *B. thuringiensis* isolation are taken from locations where insect pest is present. One gram of a homogenous soil sample is sufficient to obtain isolates. Serial dilutions of soil samples must be done and submitted to heat shock to kill all vegetative cells. A loop of the dilution containing spores is stroke onto nutrient agar. Colony will develop at 25-28°C for 24 h. Characteristic colonies are selected and observed under a microscope to check crystals and spores. If the colony is contaminated should be purified through several passes in nutrient agar. Once each typical colony has been purified and checked for crystals and spores, the new isolate is stroke onto nutrient agar and kept in refrigeration until needed; if isolate will be kept longer a lyophilization process should be conducted.

14.3.2. Culture Media

Many raw materials to prepare culture media have been used, intended to reduce cost production and favor massive production of the spore-crystal complex. Table 2. shows different culture media used for growth and sporulation of *B. thuringiensis*.

Table 2. Media for growth and sporulation of _Bacillus thuringiensis_

Culture media	Ingredients	Amount	Author
GYS Glucose- yeast-salts	glucose	0.1%	Nakata and Halvorson, 1960
	yeast extract	0.2%	
	K_2HPO_4	0.5%	
	$(NH_4)_2SO_4$	0.2%	
	$MgSO_4$	0.02%	
	$MnSO_4.H_2O$	0.005%	
	$CaCl_2·2H_2O$	0.08%	
	$FeSO_4·7H_2O$	0.00005%	
	$CuSO_4·5H_2O$	0.0005%	
	$ZnSO_4·7H_2O$	0.0005%	
G-Tris	$CaCl_2$	0.08%	Aronson _et al._, 1971
	$FeSO_4$	0.0025%	
	$CuSO_4$	0.005%	
	$ZnSO_4$	0.005%	
	$MnSO_4$	0.05%	
	$MgSO_4$	0.2%	
	$(NH_4)_2SO_4$	2%	
	glucose	2%	
	1 M Tris (pH 7.5)	0.5%	
	yeast extract	0.15%	
	K_2HPO_4	0.5%	
2XSG	Nutrient broth	16.0 g	Leighton and Doi, 1971
	KCL	2.0 g	
	$MgSO_4·7H_2O$	17.0 g	
HCO	Casaminoacids	7.0 g/l	Lecadet _et al._, 1980
	KH_2PO_4		
	$MgSO_4·7H_2O$	6.8 g/l	
	$MnSO_4·4H_2O$	0.12 g/l	
	$ZnSO_4·7H_2O$	0.0022 g/l	
	$Fe_2(SO_4)_3$	0.014 g/l	
	$CaCl_2·4H_2O$	0.02 g/l	
	Glucose	0.18 g/l	
	pH adjusted to 7.2	3.0 g/l	

A good example is the GYS medium which has been modified for many researchers to improve growth and sporulation. Modifications can be done omitting or adding some ingredients (Nickerson and Bulla, 1974; Park _et al.,_ 1998; Silveira and Molina, 2005).

Leighton and Doi (1971) modified the Schaeffer's medium that was designated as 2XSG and used for _B. subtilis_, with good results for _B. thuringiensis_. The HCO medium is a semidefined medium for growth and sporulation in _B. thuringiensis_ Berliner and many other strains, a complete medium (BP medium) could be done substituting peptone for casaminoacids, and MA18 medium which is a minimal medium for growth and sporulation could be prepared substituting casaminoacids for a mixture of 18 aminoacids: alanine, arginine, aspartic acid, cysteine, glutamic acid, glycine, histidine, isoleucine, leucine, lysine, methionine, phenylalanine, proline, tyrosine, threonine, tryptophan, serine, and valine (Lecadet *et al.*, 1980). The culture media above mentioned are frequently used in laboratory for different puroposes, however if some strains will be produced in higher amounts in bioreactors, culture media are sometimes different as they must provide nutrients and growth conditions especially favorable for massive production. Many fermentation media are used successfully. A parameter that should be carefully considered is the carbon:nitrogen ratio (C:N) because a balance between the carbon and nitrogen sources are necessary to avoid pH values lower than 5.6 which affect the cell growth (Dulmage *et al.*, 1990). According to Farrera *et al.* (1998) the C:N ratio of 7:1 yielded the highest relative Cry protein production of *B. thuringiensis* HD-73 in a culture medium composed by soybean meal, glucose, yeast extract, corn steep solids and mineral salts. A fermentation media containing molasses as carbon source and soya meal as a nitrogen source supplemented with 5.0 g/l KH_2PO_4 and K_2HPO_4 achieved maximum yield of δ-endotoxin upon large scale application for industrial production of the strain S128 of *B. thuringiensis* H-14 (Abdel-Hameed, 2001). The use of residual glucose concentration (2.72 g/l) at a constant level in a fed-batch cultivation with glucose and yeast extract fed at equal concentration increased cell density (16.0 g/l) with specific growth rate of 0.69 h of *B. thuringiensis* var. *galleriae* (Anderson and Jayaraman, 2005). Berbert-Molina *et al.* (2008) considered that glucose is the most suitable carbohydrate for *B. thuringiensis* production and its inhibitory effects can be avoided by the use of fed-batch fermentation system, however in a batch system higher glucose concentration (above 75.0 g/l) resulted in growth inhibition for *B. thuringiensis* var. *israelensis*. Many low-cost raw materials and by-products have been used as carbon and nitrogen sources in culture media for *B. thuringiensis* production such as: maize glucose, soybean flour, peanuts, molasses, liquid swine manure, coconut water, hydrolyzed orange peel, agave juice, fish flour, steep corn liquor, dehydrated blood flour, yeast extract, groundnut cake powder, wheat, brand extract, (Medrano-Roldán *et al.*, 1992; Prabakaran and Balaraman, 2006; Valicente and Mourão, 2008). Although carbon and nitrogen sources play an essential role in *B. thuringiensis* production, many substrates have been tested in order to develop more efficient and cheaper culture media. Table 3. shows several substrates for *B. thuringiensis* production, and some of them, which have been found to be useful, are derived products from industries. An interesting case is that reported by Rojas-Avelizapa *et al.* in 1999. They determined the ability of *B. thuringiensis* to grow in milled shrimp waste as the sole ingredient; this finding may expand the diversity of raw materials for crystal production and may reduce production costs as substrate is a waste material.

Table 3. Substrates used to improve *Bacillus thuringiensis* production

Substrates	Reference
Chicken feather powder	Poopathi and Abidha, 2008
Potato, sugar, and bengalgram powder	Poopathi *et al.*, 2002
Amino nitrogen (vegetable peptones and brewer's yeast)	Prabakaran and Hoti, 2008
Casaminoacids	Sachidanandham *et al.*, 1997
Brewer's yeast extract	Saksinchai *et al.*, 2001.
Gruel and fish meal	Zouari and Jaoua, 1999; Ghribi *et al.*, 2007
Barely	Devi *et al.*, 2005.

14.3.3. Inoculum Production

The inoculum preparation is of fundamental concern for every fermentation process. Its importance is due to that cell should be synchronized to be added to fermentation media and obtaining a suitable final product. Generally the inoculum is prepared from a stock culture from where a loopful of cells is seeded in a special medium that sometimes is the same than the fermentation medium. Frequently the tryptose-phosphate broth, pH 7.0 is used to synchronize the culture. This broth activates the strain and propagates it during 18 h in a rotary shaker. This is a special medium to avoid sporulation, in this way all cells are in vegetative state ready to continue growing in fermentation media, in which it is important to avoid lag phase.

A good example is the production of a strain of *B. thuringiensis* var. *israelensis*. The process was conducted in batch fermentation, and the inoculum was based on the same fermentation medium containing byproducts such as molasses, steep corn liquor, soup paste powder and several mineral salts. (Maldonado-Blanco *et al.*, 2003). On the other hand, an inoculum of a strain of *B. thuringiensis* var. *kurstaki* was prepared on tryptose-phosphate broth, and batch fermentation was carried out using a molasses based medium, soybean flour, steep corn liquor and $CaCO_3$ (Rosas-García, 2002).

14.4. FERMENTATION PROCESS

There is a number of fermentation processes developed under different conditions; most of them are conducted to obtain higher amounts of the spore-crystal complex. For this purpose different cultivations are done in different fermentation devices with any suitable culture media. Culture media could be very variable as well as fermentors, which both could be carefully selected in order to obtain higher amounts of entomopathogenic crystal protein. For example, Feng *et al.* (2001) reported a higher crude protein production using fed-batch fermentation than batch fermentation, although crystal released occurred earlier in batch fermentation than in fed-batch fermentation maintaining the same glucose concentration in the two bioprocesses. Also an improved delta-endotoxin production (7% more) was obtained by using a new medium proposed by Ghribi *et al.* (2007). The medium was composed by starch, soya bean and diluted sea water which provided 7.5 g/l NaCl. This medium contributes in this way to a significant cost reduction for large-scale production of the crystal protein.

The main function of the fermentor is to provide a suitable environment for the growth of the microorganisms. The fermentation process should be carried out under strict operating conditions, such as temperature and pH control, suitable aeration and agitation, pressure and dissolved oxygen. The kind of fermentation process could be very variable, most of the processes directed to produce the spore-crystal complex in higher amounts use the batch fermentation, which is widely accepted. In a batch reactor a material balance of components shows that the rate of accumulation of component X, given by the time derivative of the total amount of the component X in the reactor, must be equal to the net rate of formation of component X due to the chemical reactions in the vessel. Thus moles X formed by reaction is directly proportional to culture volume and inversely proportional to time (Bailey and Ollis,

1986). The batch fermentations have been widely used for *B. thuringiensis* var. *israelensis* (Avignone-Rossa *et al.,* 1992; Silveira and Molina, 2005; Tokcaer *et al.,* 2006; Berbert-Molina *et al.,* 2008), *B. thuringiensis* var. *kurstaki* HD-73 (Farrera *et al.,* 1998), and *B. thuringiensis* var. *kurstaki* HD-1 (Yezza *et al.,* 2004) among many others. For bioinsecticides production the initial culture should be pure, although the rest of the process could be carried out without taking care of sterile conditions. The fermentation vessels are usually constructed from stainless steel or glass to minimize corrosion problems. The agitator is designed to meet the mixing and aeration required, one impeller is usually required in laboratory-scale fermentation while several may be necessary in large-scale production. Typically 70 to 80% of the vessel volume is filled with the fermentation broth (Bailey and Ollis, 1986). Frequently the use of culture media for propagation of *B. thuringiensis* offers the formation of foam on the liquid surface. This is caused by the effect of agitation and aeration and if the medium contains high protein concentrations. Foam formation is especially detrimental for fermentation development as it avoids gas mass transfer, and traps many cells impeding their growth and reducing production levels. It is necessary to add antifoams to the broth in order to destabilize the foam by reducing surface tension. In the *B. thuringiensis* fermentation, oxygen requirements are low and together with the mixing system constitute the most important factor from economical, physiological, and design and equipment construction point of view (Medrano-Roldán and Galán-Wong, 1996). The mixing system is essential in a fermentation process and is represented by impellers that provide fluid circulation through the tank, and the expansion and velocity of the air passing through a fermentor impart fluid motion (Oldshue, 1960). It is important to quantify the potency that it is consumed by agitation in a fermentor due to costs can increase up to 30%. Pressure is another important parameter in fermentation process, from laboratory scale to industrial scale. The pressure will always be maintained to avoid external contamination. During *B. thuringiensis* fermentation a pressure of 1.2 atm are sufficient to maintain aseptic condition during the process. Temperature is a parameter that should be measured and controlled effectively. To control temperature fermentation vessels are cooled with cooling coils (Medrano-Roldán and Galán-Wong, 1996). To optimized the spore-crystal production it is important to maintain a temperature around 25-28°C. Higher temperatures could prevent sporulation process. In industrial fermentors a simple regulatory valve is sufficient to allow cooling water entrance (Medrano-Roldán and Galán-Wong, 1996). In the typical *B. thuringiensis* kinetic the pH values are throwing down, this makes necessary that measurement and control are effective to obtain the maximum yield of the product. A careful design of the culture media can avoid drastic changes in pH values, especially if nitrogen and carbon sources are adequately selected. The variation in pH values can be adjusted by addition of appropriate volumes of NaOH or NH_4OH if culture medium is acidic or H_2SO_4 (Medrano-Roldán and Galán-Wong, 1996) or HCl if medium is alkaline. The pH values must be maintained between 7.0 and 7.5 during all fermentation process and at the final step when crystals are formed. A pH value above 9.0 can begin dissolve crystals. Fermentation broth must be continuously monitored by sampling and observation under microscope (100X) to determine if crystal shape is still good. If crystal has begun to dissolve it would be observed with rounded edges. In this case the product should be discarded.If an 80% sporulation appears under microscope field, the fermentation will be stopped immediately and pH value adjusted to 7.0. The fermentation broth can be refrigerated if it will not be processed in that moment, for no more than 2 days (Figure. 5).

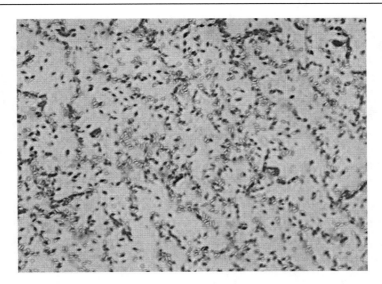

Figure 5. Spores and crystals of *Bacillus thuringiensis* grown in molasses-based medium released at the end of the fermentation process. Bypiramidal crystals are observed. Provided by the author.

After sporulation has been completed, the spore-crystal complex should be recovered as soon as possible. A very common method for this extraction is described here. The fermentation broth must be centrifuged at 10,000 rpm for 30 min in a refrigerated centrifuge. A precipitate is obtained containing a mixture of spores, crystals, and broth components, mainly soy flour (Figure. 6. A). Then the weight of the precipitate is calculated by the following equation: WP = W2 −W1, where W1 is the weight of the centrifuge bottle alone, and W2 is the weight of the bottle plus precipitate after supernatant is discarded, and WP corresponds to the weight of the precipitate obtained by the difference of both weights. The precipitate is thick and creamy and it should be resuspended in a 5% lactose solution for co-precipitation. To obtain the necessary volume of lactose solution the WP value is multiplied by a factor of 1.71, to this volume the WP value is added, in this way the total volume of lactose is obtained. The total volume of lactose is multiply by a factor of 3.34 to obtain the necessary volume of acetone for co-precipitation. All the precipitate is placed in a glass or pail with the lactose solution and acetone volumes, stirring 30 min (Figure. 6. B). Subsequently the aqueous acetone suspension settled down for 10 min, and then was filtered with suction in a Buchner funnel by using Whatman No. 1 filter paper (Figure. 6. C), washed with three volumes of acetone and left dry for 2 or 3 h (Figure. 6. D). Finally was finely ground in a mortar (Figure. 6. E), and kept in air-tight containers (Figure. 6. F) (Dulmage *et al.,* 1970). This method is very efficient and economic. The spore-crystal complex recovered in this way has a good shelf life (1-3 years), because toxic activity decreases with time. The following photographs show some important parts of this process (Figure. 6). The massive production of *B. thuringiensis* has not only been conducted for production of biological insecticides but to obtain large quantities of proteins that could serve to conduct safety-assessment studies to support registration of transgenic plants. Moreover, the recently recombinant DNA technology allows the expression of *cry* genes in alternative host organisms with many advantages that may include higher levels of expression of proteins, faster production times, economic fermentation media and lack of spore production (Gustafson *et al.,* 1997).

Figure 6. Procedure for the spore-crystal complex recovery after fermentation process. Refer text for a description of each photograph. Provided by the author.

Massive production of entomopathogenic bacteria will increase according to market needs arise. The use and application of biological insecticides based on entomopathogenic bacteria should be positioned among the most important strategies to control agriculturally insect pests. For this reason new more techniques in massive production should be developed in a near future, to resolve the coming challenges related to preserve a healthy environment.

CONCLUSION

In this world exists a great variety of bacteria that possess entompathogenic effects towards many insect pests. However, for different reasons many of them cannot be used to kill insects since they are also pathogenic to humans and animals. Up to date few bacteria have possibilities of being used for this purpose, nevertheless *B. thuringiensis* has fulfill all requirements needed.

Since its discovery as a biological control agent, a world of possibilities was opened mainly to scientific research. These studies not only increased knowledge in microbiology, chemistry, biochemistry, entomology, molecular biology and other related sciences, but contributed with such unexpected possibilities for control of insect pests that affect important crops for human life.

The application of *B. thuringiensis* as a biological insecticide involves a complex process that initiates with obtaining suitable strains until formulation preparation to be applied in the field. This time-consuming process but fascinating at the same time, has been conducted for those who have looked for better alternatives to chemical insecticide applications. It is people that look for living in a healthy environment.

Selection of suitable *B. thuringiensis* strains, bioassays performance for toxic activity determination, massive production and formulation process, are exceptionally wonderful steps that have allowed obtaining deep knowledge from a microorganism that has revolutionized world agriculture development. Its production and sales worldwide confirm acceptance from growers to improve their crops and to obtain high quality products.

REFERENCES

Abdel-Hameed, A. 2001. Stirred tank culture of *Bacillus thuringiensis* H-14 for production of the mosquitocidal δ-endotoxin: mathematical modeling and scaling-up studies. World *Journal of Microbiology and Biotechnology*, 17: 857-861.

Anderson, R.K.I. and Jayaraman, K. 2005. Impact of balanced substrate flux on the metabolic process employing fuzzy logic during the cultivation of *Bacillus thuringiensis* var. *Galleriae*. *World Journal of Microbiology and Biotechnology*, 21: 127-133.

Aronson, A.I., Angelo, N. and Holt, S.C. 1971. Regulation of extracellular protease production in *Bacillus cereus* T: Characterization of mutants producing altered amounts of protease. *Journal of Bacteriology*, 106: 1016-1025.

Asano, S., Maruyama, T., Iwasa, T., Seld, A., Takahashi, M. and Soares, Jr. G. 1993. Evaluation of biological activity of *Bacillus thuringiensis* test samples using a diet incorporation method with diamondback moth, *Plutella xylostella* (Linnaeus) (Lepidoptera: Yponomeutidae). *Applied Entomology and Zoology*, 28: 513-524.

Avignone-Rossa, C., Arcas, J. and Mignone, C. 1992. *Bacillus thuringiensis* growth, sporulation and δ-endotoxin production in oxygen limited and non-limited cultures. *World Journal of Microbiology and Biotechnology*, 8: 301-304.

Bailey, J.E. and Ollis, D.F. 1986. *Biochemical Engineering Fundamentals*. McGraw-Hill International Editions, Singapore.

Bartelt, R.J., McGuire, M.R., and Black, D.A. 1990. Feeding stimulants for the European corn borer (Lepidoptera:Pyralidae): Additives to a starch-based formulation for *Bacillus thuringiensis*. *Environmental Entomology*, 19: 182-189.

Berbert-Molina, M.A., Prata, A.M.R., Pessanha, L.G. and Silveira, M.M. 2008. Kinetics of *Bacillus thuringiensis* var. *israelensis* growth on high glucose concentrations. *Journal of Industrial Microbiology and Biotechnology*, 35: 1397-1404.

Devi, P.S.V., Ravinder, T. and Jaidev, C. 2005. Barley-based medium for the cost-effective production of *Bacillus thuringiensis*. *World Journal of Microbiology and Biotechnology*, 21:173-178.

Dulmage, H. T., Correa, J. A. and Martinez, A. J. 1970. Coprecipitation with lactose as a means of recovering the spore-crystal complex of *Bacillus thuringiensis*. *Journal of Invertebrate Pathology*, 15: 15-20.

Dulmage, H.T., Yousten, A.A., Singer, S. and Lacey, L.A. 1990. Guidelines for production of *Bacillus thuringiensis* H-14 and *Bacillus sphaericus*. UNDP/WORLD BANK/WHO Special Programme for Research and Training in Tropical Diseases, p 42.

Farrera, R.R., Pérez-Guevara, F. and de la Torre, M. 1998. Carbon:nitrogen ratio interacts with initial concentration of total solids on insecticidal crystal protein and spore production in *Bacillus thuringiensis* HD-73. *Applied Microbiology and Biotechnology*, 49: 758-765.

Feng, K.-C., Liu, B.-L., Chan, H.-S. and Tzeng, Y.-M. 2001. Morphology of a spectrum of parasporal endotoxin crystals from cultures of *Bacillus thuringiensis* ssp. *kurstaki* isolate A3-4. *World Journal of Microbiology and Biotechnology*, 17: 119-123.

Ferré, J., Real, M.D., Van Rie, J., Jansens S. and Peferoen, M. 1991. Resistance to the *Bacillus thuringiensis* bioinsecticide in a field population of *Plutella xylostella* is due to a change in a midgut membrane receptor. *Proceedings of the National Academy of the United States of America*, 88: 5119-5123.

Finney D.J. 1962. Probit Analysis. Cambridge University Press. UK.

Galán-Wong, J.L., García Salas, J.A., Elías Santos, M., Quintero Zapata, I. and Luna Olvera, H.A. 1996. Producción de *Bacillus thuringiensis*. In: Galán Wong, J.L., Rodríguez Padilla, C. and Luna Olvera, H.A. (Eds), *Avances Recientes en la Biotecnología en Bacillus thuringiensis*. Universidad Autónoma de Nuevo León, Monterrey, México, pp. 139-155.

Ghribi, D., Zouari, N., Trabelsi, H. and Jaoua, S. 2007. Improvement of *Bacillus thuringiensis* delta-endotoxin production by overcome of carbon catabolite repression through adequate control of aeration. *Enzyme and Microbial Technology*, 40: 614-622.

Gillespie, R.L., McGuire, M.R. and Shasha, B.S. 1994. Palatability of flour granular formulations to European corn borer larvae (Lepidoptera:Pyralidae). *Journal of Economic Entomology*, 87: 452-457.

Gustafson, M.E., Clayton, R.A., Lavrik, P.B., Johnson, G.V., Leimgruber, R.M., Sims, S.R. and Bartnicki, D.E. 1997. Large-scale production and characterization of *Bacillus thuringiensis* subsp. *tenebrionis* insecticidal protein from *Escherichia coli*. *Applied Microbiology and Biotechnology*, 47: 255-261.

Heimpel, A. M. 1967. A critical review of *Bacillus thuringiensis* var. *thuringiensis* Berliner and other crystalliferous bacteria. *Annual Review of Entomology*, 12: 287-322.

Johnson, D.E., McGaughey, W.H. and Barnett, B.D. 1991. Small scale bioassay for the determination of *Bacillus thuringiensis* toxicity toward *Plodia interpunctella*. *Journal of invertebrate Pathology*, 57: 159-165.

Lecadet, M.-M, Blondel, M.-O. and Ribier, J. 1980. Generalized transduction in *Bacillus thuringiensis* var. *Berliner* 1715 using bacteriophage CP-54Ber. *Journal of General Microbiology*, 121: 203-212.

Leighton, T.J. and Doi, R.H. 1971. The stability of messenger ribonucleic acid during sporulation in *Bacillus subtilis*. *The Journal of Biological Chemistry*, 246(10): 3189-3195.

Maldonado-Blanco, M.G. 2005. Control de *Aedes aegypti* con microorganismos. Simposio control biológico del mosquito transmisor del dengue *Aedes aegypti*. Edición especial no. 6-2005. *Revista Salud Pública y Nutrición*.

Maldonado-Blanco, M.G., Solís-Romero, G. and Galán-Wong, L.J. 2003. The effect of oxigen tension on the production of *Bacillus thuringiensis* subsp. *israelensis* toxin active against *Aedes aegypti* larvae. *World Journal of Microbiology and Biotechnology*, 19: 671-674.

McGuire, M.R., Galán-Wong, L.J. and Tamez-Guerra, P. 1997. Bacteria: Bioassay of *Bacillus thuringiensis* against lepidopteran larvae. In Lacey, L. (Ed), *Manual of Techniques in Insect Pathology*. Academic Press, USA. pp. 91-99.

Medrano-Roldán, H. and Galán-Wong, L.J. 1996. Bioingeniería y biotecnología en la producción de bioinsecticidas. In: Galán Wong, J.L., Rodríguez Padilla, C. and Luna Olvera, H.A. (Eds), *Avances Recientes en la Biotecnología en Bacillus thuringiensis*. Universidad Autónoma de Nuevo León, Monterrey, México, pp. 115-137.

Medrano-Roldán, H., Morales-Ramos, L. and Galán-Wong, L.J. 1992. Tecnología para la producción de insecticidas biológicos. *UBAMARI*, 25: 91-124.

Nakata, H.M. and Halvorson, H.O. 1960. Biochemical changes occurring during growth and sporulation of *Bacillus cereus*. *Journal of Bacteriology*, 80: 801-810.

Nickerson K.W. and Bulla L.A. Jr. 1974. Physiology of sporeforming bacteria associated with insects: Minimal nutritional requirements for growth, sporulation, and parasporal crystal formation of *Bacillus thuringiensis*. *Applied Microbiology*, 28:124-128.

Oldshue, J.Y.1960. Fluid Mixing in Fermentation Processes. *Industrial and Engineering Chemistry Research*, 52:60.

Park, H.-W., Ge, B., Bauer, L.S. and Federici, B.A. 1998. Optimization of Cry3A yields in *Bacillus thuringiensis* by use of sporulation-dependent promoters in combination with the STAB-SD mRNA sequence. *Applied and Environmental Microbiology*, 64: 3932-3928.

Perez, C.J., Tang, J.D. and Shelton, A.M. 1997. Comparison of leaf-dip and diet bioassays for monitoring *Bacillus thuringiensis* resistance in field populations of diamondback moth (Lepidoptera: Plutellidae). *Journal of Economic Entomology*, 90(1): 94-101.

Poopathi, S. and Abidha, S. 2008. Biodegradation of poultry waste for the production of mosquitocidal toxins. *International Biodeterioration and Biodegradation*, 62: 479-482.

Poopathi, S., Kumar, K.A., Kabilan, L. and Sekar, V. 2002. Development of low cost media for the culture of mosquito larvicides, *Bacillus sphearicus* and *Bacillus thuringiensis* serovar. *israelensis*. *World Journal of Microbiology and Biotechnology*, 18: 209-216.

Prabakaran, G. and Balaraman, K. 2006. Development of a cost-effective medium for the large scale production of *Bacillus thuringiensis* var. *israelensis*. *Biological Control*, 36: 288-292.

Prabakaran, G. and Hoti, S.L. 2008. Influence of amino nitrogen in the culture medium enhances the production of δ-endotoxin and biomass of *Bacillus thuringiensis* var. *israelensis* for the large-scale production of the mosquito control agent. *Journal of Industrial Microbiology and Biotechnology*, 35: 961-965.

Robertson, J.L., Preisler, H.K., NG, S.S., Hickle, L.A. and Gelernter, W.D. 1995. Natural variation: A complicating factor in bioassays with chemical and microbial pesticides. *Journal of Economic Entomology*, 88(1): 1-10.

Rojas-Avelizapa, L.I., Cruz-Camarillo, R., Guerrero, M.I., Rodríguez-Vázquez, R. and Ibarra, J.E. 1999. Selection and characterization of a proteo-chitinolytic strain of *Bacillus thuringiens*, able to grow in shrimp waste media. *Journal of Industrial Microbiology and Biotechnology*, 15: 299-308.

Rosas-García, N.M. 2002. Elaboración de formulados de *Bacillus thuringiensis* var. *kurstaki* y determinación de la actividad tóxica contra larvas de *Diatraea saccharalis* (Fabricius) (Lepidoptera:Pyralidae) en laboratorio y campo. Universidad Autónoma de Nuevo León, Monterrey, N.L. México.

Sachidanandham, R., Jenny, K., Fletcher, A. and Jayaraman, K. 1997. Stabilization and increased production of insecticidal crystal proteins of *Bacillus thuringiensis* subsp. *galleriae* in steady- and transient-state continuous cultures. *Applied Microbiology and Biotechnology*, 47: 12-17.

Saksinchai, S., Suphantharika, M. and Verduyn, C. 2001. Application of a simple yeast extract from spent brewer's yeast for growth and sporulation of *Bacillus thuringiensis* subsp. *kurstaki*: a physiological study. *World Journal of Microbiology and Biotechnology*, 17: 307-316.

Sarrafzadeh, M.H, Bigey, F., Capariccio, B., Mehrnia, M.-R., Guiraud, J.-P. and Navarro, J.-M. 2007. Simple indicators of plasmid loss during fermentation of *Bacillus thuringiensis*. *Enzyme and Microbial Technology*, 40: 1051-1058.

Shelton, A.M., Wyman, J.A., Cushinlli, N.L., Apfelbeck, K., Dennehy, T.J., Mahr, S.E.R. and Eigenbrode, S.D. 1993. Insecticide resistance of diamondback moth (Lepidoptera: Plutellidae) in North America. *Journal of Economic Entomology*, 86: 11-19.

Silveira, M.M. and Molina, M.A.B. 2005. Indirect estimation of *Bacillus thuringiensis* var. *israelensis* biomass concentration using oxygen balance data. *Brazilian Journal of Chemical Engineering*, 22(4): 495-500.

Tabashnik, B.E., Cushing, N.L., Finson, N. and Johnson, M.W. 1990. Field development of resistance to *Bacillus thuringiensis* in diamondback moth (Lepidoptera: Plutellidae). *Journal of Economic Entomology*, 83: 1671-1676.

Tamez-Guerra, P., Castro-Franco, R., Medrano-Roldán, H., McGuire, M.R., Galán-Wong, L.J. and Luna-Olvera, H.A. 1998. Laboratory and field comparisons of strains of *Bacillus thuringiensis* for activity against noctuid larvae using granular formulations (Lepidoptera). *Journal of Economic Entomology*, 91(1): 86-93.

Tang, J.D., Shelton, A.M., Van Rie, J., De Roeek, S., Moar, W.J., Roush R.T. and Peferoen, M. 1996. Toxicity of *Bacillus thuringiensis* spore and crystal protein to resistant diamondback moth *(Plutella xylostella)*. *Applied and Environmental Microbiology*, 62: 564-569.

Tokcaer, Z., Bayraktar, E, Mehmetoğlu, Ü., Özcengiz, G., Alaeddinoğlu. 2006. Response surface optimization of antidipteran delta-endotoxin production by *Bacillus thuringiensis* subsp. *israelensis* HD 500. *Process Biochemistry*, 41: 350-355.

Valicente, F.H. and Mourão, A.H.C. 2008. Use of by-products rich in carbon and nitrogen as a nutrient source to produce *Bacillus thuringiensis* (Berliner)-based biopesticide. *Neotropical Entomology*, 37(6): 702-708.

Yendol, W.G., Hamlen, R.A. and Rosario, S.B. 1975. Feeding behavior of gypsy moth larvae on *Bacillus thuringiensis*-treated foliage. *Journal of Economic Entomology*, 68: 25-27.

Yezza, A., Tyagi, R.D., Valèro, J.R., Surampalli, R.Y. and Smith, J. 2004. Scale-up of bipesticide production process using wastewater sludge as a raw material. *Journal of Industrial Microbiology and Biotechnology*, 31: 545-552.

Zouari, N. and Jaoua, S. 1999. The effect of complex carbon and nitrogen, salt, Tween-80 and acetate on delta-endotoxin production by a *Bacillus thuringiensis* subsp *kurstaki*. *Journal of Industrial Microbiology and Biotechnology*, 23: 497-502.

In: Microbial Insecticides: Principles and Applications   ISBN: 978-1-61209-223-2
Editors: J. Francis Borgio, K. Sahayaraj, et al.   © 2011 Nova Science Publishers, Inc

Chapter 15

BIOLOGICAL CONTROL POTENTIAL OF ENTOMOPATHOGENIC BACTERIA

*Xiaoyi Wu[1, *], Jichao Fang[2] and Huifang Wang[3]*

[1] Institute for the Control of Agrochemicals, Jiangsu Provincial Commission of Agriculture, 1909 Agro-Forestry Tower, 8 Moonlight Square, Caochangmen Street, Nanjing, Jiangsu 210036, China

[2] Institute of Plant Protection, Jiangsu Academy of Agricultural Sciences, 50 Zhongling Street, Nanjing, Jiangsu 210014, China

[3] Jiangsu Provincial Commission of Agriculture, 1316 Agro-Forestry Tower, 8 Moonlight Square, Caochangmen Street, Nanjing, Jiangsu 210036, China

ABSTRACT

This chapter summarizes the entomopathogenic bacteria that have commonly been used or have potential as biological control agents for insect pests. The widespread soil-dwelling bacterium, *Bacillus thuringiensis* Berliner (Bt), are the most primary tool used to control many insect species in Lepidoptera, Coleoptera, Diptera and a few Hymenoptera. Bt strains produce three types of toxins namely Cry, Cyt and Vip. These toxins are composed of more than 50 groups with over 400 members. Many Bt strains have been commercialized as the leading biopesticides in agriculture, forestry or for the control of vectors of many insect-borne diseases, however their use as sprayable insecticides were limited due to higher costs and lower efficacy than chemical insecticides. Limited persistence, narrow insect spectrum and poor efficacy in controlling sucking or borer insects also limited their portion in insecticide market. Instead, the transgenic Bt crops are widely adopted in the world as plant pesticide and solution for Bt dilemma. Some entomopathogenic bacteria include *Bacillus sphaericu*s, *Paenibacillus popilliae*, *Serratia spp.*, *Streptomyces avermitilis*, *Sacchroplolyspora spinosa*. Entomopathogenic nematodes symbiotic bacteria, such as *Photorhabdus spp.* and *Xenorhabdus spp.*, have also been commercially used to demonstrate their potential against some target insect pests as biocontrol agents.

* E-mail: xwu3@tigers.lsu.edu, Fax: 86-25-86263930 Phone: 86-25-86263936

15.1. INTRODUCTION

Many insects are economically important agricultural or forestry pests, or vectors of some human, animal or plant diseases. As chemical insecticides can markedly protect crop yield and reduce the risk of some insect-borne diseases, they have become the most powerful, effective and cheapest approach for controlling insect pests since the World War II. With the continuous growth in its consumption, application of chemical insecticides has become the most primary pest control tool (Gullan and Cranston, 2005). However, long-term use of insecticide leads to some deleterious effects: resistance to chemicals, destruction of non-targets organisms, pest resurgence, outbreak of secondary pest, adverse environmental effects (contamination) and dangers to human health or other animals (Gullan and Cranston, 2005; Pell *et al.*, 2010). Problems caused by chemical insecticides stimulated interest in seeking alternative measures for insect pest control. Biological control is one of the most valuable alternatives demanded for the chemical control. Classical biological control involves introducing or enhancing target insect pests natural enemies that enables restoring balance in several managed ecosystems, such as the agro-ecosystem. Despite the benefits of many natural enemies, in some cases negative effects arise with introductions of exotic biological agents. Some introduced agents failed to control their proposed target organisms, and some exacerbated pest problems and even became pest themselves. Because exotic introduction is irreversible, unpredictable and not in full controlled conditions, much worse consequences might be caused by exotic agents for the non-target organisms or native natural enemies in some failed cases. These adverse effects and uncertainties compelled people to seek for much safer and manageable agents for biological control. Microbial control comes out as a good alternative for classical biological control agent (Lacey, 2004). Some microorganisms, including bacteria, viruses and small eukaryotes (such as fungi, protozoans, and nematodes), may cause diseases in natural or cultured insect pest populations. The diseases may reduce the feeding rate and growth of insect pests, disrupt reproduction or kill insect pests directly (Stoner, 1998; Lacey, 2004; Hajek and Delalibera 2010). Among these microorganisms, several bacteria have been proved as entomopathogens for controlling target insect pests. In addition, many bacteria carried by insect-attacking nematodes such as entomopathogenic symbiotic bacteria, also cause disease or death of insects (Gullan and Cranston, 2005). In classical biological control, exotic entomopathogenic bacterium such as *Paenibacillus popilliae* (Formerly known as *Bacillus popilliae),* which causes milky disease (or called milky spore disease) in some coleopterans*,* was introduced and established for the control of the grub worm of the Japanese beetle, *Popillia japonica* (Scarabaeidae) in USA. Another bacterium *P. lentimorbis*, also has potent effect on coleopterans control (Klein, 1992; Harrison *et al.,* 2000; Capinera, 2008; Dingman, 2008).

Meanwhile, unlike all the other approaches to biological control, biological insecticides knock down the pest population in quantities instead of aiming to establish a new population of natural enemies that reaches a long-term balance with host or prey population. Most commercialized formulations of entomopathogens are applied in quantities that reduce the pest population in one generation, rather than allowing the entomopathogens to multiply and then spread among the pests, like other classical biological agents (Stoner, 1998; Gullan and Cranston, 2005).

Bacillus thuringiensis strains, the most widely applied biological insecticides, have a relatively broad spectrum of activity against larvae of several insect species in the order of Lepidoptera, Coleoptera, Diptera (aquatic larva or pupa) and a few Hymenopterans due to the abundant toxins produced by diffent *B. thringiensis* strains. Bt insecticide works by paralyzing the gut of the insect because the protein produced by Bt is the active ingredient that paralyzes the gut. However, many strains of *B. thuringiensis* can be used only as inundative microbial insecticides because of their short-lived efficacy or lack of persistence in the field, especially when exposed to ultraviolet light (Gullan and Cranston, 2005; Yu, 2008). Moreover, in many Bt products, there are no bacterial spores present, instead they exist as just a formulation containing only the active Bt proteins. Thus, the disease caused by Bt insecticides does not spread in the insect population in such a case (Stoner, 1998). During the growth period of Bt cells, a large number of additional undesirable toxins may also be produced. For example, β-exotoxin inhibits RNA polymerase enzymes by competing with ATP and is toxic to almost all insect species as well as vertebrates. It's a critical issue that may affect the purity, concentration of the desired Bt proteins, even the fate of a Bt biopesticide (Garczynski and Siegel, 2007). However, two solutions have dramatically resolved this problem. One way is by using the *Pseudomonas fluorescens*, a common bacterium that serves as an incubator that produces Bt proteins. The desired Bt protein gene is inserted into the *P. fluorescens*. As the *P. fluorescens* grows and reproduces, the imported gene codes for the desired Bt toxin and then the *P. fluorescens* can manufacture the desired Bt protein. *P. fluorescens* not only produces higher concentrations of toxins than Bt, but also protects the endotoxins that remain within the *P. fluorescens* cells ("encapsulated"), which makes collecting the endotoxins relatively easy. At the same time, the dead *P. fluorescens* cells protect the Bt endotoxins from decomposition under ultraviolet light and lengthen the time that the toxins remain active. In addition, *P. fluorescens* does not produce a large number of additional undesired toxins made by Bt that might harm non-target species (USEPA, 1996). As a kind of biological control agents, Bt toxin is the most widely used microbial insecticide taking up 90-95% of the biological insects market in the world (Gill *et al.*, 1992), but it accouts for only 2% of the insecticides in market due to several disadvantages, such as higher price, poor coverage, instability to UV light etc (Lambert and Peferoen,1992). These factors have greatly limited the utility of Bt microbial insecticides for managing insect pests in many ecosystems (Wu, 2008; Yu, 2008). The advances in biotechenology has allowed scientists to transfer Bt insecticidal protein genes into plant genomes and produce transgenic Bt plants that can directly produce insecticidal Bt proteins within their tissues and confer a whole-stage protection for the crop (Wu, 2008; Höfte *et al.*, 1986). As one of the transgenic Bt crops founders, the United States has approved and commercialized many transgenic Bt crops since 1996. During the past 13 years it has been the leading country in planting Bt crops with a size of approximately one-third of the accumulated Bt crop acreage. Transgenic Bt crop (Plant-Incorporated Protectant), as a special plant biopesticide, has become a primary tool for managing major insect pests on corn, cotton and other economically important crops. The target insect pests include the European corn borer, corn rootworm, fall armyworm, cotton bollworm, pink bollworm, tobacco budworm, ant the Colorado potato beetle, etc. More and more countries have paid attention to the microbial insecticides or transgenic Bt crops during last 13 years since the first transgenic Bt crop was commericialized in 1996. The number of countries electing to grow biotech crops including Bt crops has increased steadily from 6 in 1996, to 18 in 2003 and 25 in 2008 with the number of farmers benefited from biotech crops worldwide rising to 13.3 million or more as of 2008 (James, 2008).

In 2005, the secretariat of the European Cooperation in the field of Scientific and Technical Research (COST) launched "FA Action 862" (or called "COST 862 Action"): "Bacterial Toxins for Insect Control", of which the main objective is to increase the availability of novel and improved bacterial antagonists and/or their toxins in biological control of insect pests in agriculture, thus will effectively create economic value for biocontrol industry and the farmers (COST, 2005). The Action promotes the use of many Bt or other microbial pesticides and transgenic Bt crops in Europe. Meanwhile, lots of international organizations, including FAO, the Group of Eight (G8), European Commission and WHO, have emphasized that they highly supported the applications of the Biotech/GMO crops in the world to relieve the world food crisis (James, 2008).

Bacillus sphaericus strain, an obligate aerobic bacterium, isolated on the basis of pathogenicity for mosquito larvae, and commercially produced as larvicide for mosquito control, has some advantages over than Bti, such as longer persistence under field conditions (Lacey and Undeen, 1986; Hougard, 1990; Nicolas *et al.,* 1994; Lacey *et al.,* 2001; Federici, 2007). Two species of *Serratia* which cause amber disease in the larvae of the New Zealand grass grub, *Costelytra zealandica,* have been studied for many years, and developed for scarab control (Glare *et al.,* 1993; Jackson *et al.,* 1993, 2001).

Some entomopathogenic bacteria have been found as the symbiotic bacteria associated with entomopathogenic nematodes, for example, *Photorhabdus spp.* and *Xenorhabdus spp,* associated with the soil-dwelling nematodes, heterorhabditids and steinermatids (Burg *et al.,* 1979; Hotson, 1982; Lasota and Dybas, 1991; Salgado, 1997; Thompson *et al.,* 1997; Fang *et al.,* 2002; Toews *et al.,* 2003; Flinn *et al.,* 2004; Darriet *et al.,* 2005; Yu, 2008). In conjuction with the nematodes, these nematodes-bacteria complex are pathogenic to some host pests, and kill the pests by septicemia (Mohan *et al.,* 2003; Shahina *et al.,* 2004; Gullan and Cranston, 2005), for example, *X. nematophila* cells and their metabolic secretions have been found to be lethal to the the greater wax moth larvae, *Galleria mellonella,* the fire ants, *Solenopsis invicta,* larvae and pupae of the beet army worm, *Spodoptera exigua* and so on (Dudney, 1997; Elawad *et al.,* 1999; Mahar *et al.,* 2005), and hemocyte apoptosis induced by *P.* spp. and *X.* spp. in silkworm, *Bombyx mori,* has also been reported (Cho and Kim, 2004). One patent has been approved and commercialized in the United States for the use of toxins from *Xenorhabdus* spp. for insect control (Ensign *et al.,* 2002).

In this chapter, only some currently used major entomopathogenic bacteria is disscussed for their potential as biological control agents and some key issues encountered in their applications.

15.2. Biologic Control Potential

15.2.1. Bacillus Thuringiensis

More than 100 years have passed since *Bacillus thuringiensis* (Bt), a widespread soil dwelling bacterium was first discovered at a silk production farm in 1901 by Shigetane Ishiwatari, a Janpanese bacteriologist. This bacterium was found to kill the silkworm, *Bombyx mori* L. Ten years later, a similar bacterium was isolated from the diseased Mediterranean flour moth (*Anagasta kuehniella*) larvae, in the German province of Thuringia, by Ernst Berliner. The bacterium was named *Bacillus thuringiensis* Berliner thereafter (Beegle and Yamamoto, 1992; Gill *et al.* 1992). After the re-isolation of *B. thuringiensis* by Mattes in 1927 (Heimpel and Angus, 1960), people began to use Bt as a tool to control the European corn borer, *Ostrinia nubilalis* (Hübner) in Europe. Bt-based products began to come into use. Sporeine was the first commercial formulation produced in France in 1938. More Bt formulations were produced in 1950s in Europe. Thuricide was the first commercialized Bt formulation in USA developed in 1957 (Briggs, 1986). More and more *B. thuringiensis* strains have been discovered and isolated, and most of them are associated with diseases of some lepidopterans. In order to scientifically classify different Bt strains, a method named H-serotyping was established, based on the serological reaction to the Bt cell's flagellar antigens. Dipel, a formulation followed by a more potent *B. thuringiensis* strain HD-1 isolated in 1970s, was commercialized. HD-1 was named as serovar kurstaki, it can target lepidopterous insects. Later, *Bt* israelensis, another serovar, was isolated for its potent effect on the control of mosquito larvae, *Culex pipiens*. Serovar tenebrionis was isolated at early 1980s for anti-coleopteran activity. It is believed that about 60,000 *B. thuringiensis* isolates have been discovered and collected around the world by the end of 20th century (Stenersen, 2004; Côté, 2007). Some reports showed that *B. thuringiensis* isolates could be active against insects from other orders including Hymenoptera, Homoptera, Orthoptera, and Mallophaga. Furthermore, its effectiveness against nematodes, mites, and protozoa has also been reported (Feitelson *et al.,* 1992; Feitelson, 1993).

Since 1955, Bt Cry toxins have been widely used as important biological insecticides under various trade names, such as Able[TM], Biobit®, Dipel®, Thuricide®, and so on (NPTN 2000) to control economically important insect pests. Commercial Bt insecticidal products usually contain both dried spores and crystal delta-endotoxins while some products only contain protein toxins (Madigan and Martinko, 2005; Yu, 2008). *B. thuringiensis* insecticidal proteins are highly specific insect gut toxins, they have a superior safty record in their effects on nontarget organisms (Glare and O'Callaghan, 2000; Lacey and Siegel, 2000; Lacey *et al.,* 2001). Bt toxins ingested by their target insect pests such as Lepidopterous larvae, are solubilized in the alkaline midgut and activated by certain gut proteases, and then, binded to their specific binding sites at the brush border membrane surface of epithelium, and finally these activated toxins will form pores and cause cell lysis to lead to target insects death (Van Rie *et al.,* 1989; Slatin *et al.,* 1990; Schwartz *et al.,* 1993; Schnepf *et al.,* 1998).

Currently, main six *B. thuringiensis* strains that possess specific activity against insect species have been commericially used, including the *B. thuringiensis* var. kurstaki strain, for controlling most lepidopterous larvae with high gut pH. *B. thuringiensis* var. israelensis, is used for controlling larvae of aquatic insects such as mosquitoes and blackflies, *B. thuringiensis* var. aizawi, is used to control some lepidopterans in corn, fruits, tobacco, cotton or vegetables. *B. thuringiensis* var. morrison controls lepidopterous larvae on most crops; *B. thuringiensis* var. san diego and *B. thuringiensis* var. tenebrionis are believed to be the same strain, but they are developed by different companies for controlling Colorado potato beetle, *Leptinotarsa decemlineata*. Three types of Bt toxins can be produced by *B. thuringiensis* strains. The first type is Cry (crystal) toxin, which is the major toxin encoded by *cry* gene. The second type is called Cyt (cytolytic) toxin, which can augment the toxicity of Cry toxin against target insects. Only some *B. thuringiensis* strains produce the third type of insecticidal proteins called VIP toxin (vegetative insecticidal proteins) (Yu, 2008). Up to now, about 500 Cry, Cyt and VIP toxins have been discovered and their genes have been sequenced and submitted into genbank. Data showed that some toxins produced by the same *B. thuringiensis* strain were encoded by different genes, and their known toxicity and insecticidal spectrum were different from each other.

A given Cry protein usually has a fairly narrow range of target organisms (Table 1). An updated full list of delta-endotoxins including gene sequence information of each toxin in details is available online (Crickmore *et al.*, 2009).

Table 1. Pesticidal activity of *B.thuringiensis* Cry, Cyt and Vip proteins*

Name	Source Strain	Known Toxicity
Cry1Aa1	*B.t. kurstaki* HD-1; *B.t. aizawai* HD-68	*Heliothis virescens, Mamestra brassicae, Pseudoplusia includens* (Lepidoptera: Noctuidae); *Manduca sexta* (Lepidoptera: Sphingidae); *Pieris brassicae* (Lepidoptera: Pieridae); *Bombyx mori* (Lepidoptera: Bombycidae); (Lepidoptera: Lymantriidae); *Sciropophaga incertulas, Chilo suppressalis, Ostrinia nubilalis* (Lepidoptera: Pyralidae); *Choristoneura fumiferana* (Lepidoptera: Tortricidae); *Hyphantria cunea* (Lepidoptera: Arctiidae); *Plutella xylostella* (Lepidoptera: Plutellidae)
Cry1Ab2	*B.t. kurstaki* HD-1	*Lymantria dispar* (Lepidoptera: Lymantriidae); *Heliothis virescens, Trichoplusia ni* (Lepidoptera: Noctuidae); *Manduca sexta* (Lepidoptera: Sphingidae)
Cry1Ac1	*B.t. kurstaki* HD-73, *B.t. kurstaki* HD-244	*Bombyx mori* (Lepidoptera: Bombycidae); *Agrotis segetum, Helicoverpa zea, Heliothis virescens, Mamestra brassicae, Trichoplusia ni, Spodoptera exigua* (Lepidoptera: Noctuidae); *Ephestia kuehniella, Sciropophaga incertulas, Chilo suppressalis, Ostrinia nubilalis* (Lepidoptera: Pyralidae); *Manduca sexta* (Lepidoptera: Sphingidae); *Lymantria dispar* (Lepidoptera: Lymantriidae); *Pieris brassicae* (Lepidoptera: Pieridae)
Cry1Ad1	*B.t. aizawai* PS81I	*Trichoplusia ni, Spodoptera exigua* (Lepidoptera: Noctuidae); *Choristoneura fumiferana* (Lepidoptera: Tortricidae); *Plutella xylostella* (Lepidoptera: Plutellidae)
Cry1Ae1	*B.t. alesti*	*Heliothis virescens, Trichoplusia ni* (Lepidoptera: Noctuidae)
Cry1Af1	*B. thuringiensis* NT0423	Reported dual activity against Diptera and Lepidoptera
Cry1Ba1	*B. thuringiensis* HD-290-I; *B. thuringiensis* HD2	*Chrysomela scripta* (Coleoptera: Chrysomelidae); *Manduca sexta* (Lepidoptera: Sphingidae); *Artogeia rapae* (Lepidoptera: Pieridae)

Name	Source Strain	Known Toxicity
Cry1Bb1	*B. thuringiensis* EG5847	*Spodoptera frugiperda, Pseudoplusia includens, Trichoplusia ni* (Lepidoptera: Noctuidae); *Plutella xylostella* (Lepidoptera: Plutellidae); *Lymantria dispar* (Lepidoptera: Lymantriidae); *Ostrinia nubilalis* (Lepidoptera: Pyralidae);
Cry1Be1	*B. thuringiensis* 158C2	Strain of origin active against lepidopterans
Cry1Ca1	*B.t. entomocidus* 60.5, *B.t. aizawai* HD-229	*Sciropophaga incertulas, Chilo suppressalis* (Lepidoptera: Pyralidae); *Heliothis virescens, Spodoptera exigua, Spodoptera frugiperda, Trichoplusia ni* (Lepidoptera: Noctuidae); *Pieris brassicae* (Lepidoptera: Pieridae)
Cry1Cb1	*B.t. galleriae* HD-29	*Spodoptera exigua, Trichoplusia ni* (Lepidoptera: Noctuidae)
Cry1Da1	*B.t. aizawai* HD-68	*Plutella xylostella* (Lepidoptera: Plutellidae); *Choristoneura fumiferana* (Lepidoptera: Tortricidae); *Bombyx mori* (Lepidoptera: Bombycidae); *Lymantria dispar, Orgyia leucostigma* (Lepidoptera: Lymantriidae); *Manduca sexta* (Lepidoptera: Sphingidae); *Malacosoma disstria* (Lepidoptera: Lasiocampidae); *Lambdina fiscellaria fiscellaria* (Lepidoptera: Geometridae); *Spodoptera frugiperda* (Lepidoptera: Noctuidae)
Cry1Ea1	*B.t. darmstadiensis* HD-146	*Spodoptera littoralis, Spodoptera exempta* (Lepidoptera: Noctuidae); *Manduca sexta* (Lepidoptera: Sphingidae)
Cry1Eb1	*B.t. aizawai*	*Trichoplusia ni, Spodoptera exigua* (Lepidoptera: Noctuidae); *Plutella xylostella* (Lepidoptera: Plutellidae)
Cry1Fa1	*B.t. aizawai* EG6346	*Plutella xylostella* (Lepidoptera: Plutellidae); *Heliothis virescens, Spodoptera exigua, Spodoptera littoralis* (Lepidoptera: Noctuidae); *Ostrinia nubilalis* (Lepidoptera: Pyralidae)
Cry1Ia1	*B.t. kurstaki* INA-02, 4835	*Spodoptera littoralis* (Lepidoptera: Noctuidae); *Bombyx mori* (Lepidoptera: Bombycidae); *Plutella xylostella* (Lepidoptera: Plutellidae); *Ostrinia nubilalis* (Lepidoptera: Pyralidae); *Leptinotarsa decemlineata* (Coleoptera: Chrysomelidae)
Cry1Ib1	*B.t. entomocidus* BP465	*Plutella xylostella* (Lepidoptera: Plutellidae)
Cry1Ja1	*B. thuringiensis* EG5847	*Helicoverpa zea, Heliothis virescens, Pseudoplusia includens, Spodoptera exigua, Spodoptera frugiperda, Trichoplusia ni* (Lepidoptera: Noctuidae); *Plutella xylostella* (Lepidoptera: Plutellidae)
Cry1Jb1	*B. thuringiensis* EG5092	*Pseudoplusia includens, Trichoplusia ni* (Lepidoptera: Noctuidae); *Ostrinia nubilalis* (Lepidoptera: Pyralidae); *Plutella xylostella* (Lepidoptera: Plutellidae)

Table 1. (Continued)

Name	Source Strain	Known Toxicity
Cry1Ka1	*B.t. morrisoni* BF190	*Artogeia rapae* (Lepidoptera: Pieridae)
Cry2Aa1	*B.t. kurstaki* HD-1, HD-263	*Sciropophaga incertulas, Chilo suppressalis, Ostrinia nubilalis* (Lepidoptera: Pyralidae); *Lymantria dispar* (Lepidoptera: Lymantriidae); *Helicoverpa armigera, Heliothis virescens, Trichoplusia ni* (Lepidoptera: Noctuidae); *Aedes aegypti* (Diptera: Cuclidae)
Cry2Ab1	*B.t. kurstaki* HD1	*Manduca sexta* (Lepidoptera: Sphingidae)
Cry2Ac1	*B. thuringiensis* S	*Heliothis virescens, Trichoplusia ni* (Lepidoptera: Noctuidae); *Manduca sexta* (Lepidoptera: Sphingidae)
Cry3Aa1	*B.t. san diego, B.t. tenebrionis*	*Haltica tombacina, Leptinotarsa decemlineata, Pyrrhalta luteola* (Coleoptera: Chrysomelidae); *Hypera brunneipennis, Otiorhynchus sulcatus, Anthonomus grandis* (Coleoptera: Curculionidae); *Tribolium castaneum, Tenebrio molitor* (Coleoptera: Tenebrionidae)
Cry3Ba1	*B.t. tolworthi* EG2838	*Leptinotarsa decemlineata* (Coleoptera: Chrysomelidae)
Cry3Bb1	*B.t. kumamotoensis* EG4961	*Leptinotarsa decemlineata* (Coleoptera: Chrysomelidae)
Cry3Ca1	*B.t. san diego*	*Pyrrhalta luteola* (Coleoptera: Chrysomelidae)
Cry4Aa1	*B.t. israelensis* 4Q2-72	*Anopheles stephensi, Aedes aegypti, Culex pipiens* (Diptera: Cuclidae)
Cry4Ba1	*B.t. israelensis* 4Q2-72	*Aedes aegypti* (Diptera: Cuclidae)
Cry5Aa1	*B. thuringiensis* PS17A	*Caenorhabditis elegens, Pratylenchus* spp. (plant parasitic nematodes)
Cry5Ab1	*B. thuringiensis* PS7	*Fasciola hepatica* (liver fluke); *Caenorhabditis elegens, Pratylenchus* spp. (plant parasitic nematodes)
Cry6Aa1	*B. thuringiensis* PS52A1	*Pratylenchus* spp., *Panagrellus redivivus* (plant pathogenic nematodes)
Cry6Ba1	*B. thuringiensis* PS52A1	*Pratylenchus* spp. (plant pathogenic nematode)
Cry7Aa1	*B. thuringiensis* BTS137J	*Leptinotarsa decemlineata* (Coleoptera: Chrysomelidae)
Cry8Aa1	*B.t. kumamotoensis* PS50C	*Leptinotarsa decemlineata* (Coleoptera: Chrysomelidae)
Cry8Ba1	*B.t. kumamotoensis* PS50C	*Cotinis* spp. (Coleoptera: Scarabaeidae)
Cry8Ca1	*B.t. japonensis* strain Buibui	*Anomala cuprea* (Coleoptera: Scarabaeidae)
Cry9Aa1	*B.t. galleriae* 11-67	*Galleria mellonella* (Lepidoptera: Pyralidae)

Name	Source Strain	Known Toxicity
Cry9Ca1	*B.t. tolworthi* H9	*Agrotis segetum, Helicoverpa armigera, Heliothis virescens, Mamestra brassicae, Spodoptera exigua, Spodoptera littoralis* (Lepidoptera: Noctuidae); *Manduca sexta* (Lepidoptera: Sphingidae); *Ostrinia nubilalis* (Lepidoptera: Pyralidae); *Plutella xylostella* (Lepidoptera: Plutellidae); *Bombyx mori* (Lepidoptera: Bombycidae); *Choristoneura fumiferana* (Lepidoptera: Tortricidae)
Cry10Aa1	*B.t. israelensis* ONR60A	*Aedes aegypti* (Diptera: Cuclidae)
Cry11Aa1	*B.t. israelensis* HD-567	*Anopheles stephensi, Aedes aegypti, Culex pipiens* (Diptera: Cuclidae)
Cry11Ba1	*B.t. jegathesan* 367	*Anopheles stephensi, Aedes aegypti, Culex pipiens* (Diptera: Cuclidae)
Cry11Bb1	*B.t. medellin*	*Anopheles albimanus, Aedes aegypti, Culex quinquefasciatus* (Diptera: Cuclidae)
Cry12Aa1	*B. thuringiensis* PS33F2	*Pratylenchus* spp. (plant pathogenic nematode)
Cry13Aa1	*B. thuringiensis* PS63B	Nematodes
Cry14Aa1	*B.t. sotto* PS80JJ1	*Diabrotica* (Coleoptera:); nematodes
Cry15Aa1	*B.t. thompsoni* HD-542	*Manduca sexta* (Lepidoptera: Sphingidae)
Cry16Aa1	*Clostridium bifermentans malaysia* CH18	*Anopheles stephensi, Aedes aegypti, Culex pipiens* (Diptera: Cuclidae)
Cry19Aa1	*B.t. jegethesan*	*Anopheles stephensi, Culex pipiens* (Diptera: Cuclidae)
Cry20Aa1	*B.t. fukuokaensis*	*Aedes aegypti* (Diptera: Cuclidae)
Cry21Aa1 Cry22Aa1 Cry23Aa1	*B.t. higo*	*Culex pipiens molestus* (Diptera: Cuclidae) Hymenopterans *Tribolium castaneum* (Coleoptera: Tenebrionidae); *Popillia japonica* (Coleoptera: Scarabaeidae)
Cyt1Aa1	*B.t. israelensis* IPS82	*Anopheles stephensi, Aedes aegypti, Culex pipiens* (Diptera: Cuclidae)
Cyt1Ab1	*B.t. medellin* 163- 131	*Anopheles stephensi, Aedes aegypti, Culex pipiens* (Diptera: Cuclidae)
Cyt2Aa1 Cyt2Bb1	*B.t. kyushuensis*	*Anopheles stephensi, Aedes aegypti, Culex pipiens* (Diptera: Cuclidae) *Aedes aegypti* (Diptera: Cuclidae)
Vip1		Coleopterans
Vip 2		Coleopterans
Vip 3		Lepidopterans

*Source: Zeigler (1999), Yu (2008)

Owing to the excellent performance of the insecticidal proteins (Cry proteins) produced by *B. thuringiensis* strains against some target insect pests, *B. thuringiensis* has been used as the leading biopesticide in agriculture, forestry and other field for more than 60 years and is regarded as the most successful alternative to conventional chemical insecticides (Romeis *et al.*, 2006). Until 1996, about 20 *B. thuringiensis* strains or close species from genus *Bacillius* had been registered by the United States Environmental Protection Agency (USEPA) for commercial purpose (Table 2). Upto 1998, about 200 *B.thuringiensis*-based products had been registered in the United States alone (Schnepf *et al.*, 1998; USEPA, 2009). Though these Bt microbial insecticides are environmentally friendly with highly selectiveness and only a few cases of adverse effects on nontarget organisms, the share of Bt products in the overall insecticide market was as low as less than 2% before 1996 due to the reasons mentioned previously (Schnepf *et al.*, 1998; Ravon, 2000; Romeis *et al.*, 2006; Yu, 2008).

Table 2. Microbial pesticides registered by the U.S. EPA as of 2009*

Agents	Active ingredient(s)	Crop	Yr registered	Target pest
Bacterium	*B. popilliae*, *B. lentimorbus*		1948	Japanese beetle larva
	B. thuringiensis subsp. *kurstaki*		1961	Lepidopteran larva
	B. thuringiensis subsp. *israelensis*		1981	Dipteran larva
	B. thuringiensis subsp. Berliner		1984	Lepidopteran larva
	B. thuringiensis subsp. *tenebrionis*		1988	Coleopteran larva
	B. thuringiensis subsp. *kurstaki* EG2348		1989	Lepidopteran larva
	B. thuringiensis subsp. *kurstaki* EG2424		1989	Lepidopteran larva
	B. thuringiensis subsp. *kurstaki* EG2371		1990	Lepidopteran larva
	B. sphaericus		1991	Dipteran larva
	B. thuringiensis subsp. *aizawai* GC-91		1992	Lepidopteran larva
	B. thuringiensis subsp. *aizawai*		1992	Lepidopteran larva
	B. thuringiensis subsp. *kurstaki* BMP123		1993	Lepidopteran larva
	B. thuringiensis subsp. *kurstaki* EG7673		1995	Lepidopteran larva
	B. thuringiensis subsp. *kurstaki* EG7673		1995	Colorado potato beetle
	B. thuringiensis subsp. *kurstaki* EG7841		1996	Lepidopteran larva
	B. thuringiensis subsp. *kurstaki* EG7826		1996	Lepidopteran larva
	B. thuringiensis subsp. *kurstaki* M200		1996	Lepidopteran larva
	B. thuringiensis subsp. *israelensis* EG2215		1998	Mosquito larva
	B. thuringiensis subsp. *aizawai* NB200		2005	Lepidopteran larva

Agents	Active ingredient(s)	Crop	Yr registered	Target pest
Nonviable microbial pesticide	*B. thuringiensis* subsp. *kurstaki* delta- endotoxin in killed *P. fluorescens*		1991	Lepidopteran larva
	B. thuringiensis subsp. *san diego* delta-endotoxin in killed *P. fluorescens*		1991	Coleopteran larva
	B. thuringiensis Cry1Ac and Cry1C delta-endotoxin in killed *P. fluorescens*		1995	Lepidopteran larva
	B. thuringiensis subsp. *kurstaki* Cry1C delta-endotoxin in killed *P. fluorescens*		1996	Lepidopteran larva
Plant pesticide	*B. thuringiensis* Cry3A delta-endotoxin	Potato	1995	Colorado potato beetle
	B. thuringiensis Cry1Ab delta-endotoxin	Corn	1995	Lepidopteran larva
	B. thuringiensis Cry1Ac delta-endotoxin	Cotton	1995	Lepidopteran larva
	B. thuringiensis Cry1Ab delta-endotoxin	Corn	1996	Lepidopteran larva
	B. thuringiensis subsp. *kurstaki* delta- endotoxin from HD-1-derived plasmid vector pZ01502	Corn	1996	Lepidopteran larva
	B. thuringiensis subsp. *kurstaki* Cry1Ac delta-endotoxin	Corn	1997	Lepidopteran larva
	B. thuringiensis Cry 1F protein and the genetic material necessary for its production (plasmid PHI 8999)	Corn	2001	Lepidopteran larva
	B. thuringiensis subsp. *tolworthi* Cry9C protein and the genetic material necessary for its production (pRVA9909)	Corn	1998	lepidoptran larva
	B. thuringiensis Cry1A(c) delta-endotoxin and the genetic material necessary for its production	Cotton	2001	Lepidopteran larva
	B. thuringiensis Cry1A(b) in corn from PV CIB4431	Corn	2001	Lepidopteran larva
	B. thuringiensis Cry2Ab2 protein and the genetic material necessary for its production	*Cotton*	2002	Lepidopteran larva TBW, CBW, PBW
	B. thuringiensis Cry3Bb1 protein and the genetic material necessary for its production (vector ZMIR13L)	Corn	2003	Corn Rootworm

Table 2. (Continued)

Agents	Active ingredient(s)	Crop	Yr registered	Target pest
	B. thuringiensis Cry3Bb1 Protein and the Genetic Material in Event MON 863 Corn and *B. thuringiensis* Cry1Ab Delta-Endotoxin and the Genetic Material in Corn (Stacked)	Corn	2003	Lepidopteran larva, Corn rootworm
	B. thuringiensis subsp. *kumamotoensis* Cry3Bb1/Cry1Ab delta- endotoxin	Corn	2003	Lepidopteran larva, Corn rootworm
	B. thuringiensis var. aizawai Cry1F (pGMA281) and B. thuringiensis var. kurstaki Cry1Ac and genetic material (plasmid pMYC3006) for its production (Pyramided)	*Cotton*	2004	Lepidopteran larva
	B. thuringiensis var. *aizawai* strain PS811 Cry1F protein and the genetic material necessary for its production (plasmid PHP12537) in Event DAS-06275-8 corn	Corn	2005	Lepidopteran larva
	B. thuringiensis Cry34Ab1 and Cry35Ab1 proteins and the genetic material for the production (plasmid PHP 17662)	Corn	2005	Corn Rootworm
	B. thuringiensis Cry3Bb1 protein and the genetic material necessary for its production (vector ZMIR39) in Event MON 88017 corn	Corn	2005	Corn Rootworm
	B. thuringiensis Cry34Ab1 and Cry35Ab1 proteins and the genetic material for their production (plasmid PHP 17662) x *B. thuringiensis* Cry1F protein and the genetic material for its production (plasmid PHI8999)	Corn	2005	Lepidopteran larva and Corn Rootworm
	Modified Cry3A (m Cry3A) protein and the genetic material for its production (pZM26) in event MIR604 corn	Corn	2006	Corn rootworm
	B. thuringiensis Vip3Aa19 and modified Cry1Ab insecticidal proteins and the genetic material necessary for their production in COT102 x COT67B cotton (VipCot)	Cotton	2008	Lepidopteran larva
	B. thuringiensis Vip3Aa insecticidal proteins and the genetic material for its production	Cotton	2008	Lepidopteran larva

Agents	Active ingredient(s)	Crop	Yr registered	Target pest
	B. thuringiensis Cry1A.105 protein and the genetic material for its production	Corn	2008	Lepidopteran larva
	B. thuringiensis Cry2Ab2 protein and the genetic material for its production	Corn	2008	Lepidopteran larva
	B. thuringiensis Cry1Ab delta-endotoxin and the genetic material for its production (pZO1502) in Event *Bt*11 corn x *B. thuringiensis* Vip3Aa20 insecticidal protein and the genetic material for its production (pNOV1300) in Event MIR162 maize	Corn	2009	Lepidopteran larva
	B. thuringiensis Cry1Ab delta-endotoxin protein and the genetic material necessary for its production (pZO1502) in Event Bt11 corn x *B. thuringiensis* Vip3Aa20 insecticidal protein and the genetic material necessary for its production (pNOV1300) in Event MIR162 maize x modified Cry3A protein and the genetic material necessary for its production (pZM26) in Event MIR604 corn	Corn	2009	Lepidopteran larva and Corn rootworm

[*] Source: Schnepf *et al.*(1998), USEPA (2009)

This situation has dramatically changed since 1996. Transgenic Bt crops, a major breakthrough, known as plant pesticide, or Plant-Incorporated Protectant (PIP), have been commeriallized and planted in USA and other five contries. *Cry* gene, which encodes a certain Bt protein, was inserted into the chromosome of a certain crop plant to control their target insect pests. For example, the Bt-potato (New Leaf®), expresses the Cry3A protein, used to control Colorado patato beetle, *Leptinotarsa decemlineata*. Bt-cotton (*Bollgard*®), it produces Cry1Ac toxin that protects cotton plants from three major insect pests including the tobacco budworm, *Heliothis virescens,* the cotton bollworm, *Helicoverpa zea,* and the pink bollworm, *Pectinophora gossypiella*. Bt-corn, such as YieldGard®. It also expresses the Cry1Ab Bt protein, which manages the European corn borer, O*strinia nubilalis,* and other corn stalk borers (Wu *et al.,* 2007).

Newer generation transgenic Bt corn hybrids can express two or more Cry proteins and are able to control more than one group of insect pests. For example, YieldGard® Plus corn produces both Cry1Ab and Cry3Bb1 proteins which are efficacious against both Lepidopteran stalk-boring pests and Coleopteran corn rootworms, and shows some activity against earworms, and armyworms (USEPA, 2005). In 2007, the number of countries planting Bt crops increased to 21 with a total of > 36.8 million hectares throughout the world (James, 2007). Currently, both the Bt crops expressing single Bt gene and multi Bt genes are planted in United States and other countries. It is expected that the demand for stacked-gene varieties will increase with Bt crops with pyramided or stacked *cry* genes available for controlling more species of target insects. Single-Bt-gene varieties such as YieldGard® will likely be phased out in United States and other countries in the near future (Johnson, 2007; James, 2008; USEPA, 2009). The *vip* gene is being genetically engineered into Bt corn or Bt cotton. For example, the VipCot (COT102 x COT67B) cotton is developed via conventional breeding of COT102 (Vip3Aa19) plants crossed with COT67B (modified Cry1Ab) plants (Table 2).

15.3. RISK, SAFETY AND ECOTOXICOLOGY OF BT

B. thuringiensis has been the most widely used and leading biopesticide in the world as biological control agent for many decades. Before commercialization, most of the Bt products including transgenic crops have been carefully analyzed and assessed for any potential deleterious effects on the environment, human health and the potential ecological consequences on nontarget organisms, including natural enemies and other beneficial arthropods regarded as key concerns. These basic data are mandatory requirement for product registration in many countries, like USA. The early phase evaluation (laboratory and greenhouse assessment) can eliminate any potiential candidates with risk or fatal defects, and then further assessments via semi-field evaluation or field evaluation will determine the final fate of a potential biological control product/agent or transgenic Bt crop (Romeis *et al.,* 2006).

In recent years, some studies in labtorary have showed that Bt may have some adverse effects on nontarget or benefitial organisms directly or indirectly. However, based on the conclusions made by Romeis *et al.* (2006) after reviewing a number of pubished papers related to the risk assessment of transgenic Bt crops, a major concern has arisen to adopt Bt crops for biological control so that it does not show potential adverse effect on nontarget organisms including the natural enemies used. Studies showed that side-effects posed by Bt crops on natural enemies only occurred when Bt-susceptible, sublethally damaged herbivores were used as prey or host in experiments, with no indiction of direct toxic effects. Papers related to field studies confirmed that the there is no significant difference in abundance and activity of parasitoids and predators in Bt and non-Bt crops (Romeis *et al.,* 2006).

Monarch butterfly was the first case which aroused people's concerns on the impact of Bt-crops on nontarget insects (Losey *et al.,* 1999), however, subsequent studies concluded that there was no evidence claiming that Bt corn pollen has adverse effect on the monarch larvae, because of the low expression of Bt toxin in pollen of most commercialized transgebic Bt-crops (Sears *et al.,* 2001; Zangerl *et al.,* 2001; Wolt *et al.,* 2003).

One study showed that *B. thuringiensis* subsp. israelensis caused significant mortality and reduction in the reproductive potential of the parasitoids wasp, *Muscidifurax raptor* (Ruiu *et al.,* 2007). In another study, significant increase in mortality, prolonged developmental time and slight decrease in weight was also observed in the predator, *Chrysoperla carnea* while fed with 'Bt-contaminated' *Spodoptera littoralis* larvae (Dutton *et al.,* 2003). Growth and development of the offspring of the cotton bollworm parasitoid *Microplitis mediator* can be indirectly but significantly affected when fed with Bt cotton powder leaf diet. The parasitism rate and adult emergence of the parasitoid decreased and the abnormal pupal rate increased, especially when Bt concentration was increased in the host larvae diet (Liu *et al.,* 2005). However, in most of the cases, host/prey-quality mediated effects can be expected if susceptible herbivores ingest the toxin (Romeis *et al.,* 2006).

Moreover, some cases showed that Bt biopesticides or transgenic Bt-crops had slight or no adverse effect on nontarget arthropods. A 3-year field-scale monitoring of foliage-dwelling spiders (*Araneae*) in transgenic Bt maize fields and adjacent field margin showed that Bt maize had no consistent effect on individual numbers, species richness and guild structure of spiders (Lude and Lang, 2006). When Bt was ingested at concentrations that had minor effects on development or survival of *Spodoptera litura* (F.) and *Helicoverpa armigera* (Hübner) larvae in *Bt* -transformed potato and brassica crops, no impact was observed on two of their larval Hymenopterous parasitoids species (Walker *et al.,* 2007). No significant adverse effect was observed on the survival and fecundity of *Apanteles subandinus*, a parasitoid, and the Tasmanian lacewing, *Micromus tasmaniae*, an predator, when their host insects, potato tuberworm larvae, *Phthorimaea operculella,* or green peach aphids, *Myzus persicae*, reared on transgenic Bt potato plants (Davidson *et al.,* 2006).

StarLink corn is another public relationship crisis rather than a safety disaster caused by the Starlink corn itself. Starlink corn was once a variety of Bt corn patented by Aventis, initially registered for animals feed, not for human food components. But later it was subsequently found as food component for consumption by humans. Many people suffer from serious allergic reactions to corn products that may contain the Starlink protein. Then the sale of StarLink corn seed was discontinued, and registration for Starlink variety was voluntarily cancelled by Aventis in October 2000. However, further studies by the US CDC concluded that there was no evidence that the reactions experienced by those people were associated with hypersensitivity to the Starlink corn Bt protein. The Federal Insecticide, Fungicide, and Rodenticide Act Scientific Advisory Panel (FIFRA SAP) also pointed out that the probability of the Cry9C protein was the cause of allergic symptoms is very low (King and Gordon, 2000; CDC, 2001; FDA, 2001;FIFRA SAP, 2001;).

Concerns about the effects of thansgenic introduction from Bt crops on the genetic diversity of native landraces have also been raised after the wide adoption of these genetic engineering crops. Some reports showed that the transgene DNA from Bt-corn was detected in native maize (Quist and Chapela, 2001), however, it was claimed that these conclusions were incorrect due to misinterpretion of the data (Kaplinsky *et al.,* 2002; Metx and Fütterer, 2002). Currently, no sufficient evidence shows that transgene introduction of Bt-crops has occurred in the crop landraces and close wild relatives (Yu, 2008).

Insect resistance to transgenic Bt crops is a critical concern, which affects the Bt application and long-term success for such a good tool in insect pest management. Similiar to the conventional chemical insecticides, a number of insect species could develop resistance to *B. thuringiensis* toxins under laboratory conditions (Gelernter, 1997; Gould, 1992, 1995; Roush, 1994; Huang, 1997, 1999; Liu and Tabashnik, 1997; Sayyed and Wright, 2001; Tabashnik *et al.,* 2000, 2003; Wu *et al.,* 2009), but only few cases exhibited resistance development in the field, which includes the diamonbackmoth, *Plutella xylostella*, the cabbage looper, *Trichoplusia ni,* the sugarcane borer, *Diatraea saccharalis* and etc. (Tabashnik, 1994b; Tabashnik *et al.,* 1994, 1996,1997; Janmaat *et al.,* 2004; Janmaat and Myers, 2003, 2005; Huang *et al.,* 2007a; Tabashnik and Carrière, 2007; Wu *et al.,* 2007). Several insect resistance management (IRM) strategies have been proposed, evaluated and adopted, including the use of non-treated refuge for Bt insecticide and non-Bt crop refuge for the transgenic Bt-crops, high dosage of Bt toxins in products or expressed in Bt-crop plants, seed mixture (Bt and non-Bt cultivars), toxin mixtures, rotation or alternation of Bt toxins in pesticides or Bt-crops, and using stacked genes or pyramided genes in Bt-crops, etc. Currently, the "high dose/refuge" strategy is widely accepted than other strategies, because most of the cases showed that insect pests resistance to transgenic Bt-crops is recessive, which matches one of the four key assumptions required for the "high dose/refuge" strategy (Tabashnik, 1994a, b; Ostlie *et al.*, 1997; FIFRA SAP, 1998; Tabashnik *et al.*, 2003; Baute 2004; Huang *et al.*, 2007b; Tabashnik and Carrière, 2007).

Therefore, overall, it can be concluded that Bt biopesticides and transgenic Bt crops are much safer to environment, human health and nontarget organisms than conventional chemicals and can be continuously used as the leading biological control agents and play a very important role in sustainable agriculture.

15.4. OTHER ENTOMOPATHENIC BACTERIA

15.4.1. Bacillus Sphaericus

B. sphaericus was registered in 1991 by EPA of USA, and commercially produced for the control of some dipterans larvae, such as mosquito and black fly (Table 2). *B. sphaericus* has some advantages over *Bti* that it is more persistent in some polluted habitats and may be recycled under certain suitable conditions, but its host range is much narrower than that of *Bti* (Hougard, 1990; Nicolas *et al.*, 1994; Charles *et al.*, 1996; Rippere, 1998; Skovmand *et al.*, 2007). Meanwhile, high levels of resistance to *B. sphaericus* in some populations of mosquito, i.e., *Culex quinquefasciatus* Say, raise concerns about its long-term adoption for mosquito control (Rao *et al.*, 1995; Nielsen-Leroux *et al.*, 1997).

15.4.2. Paenibacillus Popilliae

P. popilliae is the causative agents of types A and B (respectively) milky disease, a fatal infection of Japanese beetle, *Popilla japonica,* and realted scarab larvae (Bulla *et al.*, 1978, Klein, 1992; Rippere, 1998; Rippere *et al.*, 1998; Harrison *et al.*, 2000; Capinera, 2008; Dingman, 2008; Oliver *et al.*, 2009). Efforts have been made to develop *P. popilliae* and *P. lentimorbus,* as bioinsecticides against some economically important scarabs. *P. popilliae* was the first microbial agent registered as an insecticide in the USA, in suppressing Japanese beetle populations for over 60 years (Klein and Kaya, 1995; Klein *et al.*, 2007), and its marketed product in the United States was Milky Spore (Patel *et al.*, 2000). However, *P. popilliae* spores must be produced in living host larvae or obtained from field collected infected larvae. The slow and sporadic nature of its activity has severely limited the large-scale production and decreased the potential of this bacterium for large-scale control. Only properly prepared spore powders can persist for several years (Klein and Kaya, 1995; Lacey *et al.*, 2001; Lecey, 2004; Klein *et al.*, 2007). Meanwhile, reports showed that both the Japanese beetle, and the oriental beetle, *Anomala (Exomala) orientalis*, had developed resistance to *P. popilliae* (Shelton *et al.*, 2007). Thus, how to produce *P. popilliae* in a large scale and make an appropriate strategy to manage the resistance is very important for the continuous utilization of this biological control tactic.

15.4.3. *Serratia* spp.

Serratia entomophila and *S. proteamaculans* cause amber disease in the grass grub, *Costelystra zealandica*, an important pest of pastures in New Zealand. *S. entomophila* has been developed as a biological control agent. This bacterium is easily grown and mass produced *in vitro* to very high densities, thus it's leading to rapid commercialization and successful adoption in the New Zealand as the product Invade™ against the grass grub *C. zealandica*. Even the similar bacteria have been isolated from Japanese beetle and masked chafer grubs populations in the USA, but haven't been commercially developed in USA yet. However, the amber disease may be a significant natural mortality factor in grub populations in turfgrass soils (Jackson *et al.,* 1992; Klein and Kaya, 1995; Godfray *et al.,* 1999; Tan *et al.,* 2006; Federici, 2007; Ferro, 2007; Hurst *et al.,* 2007; Klein *et al.,* 2007).

15.4.4. Streptomyces spp.

Two different series of components (As, Bs) contained in avermectins can naturally occur by *Streptomyces avermitilis* strain. These avermectins are used to control insect and mite pests of a wide range of agricultural and ornamental crops, or used for controlling fire ants. Based on the structures of these naturaly occurring components such as the Ivermectin, Emamectin benzoate, or completely synthesitic product, Milbemectin, more effective components can be produced by semisynthesis (Campbell *et al.,* 1983; Lasota and Dybas, 1991; Omura *et al.,* 2001; Kim and Goodfellow, 2002; Yu, 2008). After the prohibition of some hypertoxic pesticides, avermectin began to develop rapidly in China, and its dosages were increased gradually. In recent years, it had become a common pesticide in China and other countries. However, resistance has been developed in many insect pest populations to the pesticide after a wide and long-term use, as well as unsuitable usage of avermectin by farmers. This will speed up its substitution if there is no effective IRM strategy deployed to prevent further development of the resistance to Avermectins (Argentine *et al.,* 1992; Clark *et al.,* 1995; Qiu and Zhang, 1999; ReportLinker, 2009).

15.4.5. Saccharopolyspora Spinosa

Spinosad has a unique and novel mode of action, which kills susceptible species of insect pests by causing rapid excitation of the insect nervous system, thus it is highly valued in resistance management programs in many ecosystems (Salgado, 1997). Studies showed that it's safe to non-target organisms, including beneficial insects or predatory mites, ladybugs, green lacewings, minute pirate bugs and so on. Spinosad has been commercially registered in many countries and used to control various insect pests, such as lepidopterous caterpillars, leafminers, fruit flies, sawflies, thrips, spider mites, leaf beetle larvae, and fire ants. It is also recommended to use in IPM programs for commercial greenhouses since Spinosad does not harm most beneficial insects or natural enemies. Currently, more than 70 products containing spinosad are commercialized in USA (Ismail *et al.,* 2007; Kegley *et al.,* 2009). Howerver, some studies also suggest that long-term application of spinosad or exposure may lead to resistance in several insect species (Zhao *et al.,* 2002; Shono and Scott, 2003; Wyss *et al.,* 2003; Hsu and Feng, 2006; Mota-Sanchez *et al.,* 2006).

15.4.6. Photorhabdus spp. and Xenorhabdus spp.

The most common bacterial symbionts of entomopathogenic nematodes, such as *P. luminescens* and *X. nematophilus,* and their roles in insect pests control have been studied by scientists for many years (Ensign *et al.,* 2002; Mohan *et al.,* 2003; Cho and Kim, 2004; Shahina *et al.,* 2004; Gullan and Cranston, 2005). How to use these symbiotic bacteria and their toxins or secretions as biopesticides for insect pests control is a hotspot in recent years, and has high potential to be deployed for biological control for certain insect pests in the future (Blackburn *et al.,* 2005; Mahar *et al.,* 2005; Gullan and Cranston, 2005).

CONCLUSION

Bacteria that are currently applied for insect pest control primarily belong to the genera *Bacillus* and *Paenibacillus,* and other few species. There is no doubt that more and more entomopathogenic bacteria will be discovered and developed into biological control agents, microbial insecticides or transgenic plant pesticides (PIPs) for managing some target insect pests in the future. Generally, entomopathogenic bacteria that are to be commercialized should meet some basic and common requirements as potent biological control agents. First of all, it should be safe to human beings and non-target organisms and friendly to environment. Secondly, the potential candidate should be efficacious in controlling its target insect pests. It can be improved via other technological advances in genetic engineering or such recombinant bacteria as *Pseudomonas.* Thirdly, the potential biological control agent shall be accepted by farmers for its appropriate price, easy application and being integrated with other approches. Last, but not the least is that the IRM strategy should be taken into consideration before a novel biocontrol agent comes into use.

REFERENCES

Argentine, J. A., Clark, J. M. and Lin, H. 1992. Genetics and biochemical mechanisms of abamectin resistance in two isogenic strains of Colorado potato beetle. *Pesticide biochemistry and physiology*, 44 (3): 191-207.
Baute, T. 2004. A grower's handbook: controlling corn insect pests with Bt corn technology, pp. 1-24. Canadian Corn Coalition. Ridgetown, ON, Canada.
Beegle, C. C. and Yamamoto, T. 1992. Invitation paper (C. P. Fund): History of *Bacillus thuringiensis* Berliner research and development. *Canadian Entomologist*, 124: 587-616.

Blackburn, M. B., Domek, J. M., Gelman, D. B. and Hu, J. S. 2005. The broadly insecticidal *Photorhabdus luminescens* toxin complex a (Tca): activity against the Colorado potato beetle, *Leptinotarsa decemlineata*, and sweet potato whitefly, *Bemisia tabaci*. *Journal of Insect Science*, 5: 32-44.

Briggs, J. D. 1986. Pioneering and advanced phases of commercial use of *Bacillus thuringiensis* in North America, p. 25-35. In: A. Krieg and A. M. Huger (Eds.), *Mitt. Biol. Bundesanst*. Land Forstwirtsch. Berl. Dahlem. Vol. 233. Paul Parey, Berlin.

Bulla, L. A., Jr, Costilow, R. N. and Sharpe, E. 1978. Biology of *Bacillus popilliae*. *Advances in Applied Microbiology*, 2:1-18.

Burg, R. W., Miller, B. M., Baker, E. E., Birnbaum, J., Currie, S. A., Hartman, R., Kong, Y. L., Monaghan, R. L., Olson, G., Putter, I., Tunac, J. B., Wallick, H., Stapley, E. O., Oiwa, R. and Omura, S. 1979. Avermectins, new family of potent anthelmintic agents: producing organism and fermentation. *Antimicrobial Agents and Chemotherapy*, 15(3): 361-367.

Campbell, W. C., Fisher, M. H., Stapley, E. O., Albers-Schonberg, G. and Jacob T. A. 1983. Ivermectin: a potent new antiparasitic agent. *Science,* 221:823-828.

Capinera, J. L. 2008. *Encyclopedia of Entomology*. Springer, New York, USA.

CDC. 2001. Investigation of Human Health Effects Associated with Potential Exposure to Genetically Modified Corn. Centers for Disease Control and Prevention. Available: http://www.cdc.gov/nceh/ehhe/Cry9CReport/summary.htm. Visited on 1Nov 2009.

Charles, J. F., Nielsen-Leroux, C. and Delecluse, C. 1996. *Bacillus sphaericus* toxins: Molecular biology and mode of action. *Annual Review of Entomology*, 41: 451-472.

Cho, S. and Kim Y. 2004. Hemocyte apoptosis induced by entomopathogenic bacteria, *Xenorhabdus* and *Photorhabdus*, in *Bombyx mori*. *Journal of Asia-Pacific Entomology*, 7: 195-200.

Clark, J. M., Scott, J. G., Campos, F. and Bloomquist, J. R. 1995. Resistance to avermectins: extent, mechanisms, and management implications. *Annual Review of Entomology,* 40:1-30.

COST. 2005. Memorandum of Understanding implementing a European Concerted Research Action designated as COST Action 862 "Bacterial Toxins for Insect Control". Available: http://register.consilium.eu.int/pdf/en/05/st00/st00209.en05.pdf. Visited on 20 Oct 2009.

Côté, J. -C. 2007. How early discoveries about B. *thuringiensis* prejudiced subsequent research and use. In: Vincent, C., Goettel, M. S. and Lazarovits, G. (Eds.), *Biological Control: A global perspective - Case studies from around the world* (E-publication), CABI Publishing, Wallingford, UK, pp. 169-178.

Crickmore, N., Zeigler, D. R., Schnepf, E., Van Rie, J., Lereclus, D., Baum, J, Bravo, A. and Dean, D. H. 2009. *Bacillus thuringiensis* toxin nomenclature. Available: http://www.lifesci.sussex.ac.uk/Home/Neil_Crickmore/Bt/. Visited on 29 Oct 2009.

Darriet, F., Duchon, S. and Hougard, J. M. 2005. Spindosad: a new larvicide against insecticide-resistant mosquito larvae. *Journal of the American Mosquito Control Association*, 21(4): 495-496.

Davidson, M. M., Butler, R. C., Wratten, S. D. and Conner, A. J. 2006. Impacts of insect-resistant transgenic potatoes on the survival and fecundity of a parasitoid and an insect predator. *Biological Control*, 37(2):224-230.

Dingman, D. W. 2008. Geographical distribution of milky disease bacteria in the eastern United States based on phylogeny. *Journal of Invertebrate Pathology*, 97(2): 171-181.

Dudney, R. A. 1997. Use of *Xenorhabdus nematophilus* Im/l and 1906/1 for fire ant control. *US Patent*, No. 5616318.

Dutton, A., Klein, H., Romeis, J. and Bigler, F. 2003. Prey-mediated effects of *Bacillus thuringiensis* spray on the predator *Chrysoperl acarnea* in maize. *Biological Control*, 26(2): 209-215.

Elawad, S. A., Gowen, S. R., Hague, N. G. M. 1999. Efficacy of bacterial symbionts from entomopathogenic nematodes against the beet army worm (*Spodoptera exigua*). Test of Agrochemicals and Cultivars No. 20, *Annals of Applied Biology (Supplement)*, 134:66-67.

Ensign, J. C., Bowen, D. J., Tenor, J. L., Ciche, T. A., Petell, J. K., Strickland, J. A., Orr, G. L., Fatig, R. O., Bintrim, S. B. and ffrench-Constant, R. H., 2002. Proteins from the genus *Xenorhabdus* are toxic to jnsects on oral exposure. *US Patent*, No. 0147148 A1.

Fang, L., Subramanyam, Bh and Arthur, F. H. 2002. Effectiveness of spinosad on four classes of wheat against five stored-product insects. *Journal of Economic Entomology*, 95(3): 640-650.

FDA (Food and Drug Administration). 2001. FDA Evaluation of Consumer Complaints Linked to Foods Allegedly Containing StarLink Corn. Centers for Food Safty and Applied Nutrition. Available: http://www.epa.gov/scipoly/SAP/meetings/2001/july/fda.pdf. Visited on 1 Nov 2009.

Federici, B. A. 2007. Bacteria as biological control agents for insects: economics, engineering, and environmental safety. In: Vurro, M. and Gressel, J. (Eds.), *Novel Biotechnologies for Biocontrol Agent Enhancement and Management*. Springer, New York, USA, pp. 25-51.

Feitelson, J. S. 1993. The *Bacillus thuringiensis* family tree. In: L.Kim (Ed.), *Advanced engineered pesticides*. Marcel Dekker, Inc., New York, USA, pp. 63-71.

Feitelson, J. S., Payne, J. and Kim, L. 1992. *Bacillus thuringiensis*: insects and beyond. *Bio/Technology*, 10: 271-275.

Ferro, D. N. 2007. Bacterial pest control. In: Pimental, D. (Ed.), *Encyclopedia of Pest Management, Volume II*. CRC Press, Boca Raton, Florida, USA, pp. 30-36.

FIFRA SAP. 1998. Report of Subpanel on *Bacillus thuringiensis* (Bt) plant-pesticides and resistance management. EPA SAP Report. Available: http://www.mindfully.org/GE/FIFRA-SAP-Bt.htm. Visited: 18 Nov 2009.

FIFRA SAP. 2001. FIFRA Scientific Advisory Panel Report No. 2001-09. Available: http://www.epa.gov/scipoly/SAP/meetings/2001/july/julyfinal.pdf. Visited on 1 Nov 2009.

Flinn, P. W., Subramanyam, Bh and Arthur, F. H. 2004. Comparison of aeration and spinosad for suppressing insects in stored wheat. *Journal of Economic Entomology*, 97(4): 1465-1473.

Garczynski, S. F. and Siegel, J. P. 2007. Bacterial. In: Lacey, L. A. and Kaya, H. K. (Eds.), *Field Manual of Techniques in Invertebrate Pathology*. Springer Netherlands, pp. 175-197.

Gelernter, W. D. 1997. Resistance to microbial insecticides: The scale of the problem and how to manage it. In: Evans, H. F., *Microbial Insecticides: Novelty or Necessity? British Crop Protection Council Symposium Proceedings*, 68: 201–212.

Gill, S. S., Cowles, E. A. and Pietrantonio, P. V. 1992. The mode of action of *Bacillus thuringiensis* endotoxins. *Annual Review of Entomology*, 37: 615-636.

Glare, T. R., and O'Callaghan, M. O. 2000. *Bacillus thuringiensis: Biology, Ecology and Safety.* Wiley, New York, USA.

Glare, T. R., Corbett, G. E. and Sadler, T. J. 1993. Association of a large plasmid with amber disease of the New Zealand grass grub, *Costelytra zealandic*a, caused by *Serratia entomophila* and *Serratia proteamaculans. Journal of Invertebrate Pathology*, 62(2): 165-170.

Godfray, H. C. J., Briggs, C. J., Barlow, N. D., O'Callaghan, M., Glare, T. R. and Jackson, T. A..1999. A Model of insect-pthogen dynamics in which a pathogenic bacterium can also reproduce saprophytically. *Proceedings: Biological Sciences*, 266(1416): 233-240.

Gould, F., Anderson, A., Reynolds, A., Bumgarner, L. and Moar, W. 1995. Selection and genetic analysis of a Heliothis virescens (Lepidoptera: Noctuidae) strain with high levels of resistance to *Bacillus thuringiensis* toxins. *Journal of Economic Entomology*, 88: 1545-1559.

Gould, F., Martinez-Ramirez, A., Anderson, A., Ferré, J., Silva, F. J. and Moar, W. J. 1992. Broad-spectrum resistance to *Bacillus thuringiensis* toxins in *Heliothis virescens. Proceedings of the National Academy of Sciences of the United States of America*, 89: 7986-7988.

Gullan, P. J. and Cranston, P. S. 2005. *The Insects: An Outline of Entomology*. 3rd edition. Blackwell Publishing Ltd, MA, USA.

Hajek, A.E. and Delalibera J.I. 2010. Fungal pathogens as classical biological control agents against arthropods. *BioControl*, 55(1): 147-158.

Harrison, H., Patel, R. and Yousten, A. A. 2000. *Paenibacillus* associated with milky disease in central and South American scarabs. *Journal of Invertebrate Pathology*, 76(3): 169-175.

Heimpel, A. M. and Angus T. A. 1960. Bacterial insecticides. *Bacteriological Reviews*, 24: 266-88.

Höfte, H., Greve, H. de, Seurinck, J., Jansens, S., Mahillon, J., Ampe, C., Vandekerckhove, J., Vanderbruggen, H., Montagu, M. van and Zabeau, M. 1986. Structural and functional analysis of a cloned delta endotoxin of *Bacillus thuringiensis* Berliner 1715. *European Journal of Biochemistry*, 161: 273-280.

Hotson, I. K. 1982. The avermectins: A new family of antiparasitic agents. *Journal of the South African Veterinary Association*, 53(2): 87-90.

Hougard, J. M. 1990. Formulations and persistence of *Bacillus sphaericus* in *Culex quiquefasciatus* larval sites in tropical Africa. In: de Barjac, H. and Sutherland, D. (Eds), *Bacterial control of mosquitoes and black flies: biochemistry, genetics, and applications of Bacillus thuringiensis israelensis and Bacillus sphaericus.* Rutgers Univ. Press, New Brunswick, NJ, USA.

Hsu, J. C. and Feng, H. T. 2006. Development of resistance to spinosad in oriental fruit fly (Diptera: Tephritidae) in laboratory selection and cross-resistance. *Journal of Economic Entomology*, 99(3): 931-936.

Huang, F., Buschman, L. L., Higgins, R. A. and McGaughey, W. H. 1999. Inheritance of resistance to *Bacillus thuringiensis* toxin (Dipel ES) in the European corn borer. *Science*, 284: 965-967.

Huang, F., Higgins, R. A. and Buschman, L. L. 1997. Baseline susceptibility and changes in susceptibility to *Bacillus thuringiensis* subsp. kurstaki under selection pressure in

European corn borer (Lepidoptera: Pyralidae). *Journal of Economic Entomology*, 90: 1137-1143.

Huang, F., Leonard, B. R. and Andow, D. A. 2007a. Sugarcane borer resistance to transgenic *Bacillus thuringiensis* maize. *Journal of Economic Entomology*, 100: 164-171.

Huang, F., Leonard, B. R. and Wu, X. 2007b. Resistance of sugarcane borer to *Bacillus thuringiensis* Cry1Ab toxin. *Entomologia Experimentalis et Applicata*, 124: 117-123.

Hurst, M. R. H., Jones, S. M., Tan, B-L. and Jackson, T. A. 2007. Induced expression of the *Serratia entomophila* Sep proteins shows activity towards the larvae of the New Zealand grass grub *Costelytra zealandica. FEMS Microbiology Letters*, 275(1): 160-167.

Ismail, M. S. M., Soliman, M. F. M., Naggar, M. H. El and Ghallab, M. M. 2007. Acaricidal activity of spinosad and abamectin against two-spotted spider mites. *Experimental and Applied Acarology,* 43(2): 129-135.

Jackson, T. A., Huger, A. M. and Glare, T. R. 1993. Pathology of Amber Disease in the New Zealand Grass Grub *Costelytra zealandic*a (Coleoptera: Scarabaeidae). *Journal of Invertebrate Pathology*, 61(2): 123-130.

Jackson, T. A., Pearson, J. F., O'Callaghan, M. O., Mahanty, H. K. and Willocks, M. J. 1992. In: Jackson, T. A. and Glare, T. R. (Eds.), *Use of Pathogens in Scarab Pest Management*, Intercept, Andover, England, pp. 191-198.

Jackson,T. A., Boucias, D. G. and Thaler, J. O. 2001. Pathobiology of amber disease, caused by *Serratia spp.*, in the New Zealand grass grub, *Costelytra zealandica. Journal of Invertebrate Pathology*, 78(4): 232-243.

James, C. 2007. Global Status of Commercialized Biotech/GM Crops: 2007. ISAAA Brief No. 37. ISAAA: Ithaca, NY.

James, C. 2008. Global Status of Commercialized Biotech/GM Crops: 2008. ISAAA Brief No. 39. ISAAA: Ithaca, NY.

Janmaat, A. F. and Myers, J. 2003. Rapid evolution and the cost of resistance to *Bacillus thuringiensis* in greenhouse populations of cabbage loopers, *Trichoplusia ni. Proceedings of the Royal Society B: Biological Sciences,* 270: 2263-2270.

Janmaat, A. F. and Myers, J. H. 2005. The cost of resistance to *Bacillus thuringiensis* varies with the host plant of *Trichoplusia ni. Proceedings of the Royal Society B: Biological Sciences,* 272: 1031-1038.

Janmaat, A. F., Wang, P., Kain, W., Zhao, J. -Z., and Myers, J. 2004. Inheritance of resistance to *Bacillus thuringiensis* subsp. kurstaki in *Trichoplusia ni. Applied and Environmental Microbiology,* 70: 5859-5867.

Johnson, C. 2007. No refuge required by EPA, but one gene Bt cotton to be phased out in US. http://gmopundit.blogspot.com/2007/10/no-refuge-required-but-one-gene-bt.html.

Kaplinsky, N., Braun, D., Lisch, D., Hay, A., Hake, S. and Freeling, M. 2002. Maize transgene results in Mexico are artefacts. *Nature*, 416: 601-602.

Kegley, S.E., Hill, B. R., Orme, S., and Choi, A. H. 2009. PAN Pesticide Database, Pesticide Action Network, North America. Available: http://www. pesticideinfo.org/Detail_ Chemical.jsp?Rec_Id=PC35758. Visited on 10 Nov 2009.

Kim, S. B. and Goodfellow, M. 2002. *Streptomyces avermitilis* sp. nov., nom. rev., a taxonomic home for the avermectin-producing streptomycetes. *International Journal of Systematic and Evolutionary Microbiology*, 52(6): 2011-2014.

King, D. and Gordon, A. 2000. Contaminant found in Taco Bell taco shells. Food safety coalition demands recall (press release), vol 2001. Washington, DC: Friends of the Earth, 2000. Available: http://www.foe.org/act/getacobellpr.html.

Klein, M. G. 1981. Advances in the use of *Bacillu spopilliae* for pest control. In: Burges, H. D. (Ed.), *Microbial Control of Pests and Plant Diseases*. Academic Press, London.

Klein, M. G. and Kaya, H. K. 1995. *Bacillus* and *Serratia* speicies for scarab control. *Memorias do Instituto Oswaldo Cruz Rio de Janeiro*, 90(1): 87-95.

Klein, M. G., Grewal, P. S., Jackson, T. A. and Koppenhöfer, A. M. 2007. Lawn, turf and grassland pests. In: Lacey, L. A. and Kaya, H. K. (Eds.), *Field Manual of Techniques in Invertebrate Pathology*. Springer Netherlands, pp. 655-675.

Klein, M.G., 1992. Use of *Bacillus popilliae* in Japanese beetle control. In: T.R. Glare and T.A. Jackson (Eds.), *Use of Pathogens in Scarab Pest Management*. Intercept, Andover, England. pp. 179-189.

Lacey, L. A. 2004. Microbial control of insects. In: Capinera, J. (Ed.), *Encyclopedia of Entomology*. Kluwer Academic Publishers, Dordrecht, The Netherlands. pp. 1401-1407.

Lacey, L. A. and Siegel, J. P. 2000. Safety and ecotoxicology of entomopathogenic bacteria. In: Charles, J. F., Delécluse A. and Nielsen-le Roux C. (Eds.), *Entomopathogenic bacteria: from laboratory to field application*. Kluwer Academic Publishers, Dordrecht, Netherlands, pp. 253-273.

Lacey, L. A., and Undeen, A. H. 1986. Microbial control of black flies and mosquitoes. *Annual Review of Entomology*, 31: 265-296.

Lacey, L. A., Frutos, R., Kaya, H. K. and Vail, P. 2001. Insect pathogens as biological control agents: do they have a future? *Biological Control*, 21(3): 230-248.

Lambert, B. and Peferoen, M. 1992. Insecticidal promise of *Bacillus thuringiensis*. Facts and mysteries about a successful biopesticide. *BioScience*, 42:112-122.

Lasota, J. A. and Dybas R. A. 1991. Avermectins, a novel class of compounds: implications for use in arthropod pest control. *Annual Review of Entomology*, 36: 91-117.

Liu, X-X., Zhang, Q-W., Zhao, J-Z., Li, J-C., Xu, B-L. and Ma, X-M. 2005. Effects of Bt transgenic cotton lines on the cotton bollworm parasitoid *Microplitis mediator* in the laboratory. *Biological Control*, 35(1): 134-141.

Liu, Y. B. and Tabashnik, B. E. 1997. Inheritance of resistance to the *Bacillus thuringiensis* toxin Cry1C in the diamondback moth. *Applied and Environmental Microbiology*, 63: 2218 - 2223.

Losey, J. E., Rayor, L. S. and Carter, M. E. 1999. Transgenic pollen harms monarch larvae. *Nature*, 399: 214.

Lude, C. and Lang, A. 2006. A 3-year field-scale monitoring of foliage-dwelling spiders (*Araneae*) in transgenic Bt maize fields and adjacent field margins. *Biological Control*, 38(3): 314-324.

Madigan M. and Martinko J. 2005. Brock Biology of Microorganisms, 11th ed. Prentice Hall.

Mahar, A. N., Munir, M., Elawad, S., Gowen, S. R. and Hague, N. G. M. 2005. Pathogenicity of bacterium, *Xenorhabdus nematophila* isolated from entomopathogenic nematode (*Steinernema carpocapsae*) and its secretion against *Galleria mellonella* larvae. *Journal of Zhejiang University Science B*, 6(6): 457- 463.

Metz, M. and Fütterer, J. 2002. Suspect evidence of transgenic contamination. *Nature*, 416: 600-601.

Mohan, S., Raman, R. and Gaur, H. S. 2003. Foliar application of *Photorhabdus luminescens*, symbiotic bacteria from entomopathogenic nematode *Heterorhabditis indica*, to kill cabbage butterfly *Pieris brassicae. Current Science*, 84(11): 1397.

Mota-Sanchez, D., Hollingworth, R. M., Grafius, E. J. and Moyer, D. D. 2006. Resistance and cross-resistance to neonicotinoid insecticides and spinosad in the Colorado potato beetle, *Leptinotarsa decemlineata* (Say) (Coleoptera: Chrysomelidae). *Pest Management Science*, 62(1):30-7.

Nicolas, L., Regis, L. N. and Rios, E. M. 1994. Role of the exosporium in the stability of the *Bacillus sphaericus* binary toxin. *FEMS Microbiology Letters*, 124: 271-276.

Nielsen-Leroux, C., Pasquier, F., Charles, J. F., Sinegre, G., Gaven, B. and Pasteur, N. 1997. Resistance to *Bacillus sphaericus* involves different mechanisms in *Culex pipiens* (Diptera: Culicdae) larvae. *Journal of Medical Entomology*, 34: 321-327.

NPTN (National Pesticide Telecommunications Network). 2000. *Bacillus thuringiensis* General Fact Sheet. Oregon State University. Available: http://npic.orst.edu/factsheets/BTgen.pdf. Visited on 20 Oct 2009.

Oliver, J. B., Reding, M. E., Youssef, N. N., Klein, M. G., Bishop, B. L. and Lewis, P. A. 2009. Surface-applied insecticide treatments for quarantine control of Japanese beetle, *Popillia japonica* Newman (Coleoptera: Scarabaeidae), larvae in field-grown nursery plants. *Pest Management Science,* 65(4): 381-390.

Omura, S., Ikeda, H., Ishikawa, J., Hanamoto, A., Takahashi, C., Shinose, M., Takahashi, Y., Horikawa, H., Nakazawa, H., Osonoe, T., Kikuchi, H., Shibai, T., Sakaki, Y. and Hattori, M. 2001. Genome sequence of an industrial microorganism *Streptomyces avermitilis*: deducing the ability of producing secondary metabolites. *Proceedings of the National Academy of Sciences of the United States of America*, 98(21): 12215-12220.

Ostlie, K. R., Hutchison, W. D. and Hellmich, R. L. 1997. Bt corn and European corn borer: long term success through resistance management. North Central Region Extension Publication NCR602. Available: http://www.extension.umn.edu/distribution/ cropsystems/DC7055.html. Visit on 20 Nov 2009.

Patel, R., Piper, K., Cockerill III, F. R., Steckelberg, J. M. and Yousten, A. A. 2000. The biopesticide *Paenibacillus popilliae* has a vancomycinresistance gene cluster homologous to the enterococcal VanA vancomycin resistance gene cluster. *Antimicrobial Agents and Chemotherapy*, 44(3): 705-709.

Pell, J. K., Hannam, J. J., and Steinkraus, D. C. 2010. Conservation biological control using fungal entomopathogens. *BioControl*, 55(1): 187-198.

Qiu, L. H. and Zhang, W. J. 1999. Pest resistance to avermectins and its management. *Journal of China Agricultural University,* 4(1): 43-48.

Quist, D. and Chapela, I. H. 2001. Transgenic DNA introgressed into traditional maize landraces in Oaxaca, Mexico. *Nature*, 414: 541-543.

Rao, D. R., Mani, T. R., Rajendran, R., Joseph, A. S., Gajanana, A. and Reuben, R. 1995. Development of a high level of resistance to *Bacillus sphaericus* in a field population of *Culex quinquefasciatus* from Kochi, India. *Journal of the American Mosquito Control Association*, 11: 1-5.

Ravon, A. 2000. *Bacillus thuringiensis* application in agricultural. In: Charles, J. F., Delécluse A. and Nielsen-le Roux C. (Eds.), *Entomopathogenic bacteria: from laboratory to field application.* Kluwer Academic Publishers, Dordrecht, Netherlands, pp. 355.

ReportLinker. 2009. Research Report on Chinese Avermectin Industry, 2009-2010. Available: http://www.reportlinker.com/p0147916/Research-Report-on-Chinese-Avermectin -Industry-2009-2010.html. Visited on 10 Nov 2009.

Rippere, K. E. 1998. Systematics of the entomopathogenic bacteria *Bacillus popilliae, Bacillus lentimorbus,* and *Bacillus sphaericus.* Virginia Polytechnic Institute and State University. Blacksburg, Virginia, USA.

Rippere, K. E., Tran, M. T., Yousten, A. A., Hilu, K. H. and Klein, M. G. 1998. *Bacillus popilliae* and *Bacillus lentimorbus,*bacteria causing milky disease in Japanese beetles and related scarab larvae. *International Journal of Systematic Bacteriology,* 48: 395-402.

Romeis, J., Meissle, M. and Bigler, F. 2006. Transgenic crops expressing *Bacillus thuringiensis* toxins and biological control. *Nature Biotechnology,* 24(1): 63-71.

Roush, R.T. 1994. Managing pests and their resistance to *Bacillus thuringiensis*: can transgenic crops be better than sprays? *Biocontrol Science and Technology,* 4: 501-516.

Ruiu, L., Satta, A. and Floris, I. 2007. Susceptibility of the house fly pupal parasitoid *Muscidifurax raptor* (Hymenoptera: Pteromalidae) to the entomopathogenic bacteria *Bacillus thuringiensis* and *Brevibacillus laterosporus. Biological Control,* 43(2): 188-194.

Salgado, V. L. 1997. The mode of action of spinosad and other insect control products. *Down to Earth,* 52: 35-44.

Sayyed, A. H. and Wright, D. J. 2001. Cross-resistance and inheritance of resistance to *Bacillus thuringiensis* toxin Cry1Ac in diamondback moth (*Plutella xylostella* L.) from lowland Malaysia. *Pest Management Science,* 57: 413-421.

Schnepf, E., Crickmore, N., Van Rie, J., Lereclus, D., Baum, J., Feitelson, J., Zeigler, D. R. and Dean, D. H. 1998. *Bacillus thuringiensis* and its pesticidal crystal proteins. *Microbiology and Molecular Biology Reviews,* 62(3): 3775-3806.

Schwartz, J. L., Garneau, L., Savaria, D., Masson, L., Brousseau, R., and Rousseau, E. 1993. Lepidopteran-specific crystal toxins from *Bacillus thuringiensis* form cation- and anion-selective channels in planar lipid bilayers. *Journal of Membrane Biology,* 132: 53-62.

Sears, M. K., Hellmich, R. L., Siegfried, B. D., Pleasants, J.M., Stanley-Horn, D. E., Oberhauser, K. S.and Dively, G. P. 2001. Impact of Bt Corn Pollen on Monarch Butterfly Populations: A Risk Assessment. *Proceedings of the National Academy of Sciences of the United States of America,* 98: 11937-11942.

Shahina, F., Manzar, H. and Tabassum, K. A. 2004. Symbiotic bacteria *Xenorhabdus* and *Photorhabdus* associated with entomopathogenic nematodes in Pakistan. *Pakistan Society of Nematologists,* 22(2): 117-128.

Shelton, A. M., Wang, P., Zhao, J.-Z. and Roush, R. T. 2007. Resistance to insect pathogens and strategies to manage resistance: An update. In: Lacey, L. A. and Kaya, H. K. (Eds.), *Field Manual of Techniques in Invertebrate Pathology.* Springer Netherlands, pp. 793–811.

Shono, T. and Scott, J. G. 2003. Spinosad resistance in the housefly, Musca domestica, is due to a recessive factor on autosome 1. *Pesticide Biochemistry and Physiology,* 75: 1-7.

Skovmand, O., Kerwin, J. and Lacey, L. A. 2007. Microbial control of mosquitoes and black flies. In: Lacey, L. A. and Kaya, H. K. (Eds.), *Field Manual of Techniques in Invertebrate Pathology.* Springer Netherlands, pp. 735-750.

Slatin, S. L., Abrams, C. K., and English, L. 1990. Delta-endotoxins form cation-selective channels in planar lipid bilayers. *Biochemical and Biophysical Research Communications,* 169: 765-772.

Stenersen, J. 2004. *Chemical Pesticides: Mode of Action and Toxicology.* CRC Press. Boca Raton, FL. USA.

Stoner, K. A. 1998. Approaches to the biological control of insects. UMCE Bulletin #7144. Available: http://www.umext.maine.edu/onlinepubs/htmpubs/7144.htm. Visited on 20 Oct 2009.

Tabashnik, B. E. 1994a. Delaying insect adaption to transgenic plants: seed mixtures and refugia reconsidered. *Proceedings of the National Academy of Sciences of the United States of America,* 255: 7-12.

Tabashnik, B. E. 1994b. Evolution of resistance to *Bacillus thuringiensis. Annual Review of Entomology,* 39: 47-79.

Tabashnik, B. E. and Carrière, Y. 2007. Evolution of insect resistance to transgenic plants. In: K. Tilmon (Eds.), *Specialization, speciation, and radiation: the evolutionary biology of herbivorous insects.* University of California Press, Berkeley, CA, USA, pp. 267-279.

Tabashnik, B. E., Carrière, Y., Dennehy, T. J., Morin, S., Sisterson, M. S., Roush, R. T., Shelton, A. M. and Zhao, J. 2003. Insect resistance to transgenic Bt crops: lessons from the laboratory and field. *Journal of Economic Entomology,* 96: 1031-1038.

Tabashnik, B. E., Finson, N., Johnson, M. W. and Heckel, D. G. 1994. Cross-resistance to *Bacillus thuringiensis* toxin Cry1F in the diamondback moth. *Applied and Environmental Microbiology,* 60: 4627-4629.

Tabashnik, B. E., Liu, Y. B., de Maagd, R. A. and Dennehy, T. J. 2000. Cross-resistance of pink bollworm (*Pectiniphora gossypiella*) to *Bacillus thuringiensis* toxins. *Applied and Environmental Microbiology,* 66: 4582-4584.

Tabashnik, B. E., Liu, Y-B., Malvar, T., Heckel, D. G., Masson, L., Ballester, V., Granero, F., Mensua, J. L. and Ferré, J. 1997. Global variation in the genetic and biochemical basis of diamondback moth resistance to *Bacillus thuringiensis. Proceedings of the National Academy of Sciences of the United States of America,* 94: 12780-12785.

Tabashnik, B. E., Malvar, T., Liu,Y. B., Finson, N., Borthakur, D., Shin, B. S., Park, S. H., Masson, L., de Maagd, R. A. and Bosch, D. 1996. Cross-resistance of the diamondback moth indicates altered interactions with domain II of *Bacillus thuringiensis* toxins. *Applied and Environmental Microbiology,* 62: 2839-2844.

Tan, B., Jackson, T. A. and Hurst, M. R. H. 2006.Virulence of *Serratia* strains against *Costelytra zealandica. Applied and Environmental Microbiology,* 72(9): 6417–6418.

Thompson, G. D., Michel, K. H., Yao, R. C., Mynderse, J. S., Mosburg, C. T., Worden, T. V., Chio, E. H., Sparks T. C. and Hutchins S. H. 1997. The discovery of *Saccharopolyspora spinosa* and a new class of insect control products. *Down to Earth,* 52(1): 1-5.

Toews, M. D., Subramanyam Bh and Rowan J. 2003. Knockdown and mortality of eight stored-product beetles exposed to four surfaces treated with spinosad. *Journal of Economic Entomology,* 96(6): 1967-1973.

USEPA. 1996. *Bacillus thuringiensis* delta endotoxins encapsulated in killed *Pseudomonas fluorescens* (006409, 006410, 006457, 006462) Fact Sheet. Available: http://www.epa.gov/ oppbppd1/biopesticides/ingredients/factsheets/factsheet_006409. htm. Visited on 20 Oct 2009.

USEPA. 2005. *Bacillus thuringiensis* Cry3Bb1 protein and the genetic material necessary for its production (Vector ZMIR13L) in event MON 863 corn and *Bacillus thuringiensis* Cry1Ab delta-endotoxin and the genetic material necessary for its production in corn fact sheet. Available: http://www.epa.gov/pesticides/biopesticides/ingredients/factsheets/factsheet_ 006430-006484.htm. Visited on 20 Oct 2009.

USEPA. 2009. Biopesticide Active Ingredient Fact Sheets A, B. Available: http://www.epa.gov/ oppbppd1/biopesticides/ingredients/index_ab.htm. Visited on 20 Oct 2009.

Van Rie, J., Jansens, S., Hofte, H., Degheele, D., and Van Mellaert, H. 1989. Specificity of *Bacillus thuringiensis* d-endotoxins. Importance of specific receptors on the brush-border membrane of the mid-gut of target insects. *European Journal of Biochemistry*, 186: 239-247.

Walker, G.P., Cameron, P. J., MacDonald, F. M., Madhusudhan, V. V. and Wallace, A. R. 2007. Impacts of *Bacillus thuringiensis* toxins on parasitoids (Hymenoptera: Braconidae) of *Spodoptera litura* and *Helicoverpa armigera* (Lepidoptera: Noctuidae). *Biological Control*, 40(1): 142-151.

Wolt, J.D., Peterson, R.K.D., Bystrak, P., and Mead, T. 2003. A screening level approach for nontarget insect risk assessment: transgenic Bt corn pollen and the monarch butterfly (Lepidoptera: Danaidae). *Environmental Entomology*, 32: 237-246.

Wu, X. 2008. Resistance to *Bacillus thuringiensis* in sugarcane borer, *Diatraea saccharalis* (F.). Louisiana State University, Baton Rouge, LA, USA.

Wu, X., Huang, F., Leonard, B. R. and Moore, S. H. 2007. Evaluation of transgenic *Bacillus thuringiensis* corn hybrids against Cry1Ab-susceptible and -resistant sugarcane borer (Lepidoptera: Crambidae). *Journal of Economic Entomology*,100: 1880-1886.

Wu, X., Leonard, B. R., Zhu, Y-C., Abel, C.A., Head, G. P. and Huang, F. 2009. Susceptibility of Cry1Ab-resistant and -susceptible sugarcane borer (Lepidoptera: Crambidae) to four *Bacillus thuringiensis* toxins. *Journal of Invertebrate Pathology,* 100(1):29-34.

Wyss, C. F., Young, H. P., Shukla, J. and Roe, R. M. 2003. Biology and genetics of a laboratory strain of the tobacco budworm, *Heliothis virescens* (Lepidoptera: Noctuidae), highly resistant to spinosad. *Crop Protection*, 22(2): 307-314.

Yu, S. J. 2008. *The toxicology and biochemistry of insecticides*. CRC Press. Boca Raton, FL, USA.

Zangerl, A. R., McKenna, D., Wraight, C.L., Carroll, M., Ficarello, P., Warner, R. and Berenbaum, M. R. 2001. Effects of exposure to event 176 *Bacillus thuringiensis* corn pollen on monarch and black swallowtail caterpillars under field conditions. *Proceedings of the National Academy of Sciences of the United States of America*, 98: 11908-11912.

Zeigler, D. R. 1999. *Bacillus* Genetic Stock Center Catalog of Strains, 7th ed, Part 2: *Bacillus thuringiensis* and *Bacillus cereus. Bacillus* Genetic Stock Center, Columbus, Ohio. Available: http://www.bgsc.org/Catalogs/Catpart2.pdf. Visited on 19 Oct 2009.

Zhao, J. Z., Li, Y. X., Collins, H. L., Gusukuma-Minuto, L., R. Mau,F. L., Thompson, G. D. and Shelton, A. M. 2002. Monitoring and characterization of diamondback moth (Lepidoptera: Plutellidae) resistance to spinosad. *Journal of Economic Entomology*, 95(2): 430-436.

In: Microbial Insecticides: Principles and Applications
Editors: J. Francis Borgio, K. Sahayaraj, et al.

ISBN: 978-1-61209-223-2
© 2011 Nova Science Publishers, Inc

Chapter 16

RESTRICTION-MODIFICATION SYSTEMS IN BACILLUS THURINGIENSIS

Vladimir E. Repin[*]

Institute of chemical biology and fundamental medicine, Novosibirsk, 630090 Russia

ABSTRACT

Restriction-modification (R-M) systems are believed to have evolved to defend prokaryotic organisms from foreign DNA. However, this hypothesis may not be sufficient to explain the diversity and specificity in sequence recognition, as well as other properties, of these systems. The present work is devoted to the investigation of R-M systems in microorganisms of *Bacillus thuringiensis* species. The relativity of associating different R-M systems with certain taxons as well as the benefit of using strains with powerful R-M systems for the production of entomopathogenic preparations will be demonstrated.

16.1. INTRODUCTION

In early 1950ies, bacteriophage titration on different *E. coli* strains revealed the restriction and modification of phage DNA (Luria and Human, 1952; Bertani and Weigle., 1953). This phenomenon was supposed to be based on the existence of DNA restriction-modification (R-M) systems in bacterial cells.

By the middle of the 1960ies, as a result of systematic studies carried out by Prof. Arber and co-authors, it was shown that the enzymatic basis of R-M systems is a two-component enzyme system presented by restriction endonuclease (REase, or in some cases, R) and modifying DNA methyltransferase (MTase, or in some cases, M) (Arber and Dussoix, 1962; Arber and Linn, 1969; Roberts et al, 2003). Restriction of DNA bacteriophage is performed by restriction endonuclease through hydrolysis of phosphodiester bonds resulting in the cleavage of the DNA molecule. Modification of DNA protecting it against REase hydrolysis

---

[*] E-mail: ver@niboch.nsc.ru

is performed through methylation of nitrogen bases of certain nucleotides. Restriction and methylation enzymes, being components of the same R-M system, recognize the same specific sequence called the recognition site in the DNA. Speaking the philosophical language, here we are observing the law of unity and struggle of opposites, the struggle of shield and sword with our own eyes. R-M systems are considered to be strain-specific systems protecting the cell against the intervention of foreign genetic information in the form of a phage, plasmid or any other DNA and thus present a mechanism providing relative isolation in bacteria and archaebacteria. It's quite clear that this function of R-M systems is not a single one, and in many cases it cannot solve the problem of intervention of foreign DNA (Repin and Shelkunov, 1990). A number of experiments demonstrated the participation of restriction enzymes in a complex with DNA ligase in the recombination process *in vivo* (Chang and Cohen, 1977; Arber, 1979; Schiestl and Petes, 1991; Eddy and Gold, 1992). REases can play an important role in the evolution of both extrachromosomal and chromosomal genetic material. In addition, based on the investigation of the stabilization and incompatibility of plasmids carrying R-M systems it was supposed that R-M systems are genomic parasites like bacteriophages used for bacterial cell proliferation (Kusano *et al.,* 1995; Naito *et al.,* 1995). In the authors' opinion, the competition between different R-M systems results in the observed diversity of recognized DNA sequences. The functions of R-M systems are numerous (see reviews Levin, 1986; Repin and Shelkunov, 1990; Belfort and Roberts, 1997; Pingoud and Jeltsch, 2001; Vanyushin, 2005; Arber, 2004; Kobayashi, 2004; Karyagina, 1990), however, they have not been understood completely. We find R-M systems everywhere in the world of microorganisms, plasmids and bacteriophages. However, from the very beginning we can say that, probably, they are not obligatory cell elements. There exist microorganisms in which R-M systems were neither biochemically detected nor hypothetically predicted. Nevertheless, it should be noted that there exist cells in which the evaluated diversity by the specificity of R-M systems exceeds 20 (Kong et al, 2000). Such detection became possible after the determination of the complete nucleotide sequence for a number of microorganisms. More than one thousand bacterial and archeae genomes have been completely sequenced now. The complete sequences of these genomes have revealed a remarkable fact: more than 80% of the genomes appear to have at least one DNA R-M system and 75% of these genomes appear to contain multiple R-M systems (Table I) The column labeled 'Orphan M' contains solitary genes that share similarity with one component of an R-M system, but are missing others. The dam methylase of *E.coli* would be an example. The column labeled 'mR' contains genes that share similarity with methylation-dependent restriction systems. The present work is devoted to the investigation of R-M systems in microorganisms of *Bacillus thuringiensis* species. The relativity of associating different R-M systems with certain taxons as well as the benefit of using strains with powerful R-M systems for the production of entomopathogenic preparations will be demonstrated. The great interest for restriction endonucleases, a component of restriction- modification systems, resulted in massive search for enzymes with new specificities. Different search methods were used. One of the first methods, the so-called genetic method (Repin and Shelkunov, 1990), was based on the phenomenon of restriction of phage development in the host cell (Arber and Dussoix, 1962). Further research showed that other approaches can also be successfully applied (Nathans and Smith, 1975; Belavin, et al, 1988; Sokolov et al, 1984), and the most effective ones were determined (Repin and Degtyarev, 1992).

Table I. Potential DNA R-M systems in sequenced microbial genomes based on the computational identification of putative methylase genes and the presence of adjacent ORFs of unknown function (Kong et al, 2000)

Organism	Genome size (Mb)	Type I	Type II	Type III	Orphan M	mR	Total
Aeropyrum pernix	1.67		7				7
Aquifex aeolicus	1.55					1	1
Archaeoglobus fulgidus	2.18	1	2	1			4
Bacillus subtilis	4.21		2		1	1	4
Borrelia burgdorferi	1.44		2				2
Campylobacter jejuni	1.64	1	4			1	6
Chlamydia muridarum	1.07		no candidates				0
Chlamydia trachomatis	1.05		no candidates				0
Chlamydia pneumoniae AR39	1.23		no candidates				0
Deinococcus radiodurans	2.65		4			3	7
Escherichia coli	4.60	1			2		3
Haemophilus influenzae	1.83	2	3	1	1		7
Helicobacter pylori 26695	1.66	3	14	2	3		22
Helicobacter pylori J99	1.64	3	16	2	3		23
Methanococcus jannaschii	1.66	3	8				11
Mycobacterium tuberculosis	4.40	1	1				2
Mycoplasma genitalium	0.58	1			1		2
Mycoplasma pneumoniae	0.81	1	1				2
Neisseria meningitidis serotype A	2.18	3	7	2	1		13
Neisseria meningitidis serotype B	2.27	1	4	1	1		7
Pyrococcus abyssi	1.77	1	4				5
Pyrococcus horikoshii	1.74		3		1		4
Rickettsia prowazekii	1.10		no candidates				0
Synechocystis species	3.57		1		3	1	5
Thermatoga maritima	1.80		1				1
Treponema pallidum	1.16				1		1
Ureaplasma urealyticum	0.71	1	1		2		4

The most complete summary information on R-M systems of Bacillus thuringiensis species is collected in a public database of the Nobel Prize winner Dr. R.Roberts (http://rebase.neb.com). The following systems were found in the representatives of the species and numerous subspecies of *Bacillus thuringiensis*:

Enzyme and its prototype	Recognition site 5'-... - 3', cleavage site ↓	Citation index	Notes; (information source)
MboI; Bth1786I, Btu33I, Btu34I, Btu36I, Btu37I, Btu39I, Btu41I, Bth84I, BthCanI, Bth1997I, Bth1141I, Bth945I, Bth211I, Bth1140I, Bth213I, Bth221I, BtcI	GATC	17	*ssp. amagiensis;* *ssp. canadensis;* *ssp. novosibirsk;* *ssp. pondicheriensis;* *ssp. tohokuensis;* *ssp. tolworthi;* *ssp. yunnanensis;* (Puchkova *et al.,* 2002; Repin, *et al.,* 1991; Kuzin *et al.,* 1989).
MboI; BtkII	↓GATC	1	(Azizbekyan *et al.,* 1992)
ClaI; Bth1202I, Bth9415I, BtuI	ATCGAT	3	*ssp. monterrey;* *ssp. kumamotoensis;* (Azizbekyan, *et al.,* 1988; Puchkova, *et al.,* 2002)
EcoRII; BthDI, BthEI	CC↓WGG	2	Unpublished observations
AvaII; BthAI	G↓GWCC	1	Unpublished observations
AvaII; BtiI	GGWCC	1	(Azizbekyan, *et al.,*1984)
GsuI; Bth1795I	CTGGAG	1	Unpublished observations
Fnu4HI; BthCI	GCNG↓C	1	Neoschizomer Unpublished observations
M.BthCI	GCNGC	1	Unpublished observations
PvuII; Bth2350I	CAGCTG	1	Unpublished observations
HaeII; Btu34II	RGCGCY	1	(Repin, *et al.,*1991)
BinI; Bth617I	GGATC	1	(Puchkova, *et al.,*2002)
HpaII; BthKORF840P	CCGG	1	Putative, (Han, *et al.,*2006)
HpaII; M2.BthKORF840P	CCGG	1	Putative, (Han, *et al.,*2006)

Enzyme and its prototype	Recognition site 5'-...- 3', cleavage site ↓	Citation index	Notes; (information source)
M.BetI; M1.BthKORF840P	WCCGGW	1	Putative (Han, *et al.*, 2006).
M.BcefI; M.BthIPS78P	ACGGC	1	Putative, (Jenkinson, *et al.*,2003).
PstI; Bth9411I	CTGCAG	1	(Puchkova, *et al.*,2002)
FnuDII; BtkI	CG↓CG	1	(Azizbekyan *et al.*,1992).
I-HmuI; I-BasI	AGTAATGAG CCTAACGCT CAGCAA	1	Homing endonuclease, (Landthaler, *et al.* 2003; Landthaler, *et al.* 2006)
I-BthII; I-BthII	ATTATCCGTG ATGAGTCAA TTCA (-21/-23)	1	Homing endonuclease, (Nord and Sjoberg, 2008).
I-BmoI; I-BthORFAP	GAGTAAGAG CCCGTAGTA ATGACATGG C	1	Homing endonuclease, putative; (Nord and Sjoberg, 2008).

(where R = G or A, Y = C or T, W = A or T, N = A or C or G or T)

The table summarizes complete R-M systems with restriction endonucleases of type II; those presented only by methyltransferases; homing endonucleases, and systems predicted on the basis of the presence of known motives in the nucleotide sequence, but not biochemically isolated.

The analysis of R-M systems in *Bacillus thuringiensis* species, published or obtained from unpublished studies, shows that the specificity 5'- GATC -3' is most widespread. It is not surprising because systems with this specificity are also most often found in other bacterial taxons and, probably, are among the oldest ones (Pingoud et al, 2005). However, in this case, the number of producer strains was increased also due to a task-oriented search for this very restriction-modification system (Repin, et al, 1991). Now it is difficult to imagine such a situation, but in the USSR it was always a complex task to purchase imported enzymes. The "Enzyme base" program for production of own restriction endonucleases or their manufacturable analogues was developed and implemented as an alternative to imported enzymes. One of such enzymes was restriction endonuclease Sau3AI (neoschizomer MboI). The pathogenic strain *Staphylococcus aureus* is known to be the producer of this endonuclease; the cell wall of this strain is not readily lysed by enzymes. The detected strains of this species gave a very low enzyme yield.

According to our conception (Repin and Shelkunov, 1990), *Bacillus* genus was among the most promising ones for revealing manufacturable producers. Representatives of *Bacillus thuringiensis* species were selected for the search of isoschizomers Sau3AI. Besides the above-mentioned manufacturability and non-pathogenicity, there were also other reasons to do this: the availability of a large collection of entomopathogenic strains of different subspecies; an attempt to reduce the risk of industrial phagolysis under other equal conditions. It is known that the correlation of restriction was shown in vivo and in vitro for a number of microorganisms. Six strains carrying isoschizomers Sau3AI were detected from among 15 preliminarily selected strains. All the strains restricted the development of bacteriophages D5, K, M available in the collection by several orders (Repin, et al, 1991). Recommendations were given for insecticide production concerning preferred use of *B. thuringiensis* strains having restriction-modification systems.

We would like to note that entirely different subspecies have the same restriction-modification systems. We hypothesize that the division of *B. thuringiensis* into subspecies occurred at later stages of the species evolution i.e. we suppose that R-M systems are more ancient. Probably, the division into numerous subspecies has not been completed yet or more generalized criteria should be introduced for their reclassification. Somehow or other, our latest investigations of the microbiote of the unique ecological niche, the Valley of Geysers, Kamchatka, detected endospore-forming bacteria producing unusual parasporal crystalline inclusions usually linked with the spore. Electron microscopy revealed the presence of a common envelope (exosporium) round the spore and the crystal; this envelope probably presents an obstacle to the manifestation of insecticide properties of the studied microorganisms under experimental conditions. Experiments on insecticide activity in larvae of insects representing three orders *Hyponomeuta evonymellus* (Lepidoptera), *Aedes aegypti* (Diptera), and *Leptinotarsa decemlineata* (Coleoptera) demonstrated a weakly expressed insecticide activity or its absence. These atypical crystal- and sporiferous strains were referred to *B. thuringiensis* atypical species. The study of antagonistic properties of these strains from the Valley of Geysers with the delayed antagonism method revealed a pronounced activity with respect to pathogenic *Candida albicans* in all the studied strains (Andreeva, et al, 2008). High antiviral activity of the atypical strains against avian influenza virus A/H5N1 (Andreeva et al, 2008) was shown. Unfortunately, screening of these atypical strains did not reveal producers of restriction endonucleases among them.

The analysis of 60 strains representing 43 subspecies of *B.thuringiensis* (Table II) revealed site-specific endonuclease activity in 17 strains.

Table 2. Characteristic of *B.thuringiensis* strains by crystal-formation and endonuclease activity

Strain	Subspecies	H-antigen	Crystal presence	R-M systems presence
221	*thuringiensis*	1	+	+
213	*thuringiensis*	1	−	+
218	*finitimus*	2	+	−
204	*alesti*	3ac	+	−
273	*kurstaki*	3abc	+	−
84-F-51-46	*sumiyoshiensis*	3ad	+	−
84-I-1-13	*fukuokaensis*	3ade	+	−
617	*sotto*	4ab	−	+
56	*sotto*	4ab	+	±
256	*kenyae*	4ac	+	±

Strain	Subspecies	H-antigen	Crystal presence	R-M systems presence
69-6	*galleriae*	5ab	+	−
Can	*canadensis*	5ac	−	+
201	*entomocidus*	6	+	−
206	*subtoxicus*	6	+	−
212	*aizawai*	7	+	−
209	*morrisoni*	8ab	+	−
PG-14	*morrisoni*	8ab	+	−
941	*ostriniae*	8ac	−	−
942	*nigeriensis*	8bd	−	−
211	*tolworthi*	9	+	+
270	*darmstadiensis*	10	+	−
811	*darmstadiensis*	10	+	−
73-E-10-2	*darmstadiensis*	10	+	−
73-E-10-16	*darmstadiensis*	10	+	−
272	*toumanoffi*	11ab	+	−
74-F-16-18	*kyushuensis*	11ac	−	−
943	*kyushuensis*	11ac	−	−
271	*thompsoni*	12	−	−
1124	*pakistani*	13	+	−
1125	*israelensis*	14	+	−
749	*israelensis*	14	+	−
157	*israelensis*	14	+	−
874	*israelensis*	14	+	−
1199	*dakota*	15	+	−
944	*dakota*	15	+	−
1200	*indiana*	16	+	+
945	*tohokuensis*	17	+	+
1202	*kumamotoensis*	18ab	+	+
1203	*tochigiensis*	19	+	−
1140	*yunnanensis*	20ab	+	+
1141	*pondicheriensis*	20ac	±	+
1205	*colmeri*	21	+	−
948	*colmeri*	21	+	−
1206	*shandongiensis*	22	+	−
949	*shandongiensis*	22	+	−
1207	*japonensis*	23	+	−
9410	*japonensis*	23	+	−
9411	*neolensis*	24ab	±	+
19-97	*novosibirsk*	24ac	+	+
9412	*coreanensis*	25	+	−
9413	*silo*	26	+	−
9414	*mexicanensis*	27	+	−
9415	*monterrey*	28ab	+	+
9416	*amagiensis*	29	+	+

Table 2. (Continued)

Strain	Subspecies	H-antigen	Crystal presence	R-M systems presence
84-F-58-35	*amagiensis*	29	+	+
2B	*toguchini*	31	+	−
9417	*toguchini*	31	−	−
9418	*cameroun*	32	−	−
9419	*leesis*	33	−	−
9420	*konkukian*	34	−	−

(+) - the presence of the characteristic (property) to be determined
(-) – the absence of this characteristic (property)

Table 3. Substrate specificity of restriction endonucleases of *B.thuringiensis* strains

#	Strain	H-antigen	Subspecies	Recognition site 5'-... - 3'
	221	1	*thuringiensis*	GATC
	213	I	*thuringiensis*	GATC
	617	4ab	*sotto*	GGATCC
	can	5ac	*canadensis*	GATC
	211	9	*tolworthi*	GATC
	945	17	*tohokuensis*	GATC
	1202	18ab	*kumamotoensis*	ATCGAT
	1140	20ab	*yunnanensis*	GATC
	1141	20ac	*pondicheriensis*	GATC
	9411	24ab	*neolensis*	CTGCAG
	19-97	24ac	*novosibirsk*	GATC
	9415	28	*monterrey*	ATCGAT
	84-F-58-35	29	*amagiensis*	GATC

As a result, substrate specificity of restriction endonucleases was determined for 13 strains (Table III). In the other cases, isolation and determination of specificity was complicated by low enzyme content and the presence of a large number of nonspecific nucleases. The specificity of endonucleases 5'-ATCGAT-3', which is widespread in the bacterial world, was detected in *B.thuringiensis* with different genetic and biochemical methods. The strains were characterized by high ability to produce endonucleases as compared with the prototype ClaI, but they were significantly inferior to other isoschizomers of *Bacillus* genus in this characteristic (for example, Repin, et al, 1989). BthCI system whose restriction endonuclease is a neoschizomer recognizing the sequence GCNG↓C is a unique R-M system in *B. thuringiensis*. This is the only presently known enzyme cleaving the above sequence in a specified site. But this is not a commercial preparation, and there are no published literature data on it. Homing endonucleases were detected in *B. thuringiensis* species. One of them is I-BasI (prototype I-HmuI). I-BasI has properties that are typical of homing endonucleases, nicking the intron-minus polymerase genes in either host genome, three nucleotides downstream of the intron insertion site. The enzymes are two highly similar nicking DNA endonucleases, which are each encoded by a group I intron inserted into homologous sites within the DNA polymerase genes of *Bacillus* phages SPO1 (I-HmuI) and

Bastille(I-BasI). Although the enzymes have distinct DNA substrate specificities, both bind to an identical 25bp region of their respective intron-minus DNA polymerase genes surrounding the intron insertion site. The endonucleases appear to interact with the DNA substrates in the downstream exon 2 in a similar manner. Structural modeling analyses predict that I-BasI might make specific base contacts both upstream and downstream of the site of intron insertion. The predicted requirement for base-specific contacts in exon 1 for cleavage by I-BasI was confirmed experimentally (Landthaler *et al.*,2006). The *Bacillus thuringiensis ssp. pakistani* nrdF intron encodes a homing endonuclease, denoted I-BthII, with an unconventional GIY-(X)8-YIG motif that cleaves an intronless nrdF gene 7 nt upstream of the intron insertion site, producing 2-nt 3' extensions (Nord and Sjoberg, 2008). Interestingly that several group I introns have been previously found in strains of the *Bacillus cereus* group at three different insertion sites in the nrdE gene of the essential nrdIEF operon coding for ribonucleotide reductase. Hypothetical restriction-modification systems predicted on the basis of the presence of conservative motives but not always biochemically isolated should be especially noted. The determination of such systems became possible after complete genome sequencing. Several such systems marked in the Note of Table 1 as "putative" were predicted for *B. thuringiensis*. Two complete nucleotide sequences of *Bacillus thuringiensis* (serovars konkukian str. 97-27 and str. A1 Hakam) were published in public press, but the other data were obtained from other sources (http://rebase.neb.com). Only a few complete R-M systems or only methyltransferases were predicted in all the cases of analyzing the complete nucleotide sequence. We could explain this fact based on the supposition that a number of R-M systems are localized in numerous plasmids, including megaplasmids (>30 MDa) typical of *Bacillus thuringiensis* species. Isolation of plasmids from typical strains showed that each strain contained at least one plasmid; if it there is only one it should necessarily be a megaplasmid, and the largest recorded number of plasmids in the strain reached 13 (Reyes-Ramírez et al, 2008). Certainly, talking about R-M systems of *Bacillus thuringiensis*, we should mention the specific character of this taxon. Perhaps, in this case, we for the first time realized the narrowness of systematics by K. Woese (Woese and Fox, 1977) based on the comparison of sequences of 16s RNA genes though it was very important and provided understanding of bacterial variability. *Bacillus* genus presented a dump of species united at that time into a single genus only by the type of the cell wall and strongly differing from each other in their relation to oxygen and the ability for spore formation, which had long needed re-systematizaiton (Ash *et al.,* 1993) performed later. There appeared the genera *Alicyclobacillus* (Wisotzkey *et al.,* 1992), *Paenibacillus* (Ash *et al.,* 1993; Heyndrickx *et al.,* 1995), *Brevibacillus* (Shida *et al.,* 1996), *Aneurinibacillus* (Shida *et al.,* 1996; Heyndrickx *et al.,* 1997), *Virgibacillus* (Heyndrickx *et al.,* 1998), *Salibacillus, Gracilibacillus* (Wainù *et al.,* 1999) and others. However, within this undivided genus, a group of species called *Bacillus cereus* was notable. It included *B. thuringiensis, B. cereus* and *B. anthracis* whose biochemical characteristics were very much alike with the exception that *B. thuringiensis* was an entomopathogenic strain, *B. cereus* was a soil microorganism found everywhere and conventionally pathogenic for man, and *B.anthracis* was a highly pathogenic species, which could probably be used as biological weapons. The first of the above species was characterized by parasporal bodies and intracellular protein crystals, the third one – by two specific plasmids. The second had neither the former nor the latter. The comparison of 16s RNA sequence revealed complete coincidence in all the three species. Multilocus enzyme electrophoresis and the analysis of sequences of 9 chromosomal genes suggested that species of this group are members of the same species (Helgason, et al, 2000) differing in plasmid composition. And while from the point of view of systematics we are merely facing an oddity, great danger of effects on humans arises from incorrect determination. In our case, however, this also concerns the set of R-M systems possible for *B.thuringiensis.* By the highest standards, R-M systems of *B. cereus* and *B. anthracis* should be added to the list. And while for *B. anthracis* such systems are not numerous for quite understandable reasons, a considerable number of R-M systems differing in specificity from enzymes of *B. thuringiensis* was isolated and predicted for *B. cereus.* Summing up the table of specificities, it should be noted that *B. thuringiensis* species (even, if we add two related species, *B. cereus* and *B. anthracis*) was not a Colondike by R-M systems like *Helicobacter pylori* species. No prototype was detected, and none of the enzymes of *Bacillus thuringiensis* R-M systems became a commercially significant chemical preparation. Yes, a large number of investigations allowed us to realize that microorganism evolution has significant differences and one cannot confine himself to the determination of 16s RNA in species identification. The presence of Cry genes and the ability to synthesize endotoxins are essential in the determination of *Bacillus thuringiensis* species. But using this line of reasoning, we can go to another extreme when the existence of species is completely negated. But it is not correct. If we look at Leeuwenhoek's pictures of bacteria and their description, we will see that there exists a quite definite and relatively stable set of characteristics (i.e. gene systems) providing a stable state round which the pendulum of genetic changes is swinging. Only fixed changes in the genotype coming from other organisms are estimated at approximately 10%, sometimes the estimates are several times higher (Nelson, et al, 1999). In fact, the number of such acts of genetic intervention is much larger. R-M systems can facilitate or impede the insertion of genetic systems. Accumulation of changes due to mutation process, horizontal transfer, etc. gives rise to the transition to another unstable state, which can result in the return to the previous stable functioning of the old system, the transition into another closely related stable state or the formation of a new taxon (Repin et al, 2001). Realization of one of these possibilities depends on external conditions, the state and conditions of changes in the ecological niche, the duration of its existence and spread. Graphically it can be presented as a 3D model where balls are ecological niches with a relatively stable genetic pool.

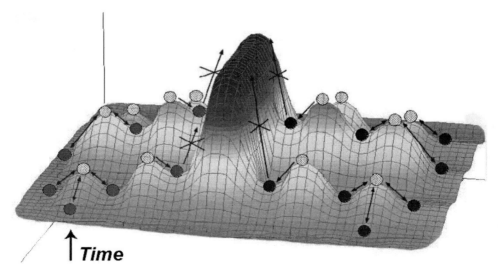

↑ **Time**

The time vector is an important component of the pattern. It means that all "hills" (unstable state) and "valleys" (stable state) can change or remain unchanged depending on the geophysical state of the planet or its parts. Naturally, sometimes this transition can take a long time, can involve several transitions and cannot be performed as a single act. We consider it necessary to make one more important observation that we draw all the conclusions based not the fact that the laws and regularities discovered by us also apply to the so-called "unculturable" microorganisms the number of which now exceeds that of culturable ones by several orders. However, microorganisms in unstable states can be found in the "unculturable" area (Repin, 2008). The possibility of transfer of large gene clusters under varying conditions was predicted and later shown in the works on gram-negative sporiferous bacteria (Andreeva et al, 2008). It is important to repeat that new ecological niches are the basis for manifestation of new properties. The last observation to be made while describing R-M systems of *Bacillus thuringiensis*: Roberts' list contains a large number of producers of restriction endonucleases (about 300) exact identification of which not always interested researchers. They confined themselves to the determination of Bacillus genus. Carriers of R-M systems were identified as Bacillus species. We believe that they include representatives of *Bacillus thuringiensis* species. It means that there exist R-M systems with a different specificity within *B. thuringiensis*. Special thanks to R.Roberts, A.Janulaitis, S. Degtyarev, R. Azizbekyan who opened me the world of the R-M systems. Particular financial support has been done by the Siberian Branch of RAS (projects ##10,117).

CONCLUSIONS

Summing up the table of specificities, it should be noted that *B. thuringiensis* species (even, if we add two related species, *B. cereus* and *B. anthracis*) was not a Colondike by R-M systems like *Helicobacter pylori* species. No prototype was detected, and none of the enzymes of *Bacillus thuringiensis* R-M systems became a commercially significant chemical preparation. And the last observation to be made while describing R-M systems of *Bacillus thuringiensis*: Roberts' list contains a large number of producers of restriction endonucleases (about 300) exact identification of which not always interested researchers. They confined themselves to the determination of Bacillus genus. Carriers of R-M systems were identified as Bacillus species. We believe that they include representatives of *Bacillus thuringiensis* species. It means that there exist R-M systems with a different specificity within *B. thuringiensis*. *Special thanks to R.Roberts, A.Janulaitis, S. Degtyarev, R. Azizbekyan who opened me the world of the R-M systems. Particular financial support has been done by the Siberian Branch of RAS (projects ##10,117).*

REFERENCES

Andreeva I.S., Pechurkina N.I., Burtseva L.I., Kalmykova G.V., Puchkova L.I., Saranina I.V., Repin V.E. 2008. Atypical *Bacillus thuringiensis* strains isolated from soil and hot springs of the Valley of geysers (Kamchatka). *Biotechnology* (Russia). (6): 41-50.

Arber W. 1979. Promotion and limitation of genetic exchange. *Science*. 205: 361-365.

Arber W. and Dussoix D. 1962. Host Specificity of DNA Producted by Escherichia Coli: I. Host controlled modification of bacteriophage lambda. *J. Mol. Biol.* 5: 18-36.

Arber, W. 2004. Genetic Variation and Molecular Evolution . In *Encyclopedia of Molecular Cell Biology and Molecular Medicine.* (Meyers, R.A. Ed., Wiley-VCH Verlag), 5: 331-352.

Arber W, Linn S. 1969. DNA modification and restriction. *Annu. Rev. Biochem.* 38: 467–500.

Ash C., Priest F. G., Collins M. D. 1993. Molecular identification of rRNA group 3 bacilli (Ash, Farrow, Wallbanks and Collins) using a PCR probe test. Proposal for the creation of a new genus *Paenibacillus.Antonie van Leeuwenhoek.* 64: 253-260.

Azizbekyan, R.R., Rebentish, B.A., Stepanova, T.V., Netyksa, E.M., Bychkova, M.A. 1984. Site specific restriction endonuclease from a *Bacillus thuringiensis* strain. *Dokl. Akad. Nauk (*Russia) 274: 742-744.

Azizbekyan, R.R., Rebentish, B.A., Netyksa, E.M. 1988. *Bacillus thuringiensis* restrictase sensitive to dam-methylation. *Biotechnology* (Russia). 4: 197-198.

Azizbekyan R.R., Rebentish B.A., Netyksa E.M., Bychkova M.A., Bolotin A.P. 1992. Site-specific restriction endonuclease from *Bacillus thuringiensis* var. *Kumantoenis. Mol. Gen. Microbiol. Virusol.* 1-2: 13-15.

Belavin, P.A., Dedkov, V.S., Degtyarev, S.K. 1988. A simple technique for detection of restriction endonucleases in bacterial colonies. *Appl.. Biochem. Microbiol.* 24: 121-124.

Belfort, M., Roberts, R.J. Homing endonucleases: keeping the house in order. *Nucleic Acids Res.* – 1997. –V.25. – P. 3379-3388.

Bertani G., Weigle J. J. Host-controlled variation in bacterial viruses. *J. Bacteriol.* – 1953. – V.65. – P.113–121.

Chang, S., Cohen, S. N. In vivo site specific genetic recombination promoted by the EcoRI restriction endonuclease. *Proc. Natl. Acad. Sci. U. S. A.* - 1977. – V.74. – P. 4811–4815

Eddy, S.R., Gold, L. Artificial mobile DNA element constructed from the EcoRI endonuclease gene. *Proc. Natl. Acad .Sci. U.S.A.* – 1992. – V.89. – P. 1544-1547.

Han, C.S., Xie, G., Challacombe, J.F., Altherr, M.R., Bhotika, S.S., Bruce, D., Campbell, C.S., Campbell, M.L., Chen, J., Chertkov, O., Cleland, C., Dimitrijevic, M., Doggett, N.A., Fawcett, J.J., Glavina, T., Goodwin, L.A., Hill, K.K., Hitchcock, P., Jackson, P.J., Keim, P., Kewalramani, A.R., Longmire, J., Lucas, S., Malfatti, S., McMurry, K., Meincke, L.J., Misra, M., Moseman, B.L., Mundt, M., Munk, A.C., Okinaka, R.T., Parson-Quintana, B., Reilly, L.P., Richardson, P., Robinson, D.L., Rubin, E., Saunders, E., Tapia, R., Tesmer, J.G., Thayer, N., Thompson, L.S., Tice, H., Ticknor, L.O., Wills, P.L., Brettin, T.S., Gilna, P. Pathogenomic Sequence Analysis of *Bacillus cereus* and *Bacillus thuringiensis* Isolates Closely Related to *Bacillus anthracis. J. Bacteriol.* – 2006. – V.188. – P.3382-3390.

Helgason E., Økstad O. A., Caugant D. A., Johansen H.A., Fouet A., Mock M., Hegna I. , Kolstø[*] A.-B. *Bacillus anthracis, Bacillus cereus*, and *Bacillus thuringiensis*. One Species on the Basis of Genetic Evidence. *Applied and Environmental Microbiology* – 2000. - V. 66, No. 6. -P. 2627-2630.

Heyndrickx M., Lebbe L., Kersters K., De Vos P., Forsyth G., Logan N.A. *Virgibacillus*: a new genus to accommodate *Bacillus pantothenticus* (Proom and Knight 1950). Emended description of *Virgibacillus pantothenticus. Int. J. Syst. Bacteriol.* – 1998. –V. 48. – P. 99-106.

Heyndrickx M., Lebbe L., Vancanneyt M., Kersters K., De Vos P., Logan N.A., Forsyth G., Nazli S., Ali N., Berkeley R.C.W. A polyphasic reassessment of the genus *Aneurinibacillus*, reclassification of *Bacillus thermoaerophilus* (Meier-Stauffer *et al.* 1996) as *Aneurinibacillus thermoaerophilus* comb. nov., and emended descriptions of *A. aneurinilyticus* corrig., and *A. migulanus*, and *A. thermoaerophilus*. *Int. J. Syst. Bacteriol.* – 1997. – V. 47. –P. 808-817.

Heyndrickx M., Vandemeulebroecke M., Scheldeman P., Hoste B., Kersters K, De Vos P., Logan N.A., Aziz A.M., Ali N., Berkeley R.C. *Paenibacillus* (formerly *Bacillus*) *gordonae* (Pichinoty *et al.* 1986) Ash *et al.* 1994 is a later subjective synonym of *Paenibacillus* (formerly *Bacillus*) *validus* (Nakamura 1984) Ash *et al.* 1994: emended description of *P. validus*. *Int. J. Syst. Bacteriol.* – 1995. – V. 45. –P. 661-669.

Jenkinson, E., Crickmore, N. Identification of a novel DNA methyltransferase activity from *Bacillus thuringiensis*. *Curr. Microbiol.* – 2003. – V.47. – P. 144-145.

Karyagina, A.S. Genetic determinants for the enzymes of bacterial restriction-modification systems. *Mol. Gen. Microbiol. Virusol.* – 1990. – V.7. – P. 3-11.

Kobayashi, I. Restriction-modification systems as minimal forms of life. (In *Nucleic Acids Mol. Biol.* (Pingoud, A. Ed., Springer-Verlag, Berlin Heidelberg). – 2004. – V. 14. – P. 19-62.

Kong H, Lin L.-F, Porter N, Stickel S, Byrd D, Posfai J, Roberts RJ. Functional analysis of putative restriction-modification system genes in the *Helicobacter pylori* J99 genome.. Nucleic Acids Res. – 2000. – V. 28. – P. 3216–3223.[

Kusano, K., Naito, T., Handa, N., and Kobayashi, I. 1995. Restriction-modification systems as genomic parasites in competition for specific sequences. *Proc. Natl. Acad. Sci. USA*, 92: 11095–11099.

Kuzin, A.I., Bolesnin, M.I., Smolyaninov, V.V., Azizbekyan, R.R. Site specific restriction endonuclease BtcI from *Bacillus thuringiensis var. canadensis*. *Mol. Gen. Microbiol. Virusol.* – 1989. – V.4. –P. 36-39.

Landthaler, M., Shen, B.W., Stoddard, B.L., Shub, D.A. I-Basl and I-Hmul: Two phage intron-encoded endonucleases with homologous DNA recognition sequences but distinct DNA specificities. *J. Mol. Biol.* – 2006. –V.358. – P. 1137-1151.

Landthaler, M., Shub, D.A. The nicking homing endonuclease I-BasI is encoded by a group I intron in the DNA polymerase gene of the *Bacillus thuringiensis* phage Bastille. *Nucleic Acids Res.* – 2003. – V.31. – P. 3071-3077.

Levin, B.R. Restriction-modification immunity and the maintenance of genetic diversity in bacterial populations. In *Evolutionary Processes and Theory* (S. Karlin, E. Nero Ed., Academic Press, New York). - 1986 – P. 669-688.

Luria, S.E., Human M. A nonhereditary, host-induced variation of bacterial viruses. *J. Bacteriol.* - 1952. – V.64. –P. 557-569.

Naito, T., Kusano, K. and Kobayashi, I. 1995. Selfish behavior of restriction-modification systems. *Science,* 267: 897–899.

Nathans, D., Smith, H.O. Restriction endonucleases in the analysis and restructuring of DNA molecules. *Annu. Rev. Biochem.* – 1975. – V.44. – P. 273-293.

Nelson[1] K. E., Clayton[1] R. A., Gill[1] S. R., Gwinn[1] M. L., Dodson[1] R.J., Haft[1] D. H., Hickey[1] E.K., Peterson[1] J. D., Nelson[1] W. C., Ketchum[1] K. A., McDonald[1] L., Utterback[1] T. R., Malek[1] J. A. , Linher[1] K. D., Garrett[1] M. M., Stewart[1] A. M., Cotton[1] M. D., Pratt[1] M. S., Phillips[1] C. A., Richardson[1] D., Heidelberg[1] J., Sutton[1] G. G., Fleischmann[1] R. D., Eisen[1]

J. A., White[1] O., Salzberg[1] S. L., Smith[1] H. O., Venter[1] J. C. , Fraser[1] C. M. Evidence for lateral gene transfer between Archaea and Bacteria from genome sequence of *Thermotoga maritima*. *Nature*. - 1999. – V.399. – P. 323-329.

Nord, D., Sjoberg, B.M. Unconventional GIY-YIG homing endonuclease encoded in group I introns in closely related strains of the *Bacillus cereus* group. *Nucleic Acids Res.* – 2008. – V.36. – P. 300-310.

Pingoud, A., Jeltsch, A. Structure and function of type II restriction endonucleases. *Nucleic Acids Res.* – 2001. –V.29. – P. 3705-3727.

Pingoud V., Sudina A., Geyer H., Bujnicki J. M., Lurz R., Lüder G., Morgan R., Kubareva E. , Pingoud A. Specificity changes in the evolution of Type II restriction endonucleases. A biochemical and bioinformatic analysis of restriction enzymes that recognize unrelated sequences. *J. Biol. Chem.* – 2005. – V. 280, No. 6. – P. 4289-4298.

Puchkova, L.I., Kalmykova, G.V., Burtseva, L.I., Repin, V.E. Entomapathogenic bacteria *Bacillus thuringiensis* as producers of restriction endonucleases. *Appl. Biochem. Microbiol.* – 2002. – V.38. – P. 140-144.

Repin, V.E., Burtseva, L.I., Burlak, V.A., Trusova, S.I. Search of restriction endonuclease producers with required specificity. *Mol. Gen. Microbiol. Virusol.* – 1991. –V.11. – P. 20-22.

Repin, V.E., Degtyarev, S.K. Comparison of express-methods used for the detection of restriction endonucleases in microorganisms. *Appl. Biochem. Microbiol.* – 1992. – V.28. – P. 152-155.

Repin, V.E., Degtyarev, S.K., Petrov, N.A., Prihodko, E.A., Bendukidze, K.A., Fodor, I.I. A *Bacillus subtilis* strain V-3667 is used as a producer of the restriction endonuclease Bsu15I giving an increased yield. *Soviet Patent Office* (1989). SU 1449583.

Repin, V.E., Shelkunov, S.N. Occurrence and function of restriction endonucleases (A Review). *Usp. Sovrem. Biol .(Russia)*. – 1990. - V.110. – P. 34-47.

Repin V.E., Torok T., Kuz'min M.I. Biodiversity of microorganisms from bottom sediments of Lake Baikal . Russian Geology and Geophysics.- 2001.- V. 42, N 2.- P. 219-221.

Reyes-Ramírez A. , Ibarra J. E. Plasmid Patterns of *Bacillus thuringiensis* Type Strains. *Appl. Environm. Microbiol.* – 2008. - V. 74, No. 1. –P. 125-129.

Roberts, R.J., Belfort, M., Bestor, T., Bhagwat, A.S., Bickle, T.A., Bitinaite, J., Blumenthal, R.M., Degtyarev, S.K., Dryden, D.T., Dybvig, K., Firman, K., Gromova, E.S., Gumport, R.I., Halford, S.E., Hattman, S., Heitman, J., Hornby, D.P., Janulaitis, A., Jeltsch, A., Josephsen, J., Kiss, A., Klaenhammer, T.R., Kobayashi, I., Kong, H., Kruger, D.H., Lacks, S., Marinus, M.G., Miyahara, M., Morgan, R.D., Murray, N.E., Nagaraja, V., Piekarowicz, A., Pingoud, A., Raleigh, E., Rao, D.N., Reich, N., Repin, V.E., Selker, E.U., Shaw, P.C., Stein, D.C., Stoddard, B.L., Szybalski, W., Trautner, T.A., Van Etten, J.L., Vitor, J.M., Wilson, G.G., Xu, S.Y.; A nomenclature for restriction enzymes, DNA methyltransferases, homing endonucleases and their genes. *Nucleic Acids Res.* – 2003. – V.31, - P. 1805-1812.

Schiestl R. H., Petes T. D. The BamHI restriction enzyme mediates integration of nonhomologous DNA into the *Saccharomyces cerevisiae* genome . *Proc. Natl. Acad. Sci. USA*. – 1991. –V.88. – P.7585-7589.

Shida O., Takagi H., Kadowaki k., Kamagata K. Proposal for two new genera, *Brevibacillus* gen. nov. and *Aneurinibacillus* gen. nov.. *Int. J. Syst. Bacteriol.* – 1996 – V. 46. –P.939-946.

Sokolov, N.N., Fitsner, A.B., Khoroshutina, E.B., Kheislere, M.Ya. Determination of restriction endonuclease activity in toluene lysates of bacterial cells. *Biull. Eksp. Biol. Med. (Russia)*. – 1984. – V. 97. – P. 163-165.

Wainù, M., Tindall, B. J., Schumann, P., Ingvorsen, K. *Gracilibacillus* gen. nov., with description of *Gracilibacillus halotolerans* gen. nov., sp. nov.; transfer of *Bacillus dipsosauri* to *Gracilibacillus dipsosauri* comb. nov., and *Bacillus salexigens* to the genus *Salibacillus* gen. nov., as *Salibacillus salexigens* comb. nov.. *Int. J. Syst. Bacteriol.* – 1999 – V. 49. –P.821-831.

Vanyushin, B.F. Enzymatic DNA methylation is an epigenetic control for genetic functions of the cell. *Biochemistry* – 2005. – V.70. – P. 598-611.

Wisotzkey, J. D., Jurtshuk, P., Jr, Fox, G. E., Deinhard, G. and Poralla, K. Comparative sequence analyses on the 16S rRNA (rDNA) of *Bacillus acidocaldarius, Bacillus acidoterrestris*, and *Bacillus cycloheptanicus* and proposal for creation of a new genus, *Alicyclobacillus* gen. nov.. *Int. J. Syst. Bacteriol.* - 1992. – V.42. – P. 263–269.

Woese C, Fox G. Phylogenetic structure of the prokaryotic domain: the primary kingdoms. Proc Natl Acad Sci USA – *1977. – V. 74 (11). – P. 5088–5090.*

PART V: INTERACTIONS

In: Microbial Insecticides: Principles and Applications   ISBN: 978-1-61209-223-2
Editors: J. Francis Borgio, K. Sahayaraj, et al.   © 2011 Nova Science Publishers, Inc

Chapter 17

MICROORGANISMS AS BIOCONTROL AGENTS AND THEIR INTERACTION WITH NEMATODES

Ebrahim Karimi [*], *Gholamreza Salehi Jouzani*
and Yadollah Dalvand
Agricultural Biotechnology Research Institute of Iran (ABRII), Mahdasht Road,
31535-1897, Karaj, Iran

ABSTRACT

Parasitic Nematodes cause numerous diseases in human beings, animals and plants placing major burdens on agricultural production and global health. Common management methods used include planting resistant crop varieties, rotating crops, incorporating soil amendments, soil solarization and applying nematicides. Biological control is other recently developed approach to controlling nematodes using microorganisms and their natural products. Because of that nematodes often occur in high numbers, it is not surprising that a wide variety of microorganisms exploit nematodes as food, i.e., as sources of carbon, nitrogen, and energy. Predators of nematodes include mites, collembola, flatworms, protozoa, and other predacious nematodes. Many exploiters of nematodes have prolonged and specialized interactions with nematodes; these microorganisms are called parasites such as fungi, bacteria, and etc. Other organisms and microorganisms may have a detrimental effect on nematodes without utilizing them as a substrate, by competing for food, space, and necessary resources. Competitors of nematodes include other nematodes, bacteria, and fungi. However, interaction between nematodes and other competitors may cause increased damage to a particular food source. Some microorganisms may antagonize nematodes by producing nematicidal or nemastatic compounds such as ammonia, certain fatty acids, and avermectins. This mode of action is referred to as antibiosis, and involves bacteria and fungi. Of these four mechanisms of antagonism (predation, parasitism, competition, and antibiosis), parasitism has received the most research effort. This chapter will focus on different microorganisms and their negative effects on nematodes. Understanding of these interactions is a branch of biocontrol science that can help us achieve scientific and friendly environmental ways in order to manage nematodes population.

[*] E-mail: ekarimi20@yahoo.com, Tel: +98-261-2703536, Fax: +261- 2704539

17.1. NEMATODES

Nematodes (= roundworms) are the most abundant and successful group of metazoans (Bird and Kaloshian, 2003). They are common in terrestrial and aquatic ecosystems and include a wide range of feeding types (Yeates *et al.,* 1993). Nematodes can feed on bacteria, fungi, algae, plants, and they can also parasitize insects, animals and humans. The range of nematode body size can vary from 0.5 mm (plant parasites) till several meters (parasites of animals). Nematodes have a huge impact on human life. Human parasites cause severe diseases (for example *Ascaris*) and plant parasites lead to huge losses in agriculture, which are estimated worldwide as much as 100 billion US dollars annually (Bird and Kaloshian, 2003). All parts of plants can be invaded by plant feeding nematodes, but the root feeders are economically the most important since they are mainly responsible for yield losses. In natural grasslands, plant parasitic nematodes have been estimated to take up around a quarter of the net primary production (Stanton, 1988).

The feeding modes of plant parasitic nematodes differ according to the plant parts they feed upon. They can feed from the outside of the roots on outer cortical cell layers, while never entering the roots with more than the feeding stylet. These are ectoparasitic nematodes and they are considered to be feeding generalists (Yeates *et al.,* 1993; van der Putten *et al.,* 2005). Plant parasitic nematodes enter and feed inside roots, are called endoparasites, some of which are feeding specialists (Yeates *et al.,* 1993). The majority of crop damage is caused by the tylenchid nematodes, particularly the approximately 60 species of the genus *Meloidogyne* (root-knot nematodes), *Heterodera* (cyst nematode), *Globodera* (cyst nematode), *Pratylenchus* (lesion nematode), *Ditylenchus* (stem and bulb nematode), *Tylenchulus* (citrus nematode), *Xiphinema* (dagger nematode), *Radopholus* (burrowing nematode), *Rotylenchulus* (reniform nematode), *Helicotylenchus* (spiral nematode), *Belonolaimus* (sting nematode) (Obannon and Nemec, 1979; Siti *et al.,* 1982; Calvet *et al.,* 1995; Jayakumar *et al.,* 2002; McK Bird, 2004).

17.2. CONTROL OF NEMATODES

The traditional methods of reducing plant parasitic numbers in agriculture include cultural practices, crop rotation and the use of chemical pesticides. Newer methods of nematode suppression include organic matter addition (Akhtar and Alam, 1993; Vawdrey and Stirling, 1997; Akhtar and Malik, 2000; Widmer *et al.,* 2002, Oka, 2010) and biocontrol practices (Kerry and Gowen ,1995; Oka *et al.,* 2000; Meyer and Roberts, 2002; Alabouvette *et al.,* 2006; van Bruggen *et al.,* 2006, Mateille *et al.,* 2009). Moreover, the studies on plants resistant against nematode infections give promising results (Strauss and Agrawal, 1999; Williamson and Kumar, 2006, Li *et al.,* 2008, Klink *et al.,* 2010).

17.2.1. Chemical Control of Nematodes

All nematicides are eventually degraded if they remain in the topsoil where there is greatest microbial activity. Once nematicides or their degradation products are flushed through the upper soil layers their persistence may be extended. It is the problem of toxic products in groundwater that has led to the prohibition of fumigant and non-fumigant nematicides in some countries. The permitted level of pesticide residues in drinking-water is different from country to country. In regions of intensive agricultural production these tolerance levels may be exceeded at certain times of the year (Kottegoda, 1985; Hague and Gowen, 1987; Hamilton *et al.,* 2003; Boobis *et al.,* 2008).

Nematicides are highly toxic compounds that have very low LD_{50} values. This is particularly important for operators of application machinery and people at risk from exposure to the chemicals during their application. The liquid formulations of some of the non-fumigant nematicides are emulsifiable concentrates. Their use should therefore be restricted to skilled operators who take adequate safety precautions. This may not always be the case where basic levels of education are poor or where operators cannot read the instructions on the labels of the products. The application of nematicides to crops too near to harvest is another risk which pesticide residue monitoring may not be sufficiently well coordinated to prevent (van Berkum and Hoestra, 1979; Bromilow, 1980; Wright, 1981; Kottegoda, 1985; Hague and Gowen, 1987). Several general purpose fumigants give excellent control of nematodes in soil. The efficacy is related to their high volatility at ambient temperatures. All fumigants have low molecular weights and occur as gases or liquids. As they volatilize, the gas diffuses through the spaces between soil particles; nematodes living in these spaces are killed.

Fumigants perform best in soils that do not have high levels of organic matter (which deactivates the toxicant) and they are free-draining but have adequate moisture. In general, fumigants are most effective in warm soils (12° to 15°C) as dispersion is temperature related. The most famous fumigants used for control of plan parasitic nematodes are D-D (1,2-dichloropropane and 1,3-dichloropropene),

1,3-Dichloropropene, Ethylene Dibromide, 1,2-Dibromo-3-Chloropropane, Methyl Bromide, Chloropicrin, Metam Sodium, Dazomet, and Methyl Isothiocyanate (MITC), Sodium Tetrathiocarbonate. In addition to fumigants, other chemicals including Carbamates (Aldicarb, Aldoxycarb, Carbofuran, Oxamyl) and Organophosphates (Ethoprop, Fenamiphos, Cadusafos, Fosthiazate, Terbufos, Fensulfothion, Phorate, thionazin, fosthietan, and isazofos) also some biochemicals including DiTera, ClandoSan and Sincocin are used for control of the nematodes (Chitwood, 2003, Giannakou and Anastasiadis, 2005, Kong   *et al.,* 2006, Karpouzas *et al.,* 2007).

17.2.2. Problems with Chemical Nematicides

17.2.2.1. Effects on Non-Target Organisms

Soil fumigants are extensively used to control plant-parasitic nematodes, weeds, fungi, and insects for planting of high value cash crops. The ideal pesticide should be toxic only to the target organisms; however, fumigants are a class of pesticide with broad biocidal activity and affect many non-target soil organisms. Soil microorganisms play one of the most critical roles in sustaining the health of natural and agricultural soil systems. The ability of soil microorganisms to recover after treatment with pesticide is critical for the development of healthy soils. Because of broad-spectrum activities of nematicides, they radically alter soil flora and fauna. Fumigant usage may result in the absence of nematode competitors, predators, and parasites in soils (Sipes and Schmitt, 1995; Ibekwe, 2004; Arias-Estévez *et al.,* 2008). The elimination of mycorrhizae by methyl bromide can result in poorer plant growth (Klein, 1996). Long-term aldicarb treatment of potato fields decreased the number of bacterial genera and species, decreased the population levels of plant growth promoting rhizobacteria, and increased total bacterial biomass compared to untreated soils (Sturz and Kimpinski, 1999). Nematicides can greatly alter the subsequent structure of nematode communities in soils; for example, *Pratylenchus* recolonized methyl bromide–treated pasture soil, replacing *Helicotylenchus* as the dominant phytoparasitic nematode (Yeates and van der Meulen, 1996). Nematodes and other organisms play a complex role in agroecosystems (Barker and Koenning, 1998); use of broad-spectrum biocides makes it difficult to exploit some of these roles (Ibekwe, 2004; Arias-Estévez *et al.,* 2008).

17.2.2.2. Environmental Contamination

One of the greater environmental problems sometimes associated with nematicide usage is groundwater contamination. Indeed, the initial detection of the nematicides DBCP and aldicarb in groundwater in the United States over 20 years ago led to the stimulation of scientific and regulatory interest in pesticide contamination of groundwater that continues to this day (Cohen, 1996). Even though DBCP usage was prohibited in 1977, groundwater contamination persists (Soutter and Loague, 2000). In 1990, the manufacturer of Temik (aldicarb) announced a voluntary halt on its sale for use on potatoes because of concerns about groundwater contamination. The following year, a train wreck released 72,000 L of metam sodium into the Upper Sacramento River and resulted in soil microbial changes that persisted for at least a year (Taylor *et al.,* 1996). When the special review of 1,3-D by the U.S. EPA was terminated, several measures for reducing potential groundwater contamination were instituted, such as prohibition of usage within 100 feet of drinking-water wells, in areas overlying karst geology, and in several states with certain soil types and where groundwater is 50 feet from the soil surface (U.S. Environmental Protection Agency, 2001). As previously indicated, 1,3-D use was suspended in California in 1990 for several years because of its detection in air distant from application sites, specifically in a school. This has resulted in the creation of 300-foot–wide buffer zones around residences for fumigation (100 feet wide if fields are drip irrigated). In addition, "township caps" limit the total amount of 1,3-D that can be used in a given area in California (Carpenter *et al.,* 2001).

17.2.2.3. Resistance to Nematicides

Resistance of field populations to nematicides has not been well characterized and is remarkably insignificant in comparison to the levels of resistance observed with mammalian parasites. Indeed, a recent National Academy of Sciences monograph stated, "Resistance of nematodes to soil fumigants has yet to be observed but systemic nematicides are relatively new and it is probably only a matter of time until resistance does appear" (Berenbaum, 2000). In one interesting study, Moens and Hendrickx (Moens and Hendrickx, 1998) evaluated populations of *Meloidogyne naasi*, *G. rostochiensis*, and *Pratylenchus crenatus* exposed to aldicarb for 15 years. Although some developmental differences were noticed between treated and control populations when challenged with aldicarb, the differences were species specific and were concluded to be not significant.

In another investigation, the free-living nematode *Rhabditis oxycerca* was bred for 400 generations in order to obtain strains adapted to reproducing on concentrations of 600- and

480-μg/ml aldicarb and oxamyl, respectively. Compared with wild type, the two mutant strains were characterized by decreased size (particularly in the tail region), tolerance of warm temperature, production of offspring, and migration in electric fields, among other characteristics. In nematicide solutions, the wild type exhibited decreased motility, electric field migration, and reproduction (K¨ampfe and Sch¨utz, 1995).

In a third study, genetically selected strains of the insect pathogen *Heterorhabditis bacteriophora* possessed 8–70-fold increased resistance to fenamiphos, avermectin, and oxamyl (Glazer *et al.,* 1997). The enhanced resistance was generally stable in the absence of further nematicide pressure; the strains have obvious potential utility in integrated pest management systems.

17.3. Crop Rotations and Cover Crops

Crop rotation to a non-host crop is often adequate by itself to prevent nematode populations from reaching economically damaging levels. However, it is necessary to positively identify the species of nematode in order to know what plants are its host(s) and non-hosts. A general rule of thumb is to rotate to crops that are not related to each other. For example, pumpkin and cucumbers are closely related and rotating between them would probably not be effective to keep nematode populations down. A pumpkin/bell pepper rotation might be more effective. Even better is a rotation from a broadleaf to a grass. Asparagus, corn, onions, garlic, small grains, Cahaba white vetch, and Nova vetch are good rotation crops for reducing root-knot nematode populations. Crotalaria, velvet bean, and grasses like rye are usually resistant to root-knot nematodes. (Yepsen, 1984; Peet, 1996; Wang, *et al.,* 2004) Rotations like these will not only help prevent nematode populations from reaching economic levels, they will also help control plant diseases and insect pests.

17.4. SOIL SOLARIZATION

Soil solarization is consistently effective only where summers are predictably sunny and warm. The basic technique entails laying clear plastic over tilled, moistened soil for approximately six to eight weeks. Solar heat is trapped by the plastic, raising the soil temperature. It is a rather recently developed technique, which has shown promise for the control of several soilborne pathogens and weeds in warm areas (Katan *et al.,* 1976; Katan, 1987). Its effectiveness against nematodes was first demonstrated in Israel and later, attempts to control nematodes by soil solarization were also made in Australia (Porter and Merriman, 1983), USA (Stapleton and DeVay, 1983; LaMondia and Brodie, 1984), Italy (Greco *et al.,* 1985), and South Africa (Barbercheck *et al.,* 1986). The use of soil solarization for control of nematodes has received increasing attention in recent years.

The efficacy of soil solarization is based on the sensitivity of nematodes to relatively high temperatures. Hot water treatment is still suggested as a method to kill nematodes within plant parts. Soaking narcissus bulbs for 3-4 hours in hot water at 44 - 45°C is suggested for the control of Ditylenchus dipsaci. The same temperature is also effective in disinfesting chrysanthemum cuttings from Aphelenchoides ritzema-bosi, but in this case a 30 minute treatment is enough. The soaking treatment can be reduced to 5 minutes by raising the water temperature to 50°C Endo (Endo, 1962) demonstrated that the time required to kill 100 percent juveniles within cysts of Heterodera gIycine is temperature dependent. He found that I second, 8 minutes, and 8 hours were required to inhibit egg hatch of the nematode at 63, 52, and 44, respectively. Similar lethal temperatures (5 min exposure at 55°C) are reported for *Globodera rostochiensis* (Mai and Lautz, 1953). It is clear that in plastic mulch conditions, temperatures higher than 50°C can be reached in the top 5 cm of the soil, but it is important to note that temperatures of 40-50°C also have been previously reported up to 10-15 cm depth in hot seasons (Katan *et al.,* 1976; Porter and Merriman, 1983; Stapleton and DeVay, 1983; Greco *et al.,* 1985; Barbercheck *et al.,* 1986).

Moreover, temperatures of 36-40°C can be reached at 20-30 cm depth in warm areas. Such temperatures, if prolonged, can be lethal to nematodes or at least may reduce their infectivity because of energy reserve depletion. Such weakened nematodes may also be more vulnerable to biotic and abiotic stresses. Further, nematode antagonists may prevail in the soil after solarization or they may colonize the solarized soils much faster than the non-solarized soils when incorporated artificially. These possible long lasting effects have not been investigated for nematodes, but they are known to exist for other soilborne pathogens. In deeper soil profiles (30-40 cm) lethal or sublethal temperatures usually are not attained, where the temperature increases only 3-4°C. However, this may render nematodes more active, which in turn increases nematode infection by Pasteuria penetrans (Walker and Wachtel, 1988) or nematode capture by some trapping fungi. Because of soil biotic and abiotic changes following solarization, the suppressing effect on nematodes may not become evident or can last for several months. Walker and Wachtel (Walker and Wachtel, 1988) found that infection of *Meloidogyne javanica* juveniles by *P. penetrans* increases for ten months after soil solarization and Stapleton and DeVay (Stapleton and DeVay, 1983) observed a reduction of *Helicotylenchus digonicus* populations only three months after solarization treatment.

The beneficial effect of nematode control is further enhanced because of the lack of or the reduced interaction of nematodes with soilborne diseases, such as those caused by *Verticillium* spp. and *Fusarium* spp. The effectiveness of soil solarization on nematode control has been ascertained by several investigators in different areas. In fields heavily infested with *Ditylenchus dipsaci* in Israel, soil solarization protected garlic bulbs throughout the growing season resulting in a greater yield increase compared with EDB and methyl bromide applications (Siti *et al.,* 1982). In Italy, control of *D. dipsaci* in a sandy soil increased with solarization periods and only 10, 6 and 2 percent nematodes were still viable after four, six, and eight weeks solarization, respectively (Greco *et al.,* 1985).

Satisfactory control has also been obtained against cyst forming nematodes. In New York State, the hatch of *Globodera rostochiensis* from solarized soil was nil, 32, and 41% of that in non-treated soil at 5, 10, and 15 cm depth, respectively (LaMondia and Brodie, 1984). Better control was achieved in Italy, where only 24-38% of eggs survived in cysts of *Heterodera carotae* from the top 30 cm soil solarized for four to eight weeks (Greco *et al.,* 1985). In another trial the yield of carrots from plots solarized for eight weeks was 4.6 kg/m, compared with 5.5 kg in plots treated with 400 1/ha DD and only 2.3 kg in the control (Greco *et al.,* unpublished).

Root-knot nematodes, *Meloidogyne* spp., are widespread throughout the world and the most damaging nematode group. Although in some solarization experiments control of these nematodes was inconsistent (Greco *et al.,* 1985; Barbercheck *et al.,* 1986), other investigations have demonstrated that excellent control is achieved by soil solarization under greenhouse conditions (Cenis, 1984; Cartia and Greco, 1987; Cartia *et al.,* 1988; Cartia *et al.,* 1989). Here, soil temperatures in the top 30 cm of mulched soil can be 3-5°C higher than in mulched soil in open air and, therefore, higher nematode mortality can be expected. In Spain (Cenis, 1984), nematode control in solarized soil was slightly less than in soil treated with methyl bromide, but yields were similar in both treatments. In Italy, several solarization experiments have been undertaken to control pathogens of solanaceous crops in the greenhouse. Tomato roots were heavily galled and rotting by the end of the growing season in control plots, while they were nearly free of nematodes in plots treated with methyl bromide. In solarized plots root infestation was intermediate because of the late nematode attack, but no rotting of the root was observed (Cartia *et al.,* 1988). Moreover, alternating methyl bromide with solarization treatments inhibited root infestation as did soil treated with methyl bromide for two consecutive years. In the same solarized plots, yield increases averaged 60 percent over the control and were similar to those obtained with methyl bromide treatments. Solarization is well documented as an appropriate technology for control of soilborne pathogens and nematodes, but the economics of purchasing and applying plastic restrict its use to high-value crops.

17.5. PLANT RESISTANCE

Generally speaking, a resistant cultivar is more effective against sedentary endoparasitic species such as root-knot and cyst nematodes than against "grazing" ectoparasitic species. Root-knot and cyst nematodes spend most of their lifecycle within the root, relying on specialized cells for feeding. Upon entering the roots of resistant cultivars, these nematodes become trapped as the feeding cells necessary for their survival fail to develop. Many crop cultivars—tomatoes and soybeans in particular—have been specifically bred for nematode resistance. The "N" designation on tomato seed packages (usually as part of "VFN") refers to nematode resistance. A few cultivars of potatoes are resistant to the golden nematode, which is a pest only in a small area of the northeastern U.S. Although most cultivars of potatoes are susceptible to infection by nematodes, some varieties tolerate infection better than others. For example, population densities of root-lesion nematodes, Pratylenchus penetrans that would affect yield in "Superior" are tolerated with little effect by "Russet Burbank." (MacGuidwin, 1993)

At the USDA's Agricultural Research Service in Charleston, South Carolina, developed two nematode-resistant varieties of bell pepper, "Charleston Belle" and "Carolina Wonder," available from commercial seed companies. (Sanchez, 1997) Charleston Belle and its susceptible parent, "Keystone Resistant Giant," were compared as spring crops to manage the southern root-knot nematode, Meloidogyne incognita in autumn-cropped cucumber and squash. Cucumber grown in plots following Charleston Belle had lower root gall severity indices than in crops following Keystone Resistant Giant. Cucumber yields were 87 percent heavier and numbers of fruit 85 percent higher in plots previously planted to Charleston Belle than to Keystone Resistant Giant. Squash grown in plots following Charleston Belle had lower root gall severity indices than those following Keystone Resistant Giant. Squash yields were 55 percent heavier and numbers of fruit 50 percent higher in plots previously planted to

Charleston Belle than to Keystone Resistant Giant. These results demonstrate that root-knot nematode-resistant bell pepper cultivars such as Charleston Belle are useful tools to manage *M. incognita* in double-cropping systems with cucurbit crops (Thies *et al.,* 2004). Nematode-tolerant or resistant cultivars of snap beans ("Harvester" and "Alabama #1"), lima beans ("Nemagreen"), and sweet potatoes ("Carolina Bunch," "Excel," "Jewel," "Regal," "Nugget," and "Carver") also exist and may be used in a similar strategy to reduce nematode levels for crops that follow.

Breeding for nematode resistance in most crops is complicated by the ability of the nematode species (primarily cyst nematodes and root-knot nematodes) to develop races or biotypes that overcome the genetic resistance factors in the crop. In order to maintain resistant crop cultivars on farms, researchers suggest that susceptible and resistant cultivars be planted in rotation. When a nematode-resistant cultivar is planted, nematode populations generally decrease, but over the course of the growing season the few nematodes in a particular population capable of overcoming this resistance begin to increase. If in the following season the farmer plants a susceptible cultivar, the overall nematode numbers will still be low enough to avoid significant yield reduction, but more importantly, the selective pressure favoring the increase of the "counter-resistant" bio types is removed. As long as the farmer has been continues to use alternate susceptible and resistant cultivars, the nematodes can be kept at non-damaging levels. Transgenic crop resistance to nematodes and other pests is being developed for numerous crops by various companies worldwide. The use of genetically modified organisms is not accepted in organic production systems.

17.6. BIOLOGICAL CONTROL

The management of nematodes is more difficult than that of other pests because nematodes mostly inhabit the soil and usually attack the underground parts of the plants (Stirling, 1991). Although chemical nematicides are effective, easy to apply, and show rapid effects, they have begun to be withdrawn from the market in some developed countries owing to concerns about public health and environmental safety (Schneider *et al.,* 2003). The search for novel, environmentally friendly alternatives with which to manage plant-parasitic nematode populations has therefore become increasingly important.

In the past 20 years three developments have occurred which have had significant effects on the prospects and opportunities for the biological control of plant-parasitic nematodes. First, several nematicides have been withdrawn from the market because of health and environmental problems associated with their production and use (Thomason, 1987). As a result of this, and increasing public concern over the use of pesticides in food production, there has been increased interest in the development of alternative methods of control, including the use of biological agents. Second, it has been demonstrated in several soils that nematophagous fungi and bacteria increase under some perennial crops, and under those grown in monocultures, and so may control some nematode pests, including cyst and root-knot nematodes (Stirling, 1991). Such nematode-suppressive soils have been reported from around the world and include some of the best documented cases of effective biological control of nematode pests. Finally, a number of commercial products based on nematophagous fungi and bacteria have been developed, but all so far have had only limited success. Their use has been based on empirical research, and it is instructive to consider what might be the key factors for a successful biological control agent for nematodes in order to identify the reasons for the general failure of the products that have been developed. Stirling (1991) identified five key aspects in setting up an experiment to evaluate a biological control agent those are presented below in a slightly modified form.

* The test organism and any organic amendment should be applied at practical application rates; 0.1 percent w/w soil is equivalent to 2.5 tonnes/ha and should represent a maximum dose. Tests should always be performed in a non-sterilized soil with a natural residual soil microflora.

* Appropriate treatments as well as an untreated control should be included if the organism is added with a substrate. These treatments should include the substrate alone, the organism alone, and the autoclaved colonized substrate. Too often, untreated controls are compared only with large applications of the organism and substrate and this does not allow separation of the effects of the agent from the effects of the substrate. In several tests reported in the literature, application of the substrate alone has decreased nematode populations to the same extent as the substrate colonized by the agent, and there is no clear evidence of biological control.

* Population densities of the agent under test should be monitored to ensure that it has survived in soil throughout the period that activity against the nematode target is required. Such monitoring may require the development of selective media, which can be a difficult and time-consuming task.

* Nematode mortality caused by the organism under test should be measured to assess whether differences between nematode population densities in treated and untreated soil relate to the levels of kill caused by the agent. Infection levels are relatively straightforward to estimate for most parasites, but repeated sampling is required to determine total kills. The effects of agents which produce toxins or have indirect effects on nematodes through competition, the modification of root exudates or the colonization of feeding cells, can only be measured by assessing their impact on nematode development.

* The impact of the soil environment, host plant and nematode should be tested as these are likely to affect the efficacy of the biological control agent, and could account for the lack of activity of potential agents in specific test conditions.

Nematodes in soil are subject to infections by some bacteria, fungi and viruses. This creates the possibility of using soil microorganisms to control plant-parasitic nematodes (Mankau, 1980; Jatala, 1986). Following some examples of different microorganisms with power of biocontrol of plant parasitic nematodes will be included.

17.7. FUNGI

17.7.1. Nematophagous Fungi

Nematophagous (nematode-destroying) fungi comprise more than 200 species of taxonomically diverse fungi that all share the ability to attack living nematodes (juveniles, adults and eggs) and use them as nutrients (Nordbring-Hertz *et al.,* 2006). Nematophagous fungi can be grouped into three categories according to their different pathogenic mechanisms: nematode-trapping fungi, parasitic fungi, and toxic fungi (Siddiqui and Mahmood, 1996; Li *et al.,* 2000).

17.7.2. Nematode-Trapping Fungi

Nematode-trapping fungi were traditionally classified into three genera based on the morphological characters of their conidia: *Arthrobotrys corda*, *Dactylella grove*, and *Monoacrosporium oudem* (Subramanian, 1963). Recent studies with internal transcribed spacer (ITS) and 18S ribosomal DNA (rDNA) sequences indicated that trapping devices are more informative than other morphological structures in delimiting genera (Liou and Tzean, 1997; Ahrén *et al.,* 1998; Scholler *et al.,* 1999; Li (Y.) *et al.,* 2005). Nematode trapping fungi form different nematode-trapping devices that include adhesive hyphae, adhesive networks, adhesive knobs or branches, and non-adhesive rings (Yang, *et al.,* 2005; Luo *et al.,* 2006 ; Zhang and Mo, 2006). The ultrastructures of the nematode-trapping devices have been extensively studied (Heintz and Pramer, 1972; Nordbring-Hertz and Stalhammar-Carlemalm, 1978; Dijksterhuis *et al.,* 1994). Although there is variation in morphology, different types of adhesive traps (branches, nets, and knobs) share some common features that clearly distinguish them from normal vegetative hyphae (Heintz and Pramer, 1972; Dijksterhuis *et al.,* 1994). One shared feature is the presence of numerous cytosolic organelles (dense bodies) within the trapping hyphal cells (Heintz and Pramer, 1972; Nordbring-Hertz and Stalhammar-Carlemalm, 1978). Another feature is the presence of extensive layers of extracellular polymers. These polymers have been considered important for the attachment of the traps to nematode surfaces (Tunlid *et al.,* 1991).

During the past 20 years, Tunlid *et al.* have studied extensively the interaction between nematophagous fungi and their hosts (nematodes), using the soil-living fungus *Arthrobotrys oligospora* as their model species (Tunlid and Jansson, 1991; Tunlid *et al.,* 1994, 1999; Åhman *et al.,* 1996, 2002). They identified that nematode-trapping fungi infect their hosts through a sequence of events. Recognition and adhesion were the first steps in the infection. However, little is known about the molecular mechanisms of recognition and adhesion. To date, only lectin has been reported to be involved in the recognition process. The interests of studying lectins in nematode-trapping fungi came from an observation that the interaction between *A. oligospora* and nematodes were mediated by a GalNAc-(N-acetyl-D-galactosamine) specific fungal lectin binding to receptors present on the nematode surface (Nordbring-Hertz and Mattiasson 1979). Similar experiments have indicated that lectins likely play a role in the adhesion to host surfaces by a number of parasitic and symbiotic fungi (Nordbring-Hertz and Chet 1986).

Recently, a gene encoding such a lectin (*A. oligospora* lectin, AOL) was deleted in *A. oligospora* by homologous recombination (Balogh *et al.,* 2003). However, the deletion mutant showed little decrease in spore (conidia) germination, saprophytic growth, and pathogenicity. This result suggested that the fungus might be capable of compensating the absence of the lectin by expressing other proteins with similar function(s) as AOL.

Nematode-trapping fungi capture nematodes by their particular hyphal structures (Yang, *et al.,* 2005; Luo *et al.,* 2006; Zhang and Mo, 2006). The nematode cuticle is a complex structure important for motility, for maintaining their morphological integrity, and for providing protection against the environment stresses and potential pathogens (Cox *et al.,* 1981). The structure and physical properties of nematode cuticles vary with life stage, as reflected by the transient expression of certain collagen genes during different life stages (Abrantes and Curtis, 2002).

As with other pathogens, the nematode-trapping fungi enter into the host through both enzyme degradation and mechanical pressure. Several extracellular hydrolytic enzymes including serine proteases and collagenases have been detected and partly identified from different nematode-trapping fungi (Schenck *et al.,* 1980; Tunlid *et al.,* 1994; Tosi *et al.,* 2001; Wang *et al.,* 2006). These studies suggested that extracellular hydrolytic enzymes are key virulence factors involved in the penetration process. After penetration, the hosts will be eventually degraded by the invading fungi. These fungi obtain nutrients from the nematodes for their growth and reproduction.

17.7.3. Parasitic Fungi

Parasitic fungi infect nematodes mainly by ingestive spores (*Harposporium spp.*) (Shimazu and Glockling, 1997) or adhesive spores (*Drechmeria coniospora*; Jansson *et al.,* 1987). Some endoparasites, e.g., *Catenaria anguillulae*, can produce zoospores that are attracted to nematodes before adhesion. The attachment is followed by encystment on the cuticle surface (Deacon and Saxena, 1997).

Parasitic fungi cannot form trapping devices, and eggshells might be the main barriers to their infections against nematode eggs. Eggshells of root-knot and cyst nematodes are composed of three layers: the outer vitelline, the middle chitin, and the inner lipo-protein layers (Khan *et al.,* 2004). The thickness of these layers varies considerably among nematode species (Blaxter and Robertson, 1998). Before penetration, spores and penetration structures such as appressoria must first adhere to the host surface (Jansson and Lopez-Llorca, 2001). Appressoria formed by the nematode egg parasite *Verticillium suchlasporium* (syn. *Pochonia rubescens*) have been studied using scanning electron microscopy (Lopez-Llorca and Claugher, 1990). Mucilaginous material between the surface of the appressoria and the eggshell was observed. This material could function as an adhesive to assist in eggshell penetration by the fungus (Lopez-Llorca and Claugher, 1990). A similar material was found in *Dactylella oviparasitica appressoria* infecting the Meloidogyne spp. eggs (Stirling and Mankau, 1979). These infectious structures were formed likely as an adaptation to concentrate the mechanical forces and enzymatic degradation in a small area to facilitate host penetration (St Leger, 1993; Lopez-Llorca *et al.,* 2002).

Extracellular hydrolytic enzymes also play important roles in the infection process of these parasitic fungi. Lopez-Llorca (1990) isolated the first pathogenic serine protease P32 from *V. suchlasporium*. Soon afterwards, Lopez-Llorca and Robertson (1992) confirmed the role of P32 in the pathogenicity of this fungus to nematode eggs by immunocytochemical localization studies.

Recently, *Lecanicillium psalliotae* (syn. *V. psalliotae*), an opportunistic fungus, was reported to parasitize the free-living nematode *Panagrellus redivivus*. *L. psalliotae* produces an alkaline serine protease that can immobilize the nematode *P. redivivus* and degrade the nematode cuticle within hours (Yang *et al.,* 2005).

17.7.4. Toxic Fungi and Other Nematophagous Fungi

Many microorganisms produce toxic metabolites, such as antibiotics, to prevent other microorganisms from competing for nutrients. Similarly, toxin-producing fungi can attack plant-parasitic nematodes by the production of nematicidal toxins (Dong *et al.,* 2006; Stadler *et al.,* 2006). The modes of action of these compounds against nematodes are diverse and complex.

Recently, a novel nematicidal mode was reported during the study of the basidiomycetous fungi *Coprinus comatus* and *Stropharia rugosoannulata* (Luo *et al.,* 2004, 2006). These two species produce a special nematode-attacking device: acanthocyte. The microscopical observations showed that some acanthae resembled a sharp sword that could cause damage to the nematode cuticle, resulting in leakage of nematode inner materials. The results suggested that mechanical force is an important virulence factor in these fungi (Luo *et al.,* 2006). There is another category for nematophagous fungi. Depending on their relative saprophytic/predacious abilities nematophagous fungi have been divided into three ecological groups (Jansson and Nordbring-Hertz, 1980):

Group 1: considered as the most saprophytic, constitutes the adhesive network formers. The nematode trapping rings structures are formed as a result of contact with nematodes. The fungi show good saprophytic growth on artificial media and are only weakly predatory.

Group 2: constitutes fungi producing other types of traps: (non-)constricting rings, adhesive knobs and adhesive branches. Trap formation is usually spontaneous in the absence of nematodes. These fungi show a higher predacious ability and weaker saprophytic ability than the group of the adhesive network formers.

Group 3: constitutes the endoparasitic fungi, which do not form any mycelium in soil. They persist in the soil as conidia or exist as zoospores. Most of these fungi are obligate nematode parasites.

17.7.5. Taxonomy and Evolution of Nematophagous Fungi

Nematophagous fungi are found in all major groups of fungi, including lower (oomycetes, chytridiomycetes, zygomycetes) and higher fungi (ascomycetes, basidiomycetes and deuteromycetes). Most nematophagous fungi, including both nematode-trapping and endoparasitic species, are deuteromycetes (asexual fungi). The taxonomic position of some of these species has been clarified by the discovery of the corresponding sexual stages of the fungus (Pfister, 1997).

For example, the sexual stages (teleomorphs) of a number of *Arthrobotrys, Monacrosporium* and *Dactylella* species (anamorphs) have been identified as *Orbilia spp.* belonging to the discomycetes (Ascomycetes). Species of the genus Nematoctonus are distinguished from all other nematode-trapping deuteromycetes, not only by being both nematode-trapping and endoparasitic but also by having hyphae with clamp connections, typical for basidiomycetes.

Consequently, several isolates of Nematoctonus have been shown to produce fruit bodies of a gilled mushroom (*Hohenbuehelia spp.*).

17.7.6. Molecular Phylogeny

The fact that species of nematode-trapping fungi are found in all major groups of fungi indicates that nematode parasitism has evolved independently several times. Molecular methods offer new possibilities to examine the evolutionary origin and the relationships of nematode trapping fungi in more detail. Analyses of ribosomal DNA (rDNA) sequences have proven to be particularly valuable for reconstructing phylogenetical relationships of fungi. Analysis of the 18S rDNA region have recently shown that a number of the common species of nematode trapping fungi, including species of the genera *Arthrobotrys, Dactylaria* and *Monacrosporium*, form a monophyletic group (clade) (Liou and Tzean, 1997; Ahren *et al.,* 1998). Notably, the phylogenetic patterns within this clade were not concordant with the morphology of the conidia and the conidiophores according to traditional classification but rather with the morphology of the infection structures. Three lineages of species were identified within the clade of nematode-trapping fungi. One lineage contains species having constricting rings, a second lineage includes non-parasitic species of the closely related genus *Dactylella*, and a third lineage has various adhesive structures (nets, hyphae and knobs) to infect nematodes. The separation of species forming constricting rings and adhesive trapping devices is well supported by their differences in morphology and trapping mechanisms. Further studies are needed to position the identified clade of nematode-trapping fungi within the ascomycetes. The above analyses suggest that trapping devices provide the most relevant morphological features for taxonomic classification of predatory anamorphic Orbiliaceae. Accordingly, Scholler *et al.* (1999) suggested that these fungi should be divided into four genera: *Arthrobotrys* forming adhesive networks, *Drechslerella* forming constricting rings, *Dactylellina* forming stalked adhesive knobs and Gamsylella species producing adhesive columns and unstalked knobs.

17.7.8. Evolution

There is a growing body of evidence to suggest that the parasitic habit of nematode-trapping fungi has evolved among cellulolytic or lignolytic fungi as a response to nutrient deficiencies in nitrogen-limiting habitats (Barron, 1992). In such environments (like soils) with a high carbon: nitrogen ratio, nematodes might serve as an important source of nitrogen during growth on carbohydrate containing substrates. Many nematode-trapping deuteromycetes are indeed good saprophytes and can utilize cellulose and other polysaccharides as carbon sources. Notably, the saprophytic ability varies among nematode-trapping fungi and is correlated with their parasitic activity. Species with high parasitic activity grow more slowly and have more special nutrient requirements than species with low parasitic activity. Thus, it appears that over evolutionary time, the more specialized parasitic species have lost some of the activity of the enzymes involved in saprophytic metabolism. The fact that several of the identified teleomorphs of nematode-trapping deuteromycetes are wood decomposers also supports the hypothesis that nematode-trapping fungi have evolved from cellulolytic or lignolytic fungi.

The phylogenetic relationships of other nematophagous fungi, including the endoparasites, are still virtually unknown. A different evolutionary history is expected within the endoparasitic fungi, which are mostly more dependent on nematodes.

17.8. ECOLOGY OF NEMATOPHAGOUS FUNGI

17.8.1. Occurrence

Nematophagous fungi have been found in all regions of the world, from the tropics to Antarctica. They have been reported from agricultural, garden and forest soils, and are especially abundant in soils rich in organic material. A simple method of obtaining nematophagous fungi is to use the so-called soil sprinkling technique, where approximately 1 g of soil is sprinkled on the surface of a water agar plate together with a suspension of nematodes added as bait. The plates are observed for 5–6 weeks under a microscope at low magnification and examined for trapped nematodes, trapping organs and conidia of nematophagous fungi.

Many soils contain 10–15 different species of nematophagous fungi. *Arthrobotrys spp.* appear to be common in most soils, with *A. oligospora* found most frequently in temperate regions and *A. musiformis* in tropical areas, although both species occur abundantly and ubiquitously. Among the lower fungi, the zygomycetes *Stylopage* spp. and *Cystopage* spp., and the chytridiomycete *Catenaria anguillulae* are often found.

In agricultural soils in temperate regions the nematode trapping fungi follow a seasonal variation, with highest densities and number of species in late summer and autumn, possibly due to the higher soil temperature and increased input of organic debris. The fungi are most frequent in the upper 20 cm of the soil and appear to be almost absent below 40 cm (Persmark *et al.,* 1996). A strict correlation between number of propagules of nematophagous fungi and number of nematodes is difficult to obtain, although in some soils a correlation exists between number of species of nematophagous fungi and the number of nematodes. This raises the question whether the parasitism of nematophagous fungi can regulate the population size of soil nematodes. Experiments in soil microcosms using the endoparasitic fungus *H. rhossoliensis* and plant parasitic nematodes have shown that the level of fungal parasitism is dependent on the nematode density, although there is a relatively long time lag in the response of the fungal population to changes in the number of nematodes (Jaffee *et al.,* 1992).

Mostly, plant-parasitic nematodes attack plant roots and, therefore, the ability of the nematophagous fungi to grow in the rhizosphere is of great importance for their capacity to control these nematodes. Many nematode trapping fungi have been found to occur more frequently in the rhizospheres of several plants, especially leguminous plants, e.g. soybean and pea, than in root-free soil. This effect could possibly be due to increased or changed root exudation in these plants. To evaluate whether trapping structures and consequently trapping of nematodes are actually more abundant in rhizosphere soil, new techniques have to be developed to examine the activity of nematophagous fungi in situ.

17.8.2. Interactions with Other Fungi and Plants

Apart from attacking nematodes, nematophagous fungi also have the capacity to infect other fungi (act as mycoparasites). Nematode-trapping fungi such as *A. oligospora* attack their host fungi by coiling of the hyphae of the nematode-trapping fungi around the host hyphae, which results in disintegration of the host cell cytoplasm without penetration of the host. It was shown that nutrient transfer took place between the nematode-trapping fungus *A. oligospora* and its host *Rhizoctonia solani* using radioactive phosphorous tracing (Olsson and Persson, 1994).

A. oligospora, *P. chlamydosporia* and other nematophagous fungi have the capacity to colonize plant roots (Bordallo *et al.,* 2002). The fungi grow inter- and intracellularly and form appressoria when penetrating plant cell walls of epidermis and cortex cells, but never enter vascular tissues. Histochemical stains show plant defence reactions, e.g. papillae, lignitubers and other cell wall appositions induced by nematophagous fungi, but these never prevented root colonization. The growth of the nematophagous fungi in plant roots is endophytic, i.e. the host remains asymptomatic. Endophytic growth of *P. chlamydosporia* in barley and wheat roots appeared to increase plant growth and reduce growth of the plant parasitic take-all fungus *Gaeumannomyces graminis* var. *tritici* (Monfort *et al.,* 2005).

Mycoparasitism and plant endophytism may be important issues for extension of the biological control potential of the nematophagous fungi.

17.8.3. Effect of Environmental Conditions on Nematophagous Fungi

Microorganisms need to be active at prevailing soil conditions in the field and to survive abiotic conditions that may occur during the day, in order to be effective and widely applicable as biological control agents against soil-inhabiting nematodes. Gronvold (1989) found a significant effect of temperature on the adhesive network development in *Arthrobotrys oligospora* (ATCC 24927): mycelium did not respond to juveniles or responded only slowly with the development of networks at temperatures below 15 °C or above 25 °C. Also *Dactylella spp.* between 20 and 24 °C captured higher proportions of nematodes captured than at lower temperatures (Feder, 1963).

The nutrients available in the environment of the fungus have effects on its metabolism and its morphognesis (Esser *et al.,* 1991). Variations of the nutrient source showed that development of adhesive networks is highly affected by nutrients available (Gronvold, 1989). For example, isolates of *A. oligospora* developed poorly networks on water agar whereas vegetative hyphae developed normally (Soprunov, 1966; Nordbring-Hertz, 1977; Jansson and Nordbring-Hertz, 1980).

Morphological responses to light has been described for many fungi (Leach, 1971). Gronvold (1989) reported that light suppresses development of ring structures in *A. oligospora* (ATCC 24927).

Another key factor which plays a role in trapping activity is the age of the hyphae. Loss of virulence of old cultures of nematode-capturing fungi was observed by Couch (1937) and Feder (1963). Such loss of virulence by nematode-capturing fungi is of special significance because it may limit their usefulness for nematode control. More recent work by Heintz (1978) showed that ageing of mycelium of *A. dactyloides* and *A. cladodes* resulted in a reduction of the ability to capture nematodes. Also loss of adhesiveness of ring structures of *Dactylella megalospora* was found within seven days (Esser *et al.,* 1991). Reduction of capture in ring structures of *A. oligospora* (ATCC 24927) also occured, when fungal colonies were kept for seven weeks at temperatures between 5 and 35 °C (Gronvold, 1989).

Products based on trapping device forming strains of the fungus *Arthrobotrys* (e.g. *Arthrobotrys superba* and *Arthrobotrys irregularis*) have been described (Cayrol *et al.,* 1993) but these strains show delayed action against infective nematodes in soil due to the time needed for induction and formation of trapping devices. This delay in action results in limited protection of the plants against direct attack of infective stages of nematodes in the early stages of plant growth. This effect of the nematicidal preparations based on trap-forming *Arthrobotrys* strains is noticed primarily in the lower reproduction rate of the nematodes and the resulting low population in the next years and not in direct protection of the plants.

It has been found surprisingly that the capture of nematode by fungi is not only realizable by complex capture structures, but can also be accomplished by not visibly differentiated vegetative hyphae.

A special strain of *A. oligospora* was identified which is able to capture nematodes with undifferentiated mycelium. This strain (CBS 289.82) combines the properties of the above mentioned group 1 and group 2 nematophagous fungi viz. good saprotrophic growth and the ability to capture nematodes without the need for induction and formation of trapping devices. This enables *A. oligospora* strain CBS 289.82 to capture the infective stages of plant-parasitic nematodes in soil without delay to due induction and formation of trapping structures. This treat distinguishes *A. oligospora* strain CBS 289.82 from other strains of *Arthrobotrys* used in nematicidal products. *A. oligospora* strain CBS 289.82 has novel qualities as nematicidal product because of its unique way of capturing nematodes and the resulting ability to effectively protect plants from direct attack by infective stages of plant-parasitic nematodes.

17.9. NEMATODE–FUNGUS INTERACTION MECHANISMS

17.9.1. Recognition and Host Specificity

The question of how nematophagous fungi recognize their prey is complex. No simple host specificity has been found in any of the nematode-trapping species, while experiments with the endoparasite *D. coniospora* have revealed somewhat higher host specificity. Nevertheless, it appears that there are recognition events in the cell–cell communication at several steps of the interaction between fungus and nematode, which might elicit a defined biochemical, physiological or morphological response. Nematodes are attracted to the mycelia of the fungi in which they may induce trap formation and they are attracted even more to fully developed traps and spores. This is followed by a 'shortrange' or contact communication: adhesion. This step may involve an interaction between a carbohydrate-binding protein (lectin) in the fungus and a carbohydrate receptor on the nematode. Recognition of the host is probably also important for the subsequent steps of the infection, including penetration of the nematode cuticle (Nordbring-Hertz *et al.,* 2006).

17.9.2. Attraction

Nematodes are attracted by compounds released from the mycelium and traps of nematode-trapping fungi, and the spores of endoparasites. Both the morphology and consequently the saprophytic/parasitic ability strongly influence the attractiveness of the fungi. Fungi that are more parasitic appear to have a stronger attraction than the more saprophytic ones; that is, the endoparasitic species infecting nematodes with conidia are more effective in attracting nematodes than the more saprophytic species with different kinds of trapping devices (Nordbring-Hertz *et al.*, 2006).

17.9.3. Adhesion

The contact and adhesion of nematodes to the traps and spores of nematophagous fungi can be observed in the electron microscope. In *A. oligospora* the three-dimensional nets are surrounded by a layer of extracellular fibrils even before the interaction with the nematodes. After contact, these fibrils become directed perpendicularly to the host surface, probably to facilitate the anchoring and further fungal invasion of the nematode. The endoparasite *D. coniospora* shows a completely different type of adhesive that seems to be composed of radiating fibrils irrespective of whether contact with the nematode has been established or not. Furthermore, the spores of *D. coniospora* adhere specifically to the sensory organs at the tip of the head of the nematode, thereby blocking nematode attraction. The chemical composition of the surface fibrils of nematophagous fungi is not known in detail but they do contain both proteins and carbohydrate-containing polymers (Nordbring-Hertz *et al.*, 2006).

17.9.4. Penetration

The adhesion of the traps to the nematode results in a differentiation of the fungi. In *A. oligospora*, a penetration tube forms and pierces the nematode cuticle. This step probably involves both the activity of hydrolytic enzymes solubilizing the macromolecules of the cuticle and the activity of a mechanical pressure generated by the penetrating growing fungus. The nematode cuticle is composed mainly of proteins including collagen, and several proteases have been isolated from nematophagous fungi that can hydrolyse proteins of the cuticle. In all cases these proteases belong to the family of serine proteases, and after obtaining data from sequencing, it has been demonstrated that they have a high homology to the subtilisin-type of serine proteases (Bonants *et al.*, 1995; Ahman *et al.*, 1996). In the endoparasite *D. coniospora*, a chymoptrypsin-like protease appears to be involved in the penetration process.

More detailed studies of the subtilisin PII produced by *A. oligospora* have indicated that this type of proteases can have a number of different functions (Ahman *et al.*, 2002). Thus, apart from being involved in penetration and digestion of the cuticle and tissues of infected nematodes, PII appears to have a nematotoxic activity.

17.9.5. Digestion and Storage of Nutrients

Following penetration, the nematode is digested by the infecting fungus. Once inside the nematode, the penetration tube of *A. oligospora* swells to form a large infection bulb. The development of the bulb and trophic hyphae occurs in parallel with dramatic changes in the ultrastructure and physiology of the fungus. The dense bodies are degraded in the trap cells and in the bulb. The bulb and the trophic hyphae typically contain normal cell organelles, endoplasmatic reticulum being particularly well developed. At later stages, lipid droplets accumulate in the trophic hyphae, which are probably involved in the assimilation and storage of nutrients obtained from the infected nematode. In contrast to the trap-forming fungi, the endoparasite *D. coniospora* does not form an infection bulb upon penetration and does not have dense bodies, which are typical for the trap-forming fungi. Along with formation of lipid droplets, another way for *A. oligospora* to store nutrients derived from the host is to produce large amounts of a lectin in the cytoplasm (Rosen *et al.,* 1997). This protein (designated *A. oligospora* lectin, AOL) is a member of a novel family of low molecular weight lectins, sharing similar primary sequences and binding properties, which have so far only been identified in a few filamentous fungi (Rosen *et al.,* 1996).

During the infection of nematodes, AOL is rapidly synthesized in *A. oligospora* once nematodes have been penetrated and digestion has started. Large amounts of AOL are accumulated in the trophic hyphae growing inside the nematode. Later, the lectin is transported from the infected nematode to other parts of the mycelium, where it can be degraded and support the growth of the fungus.

Although the mechanisms are not known, it has been suggested that AOL, like other lectins, is involved in a recognition event during the interaction with the nematodes. The fact that the AOL family of lectins binds to sugar structures that are typical of animal glycoproteins including nematodes, but not found in fungi, supports this hypothesis.

17.9.6. Constricting Rings

Although the patterns of nematode infection of other predatory fungi, which use adhesive layers for capturing nematodes (nets, hyphae or knobs), are less thoroughly studied, they appear to be largely similar to those described for *A. oligospora*. In contrast, the trapping mechanism of constricting rings is completely different. When a nematode moves into the ring, it triggers a response such that the three cells composing the ring rapidly swell inward and close around the nematode.

Other stimuli, such as touch by a needle of the inside (luminal) surface of a ring, or heat, can also trigger the closure of the trap. The reaction is rapid (0.1 s), irreversible, and is accompanied by a large increase in cell volume leading to an almost complete closure of the aperture of the trap. Following capture, the fungus produces a penetration tube that pierces the nematode cuticle. Inside the nematode a small infection bulb is formed from which trophic hyphae develop (Nordbring-Hertz *et al.,* 2006).

The mechanism by which the constricting rings are closed is not known in detail. Electron microscopy has shown that during the ring-cell expansion, the outer cell wall of the ring cells is ruptured along a defined line on the inner surface of the ring. It has been suggested that this release of wall pressure will lead to a rapid uptake of water, followed by an expansion of the elastic inner wall of the ring cells. The signal transduction pathway involved in the inflation of the ring cells has been examined in A. dactyloides (Chen *et al.,* 2001).

In this fungus it appears that the pressure exerted by a nematode on the ring activates G-proteins in the ring cells. The activation leads to an increase in cytoplasmic Ca2+, activation of calmodulin and finally the opening of water channels. The ring cells expand to constrict the ring and thus immobilize the nematode.

17.10. MYCORRHIZAE

A mycorrhiza is a symbiotic association between a fungus and the roots of a plant (Kirk et al., 2001) In a mycorrhizal association, the fungus may colonize the roots of a host plant, either intracellularly or extracellularly. They are an important part of soil life. This mutualistic association provides the fungus with relatively constant and direct access to mono or dimeric carbohydrates, such as glucose and sucrose produced by the plant in photosynthesis (Harrison, 2005) The carbohydrates are translocated from their source location (usually leaves) to the root tissues and then to the fungal partners. In return, the plant gains the use of the mycelium's very large surface area to absorb water and mineral nutrients from the soil, thus improving the mineral absorption capabilities of the plant roots (Selosse et al., 2006).

Mycorrhizae are commonly divided into ectomycorrhizas and endomycorrhizas. The two groups are differentiated by the fact that the hyphae of ectomycorrhizal fungi do not penetrate individual cells within the root, while the hyphae of endomycorrhizal fungi penetrate the cell wall and invaginate the cell membrane. Endomycorrhiza are variable and have been further classified as arbuscular, ericoid, arbutoid, monotropoid, and orchid mycorrhizae (Peterson et al., 2004). Arbuscular mycorrhizas, or AM (formerly known as vesicular-arbuscular mycorrhizas, or VAM), are mycorrhizas whose hyphae enter into the plant cells, producing structures that are either balloon-like (vesicles) or dichotomously-branching invaginations (arbuscules). Arbuscular mycorrhizae are formed only by fungi in the division Glomeromycota. Fossil evidence (Remy et al., 1994) and DNA sequence analysis (Simon et al., 1993) suggest that this mutualism appeared 400-460 million years ago, when the first plants were colonizing land. Arbuscular mycorrhizas are found in 85% of all plant families, and occur in many crop species (Wang and Qiu, 2006). The hyphae of arbuscular mycorrhizal fungi produce the glycoprotein glomalin, which may be one of the major stores of carbon in the soil. Ectomycorrhizas, or EcM, are typically formed between the roots of around 10% of plant families, mostly woody plants including the birch, dipterocarp, eucalyptus, oak, pine, and rose (Wang and Qiu, 2006) families and fungi belonging to the Basidiomycota, Ascomycota, and Zygomycota. Ectomycorrhizas consist of a hyphal sheath, or mantle, covering the root tip and a hartig net of hyphae surrounding the plant cells within the root cortex. In some cases the hyphae may also penetrate the plant cells, in which case the mycorrhiza is called an ectendomycorrhiza. Outside the root, the fungal mycelium forms an extensive network within the soil and leaf litter. Nutrients can be shown to move between different plants through the fungal network (sometimes called the wood wide web). Carbon has been shown to move from birch trees into fir trees thereby promoting succession in ecosystems (Simard et al., 1997).

17.10.1. Antagonistic Effects of AMF on Plant Parasitic Nematodes

Establishment of Arbuscular Mycorrhizae fungi (AMF) in plant root systems is considered a biological means of protection against plant diseases caused by soilborne pathogens (Azcón-Aguilar and Barea, 1996). The parasitization of plants by nematodes (mainly endoparasitic) can be influenced by the establishment of a AMF. The penetration rate of parasitic nematodes can be decreased, their development inside the root may be retarded, or the degree of damage caused by the nematode may be lowered (Dehne, 1982).

Most studies on the interactions between AMF and plant parasitic nematodes reported that root colonization by AMF increases tolerance of the host plant to *Meloidogyne* species, such as that of tomato and white clover to *Meloidogyne hapla* (Cooper and Grandison, 1986); peanut to *M. arenaria* (Carling *et al.,* 1996); banana to *M. incognita* (Jaizme-Vega *et al.,* 1997); and *Prunus* rootstocks to *M. javanica* (Calvet *et al.,* 2001). AMF can reduce the number of *Radopholus similis* on the roots and soil (Umesh *et al.,* 1988) and also they can reduce the number of *Pratylenchus goodeyi* in the roots (Jaizme-Vega and Pinochet 1997; Pinochet *et al.,* 1997).

Several mechanisms have been proposed to explain nematode suppression by AM fungi (Mehrotra, 2005). The two populations of microorganisms may compete for limited resources such as space and food supply. Support for this mechanism comes from observations of nematode-free cortical cells colonized by AM fungi (Obannon and Nemec 1979; Cooper and Grandison, 1986) and from studies in which nematode reproduction per unit root weight was retarded, or nematode development was delayed in mycorrhizal roots compared to nonmycorrhizal ones (Salawu and Estey 1979; Saleh and Sikora 1984). In an investigation, the interaction of the soybean cyst nematode *Heterodera glycines* with soybean, studied in the presence or absence of AM fungi, in a greenhouse experiment. The authors noted that the effect of AM fungi varied with time. The AM fungi suppressed the reproduction and development of the nematode through 49 days after planting but not later. They concluded that the limited plant nutrients, which were exploited by the AM fungi during the early phase of root colonization and formation of fungal structures throughout the root system could have been responsible for the suppression of the nematode (Mehrotra, 2005).

Mycorrhizal fungi may induce changes in root physiology so that the characteristics of root exudates, as well as, the composition of the cell wall is altered, making the root less attractive to colonize, difficult to penetrate and unsuited as a niche. The development of AMF colonization has been reported to induce the production of anti-fungal hydrolytic proteins (Dumas-Gaudot *et al.,* 1996) and other compounds commonly associated with the response of plants to pathogen colonization (Harisson and Dixon, 1994). Enhanced production of lignin and phenols in response to mycorrhizal inoculation has been associated with reduced reproduction of *M. javanica* and *Radopholus similes*. Exposure of nematodes to a water extract of mycorrhizal tomato roots was more toxic to *M. incognita* juveniles than exposure to water or to a water extract of non-mycorrhizal tomato root (Mehrotra, 2005). Decreased pathogenicity of *M. incognita* in mycorrhizal soybean has been attributed to elevated content of the phytoalexin, glyceollin. In an earlier investigation, Morandi *et al.* (1984) detected greater quantities of isoflavonoid phytoalexins-glyceollin, coumestrol, and daidzein in mycorrhizal soybean, compared to non-mycorrhizal one. Glyceollin has been demonstrated to immobilize *M. incognita in vitro* (Mehrotra, 2005). These observations suggest that the effects of AM fungi on suppression of nematodes may be systemic. However, inferences based on the absence of galling in mycorrhizal segments of roots and split root experiments argue for a more localized effect (Fitter and Garbaye, 1994).

Arbuscular mycorrhizal fungi could affect nematode activity indirectly through their effect on other rhizosphere populations. The reports of Rao *et al.* (1997) and Rao and Gowen (1998) suggest that AM fungi could enhance the parasitic activity of microorganisms such as, *Pasteuria penetrans* and *Verticillium chlamydosporium* against plant parasitic nematodes. Fungi that attack cysts have been sought as a means of destroying eggs of the cyst nematode before they hatch (Wilcox and Tribe, 1974). Tribe (1977) reviewed the literature on the subject and pointed out that a number of fungal species, including some *Glomus* spp. have been reported to parasitize cyst (Mehrotra, 2005). Francl and Dropkin (1985) demonstrated that *G. fasciculatum* was capable of infecting *Heterodera glycines* cysts, eggs and young females. Lastly, AM fungi could improve plant vigor through enhanced uptake of nutrients by nematode damaged roots, which would normally absorb them inefficiently. Consequently, a number of attempts have been made to mimic the nematode suppressing activity of AM fungi through P amendment of soil. However, these efforts did not yield consistent results. Heald *et al.* (1989) evaluated the influence of phosphorus and *Glomus intraradices* on the tolerance of canteloupe to *M. incognita* and concluded that the better tolerance of mycorrhizal plants was related to improved nutrient uptake. Similarly, other researchers (Camprubi *et al.,* 1993; Pinochet *et al.,* 1993; Calvet *et al.,* 1995; Pinochet *et al.,* 1995) concluded that growth enhancement of the mycorrhizal fruit trees they studied in the presence of the root-lesion nematode *Pratylenchus vulnus* was due to the increased capacity of the plant to take up nutrient and water, rather than a direct inhibition of the nematode by the fungus. Conversely, Roncadory and Hussey (1977) did not observe any change in the number of nematode eggs in cotton roots after application of high levels of P, although in some respects supplemental P mimicked mycorrhizal effect on plant growth. In a microplot study with cotton, the severity of nematode damage was increased by P amendment (Mehrotra, 2005). In contrast, the survival and reproduction of *M. hapla* in onion was not influenced by high P (MacGuidwin *et al.,* 1985). Similarly, Strobel *et al.* (1982) noted that *M. incognita* suppressed growth of non-mycorrhizal peach, irrespective of fertility levels. They also observed that damage by nematodes was less in soils with low fertility than with high fertility. Cooper and Grandison (1986) were not able to completely eliminate the inhibitory influence of AM fungi on *M. hapla*, even at P levels, which did not produce a growth response to AM fungal colonization. In a similar study, suppression of *M. incognita* in white clover by *G. mosseae* was not explained by enhanced P uptake (Mehrotra, 2005). Improved P uptake, thus dose not consistently explain alternation of nematode damage by AM fungi. If the adverse effects of plant parasitic nematodes could readily be offset through enhanced soil fertilization *per se*, the parasites would not have posed the kind of severe problem they pose, at least in the industrial countries. However, most of the studies did not monitor the flux of P to host plants, either prior to or after P amendment. The variability observed among studies may reflect differences in soil solution P status because the same amount of added P yields different quantities of P in solution in different soils depending on the P absorption capacity of the soils. Moreover, plants differ in their external and internal P requirements, depending on their mycorrhizal dependency. It is, therefore, difficult to accurately and consistently determine the impacts of P on AM fungi –nematode interaction in the absence of precise information on soil solution P and on the AM dependency of plant species under investigation (Mehrotra, 2005).

17.11. BACTERIA

Bacteria are numerically the most abundant organisms in soil, and some of them, for example members of the genera *Pasteuria*, *Pseudomonas* and *Bacillus* (Emmert and Handelsman, 1999; Siddiqui and Mahmood, 1999; Meyer, 2003), have shown great potential for the biological control of nematodes. Extensive investigations have been conducted over the last twenty years to assess their potential to control plant-parasitic nematodes.

A variety of nematophagous bacterial groups have been isolated from soil, host-plant tissues, and nematodes and their eggs and cysts (Stirling, 1991; Siddiqui and Mahmood, 1999; Kerry, 2000; Meyer, 2003). They affect nematodes by a variety of modes: for example parasitizing; producing toxins, antibiotics, or enzymes; interfering with nematode– plant-host recognition; competing for nutrients; inducing systemic resistance of plants; and promoting plant health (Siddiqui and Mahmood, 1999). These bacteria have a wide range of suppressive activities on different nematode species, including free-living and predatory nematodes as well as animal- and plant-parasitic nematodes (Mankau, 1980; Stirling, 1991; Siddiqui and Mahmood, 1999). They form a network with complex interactions among bacteria, nematodes, plants and the environment to control populations of plant-parasitic nematodes in natural conditions (Kerry, 2000).

However, only a few commercial biocontrol products from the bacteria with nematicidal potentials have been developed and used in the agriculture system (Whipps and Davies, 2000; Gardener, 2004; Schisler *et al.,* 2004). The development of biocontrol agents is often unpredictable and too variable for large-scale implementation (Meyer, 2003). No matter how well suited a commercial nematode antagonist is to a target host in a laboratory test, in order to realize ideal biocontrol effects in practice an intensive exploration of the mechanisms of the antagonist against nematode populations, and a thorough understanding of the interactions among biocontrol strains, nematode target, soil microbial community, plant and environment must be developed.

Recently, interactions among the microorganism, nematode target, plant and environment have been well reviewed and emphasized (Kerry, 2000; Barker, 2003; Dong and Zhang, 2006). Sustainable working methodologies have been proposed, including integrated pest management (IPM). The goal of IPM is to combine biocontrol and other methods, such as green manure, organic or inorganic soil amendments, resistant plant cultivars, hot-water treatment and crop rotation, so that they act synergistically on nematodes through the direct suppression of nematodes, promotion of plant growth, and facilitation of rhizosphere colonization and activity of the microbial antagonists (Akhtar, 1997; Barker and Koenning, 1998; Meyer and Roberts, 2002; Barker, 2003). For this goal to be achieved, however, accurate knowledge is needed of the ecology, biology, and mechanisms of action of the populations of nematophagous bacteria.

An increased understanding of the molecular basis of the various bacterial pathogenic mechanisms on nematodes not only will lead to a rational nematode management decision, but also could potentially lead to the development of new biological control strategies for plant-parasitic nematodes. For example, it has been recognized that the attraction between bacteria and their hosts is governed by chemotactic factors emanating from the hosts or pathogens (Zuckerman and Jasson, 1984). Knowledge of these mechanisms could be used to attract or target nematodes intentionally by modified nematicidal bacteria or to regulate nematode populations by the chemotactic factors produced by these nematophagous bacteria.

Advances in molecular biology have allowed us to obtain important information concerning molecular mechanisms of action, such as the production of nematotoxins, the signalling pathways that induce the host-plant defence mechanism, and the infection process. Such information should provide novel approaches to improving the efficacy of nematophagous bacteria for biological control applications, to increasing the expression of toxins or enzymes from the microorganisms, and to formulating commercial nematicidal agents. For example, the developing genomic-bioinformatic approach may help to solve the difficulty of culturing the nematode parasite *Pasteuria in vitro*. This may allow mass-production of spores for commercial use.

17.11.1. Parasitic Bacteria

Pasteuria Penetrans

The genus *Pasteuria* was first described by Metchnikoff in 1888 as a parasite of *Daphnia*, and was observed in 1906 on a nematode (*Dorylaimus bulbiferous*) by Cobb. Subsequently, all plant-parasitic nematodes of major economic importance have been observed to be parasitised by either *P. penetrans* or closely related species (Sayre and Starr, 1988; Sturhan, 1988; Chen and Dickson, 1998). Based on host range, life cycle and morphology, three nominal species of *Pasteuria* able to parasitise plant parasitic nematodes have been proscribed, namely: (1) *P. penetrans*, which is a parasite of *Meloidogyne spp.*, (2) *Pasteuria thornei*, which is a parasite of *Pratylenchus spp.*, and (3) *Pasteuria nishizawae*, which is a parasite of Heterodera and Globodera spp. (Chen and Dickson, 1998). Historically, there have been two factors limiting the widespread deployment of *P. penetrans* for practical nematode control: (1) absence of a robust *in vitro* culturing system for the bacterium, amenable to the mass-production of spores, and (2) the highly nematode-species-specific nature of the bacterium-host interaction, which means strains produced for one nematode species likely will not work for other species. Recently, however, major breakthroughs have been made in culturing *P. penetrans*. A company (Entomos LLC) has developed conditions for *in vitro* culture of *P. penetrans*. Nevertheless, it is clear that the culturing achievement made by Entomos is just a first step towards broad commercial use of this biocontrol agent. The ability to tightly regulate the vegetative growth-sporulation transition, for example, will be essential to large scale production. Further, the issues of host range remain to be solved.

The Pasteuria–Nematode Interaction

The *P. penetrans* bacterium produces highly durable endospores (Stirling and Wachtel, 1980; Sayre and Starr, 1985; Stirling, 1991). During migration in the soil in search of a host root, the nematode encounters *P. penetrans* endospores. Upon contact, the spores attach (by a poorly understood mechanism) and encumber the nematode. Although as many as 60 spores can be found adhering randomly along the body of a single nematode, fewer than 10 spores is more common in a natural infestation. Specific strains of *Pasteuria* have highly defined attachment profiles on different nematode populations, indicative of an initial host recognition event between *Pasteuria* and its potential host. Some *Pasteuria* strains will only attach to relatively few nematode populations, whereas others have broader attachment profiles. Obviously, attachment is an absolute prerequisite for infection, though not all attached spores necessarily germinate. Endospore-encumbered nematodes often retain the ability to invade and migrate within the root and establish normal feeding sites.

The life cycle of the bacterium on root-knot nematodes (*Meloidogyne* spp.) is initiated when endospores adhere to the nematode cuticle. Germination occurs with the production of an infection peg that penetrates the nematode cuticle after the nematode has entered the root, and in the case of the root-knot nematodes, germination and infection by the bacterium appear to be tied to the initiation of feeding by the nematode. The terminal region of the infection peg undergoes differentiation by dichotomous branching and produces a mycelial ball, or microcolony. Infected female nematodes continue to develop but microcolonies of the bacteria proliferate and prohibit infected females from producing eggs (Sayre and Starr, 1985). The bacterium undergoes sporogenesis and the females eventually die. Each infected female can produce up to 2 £ 106 endospores that are eventually released into the soil when the roots decay and the female cadavers break open (Sayre and Starr, 1985; Davies *et al.*, 1988; Chen and Dickson, 1998).

Future Prospects

The lack of efficient technology for the large-scale production of *P. penetrans* is a major impediment to the marketing of this organism as a biological control agent. It is readily apparent that mass cultivation depends on fully understanding the nutrient requirements of *P. penetrans.* A medium similar in nature to the chemical composition of the pseudocoelomic fluid of nematodes may be required to provide adequate nutrients for development of *Pasteuria* spp. However, only the pseudocoelomic fluid of large, animal-parasitic nematodes has been even partially characterized, and there are no known reports of a *Pasteuria* or *Pasteuria* like organism parasitizing these nematodes. Consequently, models of the chemical composition and physical environment of the pseudocoelomic fluid of plant-parasitic nematodes are crucial for comparing the biological and physiological differences between plant-parasitic and animal-parasitic nematodes. The clues for rearing *Pasteuria* spp. may be revealed once the chemical composition and physical makeup of the pseudocoelomic fluid of plant-parasitic nematodes is understood. With the abundant distribution of *P. penetrans* in soil (Sayre and Starr, 1988; Sturhan, 1988; Dickson *et al.,* 1994; Hewlett *et al.,* 1994), it may be possible to amplify the soil endospore densities to levels that provide biological control of nematodes (Stirling, 1991). Unfortunately, technology for the quantification of endospores in soil is not yet available, thus limiting our understanding of the ecology of endospores in soil. Cross-generic parasitism of *Pasteura* spp. has been observed (Mankau and Prasad, 1972; Mankau, 1975; Bhattacharya and Swarup, 1988; Oostendorp *et al.,* 1990; Pan *et al.,* 1993; Sharma and Davies, 1996), but there have been few investigations using alternative hosts to culture *P. penetrans.* Relatively low-cost cultivation of some nematodes on media has been developed (Friedman, 1990). If such systems could be transferred to the cultivation of *P. penetrans,* it might be possible to produce large quantities of endospores for field application.

17.11.2. Opportunistic Parasitic Bacteria

In 1946, Dollfus investigated and documented bacteria within the body cavity, gut, and gonads of nematodes (Jatala, 1986). Other reports have since suggested the association of some bacteria with the nematode cuticle. However, these studies were unable to specify whether these bacteria were parasites or saprophytes (Jatala, 1986). In fact, most nematophagous bacteria, except for obligate parasitic bacteria, usually live a saprophytic life, targetting nematodes as one possible nutrient resource. They are, however, also able to penetrate the cuticle barrier to infect and kill a nematode host in some conditions. They are described as opportunistic parasitic bacteria here, represented by *Brevibacillus laterosporus* strain G4 and *Bacillus* sp. B16.

As a pathogen, *Br. laterosporus* has been demonstrated to have a very wide spectrum of biological activities. So far, it has been reported that four nematode species (three parasitic nematodes, namely *Heterodera glycines*, *Trichostrongylus colubriformis* and *Bursaphelenchus xylophilus*, and the saprophytic nematode *Panagrellus redivius*) could be killed by various *Br. laterosporus* isolates (Oliveira *et al.,* 2004; Huang *et al.,* 2005). Among these isolates, *Br. Laterosporus* strain G4, which was isolated from soil samples in Yunnan province in China and parasitizes the nematodes *Panagrellus redivius* and *Bursaphelenchus xylophilus*, has been extensively studied (Huang *et al.,* 2005). After attaching to the epidermis of the host body, *Br. laterosporus* can propagate rapidly and form a single clone in the epidermis of the nematode cuticle. The growth of a clone can result in a circular hole shaped by the continuous degradation and digestion of host cuticle and tissue. Finally, bacteria enter the body of the host, and digest all the host tissue as nutrients for pathogenic growth (Huang *et al.,* 2005).

During bacterial infection, the degradation of all the nematode cuticle components around the holes suggests the involvement of hydrolytic enzymes (Cox *et al.,* 1981; Decraemer *et al.,* 2003; Huang *et al.,* 2005). Histopathological observations and molecular biological analyses have demonstrated that major pathogenic activity could be attributed to an extracellular alkaline serine protease, designated BLG4 (Huang *et al.,* 2005; Tian *et al.,* 2006). The most compelling evidence to support the role of protease as virulence factor was derived from studying protease-deficient mutants (Tian *et al.,* 2006). The BLG4-deficient strain BLG4-6 was only 43% as effective as the wild-type strain at killing nematodes, and showed only 22% as much cuticledegrading activity. These results also suggest that BLG4 is not the only virulence factor responsible for nematicidal activities, and that other factors such as other extracellular enzymes or toxins are probably involved (Huang *et al.,* 2005; Tian *et al.,* 2006).

Several bacterial proteases have been shown to be involved in the infection processes against nematodes. Among these, the bacterial serine protease genes from nematophagous bacteria isolated from a different area (*Br. laterosporus* strain G4, *Bacillus* sp. B16, *Bacillus* sp. RH219 and other *Bacillus* strains) have been isolated and compared (Niu *et al.,* 2005; Tian *et al.,* 2006). The amino acid sequences of these bacterial cuticle-degrading proteases have shown high sequence identity (97–99%). The consistency of these pathogenic proteases from the different nematophagous bacterial strains suggests that proteases are highly conserved in this group of bacteria.

Similar to the nemotode-pathogenic proteases from nematophagous fungi, the protease BLG4 also belongs to the family of subtilases (Segers *et al.,* 1999). Comparison of the deduced amino acid sequence of the protease BLG4 gene with other cuticle-degrading proteases from pathogenic fungi showed lower similarities than the above comparisons (35%). However, a high degree of similarity between the sequences was found in regions containing the active site residues Asp32, His64 and Ser221. The two blocks of sidechains that form the sides of the substrate-binding pockets in these serine proteases were also conserved in BLG4 as Ser125Leu126Gly127Gly128, and Ala152Ala153Gly154, respectively (Niu *et al.,* 2005; Tian *et al.,* 2006).

At present, the majority of research efforts on opportunistic nematode-parasitic bacteria have concentrated on understanding pathogenesis using free-living nematodes as targets. Such studies should allow us to identify new pathogenic factors, and to learn more about infectious processes in nematodes. It is important to understand the mechanism that controls the switch from saprotrophy to parasitism in order to formulate effective commercial nematode control agents.

17.11.3. Bacillus Thuringiensis

Bacillus thuringiensis (Bt) is a Gram-positive, spore forming, entomopathogenic bacterium characterized by the production of proteinaceous crystalline inclusions during the sporulation phase (parasporal crystals). This bacterium is a member of the genus *Bacillus*, consisting of more than twenty different Gram-positive, spore-forming bacteria. Other important species of this group are *B. anthracis*, *B. cereus* and *B. subtilis*. Bt can be differentiated from the other closely related members of the family based on their biochemical, nutritional and serological analyses and by the presence of crystalline inclusions in Bt cells, visible by light microscopy.

Bt is common to terrestrial habitats including soil, living and dead insects, granaries, and plant surfaces (Carozzi *et al.,* 1991; Smith and Couche, 1991; Meadows *et al.,* 1992) and occurs predominantly as spores dispersed in the environment. Bt spores can be disseminated widely in space and their persistence under laboratory and field conditions has been well studied (West, 1984). Vegetative cells are sensitive to ultraviolet radiations and disappear rapidly from the environment upon exposure to sunlight. Bt spores are resistant to adverse conditions such as heat and drought, thereby enabling them survive periods of stress and allowing the bacterium to re-germinate under favorable conditions (Chiang *et al.,* 1986; Benoit *et al.,* 1990).

Bt isolates have narrow specificities against various insects, but together they span a wide range of orders including Lepidoptera (beetles and moths), Diptera (mosquitoes and blackflies), Coleoptera (beetles and weevils), Hymenoptera (wasps and bees), with some isolates also active against nematodes, mites and protozoa (De Barjac and Sutherland, 1990; Becker and Margalith, 1993; Estruch *et al.,* 1996; Rang *et al.,* 2000; Donovan *et al.,* 2001; Moellenbeck *et al.,* 2001; Sayyed *et al.,* 2001; Ellis *et al.,* 2002).

Previously, only *Meloidogyne* spp. and *Pratylenchus* spp. have been reported to be sensitive to crystal protein toxin. For example, in the 1970s, the delta-endotoxin from *B. thuringiensis* was found to be active against larvae and eggs of *Meloidogyne* spp., and remarkably decreased root-knot formation on crops (Prasad *et al.,* 1972; Ignoffo and Dropin, 1977). However, some studies suggest that many other plant-parasitic nematodes besides *Meloidogyne* spp. and *Pratylenchus* spp. (i.e. *Tylenchorhynchus*, *Ditylenchus* and

Aphelenchoides spp.), are sensitive to the crystal protein toxin (Yu *et al.*, 2008). Recently, the toxicity of this protein against free-living nematodes, i.e. *Caenorhabditis elegans*, and the zooparasitic nematodes *Nippostrongylus brasiliensis* and *Ancylostoma ceylanicum*, has been established (Wei *et al.*, 2003; Cappello *et al.*, 2006; Salehi *et al.*, 2008b). More and more studies are confirming *B. thuringiensis* nematicidal activity, suggesting its usefulness in biocontrol of plant-parasitic nematodes.

17.11.3.1. Cry Proteins

Proteinaceous crystalline inclusions (parasporal crystals) are predominantly comprised of one or more proteins Crystal (Cry) and Cytolitic (Cyt) toxins, also called δ-endotoxins. Cry proteins are parasporal inclusion (Cry) proteins from Bt that exhibit experimentally verifiable toxic effect to a target organism or have significant sequence similarity to a known Cry protein. Similarly, Cyt proteins are parasporal inclusion proteins from Bt that exhibits hemolytic (Cyt) activity or has obvious sequence similarity to a known Cyt protein. These toxins are innocuous to humans, vertebrates and plants, and are completely biodegradable. Therefore, Bt is a viable alternative for the control of pests in agriculture and of important human disease vectors (Bravo *et al.*, 2005; Salehi *et al.*, 2008a,b).

Bt Cry and Cyt toxins belong to a class of bacterial toxins known as pore-forming toxins (PFT) that are secreted as water-soluble proteins undergoing conformational changes in order to insert into, or to translocate across, cell membranes of their host. There are two main groups of PFT: (i) the α-helical toxins, in which α-helix regions form the trans-membrane pore, and (ii) the β-barrel toxins, that insert into the membrane by forming a β-barrel composed of β-sheet hairpins from each monomer (Parker and Feil, 2005). The first class of PFT includes toxins such as the colicins, exotoxin A, diphtheria toxin and also the Cry three-domain toxins. On the other hand, aerolysin, α-hemolysin, anthrax protective antigen, cholesterol-dependent toxins as the perfringolysin O and the Cyt toxins belong to the β-barrel toxins (Parker and Feil, 2005). In general, PFT-producing bacteria secrete their toxins and these toxins interact with specific receptors located on the host cell surface. In most cases, PFT are activated by host proteases after receptor binding inducing the formation of an oligomeric structure that is insertion competent. Finally, membrane insertion is triggered, in most cases, by a decrease in pH that induces a molten globule state of the protein (Parker and Feil, 2005).

Cry proteins are specifically toxic to the insect orders Lepidoptera, Coleoptera, Hymenoptera and Diptera, and also to nematodes. In contrast, Cyt toxins are mostly found in Bt strains active against Diptera (Bravo *et al.*, 2007; Salehi *et al.*, 2008a; Seifinejad *et al.*, 2008). Since the cloning of the first *cry* gene by Schnepf and Whiteley (1981), more than 300 *cry* genes have been isolated from *B. thuringiensis* (Crickmore *et al.*, 1998). These Cry proteins are classified into families Cry1 to Cry54 on the basis of their amino acid sequence homology. Nowadays, parasporal proteins with nematicidal activity are grouped as Cry5, Cry6, Cry12, Cry13, Cry14 and Cry21 (Yu *et al.*, 2004; Salehi *et al.*, 2008b).

Except for Cry6, most of the nematicidal Cry proteins are large proteins (90 to 140 kDa). Most *B. thuringiensis* *cry* genes reside on large plasmids (Schnepf *et al.*, 1998). The only exception to date was reported by Loeza-Lara *et al.* (2005): a *cry*-like gene, *cry14-4*, found in a small plasmid, pBMBt1, from a *B. thuringiensis* strain. The predicted protein sequence showed low identity with the proteins CryC53 (24.6%) and Cry15Aa (27.8%). Apart from the coding sequence, there is no detailed experimental evidence available regarding the Cry14-4 protein; whether other small plasmids from *B. thuringiensis* harbor *cry* genes or not remains uncertain.

17.11.3.2. Diversity, Structure and Evolution of Cry Toxins

Cry proteins are defined as: a parasporal inclusion protein from Bt that exhibits toxic effects to a target organism, or any protein that has obvious sequence similarity to a known Cry protein (Crickmore *et al.,* 1998). Cyt toxins are included in this definition but it was agreed that proteins that are structurally related to Cyt toxins retain the mnemonic Cyt (Crickmore *et al.,* 1998). Primary sequence identity among different gene sequences is the bases of the nomenclature of Cry and Cyt proteins. Additionally, other insecticidal proteins that are not related phylogenetically to the three-domain Cry family have been identified. Among these, are binary-like toxins and Mtx-like toxins related to *B. sphaericus* toxins, and parasporins produced by Bt (Crickmore *et al.,* 1998; Salehi *et al.,* 2008a).

The members of the three-domain family, the larger group of Cry proteins, are globular molecules containing three structural domains connected by single linkers. One particular feature of the members of this family is the presence of protoxins with two different lengths. One large group of protoxins is approximately twice as long as the majority of the toxins. The C-terminal extension found in the long protoxins is dispensable for toxicity and is believed to play a role in the formation of the crystal inclusion bodies within the bacterium (de Maagd *et al.,* 2001). (Cyt toxins comprise two highly related gene families (Cyt1 and Cyt2) (Crickmore *et al.,* 1998). Cyt toxins are also synthesized as protoxins and small portions of the N-terminus and C-terminus are removed to activate the toxin (Li *et al.,* 1996).

To date, the tertiary structures of six different three-domain Cry proteins, Cry1Aa, Cry2Aa, Cry3Aa, Cry3Bb, Cry4Aa and Cry4Ba have been determined by X-ray crystallography (Li *et al.,* 1991; Grochulski *et al.,* 1995; Galitsky *et al.,* 2001; Morse *et al.,* 2001; Boonserm *et al.,* 2005, 2006). All these structures display a high degree of similarity with a three-domain organization, suggesting a similar mode of action of the Cry three-domain protein family. The N-terminal domain (domain I) is a bundle of seven α-helices in which the central helix-a5 is hydrophobic and is encircled by six other amphipathic helices; and this helical domain is responsible for membrane insertion and pore-formation. Domain II consists of three antiparallel β-sheets with exposed loop regions, and domain III is a β-sandwich (Li *et al.,* 1991; Grochulski *et al.,* 1995; Galitsky *et al.,* 2001; Morse *et al.,* 2001; Boonserm *et al.,* 2005, 2006). Exposed regions in domains II and III are involved in receptor binding (Bravo *et al.,* 2005). Domain I shares structural similarities with other PFT like colicin Ia and N and diphtheria toxin, supporting the role of this domain in pore-formation. In the case of domain II, structural similarities with several carbohydrate-binding proteins like vitelline, lectin jacalin, and lectin Mpa have been reported (de Maagd *et al.,* 2003). Domain III, shares structural similarity with other carbohydrate-binding proteins such as the cellulose binding domain of 1,4-bglucanase C, galactose oxidase, sialidase, β-glucoronidase, the carbohydrate-binding domain of xylanase U and β-galactosidase (de Maagd *et al.,* 2003). These similarities suggest that carbohydrate moieties could have an important role in the mode of action of three-domain Cry toxins. Interestingly, in the nematode *C. elegans,* mutations in *bre* genes involved in the synthesis of certain glycolipids lead to Cry5 resistance showing that glycolipids are important receptor molecules of Cry5 (Griffits *et al.,* 2005). (Cyt proteins, on the other hand, have a single a–b domain comprising of two outer layers of α-helix hairpins wrapped around a β-sheet (Li *et al.,* 1996). Cyt toxin is structurally related to volvatoxin A2, a PFT cardiotoxin produced by a straw mushroom Volvariella volvacea (Lin *et al.,* 2004).

An analysis of the phylogenetic relationships of the isolated domains of members of the three-domain Cry family revealed interesting features regarding the creation of diversity in this protein family (Bravo, 1997; de Maagd *et al.,* 2001).

Domains I and II have coevolved. The analysis of domain III sequences, revealed a different topology due to the fact that several examples of domain III swapping among toxins occurred (Bravo, 1997; de Maagd *et al.,* 2001). Some toxins with dual specificity (coleopteran, lepidopteran) are clear examples of domain III swapping among coleopteran and lepidopteran specific toxins. This suggests that domain III swapping could create novel specificities. In this regard, *in vitro* domain III swapping of certain Cry1 toxins resulted in changes in insect specificity (Bosch *et al.,* 1994; de Maagd *et al.,* 2000). The independent evolution of the three structural domains and domain III swapping among different toxins generated proteins with similar mode of action but with very different specificities (Bravo, 1997; de Maagd *et al.,* 2001).

17.11.3.3. Mode of Action

Nematicidal and insecticidal toxins of Bt are believed to share similar modes of action. Cry protein exerts its effects by forming lytic pores in the cell membrane of gut epithelial cells (Crickmore, 2005; Salehi *et al.,* 2008b). After ingestion of toxin by target nematode larvae, the crystals dissolve within the gut of the nematode, and this is followed by proteolytic activation (Crickmore, 2005). Cry toxicity is directed against the intestinal epithelial cells of the midgut and leads to vacuole and pore formation, pitting, and eventual degradation of the intestine (Marroquin *et al.,* 2000). The binding of poreforming toxin to a receptor in the epithelial cell is a major event. In order to determine host receptors, a mutagenesis screen was performed with the genetically well-characterized nematode *Caenorhabditis elegans*. This screen obtained five bre mutants that failed to internalize toxin because they lacked the receptor. The *bre* gene encodes a glycosyltransferase, which is responsible for synthesizing a carbohydrate receptor glycolipid. Convincing evidence exists for the involvement of a set of glycolipids as receptors of Bt toxins (Huffman *et al.,* 2004; Crickmore, 2005). A detailed understanding of how the Bt toxins interact with nematodes should facilitate the production of more effective Bt biocontrol agents.

17.11.4. Rhizobacteria

Rhizobacteria have also been studied for the biological control of plant-parasitic nematodes (Sikora, 1992). Aerobic endospore-forming bacteria (AEFB) (mainly *Bacillus* spp.) and *Pseudomonas* spp. are among the dominant populations in the rhizosphere that are able to antagonize nematodes (Rovira and Sands, 1977; Krebs *et al.,* 1998). Numerous *Bacillus* strains can suppress pests and pathogens of plants and promote plant growth. Some species are pathogens of nematodes (Gokta and Swarup, 1988; Li (B.) *et al.,* 2005). The most thoroughly studied is probably *Ba. subtilis* (Krebs *et al.,* 1998; Siddiqui and Mahmood, 1999; Lin *et al.,* 2001; Siddiqui, 2002). In addition, a number of studies have reported direct antagonism by other *Bacillus* spp. against plant-parasitic nematode species belonging to the genera *Meloidogyne*, *Heterodera* and *Rotylenchulus* (Gokta and Swarup, 1988; Kloepper *et al.,* 1992; Madamba *et al.,* 1999; Siddiqui and Mahmood, 1999; Insunza *et al.,* 2002; Kokalis-Burelle *et al.,* 2002; Meyer, 2003; Giannakou and Prophetou-Athanasiadou, 2004; Li (B.) *et al.,* 2005). Rhizosphere *Pseudomonas* strains also exhibit diverse pathogenic mechanisms upon interaction with nematodes (Spiegel *et al.,* 1991; Kloepper *et al.,* 1992; Kluepfel *et al.,* 1993; Westcott and Kluepfel, 1993; Cronin *et al.,* 1997a; Kerry, 2000; Jayakumar *et al.,* 2002; Andreogloua *et al.,* 2003; Siddiqui and Shaukat, 2002, 2003; Siddiqui *et al.,* 2005). The mechanisms employed by some *Pseudomonas* strains to reduce the plantparasitic nematode population have been studied. These mechanisms include the production of antibiotics and the induction of systemic resistance (Spiegel *et al.,* 1991; Cronin *et al.,* 1997a; Siddiqui and Shaukat, 2002, 2003). Other rhizobacteria reported to show antagonistic effects against nematodes include members of the genera *Actinomycetes, Agrobacterium, Arthrobacter, Alcaligenes, Aureobacterium, Azotobacter, Beijerinckia, Burkholderia, Chromobacterium, Clavibacter, Clostridium, Comamonas, Corynebacterium, Curtobacterium, Desulforibtio, Enterobacter, Flavobacterium, Gluconobacter, Hydrogenophaga, Klebsiella, Methylobacterium, Phyllobacterium, Phingobacterium, Rhizobium, Serratia, Stenotrotrophomonas* and *Variovorax* (Jacq and Fortuner, 1979; Kloepper *et al.,* 1991, 1992; Racke and Sikora, 1992; Guo *et al.,* 1996; Cronin *et al.,* 1997b; Duponnois *et al.,* 1999; Neipp and Becker, 1999; Siddiqui and Mahmood, 1999, 2001; Jonathan *et al.,* 2000; Tian and Riggs, 2000; Tian *et al.,* 2000; Meyer *et al.,* 2001; Mahdy *et al.,* 2001; Hallmann *et al.,* 2002; Insunza *et al.,* 2002; Khan *et al.,* 2002; Mena and Pimentel, 2002; Meyer, 2003).

The rhizobacteria usually comprise a complex assemblage of species with many different modes of action in the soil (Siddiqui and Mahmood, 1999). Rhizobacteria reduce nematode populations mainly by regulating nematode behaviour (Sikora and Hoffmann-Hergarten, 1993), interfering with plant–nematode recognition (Oostendorp and Sikora, 1990), competing for essential nutrients (Oostendorp and Sikora, 1990), promoting plant growth (El-Nagdi and Youssef, 2004), inducing systemic resistance (Hasky-G"unther *et al.,* 1998), or directly antagonising by means of the production of toxins, enzymes and other metabolic products (Siddiqui and Mahmood, 1999).

Most rhizobacteria act against plant-parasitic nematodes by means of metabolic by-products, enzymes and toxins. The effects of these toxins include the suppression of nematode reproduction, egg hatching and juvenile survival, as well as direct killing of nematodes (Zuckerman and Jasson, 1984; Siddiqui and Mahmood, 1999). Ammonia produced by ammonifying bacteria during decomposition of nitrogenous organic materials can result in reduced nematode populations in soil (Rodriguez-Kabana, 1986). *Pseudomonas fluorescens* controlled cyst nematode juveniles by producing several secondary metabolites such as 2,4-diacetylphloroglucinol (DAPG) (Cronin *et al.,* 1997a; Siddiqui and Shaukat, 2003). Mena *et al.* reported that *Corynebacterium paurometabolu* inhibited nematode egg hatching by producing hydrogen sulphide and chitinase (Mena and Pimentel, 2002). Some other rhizobacteria reduce deleterious organisms and create an environment more favourable for plant growth by producing compounds such as antibiotics or hydrogen cyanide (Zuckerman and Jasson, 1984).

Recently, rhizobacteria-mediated induced systemic resistance (ISR) in plants has been shown to be active against nematode pests (van Loon *et al.,* 1998; Ramamoorthy *et al.,* 2001). Plant growth-promoting rhizobacteria (PGPR) can bring about ISR by fortifying the physical and mechanical strength of the cell wall by means of cell-wall thickening, deposition of newly formed callose, and accumulation of phenolic compounds. They also change the physiological and biochemical ability of the host to promote the synthesis of defence chemicals against the challenge pathogen (e.g. by the accumulation of pathogenesis-related proteins, increased chitinase and peroxidase activity, and synthesis of phytoalexin and other secondary metabolites) (van Loon *et al.,* 1998; Siddiqui and Mahmood, 1999; Ramamoorthy *et al.,* 2001). Bacterial determinants of ISR include lipopolysaccharides (LPSs), siderophores and salicylic acid (SA) (van Loon *et al.,* 1998; Ramamoorthy *et al.,* 2001). *Rhizobium etli* G12 has been repeatedly demonstrated to be capable of suppressing early infection by the potato cyst nematode *Globodera pallida* and the root-knot nematode *Meloidogyne incognita* (Hallmann *et al.,* 2001). LPS was identified as an inducing agent of the systemic resistance. The mechanism involved in resistance development seems to be directly related to nematode recognition and penetration of the root (Reitz *et al.,* 2000, 2001; Mahdy *et al.,* 2001). However, Siddiqui *et al.* (Siddiqui and Shaukat, 2004) found that SA-negative or SA-overproducing mutants induced systemic resistance to an extent similar to that caused by the wild-type bacteria in tomato plants. They concluded that fluorescent pseudomonads induced systemic resistance against nematodes by means of a signal transduction pathway, which is independent of SA accumulation in roots.

Except for the nematophagous fungi and actinomycetes, rhizobacteria are the only group of microorganisms in which biological nematicides have been reported. Deny is a commercial biocontrol nematode product based on a natural isolate of the bacterium *Burkholderia cepacia*. This bacterium has been shown to reduce egg hatching and juvenile mobility (Meyer and Roberts, 2002). There are two commercial bionematicidal agents based on *Bacillus* species. Through a PGPR research program of the ARS (Agriculture Research Service, USA), a commercial transplant mix (Bio YieldTM, Gustafson LLC) containing *Paenobacillus macerans* and *Bacillus amyloliquefaciens* has been developed to control plant-parasitic nematodes on tomato, bell pepper and strawberry (Meyer, 2003). Another product, used in Israel, is BioNem, which contains 3% lyophilized *Bacillus firmus* spores and 97% nontoxic additives (plant and animal extracts) to control root-knot nematodes as well as other nematodes (Giannakou and Prophetou-Athanasiadou, 2004). In extensive testing on vegetable crops (tomato, cucumber, pepper, garlic and herbs), BioNem preplant applications significantly reduced nematode populations and root infestation (galling index), resulting in an overall increase in yield (Giannakou and Prophetou-Athanasiadou, 2004). BioNem showed a higher effectiveness against root-knot nematodes in the field than did *Pas. penetrans*. However, the excellent biocontrol effects of BioNem can be partially attributed to the stimulating effect that the animal and plant additives contained in the bio-nematicide formulation have on the microbial community of the rhizosphere. Previous studies have shown that the addition of manure or other organic amendments stimulate the activity of the indigenous soil microbial community (Giannakou and Prophetou-Athanasiadou, 2004).

17.11.5. Endophytic Bacteria

Endophytic bacteria have been found internally in root tissue, where they persist in most plant species. They have been found in fruits and vegetables, and are present in both stems and roots, but do no harm to the plant (McInory and Kloepper, 1995; Hallmann *et al.,* 1997, 1999; Azevedo *et al.,* 2000; Hallmann, 2001; Surette *et al.,* 2003). They have been shown to promote plant growth and to inhibit disease development and nematode pests (Sturz and Matheson, 1996; Hallmann *et al.,* 1999; Azevedo *et al.,* 2000; Munif *et al.,* 2000; Shaukat *et al.,* 2002; Sturz and Kimpinski, 2004). For example, Munif *et al.,* (2000) screened endophytic bacteria isolated from tomato roots under greenhouse conditions. They found antagonistic properties towards *M. incognita* in 21 out of 181 endophytic bacteria. Several bacterial species have also been found to possess activity against root-lesion nematode (*Pratylenchus penetrans*) in soil around the root zone of potatoes. Among them, *M. esteraomaticum* and *K. varians* have been shown to play a role in root-lesion nematode suppression through the attenuation of host proliferation, without incurring any yield reduction (Munif *et al.,* 2000). Despite their different ecological niches, rhizobacteria and endophytic bacteria display some of the same mechanisms for promoting plant growth and controlling phytopathogens, such as competition for an ecological niche or a substrate, production of inhibitory chemicals, and induction of systemic resistance (ISR) in host plants (Hallmann, 2001; Compant *et al.,* 2005).

17.11.6. Symbionts of Entomopathogenic Nematodes

Xenorhabdus spp. and *Photorhabdus* spp. are bacterial symbionts of the entomopathogenic nematodes *Steinernema* spp. and *Heterorhabdus* spp., respectively (Paul *et al.,* 1981). They have been thought to contribute to the symbiotic association by killing the insect and providing a suitable nutrient environment for nematode reproduction (Boenare *et al.,* 1997). In recent years, a potentially antagonistic effect of the symbiotic complex on plant-parasitic nematodes has been reported (Bird and Bird, 1986; Grewal *et al.,* 1997, 1999; Perry *et al.,* 1998; Lewis *et al.,* 2001). Further investigation demonstrated that the symbiotic bacteria seemed to be responsible for the plant-parasitic nematode suppression via the production of defensive compounds (Samaliev *et al.,* 2000). To date, three types of secondary metabolites have been identified as the nematicidal agent: ammonia, indole and stilbene derivative (Hu *et al.,* 1995, 1996, 1997, 1999). They were toxic to second-stage juveniles of root-knot nematode (*M. incognita*) and to fourth-stage juveniles and adults of pine-wood nematode (*Bu. xylophilus*), and inhibited egg hatching of *M. incognita* (Hu *et al.,* 1999).

17.11.7. Understanding Bacterial Pathogenesis in Nematodes at the Molecular Level

Molecular Genetic Techniques Used in Studying Bacterial Pathogenesis in Nematodes
A number of bacteria have been shown to exhibit a variety of effects on nematodes in natural environments and laboratory conditions. However, studies on the mechanisms of bacterial pathogenicity have lagged behind those assessing their roles in biological control and resource potential. Over the past few years, a number of molecular genetic methods in bacterial pathogenicity have been developed, and it is now possible to introduce these successful techniques to the study of bacterial pathogenesis in plant-parasitic nematodes (Hensel and Holden, 1996; Aballav and Ausube, 2002; Tan, 2002; Barker, 2003). Although

some technologies have been reported not to be successful in studying plant-parasitic nematodes, knowledge from studying bacterial pathogens of *C. elegans* and other animal pathogens may enhance knowledge of bacterial pathogenesis in plant-parasitic nematodes, and provide a basic methodology for studies on plant-parasitic nematodes.

Reverse genetics is a common approach in identifying and determining functions of virulence determinants. This method involves the isolation of virulence proteins involved in pathogenicity, and cloning of the corresponding genes. The functions of virulence proteins are further confirmed by their expression in other organisms, by the inactivation of the gene in a wild-type strain, or by immunological techniques (Huang *et al.,* 2005; Tian *et al.,* 2006). For example, studies on the bacterial proteases of *Br. laterosporus* G4 serving as pathogenic factors in nematode infection used reverse genetics methods (Huang *et al.,* 2005; Tian *et al.,* 2006). However, the path of discovery from proteins to genes is very labour-intensive.

Mutational analysis is another popular technique for identifying pathogenic determinants. This tool can be divided into directed and random mutagenesis. In directed mutagenesis, a putative virulence determinant encoding a gene postulated to be responsible for a certain pathogenic trait is disrupted or replaced to construct a mutant strain. The mutant and the wild-type strain are then compared to determine the importance of the suspected virulence determinant. Siddiqui *et al.* (2005) constructed mutants of the Gac-controlled *aprA*, which encodes a major extracellular protease in *Ps. fluorescens* CHA0, by inserting a suicide plasmid into the site of the chromosomal *aprA* gene. The mutant showed significantly reduced biocontrol activity against *M. incognita* during tomato and soybean infection (Siddiqui *et al.,* 2005). Much current research is instead, however, based on the use of random mutagenesis. *Pseudomonas* sp. BG33R can suppress multiplication of *M. xenoplax* and inhibit egg hatching. To investigate the pathogenic factors, Wechter *et al.* (2001, 2002) utilized Tn5 transposonmediated mutagenesis to construct a mutant library and generate five BG33R mutants that lacked ovicidal activity. ORF analysis and amino acid comparative database searches of the Tn5 insertion sites in the five mutants revealed a high degree of homology to several putative regulatory genes (Wechter *et al.,* 2001, 2002). It is time-consuming to identify a mutant with attenuated virulence within a large population of mutants. In future, signature-tagged mutagenesis (STM) may be introduced to allow mutants to be differentiated from each other by the tagging of a unique sequence for every individual transposon (Hensel *et al.,* 1995).

Comparative genomics can identify pathogenic genes by comparing genomic sequences of pathogenic and nonpathogenic strains, or other sequences from strains of interest of the same genus. Similarly, a genomic-bioinformatic approach might further define the evolutionary relationships among the various pathogenic and nonpathogenic bacteria (Hensel and Holden, 1996). For example, a comparison of the genomes of the obligate nematode parasite *Pas. penetrans* with those of other closely related bacteria, such as *Bacillus anthracis* and *Bacillus cereus* (facultative mammalian pathogen), and *Bacillus haladurans* and *Ba. subtilis* (free-living), have shown significant colinearity in larger contiguous sequences among these species. Amino acid level analysis using concatenation of 40 housekeeping genes revealed that *Pas. penetrans* is more closely related to the saprophytic species *Ba. haladurans* and *Ba. subtilis* than to the pathogenic species *Ba. anthracis* and *Ba. cereus* (Bird *et al.,* 2003; Preston *et al.,* 2003; Charles, 2005; Charles *et al.,* 2005; Davies, 2005). A genomic-bioinformatic approach will also be useful for studying the processes of host recognition and attachment. Collagen is a filamentous protein that contains a G-x-y repeated structure. These proteins were thought to be restricted to animals; however, collagen-like proteins were recently identified in the genome of *Pas. penetrans*. They are similar to those in other species of bacilli, and are likely to be responsible for endospore attachment. Four separate nucleotide sequences, Pcl.C1, Pcl.C336, Pcl.C374 and Pcl.C384, were identified in the *Pas. penetrans* genome. Other proteins containing collagen-like sequences from other bacilli were obtained from the NCBI public database. A preliminary analysis of these collagens has shown that *Pasteuria* collagens are most closely related to *Ba. thuringiensis* and *Ba. cereus* collagens rather than to those in *Ba. anthracis* (Charles, 2005; Davies, 2005; Davies and Opperman, 2006).

Some techniques have not yet been employed in the study of bacterial infection against plant-pathogenic nematodes but may prove useful, for example *in vivo* expression technology (IVET), differential fluorescence induction (DFI), subtractive hybridization and differential display etc. All these techniques are able to monitor bacterial gene expression during infection in a living organism. IVET has allowed the identification of hundreds of *in vivo* induced (ivi) genes in bacterial pathogens (Hensel and Holden, 1996). The DFI technique can be applied to more complex environments for easy isolation of GFP-expressing bacteria. However, these approaches may miss some virulence genes whose promoters do not express during certain stages of infection, or genes that are expressed only *in vitro* (Valdivia and Ramakrishnan, 2000). Subtractive hybridization and differential display approaches are techniques based on the comparison of mRNA profiles (Ogawa *et al.,* 2000; Harakava and Gabriel, 2003). The ability to synthesize cDNA from RNA populations isolated from infected hosts permits differential screening to identify genes that are specifically expressed during infection.

The subtractive hybridization and differential display approaches that have been developed have been used to study nematophagous fungi. Recently, Ahren *et al.* (2005) compared the gene expression patterns in traps and in the mycelium of the nematode-trapping fungus, *Monacrosporium haptotylum* . Despite the fact that the knobs and mycelium were grown in the same medium, there were substantial differences in the patterns of genes expressed in the two cell types. A number of the genes that were differentially expressed in trap cells are known to be regulated during the development of infection structures in plant-pathogenic fungi (Ahren *et al.,* 2005). Therefore, the techniques used to differentiate bacterial gene expression during infection are useful tools for studying stage-specific functional genes. For example, studies on the infection processes of nematodes revealed that a series of enzymes such as protease, collagenase, chitinase, lipase etc. are involved in bacterial penetration of the nematode cuticle (Cox *et al.,* 1981; Morton *et al.,* 2004; Huang *et al.,* 2005). However, which enzymes are involved in infection and when these pathogenic factors are expressed remain largely unknown. IVET, DFI and subtractive hybridization and differential display are appealing methods to answer these questions owing to their ability to monitor gene expression during infection or directly to measure transcription levels of genes.

Developing Available Models for Studying Bacterial Pathogenesis in Plant-Parasitic Nematodes

At present there is limited knowledge of the genetics of the interactions between nematode hosts and their pathogens. It is necessary to develop an alternative model for obligate bacterial parasites to understand bacterial pathogenesis in plant-parasitic nematodes at a molecular level. Unlike *Pasteuria*, opportunistic parasitic bacteria can be easily cultured and manipulated for genetic studies, so they can be used as models to gain an understanding of bacterial infection processes in nematodes. During a study of the infection of *Brevibacillus laterosporus* against freeing-living nematodes (*Panagrellus redivius*) and pine-wood nematodes (*Bursaphelenchus xylophilus*), the extracellular protease BLG4 that served as a pathogenic factor during infection was first identified using the free-living nematode *Pan. redivius* as a model. Subsequently, its role in infection against the parasitic nematode *Bu. xylophilus* was confirmed, indicating that it is feasible to identify pathogenic factors and define their roles in the infection of plant-parasitic nematodes using an easily tractable *Br. Laterosporus–Pan. redivius* model (Huang *et al.,* 2005; Tian *et al.,* 2006). Furthermore, *Br. laterosporus* strain G4 and its spores can also attach to nematode cuticles. *Br. laterosporus– Pan. redivius* could be used as a model to understand the recognition mechanism between Pasteuria spores and parasitic nematode cuticles in future research.

Another reference for plant-parasitic nematode–pathogen interactions is the use of *C. elegans* as a high-throughput screening model to facilitate the identification of virulence determinants (Davies, 2005). To date, there are some 20 species of bacteria that are known to be pathogens of *C. elegans*, of which six are Gram-positive and the remainder are Gram-negative (Couillault and Ewbank, 2002; Ewbank, 2002). *Caenorhabditis elegans* is currently being used as a model for defining bacterial virulence factors and nematode defence response factors (Ewbank, 2002; Gravato-Nobre and Hodgkin, 2005; Gravato-Nobre *et al.,* 2005). During the identification of the bacterial virulence factors that are required for the killing of *C. elegans* by the human opportunistic pathogen *P. aeruginosa* PA14, a random insertion library was generated using Tn5-based transposon mutagenesis. Following mutagenesis, mutants were analysed either individually or in pools for attenuated or increased virulence. By this means, five structural genes involved in 'fast killing' and eight involved in 'slow killing' were identified (Aballav and Ausube, 2002; Tan, 2002).

The major question is whether the pathogenic factors identified in these models can be used to explain pathogenesis in plant-parasitic nematodes. Until recently, it was believed that animals did not share similar virulence factors. However, the existence of a universal virulence factor has been clearly demonstrated in the case of *P. aeruginosa*. Among eight bacterial mutants with reduced pathogenicity against *C. elegans*, six in an insect model and seven in a mouse model also showed attenuated virulence (Couillault and Ewbank, 2002). Moreover, the enzyme-mediated infection in the *Br. laterosporus–Pan. redivius* model has been extensively studied and confirmed to be similar to the fungal penetration of plant-parasitic nematode cuticles. It is therefore feasible to understand pathogenic mechanisms in plant parasitic nematodes using tractable models such as the *Br. laterosporus–Pan. redivius* model or the bacterium–*C. elegans* model. These models of pathogenicity have been intensively studied, including the stages of attraction and attachment between bacteria and their hosts, entry into the host through nematode stoma or penetration of the nematode body wall, and parasitism or toxin-mediated host death (Tan, 2002; Huang *et al.,* 2005).

17.11. AMOEBAE

Amoebae are protists which move and feed using pseudopodia. Amoebae (Sarcodina: Rhizopodea) occur in large numbers in almost all soil types. A variety of microflora such as bacteria, fungal spores, algae, and microfauna consisting principally of protozoans and minute invertebrates are ingested by amoebae (Esser, 1983). In 1952, a predacious amoeboid organism (Weber *et al.,* 1952) which was subsequently identified as *Theratromyxa weberi* was observed attacking and ingesting larvae of the golden nematode, *Globodera rostochiensis*. Subsequent publications dealt with amoebae predacious on phytoparasitic nematodes (Paramonov, 1954; Winslow and Williams, 1957; Sayre, 1973). In Florida, *Thecamoeba* sp. has been observed preying on nematodes.

The following nematodes have been reported as prey of *Theratromyxa weberi* (Weber *et al.,* 1952; Winslow and Williams, 1957; Sayre, 1973): *Aphelenchoides rutgersi* Hooper and Meyers, 1971; *Aphelenchus avenae* Bastian, 1865; *Ditylenchus dipsaci* (Kuhn, 1857) Filipjev, 1936; *Globodera rostochiensis*; *Hemicycliophora* sp.; *Heterodera schachtii* **A.** Schmidt, 1871; *H. trifolii* Goffart, 1932; *Meloidogyne incognita* (Kofoid and White, 1919; Chitwood, 1949); *Pratylenchus pratensis* (deMan, 1880; Filipjev, 1936); and *Rotylenchulus reniformis* (Linford and Oliveira, 1940). In Florida, *Pratylenchus* sp. and several unidentifiable nematodes have served as prey for *Thecamoeba* sp. *Theratromyxa weberi*, upon contacting its prey (principally larvae) flows over the nematode body and assimilates it entirely. After ingestion, an enfolding occurs that results in a smaller digestive cyst in approximately two hours. After 23 hours, the nematode is digested, and the protoplasm within the digestive cyst cleaves, and within 15-20 minutes some 4-10 amoebae emerge from the digestive cyst (Sayre, 1973).

Nematodes have been observed several times to be held in a pellicle fold of *Thecamoeba* sp. In one case, a nematode escaped from the fold. One occasionally observes an undigested nematode inside the body of an amoeba. Two nematodes were observed that were ingested lengthwise into the pellicle of *Thecamoeba* sp. The nematode body inside the pellicle was severely distorted as if compressed during entry. The pellicle in both cases formed a lip-like tube about the nematode body. Amoebae failed to control root-knot nematodes infecting tomatoes under greenhouse conditions (Sayre, 1973). It is unlikely that amoebae will be effective predators of nematodes since most nematodes are muscular and active compared to the slow-moving amoebae that lack muscles. Amoebae are also very sensitive to adverse soil conditions, and will encyst or perish under conditions in which nematodes may thrive.

17.12. VIRUSES

There is no report for viral infection of nematodes as a potential biological control agent. It has been reported that *Thaumamermis cosgrovei* is infected by *Iridovirus* (Williams, 2008). This nematode belongs to Mermithidae, a family of nematode worms that are endoparasites in arthropods and are not considered as a pest.

CONCLUSION

In this chapter we tried to give an overview on application of different microorganisms as biocontrol agents of plant parasite nematodes. In spite of that the research on biocontrol of nematodes has started prior to 1975, this area is young yet, but it is important to note that finding new sources as biocontrol agents of nematodes encouraged researchers and companies to have more focus in this area of research.

The research in the recent two decades was typified by attempts to replace nematicides with antagonists. Thus far, few of these efforts have resulted ineffective biological control and the research has done little to increase our understanding of how biological control may or may not be achieved. Our greatest need is for sound, in-depth biological information on how organisms, populations, and communities operate in the environment. The next phase will be typified by basic investigations of organism, population, and community ecology. Microbial biological nematode control agents do not leave harmful residues, substantially reduced impact on non-target species, when locally produced, may be cheaper than chemical pesticides and in the long-term may be more effective than chemical pesticides. However, they have some disadvantages such as:

a) high specificity, which will require an exact identification of the pest/pathogen and may require multiple pesticides to be used.

b) Often slow speed of action thus making them unsuitable if a pest outbreak is an immediate threat to a crop.

c) Often variable efficacy due to the influences of various biotic and abiotic factors.

Where microbial control of nematodes appears to be occurring, it is essential that the mechanism of antagonism be established for the system in question. This is relatively easy for parasites and predators, but quite difficult for antibiosis and competition. In addition, we must be able to quantify the antagonist and antagonist activity. Without this information, we cannot understand and remedy the inconsistency of results that is characteristic of biological control research. Quantitative assays are lacking for most antagonists of nematodes. Biological control occurs when a population or community of antagonists suppresses a population of nematodes. Quantitative assays and models must be developed in order to understand the interaction of nematode populations and biological control agents. Control of nematodes via microorganisms can be optimized by basing management decisions on the relationship between biology and activity of these antagonists and nematodes response. Lacking a well informed management plan, arbitrary selection of control practices can be costly and ineffective. Microorganisms that have biocontrol potential for nematodes show different reactions against them. Some of these differences have been related to genetic properties of microorganisms; so they need to improve via biotechnology. Advances in the techniques of biotechnology introduce additional possibilities of biological engineering of nematode antagonists. The strategy would involve selection or induction of variants of potential microbial agents with characteristics that enhance their effectiveness; it would also require development of cultural practices that promote the growth of beneficial biota in the soil. Increased effectiveness might involve the introduction of a microorganism that is a more successful competitor, a more aggressive predator, a more virulent parasite, or have enhanced survival characteristics. The approach may involve mutagenesis and screening, or the direct insertion of genes for the desired characteristic into the genome of the organism. There is an important arena of basic and applied research opportunities in the genetic tailoring organisms for effectiveness in specific environmental and cultural situations. Totally, it seems that via researches highly concerned on biocontrol of plant parasitic nematodes using useful microorganisms, soon we will see the new bioproducts for biocontrol of nematodes.

REFERENCES

Aballav, A. and Ausube, F.M. 2002. *Caenorhabditis elegans* as a host for the study of host – pathogen interactions. *Curr. Opin. Microbiol*, 5: 97–101.

Abrantes, I.M.D. and Curtis, R.H.C. 2002. Immunolocalization of a putative cuticular collagen protein in several developmental stages of *Meloidogyne arenaria*, *Globodera pallida* and *G. rostochiensis*. *J. Helminthol*, 76: 1–6.

Ahman, J., Ek, B., Rask, L. and Tunlid, A. 1996. Sequence analysis and regulation of a gene encoding a cuticle-degrading serine protease from the nematophagous fungus *Arthrobotrys oligospora*. *Microbiology*, 142: 1605–1616.

Ahman, J., Johanson, T., Olsson, M., Punt, P.J., van den Hondel, C.A.M.J.J. and Tunlid, A.S. 2002. Improving the pathogenicity of a nematode trapping fungus by genetic engineering of a subtilisin with nematotoxic activity. *Applied and Environmental Microbiology*, 689: 3408–3415.

Ahren, D., Tholander, M., Fekete, C., Rajashekar, B., Friman, E., Johansson, T. and Tunlid, A. 2005. Comparison of gene expression in trap cells and vegetative hyphae of the nematophagous fungus *Monacrosporium haptotylum*. *Microbiology*, 151: 789–803.

Ahren, D., Ursing, B.M. and Tunlid, A. 1998. Phylogeny of nematodetrapping fungi based on 18S rDNA sequences. *FEMS Microbiology Letters*, 158: 179–184.

Akhtar, M. 1997. Current options in integrated management of plant-parasitic nematodes. *Integrated Pest Managment Review*, 2: 187–197.

Akhtar, M. and Alam, M.M. 1993. Utilization of waste materials in nematode control – a review. *Bioresource Technology*, 45: 1-7.

Akhtar, M. and Malik, A. 2000. Roles of organic soil amendments and soil organisms in the biological control of plant-parasitic nematodes: a review. *Bioresource Technology*, 74: 35-47.

Alabouvette, C., Olivain, C. and Steinberg, C. 2006. Biological control of plant diseases: the European situation. *European Journal of Plant Pathology*, 114: 329-341.

Andreogloua, F.I., Vagelasa, I.K., Woodb, M., Samalievc, H.Y. and Gowena, S.R. 2003. Influence of temperature on the motility of *Pseudomonas oryzihabitans* and control of *Globodera rostochiensis*. *Soil Biol. Biochem*, 35: 1095–1101.

Arias-Estévez, M., López-Periago, E., Martínez-Carballo, E., Simal-Gándara, J., Mejuto, J.C. and García-Río, L. 2008. The mobility and degradation of pesticides in soils and the pollution of groundwater resources. *Agriculture, Ecosystems and Environment*, 123 (4): 247-260.

Azcón-Aguilar, C., Barea, J.M., 1996. Arbuscular mycorrhizas and biological control of soil-borne plant pathogens – an overview of the mechanisms involved. *Mycorrhiza*, 6: 457–64.

Azevedo, J.L., Maccheroni, W.J., Pereira, J.O. and de Arau'jo, W.L. 2000. Endophytic microorganisms: a review on insect control and recent advances on tropical plants. *Electronic Journal of Biotechnol*, 3(1): 40–65.

Balogh, J., Tunlid, A. and Rosen, S. 2003. Deletion of a lectin gene does not affect the phenotype of the nematode-trapping fungus *Arthrobotrys oligospora*. *Fungal Genet. Biol*, 39: 128–135.

Barbercheck, M.E. and Von Broembsen, S.L. 1986. Effects of soil solarization on plant-parasitic nematodes and Phytophthora cinnamoni in South Africa. *Plant Disease*, 70: 945-950.

Barker, K.R. 2003. Perspectives on plant and soil nematology. *Annu. Rev. Phytopathol*, 41: 1–25.

Barker, K.R. and Koenning, S.R. 1998. Developing sustainable systems for nematode management. *Annu. Rev. Phytopathol*, 36: 165–205.

Barron, G.L. 1992. Lignolytic and cellulolytic fungi as predators and parasites. In: Carroll, G.C. and Wicklow, D.T. (eds), *The Fungal Community, Its Organization and Role in the Ecosystems*, New York, Marcel Dekker, pp. 311–326.

Becker, N., and Margalith, J. 1993. Use of *Bacillus thuringiensis* israeliensis against mosquitoes and blackflies. In: Entwistle, P.F., Cory, P.F., Bailey, M.J. and Higgs, S. (Eds.), *Bacillus thuringiensis, an environmental biopesticide: theory and practice*. J. Wiley and Sons, New York, pp. 145-170.

Benoit, T.G., Wilson, G.R., Bull, D.L., and Aronson, A.I. 1990. Plasmid associated sensitivity of *Bacillus thuringiensis* to UV light. *Applied and Environmental Microbiology*, 56: 2282-2286.

Berenbaum, M. 2000. Committee on the Future Role of Pesticides, National Academy of Sciences, *The Future Role of Pesticides in U.S. Agriculture*, National Academy Press, Washington, D.C. p. 48.

Bhattacharya, D. and Swarup, G. 1988. *Pasteuria penetrans*, a pathogen of the genus *Heterodera*, its effect on nematode biology and control. *Indian Journal of Nematology*, 18: 61-70.

Bird, A.F. and Bird, J. 1986. Observations on the use of insect parasitic nematodes as a means of biological control of rootknot nematodes. *International Journal for Parasitology*, 16: 511–516.

Bird, D.M. and Kaloshian, I. 2003. Are roots special? Nematodes have their say. *Physiological and Molecular Plant Pathology*, 62: 115-123.

Bird, D.M., Opperman, C.H. and Davies, K.G. 2003. Interaction between bacteria and plant-parasitic nematodes: now and then. *International Journal for Parasitology*, 33: 1269–1276.

Blaxter, M.L. and Robertson, W.M. 1998. The cuticle. In: Perry, R.N., Wright, D.J. (eds), *The physiology and biochemistry of free-living and plant-parasitic nematodes*. CABI, Wallingford, UK, pp. 25–48.

Boenare, N.E., Givaudan, A., Brehelin, M. and Laumond, C. 1997. Symbiosis and pathogenicity of nematode–bacterium complex. *Symbiosis*, 22: 21–45.

Bonants, P.J.M., Fitters, P.F.L., Thijs, H. *et al.* 1995. A basic serine protease from *Paecilomyces lilacinus* with biological activity against *Meloidogyne hapla* eggs. *Microbiology*, 141: 775–784.

Boobis, A.R., Ossendorp, B.C., Banasiak, U., Hamey, P.Y., Sebestyen, I. and Moretto, A. 2008. Cumulative risk assessment of pesticide residues in food. *Toxicology Letters*, 180 (2): 137-150.

Boonserm, P., Davis, P., Ellar, D.J. and Li, J. 2005. Crystal structure of the mosquito-larvicidal toxin Cry4Ba and its biological implications. *J. Mol. Biol*, 348: 363–382.

Boonserm, P., Mo, M., Angsuthanasombat, C. and Lescar, J. 2006. Structure of the functional form of the mosquito larvicidal Cry4Aa toxin from *Bacillus thuringiensis* at a 2.8-A° resolution. *J. Bacteriol*, 188: 3391–3401.

Bordallo, J.J., Lopez-Llorca, L.V., Jansson, H-B. *et al.* 2002. Effects of eggparasitic and nematode-trapping fungi on plant roots. *New Phytologist*, 154: 491–499.

Bosch, D., Schipper, B., van der Kleij, H., de Maagd, R.A. and Stiekema, J. 1994. Recombinant *Bacillus thuringiensis* insecticidal proteins with new properties for resistance management. *Biotechnology*, 12: 915–918.

Bravo, A. 1997. Phylogenetic relationships of *Bacillus thuringiensis* d-endotoxin family proteins and their functional domains. *J. Bacteriol*, 179: 2793–2801.

Bravo, A., Gill, S.S. and Sobero'n, M. 2007. Mode of action of *Bacillus thuringiensis* Cry and Cyt toxins and their potential for insect control. *Toxicon*, 49: 423–435.

Bravo, A., Gill, S.S., Sobero'n, M. 2005. *Bacillus thuringiensis* Mechanisms and Use In: *Comprehensive Molecular Insect Science*. Elsevier, B.V., Amsterdam, pp. 175–206.

Bromilow, R.H. 1980. Behavior of nematicides in soil and plants, In: *Factors affecting the application and use of nematicides in Western Europe*. Workshop, Nematology Group Association of Applied Biologists, pp. 87-107.

Calvet, C., Pinochet, J., Camprubi, A., and Fernez, C. 1995. Increased tolerance to the root lesion nematode *Pratylenchus vulnus* in micropropagated BA-29 Quince rootstock. *Mycorrhiza*, 5: 253-258.

Calvet, C., Pinochet, J., Hernández Dorrego, A., Estaún, V. and Camprubí, A. 2001. Field microplot performance of the peach–almond hybrid GF-677 after inoculation with arbuscular mycorrhizal fungi in a replant soil infested with root-knot nematodes. *Mycorrhiza*, 10: 295–300.

Camprubi, A., Pinochet, J., Calvet, C. AND Estaun , V. 1993. Effect of the root lesion nematode *Prathylenchus vulnus* and the vesicular-arbuscular mycorrhizal funfus *Glomus mosseae* on the growth of three pulm rootstocks. *Plant Soil*, 153: 223-229.

Cappello, M., Bungiro, R.D., Harrison, L.M., *et al.* 2006. A purified *Bacillus thuringiensis* crystal protein with therapeutic activity against the hookworm parasite *Ancylostoma ceylanicum*. *Proceedings of the National Academy of Sciences of the USA*, 103(41): 15154–15159.

Carling, D.E., Roncadori, R.W. and Hussey, R.S. 1996. Interactions of arbuscular mycorrhizae, *Meloidogyne arenaria*, and phosphorus fertilization on peanut. *Mycorrhiza*, 6: 9–13.

Carozzi, N.B., Kramer, V.C., Warren, W., Evola, S. and Koziel, M.G. 1991. Prediction of insecticidal activity of *Bacillus thuringiensis* strains by polymerase chain reaction product profiles. *Applied and Environmental Microbiology*, 57: 3057–3061.

Carpenter, J. Lynch, L. and Trout, T. 2001. *Calif. Agric.* 55(3): 12–18.

Cartia, G. and Greco, N. 1987. Effetti della solarizzazione del suolo su una coltura di peperone in serra. *Colture Protette*, 16(5):61-65.

Cartia, G., Greco, N. and Cipriano, T. 1989. Effect of soil solarization and fumigants on soil-borne pathogens of pepper in greenhouse. *Acta Horticulturae*, 255: 111-116.

Cartia, G., Greco, N., and Cirvilleri, G. 1988. Solarizzazione e bromuro di mettle nella difesa dai parassiti del pomodoro in ambiente protetto. *Proc. Giornate Fitopatologiche*, Lecce, Italy, 16-20 May, 1: 437488.

Cayrol, J.C., Djian-Caporalino C. and Panchaud-Mattei E. 1993. Les biopesticides a I'assaut des nematodes du sol'. *La Recherche*, 24(250): 78-80.

Cenis, J.L. 1984. Control of the nematode Meloidogyne javanica by soil solarization. *Proc. 6th Congr. Union Phytopath*. Mediterr., Cairo, Egypt, p. 132.

Charles, L. 2005. Phylogenetic studies of *Pasteuria penetrans* looking at the evolutionary history of housekeeping genes and collagenlike motif sequences. MSc thesis, North Carolina State University, Raleigh, NC.

Charles, L., Carbone, I., Davis, K.G., Bird, D., Burke, M., Kerry, B.R. and Opperman, C.H. 2005. Phylogenetic analysis of *Pasteuria penetrans* by use of multiple genetic loci. *J. Bacteriol*, 187: 5700–5708.

Chen, T.H., Hsu, C.S., Tsai, P.J., Ho, Y.F. and Lin, N.S. 2001. Heterotrimeric G-protein and signal transduction in the nematode-trapping fungus *Arthrobotrys dactyloides*. *Planta (Berlin)*, 212: 858–863.

Chen, Z.X. and Dickson, D.W. 1998. Review of *Pasteuria penetrans*: biology, ecology, and biological control potential. *J. Nematol*, 30: 313–340.

Chiang, A.S., Yen, D.F., and Pang, W.K. 1986. Germination and proliferation of *Bacillus thuringiensis* in the gut of rice moth larva. *Journal of Invertebrate Pathology*, 48: 96-99.

Chitwood, D.J. 2003. Nematicides. In: Plimmer, J.R., (Ed.), *Encyclopedia of Agrochemicals*. Vol. 3. New York, NY: John Wiley and Sons. p. 1104-1115.

Cohen, S.Z. 1996. *J. Environ. Sci. Health* B31: 345–352.

Compant, S., Duffy, B., Nowak, J., Clement, C. and Barka, E.A. 2005. Use of plant growth-promoting bacteria for biocontrol of plant diseases: principles, mechanisms of action, and future prospects. *Applied and Environmental Microbiology*, 71: 4951–4959.

Cooper, K.M., Grandison, G.S., 1986. Interaction of vesicular–arbuscular mycorrhizal fungi and root-knot nematode on cultivars of tomato and white clover susceptible to *Meloidogyne hapla*. *Annals of Applied Biology*, 108: 555–66.

Couch, J.N. 1937. The formation and operation of the traps in the nematode-catching fungus, *Dactylella bembicoides*. *J. Elisha Mitchell Sci. Soc*, 53: 301-309.

Couillault, C. and Ewbank, J.J. 2002. Diverse bacteria are pathogens of *Caenorhabditis elegans*. *Infect Immun*, 70: 4705–4707.

Cox, G.N., Kusch, M. and Edgar, R.S. 1981. Cuticle of *Caenorhabditis elegans* its isolate and partial characterization. *J. Cell Biol*, 90:7–17.

Crickmore, N. 2005. Using worms to better understand how Bacillus thuringiensis kills insects. *Trends in Microbiology* 13: 347-350.

Crickmore, N., Zeigler, D.R. , Feitelson, J., Schnepf, E., Van Rie, J., Lereclus, D., Baum, J. and Dean, D.H. 1998. Revision of the nomenclature for the *Bacillus thuringiensis* pesticidal crystal proteins. *Microbiol. Mol. Biol. Rev*, 62: 807–813.

Cronin, D., Moenne-Loccoz, Y., Dunne, C. and O'Gara, F. 1997b. Inhibition of egg hatch of the potato cyst nematode *Globodera rostochiensis* by chitinese-producing bacteria. *Eur. J. Plant Pathol*, 103: 433–440.

Cronin, D.,Moenne-Loccoz, Y., Fenton, A., Dunne, C., Dowling, D.N. and O'gara, F. 1997a. Role of 2, 4-diacetylphloroglucinol in the interaction of the biocontrol *Pseudomonas* strain F113 with the potato cyst nematode *Globodera rostochiensis*. *Applied and Environmental Microbiology*, 63: 1357–1361.

Davies, K.G. 2005. Interactions between nematodes and microorganisms: bridging ecological and molecular approaches. *Adv. Appl. Microbiol*, 57: 53–78.

Davies, K.G. and Opperman, C.H. 2006. A potential role for collagen in the attachment of *Pasteuria penetrans* to nematode cuticle. Multitrophic Interactions in the Soil, IOBC/wprs Bulletin, (Raaijmakers, J.M. and Sikora, R.A., eds), 29: 11–16.

Davies, K.G., Kerry, B.R. and Flynn, C.A. 1988. Observations on the pathogenicity of *Pasteuria penetrans*, a parasite of root-knot nematodes. *Ann. Appl. Biol*, 112: 491–501.

De Barjac, H., and Sutherland, D.J. 1990. *Bacterial control of mosquitoes and blackflies.* Rutger University Press, Newbrunswick, N. J.

De Maagd, R.A., Bravo, A. and Crickmore, N. 2001. How *Bacillus thuringiensis* has evolved specific toxins to colonize the insect world. *Trends Genet*, 17: 193–199.

De Maagd, R.A., Bravo, A., Berry, C., Crickmore, N. and Schnepf, H.E. 2003. Structure, diversity and evolution of protein toxins from spore-forming entomopathogenic bacteria. *Ann. Rev. Genet*, 37: 409–433.

De Maagd, R.A., Weemen-Hendriks, M., Stiekema, W. and Bosch, D. 2000. Domain III substitution in *Bacillus thuringiensis* delta-endotoxin Cry1C domain III can function as a specific determinant for *Spodoptera exigua* in different, but not all, Cry1–Cry1C hybrids. *Applied and Environmental Microbiology*, 66: 1559–1563.

Deacon, J.W, and Saxena, G. 1997. Orientated zoospore attachment and cyst germination in *Catenaria anguillulae*, a facultative endoparasite of nematodes. *Mycological Research*, 101: 513–522.

Decraemer, W., Karanastasi, E., Brown, D. and Backeljau, T. 2003. Review of the ultrastructure of the nematode body cuticle and its phylogentic interpretation. *Biol. Rev*, 78: 465–510.

Dehne, H.W. 1982. Interaction between vesicular-arbuscular mycorrhizal fungi and plant pathogens. *VA Mycorrhizae and Plant Disease Research*, 72(8): 1115-1119.

Dickson, D.W., Oostendorp, M., Giblin-Davis, R. and Mitchell, D.J. 1994. Control of plant-parasitic nematodes by biological antagonists. In: Rosen, D., Bennett, F.D. and Capinera, J.L. (Eds.), *Pest management in the subtropics. Biological control—a Florida perspective.* Andover, Ug: Intercept, pp. 575-601.

Dijksterhuis, J., Veenhuis, M., Harder, W. and Nordbring-Hertz, B. 1994. Nematophagous fungi: physiological aspects and structure–function relationships. *Adv. Microb. Physiol*, 36:111–143.

Dong, J.Y., Zhou, Y., Li, R., Zhou, W., Li, L., Zhu, Y., Huang, R. and Zhang, K. Q. 2006. New nematicidal azaphilones from the aquatic fungus *Pseudohalonectria adversaria* YMF1.01019. *FEMS Microbiology Letters*, 264:65–69.

Dong, L.Q. and Zhang, K.Q. 2006. Microbial control of plantparasitic nematodes: a five-party interaction. *Plant Soil*, 288: 31–45.

Donovan, W.P., Donovan, J.C., and Engleman, J.T. 2001. Gene knockout demonstrates that vip3A contributes to the pathogenesis of *Bacillus thuringiensis* towards *Agrotis ipsilon* and *Spodoptera exigua*. *Journal of Invertebrate Pathology*, 78: 45-51.

Dumas-Gaudot, S.S., Dassi, B., Pozo, M.J., Gianinazzi-Pearson, V. and Gianinazzi, S. 1996. Plant hydrolytic enzymes (chitinase and B-1, 3-glucase) in root reactions to pathogenic and symbiotic microorganisms. *Plant Soil*, 185: 211-221.

Duponnois, R., Ba, A.M. and Mateille, T. 1999. Beneficial effects of *Enterbacter cloacae* and *Pseudomonas mendocina* for biocontrol of Meloidogyne incognita with the endospore-forming bacterium *Pasteuria penetrans*. *Nematol*, 1: 95–101.

Ellis, R.T., Stockhoff, B.A., Stamp, L., Schnepf, H.E., Schwab, G.E., Knuth, M., *et al.* 2002. Novel *Bacillus thuringiensis* binary insecticidal crystal proteins active on western corn rootworm, *Diabrotica virgifera virgifera* LeConte. *Applied and Environmental Microbiology*, 68: 1137-1145.

El-Nagdi, W.M.A. and Youssef, M.M.A. 2004. Soaking faba bean seed in some bio-agent as prophylactic treatment for controlling *Meloidogyne incognita* root-knot nematode infection. *Journal of Pest Science*, 77: 75–78.

Emmert, E.A.B. and Handelsman, J. 1999. Biocontrol of plant disease: a (Gram1) positive perspective. *FEMS Micriobiology Letters*, 171: 1–9.

Endo, B.Y. 1962. Lethal time-temperature relations for Heterodera glycines. *Phytopathology*, 52:992-997.

Esser, R.P. 1983. Amoebic predation upon nematodes. *Nematology Circular*, No. 98.

Esser, R.P., El-Gholl, N.E. and Price, M. 1991. Biology of *Dactylella megalospora* Drechs., a nematophagous fungus. *Soil Crop Sci. Sco. Fla. Proc*, 50: 173-180.

Estruch, J.J., Warren, G.W., Mullins, M.A., Nye, G.J., Craig, J.A., and Koziel, M.G. 1996. Vip3A, a novel *Bacillus thuringiensis* vegetative insecticidal protein with a wide spectrum of activities against lepidopteran insects. *Proceedings of the National Academy of Sciences*, USA, 93: 5389-5394.

Ewbank, J.J. 2002. Tackling both sides of the host–pathogen equation with *Caenorhabditis elegans*. *Microbes Infect*, 4: 247–256.

Feder, W.A. 1963. A comparison of nematode-capturing efficencies of five *Dactylelia*-species at four temperatures. *Mycopath. Mycol. Appl*, 19: 99-104. Genstat 5 Committee (1987). Genstat 5 reference manual, Clarendon Press, Oxford.

Fitter, A.H. and Garbaye, J. 1994. Interaction between mycorrhizal fungi and other soil microorganisms. *Plant Soil*, 159: 123-132.

Francl, L.J. and Dropkin, V.H. 1985. *Glomus fasciculatum*, a weak pathogen of *Heterodera glycines*. *J. Nematol*, 7: 470-475.

Friedman, M.J. 1990. Commercial production and development. In: Gaugler, R. and Kaya, H.K. (Eds.), *Entomopathogenic nematodes in biological control*. Boca Raton, FL: CRC Press, pp. 153-172.

Galitsky, N., Cody, V., Wojtczak, A., Ghosh, D., Luft, J. R., Pangborn, W. and English, L. 2001. Structure of the insecticidal bacterial d-endotoxin CryBb1 of *Bacillus thuringiensis*. Acta Cryst. allogr D 57: 1101–1109.

Gardener, B.B.M. 2004. Ecology of *Bacillus* and *Paenibacillus* spp. in agricultural systems. *Phytopathol*, 94: 1252–1258.

Giannakou, I.O. and Prophetou-Athanasiadou, D. 2004. A novel non-chemical nematicide for the control of root-knot nematodes. *Appl. Soil. Ecol*, 26: 69–79.

Giannakou, I.O. and Anastasiadis, I. 2005. Evaluation of chemical strategies as alternatives to methyl bromide for the control of root-knot nematodes in greenhouse cultivated crops. *Crop Protection*, 24 (6): 499-506.

Glazer, I. Salame, L. and Segal, D. 1997. *Biocontrol Sci. Technol.* 7: 499–512.

Gokta, N. and Swarup, G. 1988. On the potential of some bacterial biocides against root-knot cyst nematodes. *Indian Journal of Nematology*, 18: 152–153.

Gravato-Nobre, M.J. and Hodgkin, J. 2005. *Caenorhabditis elegans* as a model for innate immunity to pathogens. *Cellular Microbiol*, 7: 741–751.

Gravato-Nobre, M.J., Nicholas, H.R., Nijland, R., O'Rourke, D., Whittington, D.E., Yook, K.J. and Hodgkin, J. 2005. Multiple genes affect sensitivity of *Caenorhabditis elegans* to the bacterial pathogen *Microbacterium nematophilum*. *Genetics*, 171: 1033–1045.

Greco, N., Brandonisio, A. and Elia, F. 1985. Control of *Ditylenchus dipsaci, Heterodera carotae* and *Meloidogyne javanica* by solarization. *Nematol. Medit*, 13:191-197.

Grewal, P.S., Lewis, E.E. and Venkatachari, S, 1999. Allelopathy: a possible mechanism of suppression of plant-parasitic nematodes by entomopathogenic nematodes. *Nematol*, 1: 735–743.

Grewal, P.S., Martin, W.R., Miller, R.W. and Lewis, E.E. 1997. Suppression of plant-parasitic nematode populations in turfgrass by application of entomopathogenic nematodes. *Biocontrol Sci Technol*, 7: 393–399.

Griffits, J.S., Haslam, S.M., Yang, T., Garczynski, S.F., Mulloy, B., Morris, H., Cremer, P.S., Dell, A., Adang, M.J. and Aroian, R.V. 2005. Glycolipids as receptors for *Bacillus thuringiensis* crystal toxin. *Science*, 307: 922–925.

Grochulski, P., Masson, L., Borisova, S., Pusztai-Carey, M., Schwartz, J.L., Brousseau, R. and Cygler, M. 1995. *Bacillus thuringiensis* CryIA(a) insecticidal toxin: crystal structure and channel formation. *J. Mol. Biol*, 254: 447–464.

Gronvold, J. 1989. Induction of nematode-trapping organs in the predacious fungus, *Athrobotrys oligospora* (Hyphomycetales) by infective larvae of *Ostertagia ostertagi* (Trichostrongylidae). *Acta Vet. Scand*. 30: 77-87.

Guo, R.J., Liu, X.Z. and Yang, H.W. 1996. Study for application of rhizobacteria to control plant-parasitic nematode. *Chinese Biol. Control*, 12: 134–137 (abstract).

Hague, N.G.M. and Gowen, S.R. 1987. Chemical control of nematodes. In: Brown, R.H. and Kerry, B.R. (Eds.), *Principles and practice of nematode control in crops*. Academic Press. pp. 131-178.

Hallmann, J. 2001. Plant interactions with endophytic bacteria. In: Jeger, M.J. and Spence, N.J. (Eds.), *Biotic Interactions in Plant-Pathogen Interactions*. CAB International, Wallingford, UK. pp. 87–119.

Hallmann, J., Faupel, A., Krachel, A. and Berg, G. 2002. Occurrence and biocontrol potential of potato-associated bacteria. *Nematol*, 4: 285 (abstract).

Hallmann, J., Quadt-Hallmann, A., Mahaffee, W.F. and Kloepper, J.W. 1997. Bacterial endophytes in agricultural crops. *Can. J. Microbiol*, 43: 895–914.

Hallmann, J., Quadt-Hallmann, A., Miller, W.G., Sikora, R.A. and Lindow, S.E. 2001. Endophytic colonization of plants by the biocontrol agent *Rhizobium etli* G12 in relation to *Meloidogyne incognita* infection. *Phytopathol*, 91: 415–422.

Hallmann, J., Rodriguez-Kabana, R. and Kloepper, J.W. 1999. Chitinmediated changes in bacterial communities of the soil, rhizosphere and within roots of cotton in relation to nematode control. *Soil Biol. Biochem*, 31: 551–560.

Hamilton, D.J., Mbrus, A., Dieterle, R.M., Felsot, A.S., Harris, C.A., Holland, P.T., Katayama, A., Kurihara, N., Linders, J. Uunsworth, J.and Wong, S.S. 2003. Regulatory limits for pesticide residues in water (IUPAC technical teport). *Pure Appl. Chem.*, 75 (8): 1123–1155.

Harakava, R. and Gabriel, D.W. 2003. Genetic differences between two strains of *Xylella fastidiosa* revealed by suppression subtractive hybridization. *Applied and Environmental Microbiology*, 69: 1315–1319.

Harisson, M.J. and Dixon, R.A. 1994. Spatial patterns of expression of flavonoid/isofavonoid pathway genes during interaction between roots of *Medicago truncatula* and the mycorrhizal fungus *Glomus versiforme*. *Plant J*, 6: 9-20.

Harrison, M.J. 2005. *"Signaling in the arbuscular mycorrhizal symbiosis"*. *Annu. Rev. Microbiol.* 59: 19–42.

Hasky-G"unther, K., Hoffmann-Hergarten, S. and Sikora, R.A. 1998. Resistance against the potato cyst nematode *Globodera pallida* systemically induced by the rhizobacteria *Agrobacterium radiobacter* (G12) and *Bacillus sphaericus* (B43). *Fund. Appl. Nematol*, 21: 511–517.

Heald, .C.M., Bruton, B.D. and Davis, R.M. 1989. Influence of *Glomus intraradices* and soil phosphorus on *Meloidogyne incognita* infecting *Cucumis melo*. *J. Nematol*, 21: 69-73.

Heintz, C.E. 1978. Assessing the predacity of nematode trapping fungi in vitro. *Mycologia*, 70 : 1086-1100.

Heintz, C.E. and Pramer, D. 1972. Ultrastructure of nematode-trapping fungi. *J. Bacteriol*, 110:1163–1170.

Hensel, M. and Holden, D.W. 1996. Molecular genetic approaches for the study of virulence in both pathogenic bacteria and fungi. *Microbiol*, 142: 1049–1058.

Hensel, M., Shea, J.E., Gleeson, C., Jones, M.D., Dalton, E. and Holden, D.W. 1995. Simultaneous identification of bacterial virulence genes by negative selection. *Science*, 269: 400–403.

Hewlett, T.E., Cox, R., Dickson, D.W. and Dunn, R.A. 1994. Occurrence of *Pasteuda* spp. in Florida. *Journal of Nematology*, 26:616-619.

Hu, K., Li, J. and Webster, J.M. 1995. Mortality of plant-parasitic nematodes caused by bacterial (*Xenorhabdus* spp. and *Photorhabdus luminescens*) culture media. *J. Nematol*, 27: 502–503.

Hu, K., Li, J. and Webster, J.M. 1996. 3, 5-Dihydroxy-4-isopropylstilbene: a selective nematicidal compound from culture filtrate of Photorhabdus luminescens. *Can. J. Plant Pathol*, 18: 104.

Hu, K., Li, J. and Webster, J.M. 1997. Quantitative analysis of a bacteria-derived antibiotic in nematode-infected insects using HPLC-UV and TLC-UV methods. *J. Chromatogr*, B703: 177–183.

Hu, K., Li, J. and Webster, J.M. 1999. Nematicidal metabolites produced by *Photorhabdus luminescens* (Enterobacteriaceae), bacterial symbiont of entomopathohenic nematodes. *Nematol*, 1: 457–469.

Huang, X.W., Tian, B.Y., Niu, Q.H., Yang, J.K., Zhang, L.M. and Zhang, K.Q. 2005. An extracellular protease from *Brevibacillus laterosporus* G4 without parasporal crystal can serve as a pathogenic factor in infection of nematodes. *Res. Microbiol*, 156: 719–727.

Huffman, D.L., Abrami, L., Sasik, R., Corbeil, J., Van Der Goot, F. G. and Aroian, R.V. 2004. Mitogen-activated protein kinase pathways defend against bacterial pore-forming toxins. *Cell Biology*, 101 (30): 10995–11000.

Ibekwe, A.M. 2004. Effects of Fumigants on Non-Target Organisms in Soils. *Advances in Agronomy*, 83: 1-35.

Ignoffo, C.M. and Dropin, V.H. 1977. Deleterious effects of thermostable toxin of *Bacillus* toxin of *Bacillus thuringiensis* on species of soil-inhabiting myceliophagous and plant-parasitic nematodes. *Journal of the Kansas Entomology Society*, 50(3): 394–398.

Insunza, V., Alstrom, S. and Eriksson, K.B. 2002. Root bacteria from nematicidal plants and their biocontrol potential against trichodorid nematodes in potato. *Plant Soil*, 241: 271–278.

Jacq, V.A. and Fortuner, R. 1979. Biological control of rice nematodes using sulphate reducing bacteria. *Revue N'ematol*, 2: 41–50.

Jaffee, B., Phillips, R., Muldoon, A. and Mangel, M. 1992. Density-dependent host–pathogen dynamics in soil microcosms. *Ecology*, 73: 495–506.

Jaizme-Vega, M.C. and Pinochet, J. 1997. Growth response of banana to three mycorrhizal fungi in *Pratylenchus goodeyi* infested soil. *Nematropica*, 27: 69-76.

Jaizme-Vega, M.C., Tenoury, P., Pinochet, J.., Jaumot, M., 1997. Interactions between the root-knot nematode *Meloidogyne incognita* and *Glomus mosseae* in banana. *Plant and Soil*, 196: 27–35.

Jansson, H.B. and Lopez-Llorca, L.V. 2001. Biology of nematophagous fungi. In: Misra, J.K. Horn, B.W. (Eds.), *Mycology: Trichomycetes, other fungal groups and mushrooms*. Science Publishers, Enfield, CT, USA, pp. 145–173.

Jansson, H.B. and Nordbring-Hertz, B. 1980. Interactions between nematophagous fungi and plant-parasitic nematodes: attraction, induction of trap formation and capture. *Nematologica*, 26: 383-389.

Jansson, H.B., Dackman, C. and Zuckman, B.M. 1987. Adhesion and infection of plant parasitic nematodes by the fungus *Drechmeria coniospora*. *Nematologica*, 33:480–487.

Jatala, P. 1986. Biological control of plant-parasitic nematodes. *Annu. Rev. Phytopathol*, 24: 453–489.

Jayakumar, J., Ramakrishnan, S. and Rajendran, G. 2002. Bio-control of reniform nematode, *Rotylenchulus reniformis* through fluorescent Pseudomonas. *Pesttol*, 26: 45–46.

Jonathan, E.I., Barker, K.R., Abdel-Alim, F.F., Vrain, T.C. and Dickson, D.W. 2000. Biological control of *Meloidogmein cognition* tomato and banana with rhizobacteria actinomycetes, and *Pasteuria penetrans*. *Nematropica*, 30: 231–240.

Kampfe, L. and Sch" utz, H. 1995. *Nematologica*, 41: 449–467.

Karpouzas, D.G., Pantelelis, I., Menkissoglu-Spiroudi, U., Golia, E. and Tsiropoulos, N.G. 2007. Leaching of the organophosphorus nematicide fosthiazate. *Chemosphere*, 68 (7): 1359-1364.

Katan, J. 1987. Soil solarization. In: Chet, I. and Ed., J. (Eds.), *Innovative Approaches to Plant Disease Control*. Wiley and Sons, New York, pp.77-105.

Katan, J., A. Greenberger, H. Alon, and A. Grinstein. 1976. Solar heating by polyethylene mulching for the control of diseases caused by soilborne pathogens. *Phytopathology*, 66:683-688.

Kerry, B.R. 2000. Rhizosphere interactions and exploitation of microbial agents for the biological control of plant-parasitic nematodes. *Annu. Rev. Phytopathol*, 38: 423–441.

Kerry, B.R. and Gowen, S.R. 1995. Biological control of plant parasitic nematodes. *Nematologica*, 41: 362-363.

Khan, A., Williams, K.L. and Nevalainen, H.K. M. 2004. Effects of *Paecilomyces lilacinus* protease and chitinase on the eggshell structures and hatching of *Meloidogyne javanica* juveniles. *Biol. Control*, 31: 346–352.

Khan, M.R., Kounsar, K. and Hamid, A. 2002. Effect of certain rhizobacteria and antagonistic fungi on root-nodulation and root-knot nematode disease of green gram. *Nematologia Mediterranea*, 31: 85–89.

Kirk, P.M., Cannon, P.F., David, J.C. and Stalpers, J. 2001. *Ainsworth and Bisby's Dictionary of the Fungi*. 9th ed. CAB International, Wallingford, UK.

Klein, L. 1996. In: Bell, C. H., Price, N. and Chakrabarti, B. (Eds.), *The Methyl Bromide Issue*, Wiley, Chichester, U.K., pp. 191–235.

Klink, V.P., Hosseini, P., Matsye, P.D., Alkharouf, N.W. and Matthews, B.F. 2010. Syncytium gene expression in *Glycine max*[PI 88788] roots undergoing a resistant reaction to the parasitic nematode *Heterodera glycines*. *Plant Physiology and Biochemistry*, 48 (2-3): 176-193.

Kloepper, J.W., Rodrıguez-Kabana, R., McInroy, J.A. and Collins, D.J. 1991. Analysis of populations and physiological characterization of microorganisms in rhizospheres of plants with antagonistic properties to phytopathogenic nematodes. *Plant Soil,* 136: 95–192.

Kloepper, J.W., Rodriguez-Kabana, R., Mcinroy, J.A. and Young, R.W. 1992. Rhizosphere bacteria antagonistic to soybean cyst (*Heterodera glycines*) and root-knot (*Meloidogyne incognita*) nematodes: identification by fatty acid analysis and frequency of biological control activity. *Plant Soil*, 139: 75–84.

Kluepfel, D.A., McInnis, T.M. and Zehr, E. 1993. Involvement of rootcolonizing bacteria in peach orchard soils suppressive to the nematode *Criconemella xenoplax*. *Phytopathol*, 83: 1240–1245.

Kokalis-Burelle, N., Vavrina, C.S., Rosskopf, E.N. and Shelby, R.A. 2002. Field evaluation of plant growth-promoting rhizobacteria amended transplant mixes and soil solarization for tomato and pepper production in Florida. *Plant Soil*, 238: 257–266.

Kong, J.O. Lee, S.M., Moon, Y.S., Lee, S.G. and Ahn, Y.J. 2006. Nematicidal Activity of Plant Essential Oils against *Bursaphelenchus xylophilus* (Nematoda: Aphelenchoididae). *Journal of Asia-Pacific Entomology*, 9 (2): 173-178.

Kottegoda, M.B. 1985. Safety in use of pesticides and medical treatment. *Chemistry and Industry*, (16 September): 623-625.

Krebs, B., Hoeding, B., Kuebart, S., Workie, M.A., Junge, H., Schmiedeknecht, G., Grosch, R., Bochow, H. and Hevesi, M. 1998. Use of *Bacillus* subtilis as biocontrol agent. I. Activities and characterization of *Bacillus subtilis* strains. *Zeitschrift Pflanzenkrankh Pflanzenschutz*, 105: 181–197.

LaMondia, J.A. and Brodie, B.B. 1984. Control of Globodera rostochiensis by solar heat. *Plant Disease*, 68:474-476.

Leach, C. 1971. A practical guide to the effects of visible and ultraviolet light on fungi. In: Booth, C. (Ed.), *Methods of microbiology*. Academic Press, London, pp. 609-664.

Lewis, E.E., Grewal, P.S. and Sardanelli, S. 2001. Interactions between *Steinernema feltiae–Xenorhabdus bovienii* insect pathogen complex and root-knot nematode *Meloidogyne incognita*. *Biological Control*, 21: 55–62.

Li, B., Xie, G.L., Soad, A. and Coosemans, J. 2005. Suppression of *Meloidogyne javanica* by antagonistic and plant growth promoting rhizobacteria. *J Zhejiang Univ Sci*, 6B: 496–501.

Li, J., Carrol, J. and Ellar, D.J. 1991. Crystal structure of insecticidal d-endotoxin from *Bacillus thuringiensis* at 2.5A ° resolution. *Nature*, 353: 815–821.

Li, J., Pandelakis, A.K. and Ellar, D.J. 1996. Structure of the mosquitocidal d-endotoxin CytB from *Bacillus thuringiensis* sp. *kyushuensis* and implications for membrane pore formation. *J. Mol. Biol*, 257: 129–152.

Li, T.F., Zhang, K.Q. and Liu, X.Z. 2000. *Taxonomy of nematophagous fungi (Chinese)*. Science Press, Beijing, People's Republic of China.

Li, X.Q., Tan, A., Voegtline, M., Bekele, S., Chen, C.S. and Aroian, R.V. 2008. Expression of Cry5B protein from *Bacillus thuringiensis* in plant roots confers resistance to root-knot nematode. *Biological Control*, 47 (1): 97-102.

Li, Y., Kevin, D.H., Jeewon, R., Cai, L., Vijaykrishna, D. and Zhang, K.Q. 2005. Phylogenetics and evolution of nematode-trapping fungi (Orbiliales) estimated from nuclear and protein coding genes. *Mycologia*, 97: 1034–1046.

Lin, D., Qu, L.J., Gu, H. and Chen, Z. 2001. A 3.1-kb genomic fragment of *Bacillus subtilis* encodes the protein inhibiting growth of Xanthomonasoryzae pv. oryzae. *J. Appl. Microbiol*, 91: 1044–1050.

Lin, S.C., Lo, Y.C., Lin, J.Y. and Liaw, Y.C. 2004. Crystal structures and electron micrographs of fungal valvotoxin A2. *J. Mol. Biol*, 343: 477–491.

Liou, G.Y. and Tzean, S.S. 1997. Phylogeny of the genus *Arthrobotrys* and allied nematode-trapping fungi based on rDNA sequences. *Mycologia*, 89: 876–884.

Loeza-Lara, P.D., Benintende, G., Cozzi, J., Ochoa-Zarzosa, A., Baizabal-Aguirre, V.M., Valdez-Alarcon, J.J. and Lopez-Meza, J.E. 2005. The plasmid pBMBt1 from *Bacillus thuringiensis* subsp. *darmstadiensis* (INTA Mo14-4) replicates by the rolling-circle mechanism and encodes a novel insecticidal crystal protein-like gene. *Plasmid*, 54: 229–240.

Lopez-Llorca, L.V. 1990. Purification and properties of extracellular proteases produced by the nematophagous fungus *Verticillium suchlasporium*. *Can. J. Microbiol*, 36: 530–537.

Lopez-Llorca, L.V. and Claugher, D. 1990. Appressoria of the nematophagous fungus *Verticillium suchlasporium*. *Micron. Microsc. Acta*, 21: 125–130.

Lopez-Llorca, L.V. and Robertson, W.M. 1992. Immumocytochemical localization of a 32-kDa protease from the nematophagous fungus *Verticillium suchlasporium* in infected nematode eggs. *Exp. Mycol*, 16: 261–267.

Lopez-Llorca, L.V., Olivares-Bernabeu, C., Salinas, J., Jansson, H.B. and Kolattukudy, P.E. 2002. Prepenetration events in fungal parasitism of nematode eggs. *Mycological Research*, 106: 499–506.

Luo, H., Li, X., Li, G.H., Pan, Y.B. and Zhang, K.Q. 2006. Acanthocytes of *Stropharia rugosoannulata* function as a nematode-attacking device. *Applied and Environmental Microbiology*, 72: 2982–2987.

Luo, H., Mo, M.H., Huang, X.W., Li, X. and Zhang, K.Q. 2004. *Coprinus comatus*: a basidiomycete fungus forms novel spiny structures and infects nematodes. *Mycologia*, 96: 1218–1225.

MacGuidwin, A.E. 1993. Management of Nematodes. In: Randell C. Rowe (Ed.), *Potato Health Management*. APS Press, St. Paul, MN, pp. 159–166.

MacGuidwin, A.E., Bird, G.W. and G.R. Safir. 1985. Influence of *Glomus fasciculatum* on *Meloidogyne hapla* infecting *Allium cepa*. *Journal of Nematology*, 17: 389-395.

Madamba, C.P., Camaya, E.N., Zenarosa, D.B. and Yater, H.M. 1999. Screening soil bacteria for potential biocontrol agents against the root-knot nematode, *Meloidogyne* spp. *The Philippine Agriculturist*, 82: 113–122.

Mahdy, M., Hallmann, J. and Sikora, R.A. 2001. Influence of plant species on the biological control activity of the antagonistic rhizobacterium *Rhizobium etli* strain G12 toward the rootknot nematode *Meloidogyne incognita. Meded. Rijksuniv. Gent. Fak. Landbouwkd Toegep. Biol. Wet*, 66: 655–662.

Mai, W.F. and Lautz W.H. 1953. Relative resistance of free and excysted larvae of the golden nematode *Heterodera rostochiensis* Wollenweber to D-D mixture and hot water. *Proc. Helminthol. Soc.*, Washington, D.C., 20:1-7.

Mankau, R. 1975. *Bacillus penetrans* n. comb. Causing a virulent disease of plant-parasitic nematodes. *Journal of Invertebrate Pathology*, 26: 333-339.

Mankau, R. 1980. Biological control of nematodes pests by natural enemies. *Annual Review of Phytopathology*, 18: 415–440.

Mankau, R. and Prasad, N. 1972. Possibilities and problems in the use of a sporozoan endoparasite for biological control of plant-parasitic nematodes. *Nematropica*, 2:7-8.

Marroquin L.D., Elyassnia D., Griffitts J.S., Feitelson J.S. and Aroian R.V. 2000. *Bacillus thuringiensis* (Bt) toxin susceptibility and isolation of resistance mutants in the nematode *Caenorhabditis elegans. Genet*, 155: 1693–1699.

Mateille, T., Fould, S., Dabiré, K.R., Diop, M.T. abd Ndiaye, S. 2009. Spatial distribution of the nematode biocontrol agent *Pasteuria penetrans* as influenced by its soil habitat. *Soil Biology and Biochem*istry, 41 (2):303-308.

McInory JA and Kloepper JW (1995) Survey of indeginous bacterial endophytes from cotton and sweet corn. *Plant Soil* 173: 337–342.

McK Bird, D. 2004. Signaling between nematodes and plants. *Current Opinion in Plant Biology*, 7:372–376.

Meadows, M.P., Ellis, D.J., Butt, J., Jarrett, P., and Burges, H.D. 1992. Distribution, frequency, and diversity of *Bacillus thuringiensis* in an animal feed. *Applied and Environmental Microbiology*, 58: 1344–1350.

Mehrotra, V.S. 2005. Mycorrhiza: Role and Applications. Allied Publishers Limited, New Dehli, pp. 193-203.

Mena, J. and Pimentel, E. 2002. Mechanism of action of *Corynebacterium pauronetabolum* strain C-924 on nematodes. *Nematol*, 4: 287 (abstract).

Meyer, S.L.F. 2003. United States Department of Agriculture–Agricultural Research Service research programs on microbes for management of plant-parasitic nematodes. *Pest Management Science*, 59: 665–670.

Meyer, S.L.F. and Roberts, D.P. 2002. Combinations of biocontrol agents for management of plant-parasitic nematodes and soilborne plant-pathogenic fungi. *Journal of Nematology*, 34: 1-8.

Meyer, S.L.F., Roberts D.P., Chitwood D.J., Carta L.K., Lumsden R.D. and Mao W. 2001. Application of *Burkholderia cepacia* and *Trichoderma virens*, alone and in combinations, against *Meloidogyne incognita* on bell pepper. *Nematropica*, 31: 75–86.

Moellenbeck, D.J., Peters, M.L., Bing, J.W., Rouse, J.R., Higgins, L.S., Sims, L., *et al.* 2001. Insecticidal proteins from *Bacillus thuringiensis* protect corn from corn rootworms. *Nature Biotechnology*, 19: 668-672.

Moens, M. and Hendrickx, G. 1998. *Fund. Appl. Nematol.* 21: 199–204.

Monfort, E., Lopez-Llorca, L.V., JanssonH-B, *et al.* 2005. Colonisation of seminal roots of wheat and barley by egg-parasitic nematophagous fungi and their effects on *Gaeumannomyces graminis* var. *tritici* and development of root-rot. *Soil Biology and Biochemistry*, 37: 1229–1235.

Morandi, D., Bailey, J.A. and Gianinazzi-Person, V. 1984. Isoflavonoid accumulation in soybean roots infected with vesicular-arbuscular mycorrhizal fungi; Physiol. *Plant Pathol.* 24 357–364.

Morse, R.J., Yamamoto, T. and Stroud, R.M. 2001. Structure of Cry2Aa suggests an unexpected receptor binding epitope. *Structure*, 9: 409–417.

Morton, C.O., Hirsch, P.R. and Kerry, B.R. 2004. Infection of plantparasitic nematodes by nematophagous fungi – a review of the application of molecular biology to understand infection processes and to improve biological control. *Nematologica*, 6: 161–170.

Munif, A., Hallmann, J. and Sikora, R.A. 2000. Evaluation of the biocontrol activity of endophytic bacteria from tomato against *Meloidogyne incognita*. *Mededelingen Faculteit Landbouwkundige, Universiteit Gent*, 65: 471–480.

Neipp, P.W. and Becker, J.O. 1999. Evaluation of biocontrol activity of rhizobacteria from Beta Vulgaris against *Heterodera schachtii*. *Journal of Nematology*, 31: 54–61.

Niu, Q.H., Huang, X.W., Tian, B.Y., Yang, J.K., Liu, J., Zhang, L. and Zhang, K.Q. 2005. *Bacillus* sp. B16 kills nematodes with a serine protease identified as a pathogenic factor. *Appl. Microbiol. Biotechnol*, 69: 722–730.

Nordbring-Hertz, B. 1977. Nematode-induced morphogenesis in the predacious fungus, *Arthrobotrys oligospora*, *Nematologica*, 23: 443-451.

Nordbring-Hertz, B. and Chet, I. 1986. Fungal lectins and agglutinins. In: Mirelman D (ed), *Microbial lectins and agglutinins: properties and biological activity*. Wiley, New York, NY, pp. 393–408.

Nordbring-Hertz, B. and Mattiasson, B. 1979. Action of a nematodetrapping fungus shows lectin-mediated host–microorganism interaction. *Nature*, 281:477–479.

Nordbring-Hertz, B. and Stalhammar-Carlemalm, M. 1978. Capture of nematodes by *Arthrobotrys oligospora*, an electro microscope study. *Can. J. Bot*, 56: 1297–1307.

Nordbring-Hertz, B., Jansson, H.B. and Tunlid, A. 2006. Nematophagous fungi. In: *Eecyclopedia of Life Sciences*. John Wiley and Sons.

Nordbring-Hertz, B., Jansson, H.B., Friman, E. Persson, Y., Dackman, C., Hard, T., Poloczek, E. and Feldman, R. 1995. Nematophagous Fungi. Film No C1851 Gottingen, Germany: Institut fur den Wissenschaftlichen Film.

Obannon, J.H. and Nemec, S. 1979. The response of Citrus limon seedlings to a symbiont, *Glomus etunicatus*, and a pathogen, *Radopholus similes*. *Journal of Nematology*, 11: 270-275.

Ogawa, H., Fukushima, K., Sasaki, I. and Matsuno, O. 2000. Identification of genes involved in mucosal defense and inflammation associated with normal enteric bacteria. *Am. J. Physiol Gastrointest Liver Physiol*, 279: 492–499.

Oka, Y. 2010. Mechanisms of nematode suppression by organic soil amendments. *Applied Soil Ecology*, 44 (2): 101-115.

Oka, Y., Koltai, H., Bar-Eyal, M., Mor, M., Sharon, E., Chet, I. and Spiegel, Y. 2000. New strategies for the control of plant-parasitic nematodes. *Pest Management Science*, 56: 983-988.

Oliveira, E.J., Rabinovitch, L., Monnerat, R.G., Passos, L.K.J. and Zahner, V. 2004. Molecular characterization of *Brevibacillus laterosporus* and its potential use in biological control. *Applied and Environmental Microbiology*, 70: 6657–6664.

Olsson, S. and Persson, Y. 1994. Transfer of phosphorus from *Rhizoctonia solani* to the mycoparasite *Arthrobotrys oligospora*. *Mycological Research*, 98: 1065–1068.

Oostendorp, M. and Sikora, R.A. 1990. In-vitro interrelationships between rhizosphere bacteria and *Heterodera schachtii*. *Rev. Nematol*, 13: 269–274.

Oostendorp, M., Dickson, D.W. and Mitchell, D.J. 1990. Host range and ecology of isolates of *Pasteuria* spp. from the southeastern United States. *Journal of Nematology*, 22:525-531.

Pan, C., Lin, J., Ni, Z. and Wang, S. 1993. Study on the pathogenic bacteria parasitizing root-knot nematodes discovered in China and their application to biological control. *Acta Microbiologica Sinica*, 33: 313-316.

Paramonov, A.A. 1954. An amoeboid organism destroying infective larvae of the root-knot nematode. Tr. Gelmintologicheskoi Laboratorii Akad. Nauk U.S.S.R. 7:50-54.

Parker, M.W., Feil, S.C. 2005. Pore-forming protein toxins: from structure to function. *Progr. Biophys. Mol. Biol*, 88: 91–142.

Paul, V.J., Frautschy, S., Fenical, W. and Nealson K.H. 1981. Antibiotics in microbial ecology: isolation and structure assignment of several new antibacterial compounds from the insectsymbiotic bacteria *Xenorhabdus* spp. *J. Chem. Ecol*, 7: 589–597.

Peet, M. 1996. *Sustainable Practices for Vegetable Production in the South*. Focus Publishing, Newburyport, MA. p. 75–77.

Perry, R.N., Hominick, W.M., Beane, J. and Briscoe, B. 1998. Effects of the entomopathogenic nematodes, *Steinernema feltiae* and *S. carpocapsae* on the potato cyst nematode, *Globodera rostochiensis*, in pot trials. *Biocontrol Sci. Technol*, 8: 175–180.

Persmark, L., Banck, A. and Jansson, H.B. 1996. Population dynamics of nematophagous fungi and nematodes in an arable soil: vertical and seasonal fluctuations. *Soil Biology and Biochemistry*, 28: 1005–1014.

Peterson, R.L., Massicotte, H.B. and Melville, L.H. 2004. *Mycorrhizas: anatomy and cell biology*. National Research Council Research Press.

Pfister, D.H. 1997. Castor, Pollux and life histories of fungi. *Mycologia*, 89: 1–23.

Pinochet, J. Camprubi, A. and Calvet, C. 1993. Effect of the root lesion nematode *Pratylenchus vulnus* and the mycorrhizal fungus *Glomus mosseae* on the growth of EMLA-26 apple rootstock. *Mycorrhiza*, 4: 79-83.

Pinochet, J., Calvet, C., Camprubi, A., and Fernandez, C. 1995. Interaction between the root lesion nematode *Pratylenchus vulnus* and the mycorrhizal association of *Glomus intraradices* and Santa Lucia cherry rootstock. *Plant and Soil*, 170: 323-329.

Pinochet, J., Fernadez, C., Jaizme, M.C. and Tenuory, P. 1997. Microproporgated banana infected with *Meloidogyne javanica* responds to *Glomus intraradices* and phosphorus. *HortScience*, 32: 101-103.

Porter, 1.J. and Merriman, P.R. 1983. Effect of solarization of soil on nematode and fungal pathogens at two sites in Victoria. *Soil Biol. Biochem*, 15: 39-44.

Prasad, S.S.V., Tilsk, K.V. R. and Gollakota, K.G. 1972. Role of *Bacillus thuringiensis* var. *thuringiensis* on the larval survivability and egg hatching of *Meloidogyne* spp. the causative agent of root-knot disease. *Journal of Invertebrate Pathology*, 20(3): 377–378.

Preston, J.F., Dickson, D.W., Maruniak, J.E., Nong, G., Brito, J.A., Schmidt, L.M. and Giblin-Davis, R.M. 2003. *Pasteuria* spp.: systematics and phylogeny of these bacterial parasites of phytopathogenic nematodes. *Journal of Nematology*, 35: 198–207.

Racke, J. and Sikora, R.A. 1992. Isolation, formulation and antagonistic activity of rhizobacteria toward the potato cyst nematode Globodera pallida. *Soil Biol. Biochem*, 24: 521–526.

Ramamoorthy, V., Viswanathan, R., Raguchander, T., Prakasam, V. and Samiyappan, R. 2001. Induction by systemic resistance by plant growth promoting rhizobacteria in crop plants against pests and diseases. *Crop Protection*, 20: 1–11.

Rang, C., Lacey, L.A., and Frutos, R. 2000. The crystal proteins from *Bacillus thuringiensis* subsp. *thompsoni* display a synergistic activity against the codling moth, *Cydia pomonella*. *Current Microbiology*, 40: 200-204.

Rao, M.S. and Gowen, S.R. 1998. Biomanagement of *Meloidogyne incognita* on tomato by integrating *Glomus deserticola* of *Pasteurian penetrans*. *Journal of Plant Diseases Protection*, 105: 49-52.

Rao, M.S., Kerry, B.R., Gowen, S.R. Bourne, J.M. and Reddy, P.P. 1997. Management of *Meloidogyne incognita* in tomato nurseries by integration of *Glomus deserticola* with *Verticillium chlamydosporium*. *Journal of Plant Diseases Protection*, 104: 410-422.

Reitz, M., Hoffmann-hergarten, S., Hallmann, J. and Sikora, R.A. 2001. Induction of systemic resistance in potato by rhizobacterium *Rhizobium etli* strain G12 is not associated with accumulation of pathogenesis-related proteins and enhanced lignin biosynthesis. *Z Pflkrankh* Pflschutz, 108: 11–20.

Reitz, M., Rudolph, K., Schr̈oder, L., Hoffmann-Hergarten, S., Hallmann, J. and Sikora, R.A. 2000. Lipopolysaccharides of *Rhizobium etli* strain G12 act in potato roots as an inducing agent of systemic resistance to infection by the cyst nematode *Globodera pallida*. *Applied and Environmental Microbiology*, 66: 3515–3518.

Remy, W., Taylor, T.N., Hass, H., Kerp, H. 1994. "4 hundred million year old vesicular-arbuscular mycorrhizae". Proc. National Academy of Sciences, 91: 11841–11843.

Rodriguez-Kabana, R. 1986. Organic and inorganic nitrogen amendments to soil as nematode suppressants. *Journal of Nematology*, 18: 524–526.

Roncadory, R.W. and Hussey, R.S. 1977. Interaction of the endomycorrhizal fungus *Gigaspora margarita* and root-knot nematode on cotton. *Phytopath*, 67: 1507-1511.

Rosen, S., Kata, M., Persson, Y., Lipniunas, P.H., Wikstrom, C., Van den Hondel, C., Van den Brink, J., Rask, L., Heden, L.O. and Tunlid, A. 1996. Molecular characterization of a saline-soluble lectin from a parasitic fungus. Extensive sequence similarity between fungal lectins. *European Journal of Biochemistry*, 238: 822–829.

Rosen, S., Sjollema, K., Veenhuis, M. and Tunlid, A. 1997. A cytoplasmic lectin produced by the fungus *Arthrobotrys oligospora* functions as a storage protein during saprophytic and parasitic growth. *Microbiology*, 143: 2593–2604.

Rovira, A.D. and Sands, D.C. 1977. *Fluorescent pseudomonas* – a residual component in the soil microflora. *J. Appl. Bacteriol*, 34: 253–259.

Salawu, E.O. and Estey, R.H. 1979. Observations on the relationship between a vesicular-arbuscular fungus, a fungivorus nematode and the growth of soybeans. *Phytoprotection*, 60: 99-102.

Saleh, H. and Sikora, R.A. 1984. Relationship between *Glomus fasciculatum* root colonization of cotton and its effect on *Meloidogyne incognita*. *Nematologica*, 30: 230-237.

Salehi Jouzani, G., Goldenkova, I.V. and Piruzian, E.S. 2008a. Expression of hybrid cry3aM-licBM2 genes in transgenic potatoes (*Solanum tuberusom*). *Journal of Plant Cell Tissue and Organ Culture*, 92(3): 321-325.

Salehi Jouzani, G., Seifinejad, A., Saeedizadeh, A., Nazarian, A., Yousefloo, M., Soheilivand, S., Mousivand, M., Jahangiri, M., Yazdani, M., Maali Amiri, M. and Akbari, M. 2008b.

Molecular detection of nematicidal crystalliferous *Bacillus thuringiensis* strains of Iran and evaluation of their toxicity on free living and plant parasitic nematodes. *Canadian Journal of Microbiology*, 54(10): 812–822.

Samaliev, H.Y., Andreoglou, F.I., Elawad, S.A., Hague, N.G.M. and Gowen, S.R. 2000. The nematicidal effects of the bacteria *Pseudomonas oryzihabitans* and *Xenorhabdus nematophilus* on the root-knot nematode *Meloidogyne javanica*. *Nematologica*, 2: 507–514.

Sanchez, P. 1997. For pepper growers, built-in nematode resistance. *Agricultural Research*, October. p. 12–13.

Sayre, R.M. 1973. *Theratromyxa weberi*, an amoeba predatory on plant-parasitic nematodes. *Journal of Nematology*, 5: 258-264.

Sayre, R.M. and Starr, M.P. 1985. *Pasteuria penetrans* (ex Thorne 1940) nom.rev., comb.n., sp.n., a mycelial and endospore-forming bacterium parasitic in plant nematodes. *Proc. Helminthol. Soc.* Wash, 52: 149–165.

Sayre, R.M. and Starr, M.P. 1988. Bacterial diseases and antagonists of nematodes. In: Poinar, G. O., Jansson, H.B. (Eds.), *Diseases of Nematodes*, Vol. 1. CRC Press, Boca Raton, FL, pp. 69–101.

Sayyed, A.H., Crickmore, N., and Wright, D.J. 2001. Cyt1Aa from *Bacillus thuringiensis* subsp. *israelensis* is toxix to the diamond back moth, *Plutella xylostella*, and synergizes the activity of Cry1Ac towards a resistant strain. *Applied and Environmental Microbiology*, 67: 5859-61.

Schenck, S., Chase, T.J., Rosenzweig, W.D. and Pramer, D. 1980. Collagenase production by nematode-trapping fungi. *Applied and Environmental Microbiology*, 40: 567–570.

Schisler, D.A., Slininger, P.J., Behle, R.W. and Jackson, M.A. 2004. Formulation of *Bacillus* spp. for biological control of plant diseases. *Phytopathol*, 94: 1267–1271.

Schneider, S.M., Rosskopf, E.N., Leesch, J.G., Chellemi, D.O., Bull, C.T. and Mazzola, M. 2003. Research on alternatives to methyl bromide: pre-plant and post-harvest. *Pest Management Science*, 59: 814–826.

Schnepf, E., Crickmore, N., Van Rie, J., Lereclus, D., Baum, J., Feitelson, J., Zeigler, D.R. and Dean, D.H. 1998. *Bacillus thuringiensis* and its pesticidal crystal proteins. *Microbiol. Mol. Biol. Rev*, 62: 775–806.

Schnepf, H.E. and Whiteley, H.R. 1981. Cloning and expression of the *Bacillus thuringiensis* crystal protein gene in *Escherichia coli*. *Proc. National Academy of Sciences*, USA, 78: 2893–2897.

Scholler, M., Hagedorn, G. and Rubner, A. 1999. A reevaluation of predatory orbiliaceous fungi II A new generic concept. Sydowia 51(1): 89–113.

Segers, R., Butt, T.M., Carder, J.H., Keen, J.N., Kerry, B.R. and Peberdy, J,F. 1999. The subtilisins of fungal pathogens of insects, nematodes and plants: distribution and variation. *Mycological Research*, 103: 295–402.

Seifinejad, A., Salehi Jouzani G., Hosseinzadeh, A. and Abdmishani, C. 2008. Characterization of Lepidoptera-active *cry* and *vip* genes in Iranian *Bacillus thuringiensis* strain collection. *Journal of Biological Control*, 44 : 216–226.

Selosse, M.A., Richard, F., He, X., Simard, S.W. 2006. "Mycorrhizal networks: des liaisons dangereuses?". *Trends Ecol. Evol, 21: 621–628.*

Sharma, S.B., and Davies, K.G. 1996. Characterization of *Pasteuria* isolated from *Heterodera cajani* using morphology, pathology, and serology of endospores. *Systematic and Applied Microbiology*: 19: 106-112.

Shaukat, S.S., Siddiqui, I.A., Hamid, M., Khan, G.H. and Ali. S.A. 2002. In vitro survival and nematicidal activity of *Rhizobium*, *Bradyrhizobium* and *Sinorhizobium*. I. the influence of various NaCl concentrations. *Pakistan J. Biol. Sci*, 5: 669–671.

Shimazu, M. and Glockling, S.L. 1997. A new species of *Harposporium* with two spore types isolated from the larva of a cerambycid beetle. *Mycological Research*, 101: 1371–1376.

Siddiqui, I.A. 2002. Suppression of *Meloidogyne javanica* by *Pseudomonas aeruginosa* and *Bacillus subtilis* in tomato. *Nematologia Mediterranea*, 30: 125–130.

Siddiqui, I.A. and Shaukat, S.S. 2002. Rhizobacteria-mediated induction of systemic resistance (ISR) in tomato against *Meloidogyne javanica*. *J. Phytopathology-phytopathologische Zeitschrift*, 150: 469–473.

Siddiqui, I.A. and Shaukat, S.S. 2003. Suppression of root-knot disease by *Pseudomonas fluorescens* CHA0 in tomato: importance of bacterial secondary metabolite 2,4-diacetylphloroglucinol. *Soil Biol. Biochem*, 35: 1615–1623.

Siddiqui, I.A. and Shaukat, S.S. 2004. Systemic resistance in tomato induced by biocontrol bacteria against the root-knot nematode, *Meloidogyne javanica* is independent of salicylic acid production. *J. Phytopathol*, 152: 48–54.

Siddiqui, I.A., Haas, D. and Heeb, S. 2005. Extracellular protease of *Pseudomonas fluorescens* CHA0, a biocontrol factor with activity against the root-knot nematode *Meloidogyne incognita*. *Applied and Environmental Microbiology*, 71: 5646–5649.

Siddiqui, Z.A. and Mahmood, I. 1996. Biological control of plant parasitic nematodes by fungi: a review. *Bioresource Technology*, 58: 229–239.

Siddiqui, Z.A. and Mahmood, I. 1999. Role of bacteria in the management of plant parasitic nematodes: a review. *Bioresource Technology*, 69: 167–179.

Siddiqui, Z.A. and Mahmood, I. 2001. Effects of rhizobacteria and root symbionts on the reproduction of *Meloidogyne javanica* and growth of chickpea. *Bioresource Technology*, 79: 41–46.

Sikora, R.A. 1992. Management of the antagonistic potential in agriculture ecosystems for the biological control of plant parasitic nematodes. *Annual Review of Phytopathology*, 30: 245–270.

Sikora, R.A. and Hoffmann-Hergarten, S. 1993. Biological control of plant parasitic nematodes with plant-health promoting rhizobacteria. In: Lumsden, P.D. and Vaugh, J.L., (Eds.), *Biologically based technology*. ACS Symposium series, USA. pp. 166–172.

Simard, S.W. , Perry, D.A., Jones, M.D., Myrold, D.D., Durall D.M. and Molina, R. 1997. Net transfer of carbon between ectomycorrhizal tree species in the field. *Nature*, 388: 579-582.

Simon, L., Bousquet, J., Lévesque, R.C . and Lalonde, M. 1993. Origin and diversification of endomycorrhizal fungi and coincidence with vascular land plants. *Nature*, 363: 67-69.

Sipes, B.S. and Schmitt, D.P. 1995. *Suppl. J. Nematol*, 27: 639–644.

Siti, E., Cohn, E., Katan, J. and Mordechai, M. 1982. Control of Ditylenchus dipsaci in garlic by bulb and soil treatments. *Phytoparasitica*, 10: 93-100.

Smith, R.A., and Couche, G.A. 1991. The phylloplane as a source of *Bacillus thuringiensis* variants. *Applied and Environmental Microbiology*, 57: 311–315.

Soprunov, F.F. 1966. Predacious hyphomycetes and their application in the control of pathogenic nematodes. (Translation from Russian) *Israel program for scientific translations*, Washington, p. 292.

Soutter, L.A. and Loague, K. 2000. *J. Environ.* Qual, 29: 1794– 1805.

Spiegel, Y., Cohn, E., Galper, S., Sharon, E. and Chet, I. 1991. Evaluation of a newly isolated bacterium, *Pseudomonas chitinolytica* sp. nov., for controlling the root-knot nematode *Meloidogyne javanica*. *Biocontrol Sci. Technol*, 1: 115–125.

St Leger, R.J. 1993. Biology and mechanism of insect-cuticle invasion by deuteromycete fungal pathogens. In: Beckage, N.E., Thompson, S.N., Federici, B.A. (Eds.), *Parasites and pathogens of insects, vol 2. Pathogens*. Academic, San Diego, pp. 211–229.

Stadler, M., Quang, D.N., Tomita, A., Hashimoto, T. and Asakawa, Y. 2006. Changes in secondary metabolism during stromatal ontogeny of *Hypoxylon fragiforme*. *Mycological Research*, 10: 811–820.

Stanton, N.L. 1988. The underground in grasslands. *Annual Review of Ecology and Systematics*, 19, 573-589.

Stapleton, J.J. and DeVay, J.E. 1983. Response of phytoparasitic and free-living nematodes to soil solarization and 1, 3 - dichloropropene in California. *Phytopathology*, 73: 1429-1436.

Stirling, G.R. 1991. *Biological Control of Plant Parasitic Nematode: Progress, Problems and Prospects*. CAB International, Wallington, UK.

Stirling, G.R. and Mankau, R. 1979. Mode of parasitism of *Meloidogyne* and other nematode eggs by *Dactylella oviparasitica*. *Journal of Nematology*, 11: 282–288.

Stirling, G.R. and Wachtel, M.F. 1980. Mass production of *Bacillus penetrans* for the biological control of root-knot nematodes. *Nematologica*, 26: 308–312.

Strauss, S.Y. and Agrawal, A.A. 1999. The ecology and evolution of plant tolerance to herbivory. *Trends in Ecology and Evolution*, 14: 179-185.

Strobel, N.E., Hussey, R.S. and Roncadori, R.W. 1982. Interactions of vesicular arbuscular mycorrhizal fungi, *Meloidogyne incognita* and soil fertility on peach. *Phytopathology*, 72: 690–694.

Sturhan, D. 1988. New host and geographical records of nematodeparasitic bacteria of the Pasteuria penetrans group. *Nematologica*, 34: 350–356.

Sturz, A.V. and Kimpinski, J. 2004. Endoroot bacteria derived from marigolds (*Tagetes* spp.) can decrease soil population densities of root-lesion nematodes in the potato root zone. *Plant Soil*, 262: 241–249.

Sturz, A.V. and Kimpinski, J.1999. *Plant Pathology*, 48: 26–32.

Sturz, A.V. and Matheson, B.G. 1996. Populations of endophytic bacteria which influence host-resistance to *Erwinia*-induced bacterial soft rot in potato tubers. *Plant Soil*, 184: 265–271.

Subramanian, C.V. 1963. *Dactylella, Monacrosporium* and *Dactylina. J. Indian Bot. Soc*, 42: 289–291.

Surette, M.A., Sturz, A.V., Lada, R.R. and Nowak, J. 2003. Bacterial endophytes in processing carrots (*Daucus carota* L. var. sativus): their localization, population density, biodiversity and their effects on plant growth. *Plant Soil*, 253: 381–390.

Tan, M.W. 2002. Identification of host and pathogen factors involved in virulence using a *Caenorhabditis elegans. Method Enzomol*, 358: 13–28.

Taylor, G.E. *et al.* 1996. *Environ. Toxicol. Chem*, 15: 1694–1701.

Thies, J.A., Davis, R.F., Mueller, J.D., Fery, R.L., Langston, D.B. and Miller, G. 2004. Double-cropping cucumbers and squash after resistant bell pepper for root-knot nematode management. *Plant Disease*, 88(6): 589-593.

Thomason, I.J. 1987. Challenges facing nematology: environmental risks with nematicides and the need for new approaches. In: Veech, J.A. and Dickson, D.W. (Eds.), *Vistas on nematology*, Hyattsville, USA, Society of Nematologists, pp. 469-476.

Tian, B.Y., Li, N., Lian, L.H., Liu, J.W., Yang, J.K. and Zhang, K.Q. 2006. Cloning, expression and deletion of the cuticle-degrading protease BLG4 from nematophagous bacterium *Brevibacillus laterosporus* G4. *Arch. Microbial*, 186: 297–305.

Tian, H. and Riggs, R.D. 2000. Effects of rhizobacteria on soybean cyst nematodes, *Heterodera glycines*. *Journal of Nematology*, 32: 377–388.

Tian, H., Riggs, R.D. and Crippen, D.L. 2000. Control of soybean cyst nematode by chitinolytic bacteria with chitin substrate. *Journal of Nematology*, 32: 370–376.

Tosi, S., Annovazzi, L., Tosi, I., Iadrola, P. and Caretta, G. 2001. Collagenase production in an antarctic strain of *Arthrobotrys tortor* Jarowaja. *Mycopathologia*, 153: 157–162.

Tribe, H.T. 1977. Pathology of cyst-nematodes. *Biological Reviews*, 52: 477-. 507.

Tunlid, A. and Jansson, S. 1991. Proteases and their involvement in the infection and immobilization of nematodes by the nematophagous fungus *Arthrobotrys oligospora*. *Applied and Environmental Microbiology*, 57: 2868–2872.

Tunlid, A., Åhman, J. and Oliver, R.P. 1999. Transformation of the nematode-trapping fungus Arthrobotrys oligospora. *FEMS Microbiol Lett*, 173: 111–116.

Tunlid, A., Johansson, T., and Nordbring-Hertz, B. 1991. Surface polymers of the nematode-trapping fungus Arthrobotrys oligospora. *J. Gen. Microbiol*, 137: 1231–1240.

Tunlid, A., Rosen, S., Ek, B. and Rask, L. 1994. Purification and characterization of an extracellular serine protease from the nematode-trapping fungus *Arthrobotrys oligospora*. *Microbiology*, 140: 1687–1695.

U.S. Environmental Protection Agency. 2001. *Fed. Reg*, 66: 58468–58472.

Umesh, K.C., Krishnappa, K. and Bagyaraj, D.J. 1988. Interaction of burrowing nematode, *Radopholus similis* (Cobb, 1893) Thorne1949, and VA mycorrhiza, *Glomus fasiciulatum* (Thaxt.) Gerd. and Trappe in banana (*Musa acuminate* Colla.). *Indian Journal of Nematology*, 18: 6-11.

Valdivia, R.H. and Ramakrishnan, L. 2000. Applications of gene fusions to green fluorescent protein and flow cytometry to the study of bacterial gene expression in host cells. *Method Enzymol*, 326: 47–73.

Van Berkum, J.A. and Hoestra, H. 1979. Practical aspects of the chemical control of nematodes in soil, p. 53-154. In: Mulder, D. (Ed.), *Soil disinfestation*. Amsterdam, the Netherlands, Elsevier.

Van Bruggen, A.H.C., Semenov, A.M., van Diepeningen, A.D., de Vos, O.J. and Blok, W.J. 2006. Relation between soil health, wave-like fluctuations in microbial populations and soil-born disease management. *European Journal of Plant Pathology*, 115: 105-122.

Van der Putten, W.H., Yeates, G.W., Duyts, H., Reis, C.S. and Karssen, G. 2005. Invasive plants and their escape from root herbivory: a worldwide comparison of the rootfeeding nematode communities of the dune grass *Ammophila arenaria* in natural and introduced ranges. *Biological Invasions*, 7: 733-746.

Van Loon, L.C., Bakker, P.A.H.M. and Pieterse, C.M. 1998. Systemic resistance induced by rhizosphere bacteria. *Annual Review of Phytopathology*, 36: 453–483.

Vawdrey, L.L. and Stirling, G.R. 1997. Control of root-knot nematode (*Meloidogyne javanica*) on tomato with molasses and other organic amendments. *Australasian Plant Pathology*, 26: 179-187.

Veenhuis, M., Nordbring-Hertz, B. and Harder, W. 1985. An electron microscopical analysis of capture and initial stages of penetration of nematodes by *A. oligospora*. *Antonie van Leeuwenhoek*, 51: 385–398.

Walker, G.E. and Wachtel, M.F. 1988. The influence of soil solarization and non-fumigant nematicides on infection of Meloidogyne javanica by Pasteuria penetrans. *Nematologica*, 34: 477-483.

Wang, B. and Qiu, Y.L. 2006. "Phylogenetic distribution and evolution of mycorrhizas in land plants". Mycorrhizahello, 16 (5): 299–363.

Wang, K.H., McSorley, R. and Gallaher, R.N. 2004. Effect of Crotalaria juncea amendment on squash infected with Meloidogyne incognita. *Journal of Nematology*, 36(3): 290-296.

Wang, R.B., Yang, J.K., Lin, C. and Zhang, K.Q. 2006. Purification and characterization of an extracellular serine protease from the nematode-trapping fungus *Dactylella shizishanna*. *Lett. Appl. Microbiol*, 42: 589–594.

Weber, A.P., Zwillenberg, L.O. and Van Der Laan, P.A. 1952. A predacious amoeboid organism destroying larvae of the potato-root eelworm and other nematodes. *Nature*, 169: 834-835.

Wechter, W.P., Begum, D., Presting, G., Kim, J.J., Wing, R.A. and Kluepfel, D.A, 2002. Physical mapping BAC-end sequence analysis, and marker tagging of the soilborne nematicidal bacterium, *Pseudomonas synxantha* BG33R. *OMICS*, 6: 11–21.

Wechter, W.P., Glandorf, D.C.M., Derrick, W.C., Leverentz, B. and Kluepfel, D.A. 2001. Identification of genetic loci in a rhizosphere-inhibiting species of *Pseudomonas* involved in expression of a phytoparasitic nematode ovicidal factor. *Soil Biol. Biochem*, 33: 1749–1758.

Wei, J.Z., Hale, K., Carta L., *et al.* 2003. *Bacillus thuringiensis* crystal protein that target nematodes. *Proceedings of the National Academy of Sciences of the USA*, 100(5): 2760–2765.

West, A.W. 1984. Fate of the insecticidal, proteinaceous paraspoarl crystal of *Bacillus thuringiensis* in soil. *Soil Biol. Biochem*, 16: 357-360.

Westcott, S.W. and Kluepfel, D.A. 1993. Inhibition of *Criconemella xenoplax* egg hatch by *Pseudomonas aureofaciens*. *Phytopathology*, 83: 1245–1249.

Whipps, J.M. and Davies,, K.G. 2000. Success in biological control of plant pathogens and nematodes by microorganisms. In: Gurr, G. and Wratten, S., (Eds.), *Biological control: measures of success*. Kluwer Academic Publishers, Dordecht, The Netherlands, pp. 231–269.

Widmer, T.L., Mitkowski, N.A. and Abawi, G.S. 2002. Soil organic matter and management of plant-parasitic nematodes. *Journal of Nematology*, 34: 289-295.

Wilcox, J. and Tribe, H.T. 1974. Fungal parasitism in cysts of *Heterodera glycines*. I. Preliminary investigations. *Trans Br. Mycol. Soc*, 62: 585-594.

Williams, T. 2008. Natural invertebrate hosts of iridoviruses (Iridoviridae). *Neotropical Entomology*, 37(6): 615-632.

Williamson, V.M. and Kumar, A. 2006. Nematode resistance in plants: the battle underground. *Trends in Genetics*, 22: 396-403.

Winslow, R.D., and Williams, T.D. 1957. Amoeboid organisms attacking larvae of the potato root eelworm (*Heterodera rostochiensis* Woll.) in England and the beet eelworm (5. schachtii Schm.) in Canada. *Tiidschr. Plantenziekten*, 63: 242-243.

Wright, D.J. 1981. Nematicides: mode of action and new approaches to chemical control, , In: Zuckerman, B.M. and Rohde, R.A. (Eds.), *Plant-parasitic nematodes.* London and New York, Academic Press, pp. 421-449.

Yang, J.K., Huang, X.W., Tian, B.Y., Wang, M., Niu, Q.H. and Zhang, K.Q. 2005. Isolation and characterization of a serine protease from the nematophagous fungus, *Lecanicillium psalliotae*, displaying nematicidal activity. *Biotechnology Letters*, 27: 1123–1128.

Yang, J.K., Tian, B.Y., Liang, L.M. and Zhang, K.Q. 2007. Extracellular enzymes and the pathogenesis of nematophagous fungi. *Appl. Microbiol. Biotechnol*, 75: 21–31.

Yeates, G.W. and van der Meulen, H. 1996. *Biol. Fert. Soils*, 21: 1–6.

Yeates, G.W., Bongers, T., de Goede, R.G.M., Freckman, D.W. and Georgieva, S.S. 1993. Feeding habits in soil nematode families and genera - an outline for soil ecologists. *Journal of Nematology*, 25: 315-331.

Yepsen, R.B. Jr. (Ed.), 1984. *The Encyclopedia of Natural Insect and Disease Control.* Rev. ed. Rodale Press, Emmaus, PA, pp. 267–271.

Yu, Z.Q., Wang, Q.L., Liu, B., Zou, X., Yu Z.N. and Sun, M. 2008. *Bacillus thuringiensis* crystal protein toxicity against plant-parasitic nematodes. *Chinese Journal of Agricultural Biotechnology*, 5(1): 13–17.

Yu, Z.Q., Zhou, Y., Sun, M. and Yu, Z.N. 2004. Progress of research on activity of *Bacillus thuringiensis* against plant-parasitic nematodes. *Acta Phytophylacica Sinica*, 31(4): 418–424.

Zhang, K.Q. and Mo, M.H. 2006. *Flora fungorum sinicorum, vol 33. Arthrobotrys et genera cetera cognate (Chinese).* Science Press, Beijing, People's Republic of China.

Zuckerman, B.M. and Jasson, H.B. 1984. Nematode chemotaxis and possible mechanisms of host/prey recognition. *Annual Review of Phytopathology*, 22: 95–113.

In: Microbial Insecticides: Principles and Applications ISBN: 978-1-61209-223-2
Editors: J. Francis Borgio, K. Sahayaraj, et al. © 2011 Nova Science Publishers, Inc

Chapter 18

CONTROL OF ARTHROPOD PESTS OF TROPICAL TREE FRUIT WITH ENTOMOPATHOGENS

*Claudia Dolinski[1] * and Lawrence A. Lacey[2]*

[1]Universidade Estadual do Norte Fluminense Darcy Ribeiro/CCTA/LEF, Av. Alberto
Lamego, 2000, Pq. Califórnia Campos dos Goytacazes, RJ, Brazil, 28015-620
[2]Yakima Agricultural Research Laboratory, USDA-ARS, 5230 Konnowac Pass Rd.
Wapato, WA, 98951, USA

ABSTRACT

A plethora of arthropods attack fruit crops throughout the tropics and sub-tropics. The predominant method for controlling most of these pests is the application of broad-spectrum chemical pesticides. Growing concern over the negative environmental effects has encouraged development of alternative internative interventions with with an integrated pest management (IPM) strategy. Inundatively and inoculatively applied microbial control agents (virus, bacteria, fungi, and entomopathogenic nematodes) have been developed as alternative control methods of a wide variety of arthropods including tropical fruit pests. These appear to be ready made components for IPM in that they are safe for applicators, natural enemies of pest species and the food supply. The majority of the research and applications in tropical fruit agroecosystems has been conducted in citrus, banana, coconut and oil palms, and mango. Successful microbial control initiatives of citrus pests and mites have been reported. Microbial control of arthropod pests of banana includes banana weevil, *Cosmopolites sordidus* (with EPNs and fungi) among others. One of the most successful uses of an entomopathogen for classical biological control of an insect pest is reported for the palm rhinoceros beetles, *Oryctes* spp. with a non-occluded virus. Key pests of mango that have been controlled with microbial control agents include several fruit flies (Tephritidae) (with EPNs and fungi), and other pests. Also reported in this review is the successful microbial control of a limited number of arthropod pests of guava, papaya and pineapple. Microbial agents can provide effective control of several key pests of tropical fruit. The challenge will

* E-mail: Claudia.dolinski@censanet.com.br, Phone: (55) (22) 27261658; Fax: (55) (22) 27261549

be to find successful combinations of entomopathogens, predators, and parasitoids along with other interventions to produce effective and sustainable pest management.

18.1. INTRODUCTION

A multitude of insect and mite species are pests of tree fruit world wide (Lacey and Shapiro-Ilan, 2008). The predominant method for controlling most of these pests is the application of chemical pesticides, which has generated complex problems including: insecticide resistance; outbreaks of secondary pests normally held in check by natural enemies; safety risks for humans and wild and domestic animals; contamination of ground water and riparian habitats; and decrease in biodiversity. Growing concern over the environmental effects of pesticides has encouraged the development of alternatives to broad-spectrum pesticides. Natural pathogens of arthropods often play an important role in the regulation of insect and mite populations in agroecosystems (Ignoffo, 1985; Pell, 2007, Steinkraus, 2007). However, their main impact on pests may occur after economic thresholds are surpassed. Inundatively or inoculatively applied microbial control agents (viruses, bacteria, fungi, and nematodes) have been developed as alternative control methods for a wide variety arthropod pests (Alves, 1998a, Lacey *et al.,* 2001, Lacey and Kaya, 2007; Alves and Lopes, 2008), which include pests of tropical tree fruit. For a comprehensive review of microbial control in temperate orchards see Lacey and Shapiro-Ilan, (2008). In this review we will explore the use of entomopathogens for controling insects and mites of tropical fruit.

18.2. CANDIDATE ENTOMOPATHOGENS

18.2.1. Viruses

Many viruses have been identified from hundreds of arthropod species. For most of them, those with particles occluded in protein bodies (OBs) [Baculoviridae, Entomopoxviridae, Reoviridae (Cypoviruses)] have been used successfully in microbial control programs. The Baculoviridae (nucleopolyhedroviruses and granuloviruses) are the most studied and used as microbial control agents (Miller, 1997; Hunter-Fujita *et al.,* 1998; Moscardi, 1999; Cory and Evans, 2007; Sosa-Gómez *et al.,* 2008). They are normally transmitted *per os* and gain access to host tissues via the midgut where the OBs that surround the virus rods are dissolved. The currently unclassified virus of *Oryctes rhinoceros* L. is the most successfully used non-occluded virus. Viruses comprise some of the most host-specific entomopathogens but their main drawbacks are the requirement for *in vivo* production and their sensitivity to ultra-violet degradation.

18.2.2. Bacteria

Although several species of bacteria have been used as microbial control agents of a variety of insects, only *Bacillus thuringiensis* Berliner (Bt) has been used for practical pest control in tropical tree fruit agroecosystems. It is the most widely used inundatively applied microbial control agent (Glare and O'Callaghan, 2000). Several isolates of Bt are commercially produced with activity against Lepidoptera, Coleoptera, and Diptera. The safety of Bt for applicators and vertebrate and invertebrate non-target organisms is well documented (Glare and O'Callaghan, 2000; Lacey and Siegel, 2000). Its insecticidal activity is associated with delta-endotoxins located in parasporal inclusion bodies (or parasporal crystals) that are produced at sporulation and must be ingested by the target organism in order to be active. Limiting factors of Bt is a fairly short half-life on foliage in the field due to solar inactivation or degradation in organically enriched aquatic habitats.

18.2.3. Fungi

The majority of fungi that naturally regulate insect and mite populations are in the order Hypocreales (Ascomycetes), including the vast majority of conidial entomopathogens in more than two dozen genera formerly classified among the Hyphomycetes and Entomophthorales. Production of the latter ranges from difficult to impossible, hence, they have not been commercially produced or applied inundatively on a large scale. On the other hand, several asexually reproducing species in the Hypocreales are amenable to mass production and commercialization. The most studied for control of insects and mites are in the genera *Beauveria*, *Metarhizium*, *Isaria* (formerly *Paecilomyces*), *Aschersonia*, *Hirsutella*, and *Lecanicillium* (formerly *Verticillium*) (McCoy *et al.,* 1988; Alves, 1998b; Inglis *et al.* 2001 Goettel *et al.,* 2005, Ekesi and Maniania, 2007, Alves *et al.,* 2008). Some notable successes with certain species are reported for a number of arthropod pests of tropical fruit (Alves *et al.,* 2008). Because the normal route of invasion is through the cuticle, fungi are especially suitable microbial control agents for sucking insects (*e.g.* Hemiptera).

Phylogenetic studies now confirm that the Microsporidia are highly derived organisms correctly placed among the lower fungi rather than extremely ancient and simple organisms classified with the Protozoa (Hirt *et al.,* 1999). Although many microsporidians are common pathogens of arthropods, few have been included in microbial control programs because certain fundamental characteristics (complex life cycles, obligate parasitism, and chronic rather than acute effects) inhibit their use (Solter and Becnel, 2007).

18.2.4. Entomopathogenic Nematodes (EPNs)

Nematodes in the families Steinernematidae and Heterorhabditidae are effective control agents of dozens of insect species in soil and cryptic habitats (Kaya and Gaugler, 1993; Georgis *et al.,* 2006; Shapiro-Ilan *et al.,* 2005). These nematodes are associated with symbiotic bacteria (*Xenorhabdus* spp. and *Photorhabdus* spp.), which are housed in the intestine of the infective juveniles (IJs) (sometimes referred to as the Dauer stage). The IJ, the only free-living stage, occurs in the soil and searches for an insect host. Upon finding a host, it enters through the mouth, anus or spiracles and penetrates into the body cavity. In the case of heterorhabditids, IJ penetration can occur directly through soft cuticles. In the body cavity, the IJ releases mutualistic bacterial cells, which multiply rapidly and kill its insect host, usually within 48 h. In addition, the bacterial cells digest host tissues and produce antibiotics that protect the host cadaver from saprophytes and scavengers, and allow the nematodes to develop and reproduce. The nematodes feed on the mutualistic bacterial cells and on degraded host tissues. Depending on host size, there may be one to three nematode generations in the host cadaver. When host nutrients are depleted, the pre-IJs sequester the mutualistic bacterial cells in their intestines. The resulting IJs leave the host and search for new hosts. In the absence of a host, IJs can persist for months in moist soil. However, the IJs have their own natural enemies (*i.e.* nematophagous fungi, predatory mites and other soil predators) and must also contend with abiotic factors such as temperature extremes, low soil moisture, and ultraviolet radiation that affect their survival.

Several EPN species are commercially produced and available for large-scale application. For small-scale experimental testing, EPNs can be produced *in vivo* and on artificial media (Kaya and Stock, 1997).

18.3. RESEARCH AND APPLICATION OF ENTOMOPATHOGENS FOR CONTROL OF ARTHROPOD PESTS OF TROPICAL FRUITS

The literature on microbial control agents of tropical fruit pests has, for the most part, concentrated on key pests of a few major crops (*e.g.* citrus and banana). In this review we present information on the use of microbial control of arthropod pests of citrus, banana, coconut and related palm fruit, mango, guava, papaya, pineapple and apple.

18.3.1. Citrus

Because of the diversity of cultivars and climates in which they are grown, citrus is perhaps the most widely distributed tree fruit crop ranging from tropical and subtropical climates to temperate habitats around the world. Consequently a huge range of arthropod pests is reported from citrus varieties (Smith and Peña, 2002). Successful microbial control of several pests has been reported using fungi, bacteria, viruses and EPNs.

Citrus Rust Mite (CRM), Phyllocoptruta Oleivora (Ashmead) (Acari: Eriophyidae)

This is a major pest of citrus in several countries, including Brazil and the USA. Studies on the use of fungi for its control are limited number. Alves *et al.* (2005) assessed the pathogenicity of five concentrations of *Beauveria bassiana* (Bals.) Vuill. ranging from 10^6 to 10^8 condia/ml under laboratory conditions (25°C, 12h photophase, 98% RH). Mortality was time and dosage dependent and ranged from 24% to 91% with an LC$_{50}$ of 4.23 x 10^6 conidia/ml five days after treatment. LT$_{50}$ at the highest concentration was 2.74 days. *Hirsutella thompsonii* Fisher (Entomophthorales: Moliniaceae) is infectious for CRM and several other mite pests of greenhouse crops, coconut and turf (Samson *et al.,* 1980; McCoy, 1996; Boucias *et al.,* 2007). In the early 1980's it was mass-produced and formulated by Abbott Laboratories (Chicago, IL, USA) and registered for CRM control in the USA (McCoy and Couch , 1982). Under optimal conditions, *H. thompsonii* can control CRM within one to two weeks (McCoy *et al.,* 2007). Field applications of mycelia led to production of conidia within 48h and provided suppression of CRM for up to 14 weeks (McCoy *et al.,* 1971).

McCoy *et al.* (2007) provided protocols for the field evaluation of *H. thompsonii* and other fungi intended for CRM control. Unfortunately, commercial development of the fungus was discontinued by Abbott Laboratories.

Citrus Red Mite, Panonychus Citri (Mcgregor) (Acari: Tetranychidae)

This mite can be a serious pest of citrus in certain locations. In southern China, Shi and Feng (2006) evaluated the efficacy of four rates of *B. bassiana* (ranging from 1.2 x 10^{12} to 3.0 x 10^{13} conidia/ha) and a combination of the fungus with a low rate of pyridaben for control of *P. citri* in orange groves. All of the *B. bassiana* application rates produced significant mortality in *P. citri* and the combinations with pyridaben led to better control. Two applications of $\geq$ 1.5 x 10^{13} conidia/ha plus low rate pyridaben with a 15 day spray interval resulted in good control of *P. citri* for 35 days with mite density declines of 74-91%. Where *P. citri* is a pest in arid regions, it is not recommendable to use fungi against this pest. McCoy *et al.* (2007) summarized research and results of applications of a non-occluded virus found in *P. citri*. The virus has apparently been responsible for decimating epizootics in *P. citri* populations in Arizona and California (Reed, 1981). Shaw *et al.* (1968) reported control of the mite after application of triturated mites infected with the virus. A major limitation of this virus is the need for its mass production in *P. citri*.

Broad Mite (BM) or White Mite, Polyphagotarsonemus Latus (Banks) (Acari: Tarsonemidae)

This is a cosmopolitan pest of a variety of plants including citrus, papaya, and mango. It is abundant during warm and humid condition and thus an ideal candidate for control with fungi. However, very little research on microbial control has been conducted on this pest. Cabrera *et al.* (1987) reported natural infection of BM with *H. thompsonii*. Peña *et al.* (1996) assessed the infectivity of *B. bassiana*, *H. thompsonii*, and *Isaria fumosorosea* (Wize) Brown and Smith conidia in laboratory bioassays on bean leaves. The LC_{50} values for the fungi were 1.16 x 10^6, 2.39 x 10^3, and 1.29 x 10^5 conidia/ml, respectively. Mortality due to *B. bassiana* was most rapid in mite densities between 65 and 125 mites/leaf. The efficacy of *B. bassiana* and *I. fumosorosea* and other agents were also evaluated against BM in a greenhouse test on potted bean plants. Treatments with *B. bassiana* were the most efficacious and persistent, and resulted in 88% mortality.

False Spider Mite (FSM), Brevipalpus Phoenicis (Geijskes) (Acari: Tenuipalpidae)

FSM, also known as the red and black flat mite, is a polyphagous widely distributed tropical-subtropical species that has been reported from several hundred plant hosts including citrus, banana, macadamia, orchid, papaya, passion fruit, coffee and tea. FSM is an important citrus pest as it is a vector of the citrus leprosis virus (Childers *et al.,* 2003). Rossi-Zalaf and Alves (2006) assessed the activity of 52 isolates of fungi including *B. bassiana*, *Metarhizium anisopliae* (Metsch.) Sorokin, *Paecilomyces* spp., *H. thompsonii*, *Lecanicillium* spp. and others. The most active isolates were all *H. thompsonii*, causing 90-100% mortality six days after treatment. All other species of fungi produced less than 30% mortality six days after treatment. The authors observed conidiogenesis of *H. thompsonii* with development of mycelium and condiophores emerging from the posterior and anterior parts of mites 120h after spraying with conidia.

Whiteflies and Blackflies (Hemiptera: Aleyrodidae)

About 30 species of whiteflies and blackflies have been reported attacking citrus worldwide (Smith and Peña, 2002), six of which are considered major pests. Fransen (1990), Lacey *et al.* (1996, 2008a), and Faria and Wraight (2001) summarized the literature on fungi reported from whiteflies. Most of research on microbial control of whiteflies in citrus has focused on *Aschersonia* spp., which has produced spectacular epizootics in conditions of high humidity and rainfall (Figure 1.a). In Florida, USA, Fawcett (1944) reported epizootics in *Dialeurodes citri* (Ashmead) and *D. citrifolii* (Morgan) in citrus groves during the summer when high humidity promoted conidial sporulation and host infection, and frequent rains enabled effective dispersal of the conidia. Meyerdirk *et al.* (1980) observed *A. aleyrodis* Ashby infecting *D. citri* in Texas. Elizondo and Quezada (1990) published on the distribution of the citrus backfly, *Aleurocanthus woglumi* Ashby, and its natural enemies in four localities in Costa Rica. In addition to parasitoids and predators, mortality due to *A. aleyrodis* was significant. The same agent was unable to control *A. woglumi* following its introduction into and spread through El Salvador (Quezada, 1974). Outbreaks of disease caused by *Aschersonia* spp. among populations of *Bemisia giffardi* (Kotinsky) on citrus in Taiwan were observed (Yen and Tsai, 1969). In Brazil, the occurrence of *Aschersonia* spp. on whiteflies is very common in all areas where citrus is grown and coincides with the periods of greatest rainfall (Alves, 1998b). In addition to natural occurrence, *Aschersonia* spp. has been successfully applied against whiteflies in citrus groves in the Georgia, China and Japan (Ponomarenko *et al.*, 1975, Gao *et al.*, 1985). In field tests in Japan *Aschersonia* sp. was pathogenic to nymphs, pupae and eggs of *D. citri*, and the mortality increased with the concentration of conidia (Uchida, 1970). Ponomarenko *et al.* (1975) introduced several isolates of *Aschersonia* from six countries into orange groves near Adzharia, Georgia, being *A. placenta* Berk and Br. from Vietnam and China the most effective leading to up to 90% parasitism in favorable weather.

Figure 1. A) Citrus whiteflies infected with *Aschersonia aleyrodis*. Photo coutesy of G. Xiong; B) Adult guava weevil infected with *Beauveria bassiana*. Photo courtesy of R. Sammuels; C) Inoculation of palm rhinoceros beetle with Oryctes virus. Photo by C. Prior; D) *Galleria mellonella* larvae cadavers covered with commercial calcitic calcareum formulated as a powder and as an aqueous suspension (10%); gelatin capsules (100% gelatin, size 000); talc formulated as a powder and as an aqueous suspension (10%). Photo by C. Dolinski.

Scales (Hemiptera: Coccidae and Diaspididae)

These are regarded as the most abundant and injurious citrus pests (Smith and Peña, 2002). Hall (1981) stated that scales are among the most frequently reported hosts for *L. lecanii*. El-Choubassi *et al.* (2001) observed up to 49% infection of the diaspidid *Parlatoria ziziphi* (Lucas) by *A. aleyrodis* and *A. goldiana* Sacc. and Ellis in Cuba. Gravena *et al.* (1988) noted that the main control agent of a diaspidid, *Selenaspidus articulatus* (Morgan), in a grove near São Paulo, Brazil, was *A. aleyrodis*. Yen and Tsai (1969) observed fungal infections caused by *Podonectria coccicola* (Ellis and Everhart) Petch, *Pseudomicrocera henningsii* (Koord.) Petch, and *Sphaerostilbe aurantiicola* (B. and Br.) Petch in the coccids *Chrysomphalus aonidum* (L.), *P. ziziphus* and *Lepidosaphes beckii* (Newm.) in citrus groves in Taiwan. In South Africa, Moore (2002) reported several fungi attacking four scale pests of citrus. The green scale, *Coccus viridis* (Green), is a pest of citrus and other plants in tropical regions of the world. Entomopathogenic fungi including *L. lecanii*, and *Aschersonia cubensis*, have been documented attacking *C. viridis* and some have been credited with playing an important role in the natural limitation of the scale on citrus. *L. lecanii* was the most frequently observed. Attempts to artificially infect *C. viridis* with the fungus were unsuccessful (Fredrick, 1943; Dekle and Fasulo, 2001). No publications on the applied use of fungi for successful control of scale in citrus were found in the literature

Aphids (Hemiptera: Aphididae)

Aphids, particularly those in the genus *Toxoptera*, are important pests of citrus, especially due to their roles as vectors of diseases such as citrus tristeza. Fungi are important natural enemies of aphids in warm and humid conditions (Hall, 1981; Latgé and Papierok, 1988; Humber, 1997; Eilenberg and Pell. 2007; Steinkraus, 2007) but studies on their use in citrus groves have been limited. Poprawski *et al.* (1999) demonstrated good potential for control of the brown citrus aphid, *Toxoptera citricida* (Kirkaldy), with *B. bassiana*. Application of the fungus in field trials at 2.5 x 10^{13} and 5.0 x 10^{13} conidia/ha resulted in 79.8% and 94.4% control, respectively.

Asian Citrus Psyllid, Diaphorina Citri Kuwayama (Hemiptera: Psyllidae)

This psyllid is a serious pest of citrus in the tropics and subtropics due to the transmission of a bacterium (*Candidatus* Liberibacter sp.) that causes a disease referred to as citrus greening. *D. citri*, has been found naturally infected with *Hirsutella citriformis* Speare and *I. fumosorosea* in Florida citrus groves (Boucias *et al.,* 2007; Meyer *et al.,* 2007; 2008) There are several reports of these and other fungi in *D. citri* and other psyllids in Cuba, Guadeloupe, Malaysia, New Caladonia, The Philippines, and Taiwan. However, there have have been only a few applied studies on the potential of Hypocreales fungi for control of psyllids. (Puterka, 1999) and Puterka *et al.* (1994), but no substantive studies have been conducted on the field activity of applied fungi for control of *D. citri*.

Diaprepes Root Weevil, Diaprepes Abbreviatus (L.) (Coleoptera: Curculionidae)

This weevil is native to the Caribbean and has become a major pest in Florida citrus since its introduction was first reported in 1964 (Woodruff, 1964). Annual losses and cost of control in Florida citrus are thought to exceed $72 million (Peña *et al.* 2000).

Notching along the margins of young leaves is a typical sign of feeding by the adult. Eggs are laid on older leaves and, after hatching, larvae drop to the ground, enter the soil, and feed on roots for most of the year.

Injury caused by the weevil appears to be cumulative; root damage impedes the plant to take up water and nutrients, and can result in tree mortality (Syvertsen and McCoy, 1985). In addition, this injury provides an avenue for fungal root rot infections by *Phythophora* spp. (Graham *et al.*, 2003). A single larva can kill young trees, whereas several larvae can cause decline of older, established trees.

Since larvae are below ground, it is difficult to detect them before decline of above ground portions of the host plant is observed (Simpson *et al.*, 1996). EPNs (Steinernematidae and Heterorhabditidae) are effective in soil habitats. Because *D. abbreviatus* larvae are vulnerable while entering the soil, numerous studies have been conducted on the use of EPNs for their control.

In Florida, EPNs have been marketed for weevil control for over 15 years. Currently, two commercially available species, *Steinernema riobrave* Cabanillas, Poinar and Raulston and *Heterorhabditis indica* Poinar, Karunakar and David are used for control of the weevil. These nematodes appear to be most effective at high temperatures ($27 \pm 2°C$) in coarse sandy soils (Stuart *et al.*, 2008). Larval mortality of over 90% has been reported for field trials with *S. riobrave* when applied at 1.2×10^{10} infective juveniles (IJs)/ha (McCoy *et al.*, 2002, 2007). Clay soils can greatly limit EPN activity for control of *D. abbreviatus* and other weevils (Jenkins *et al.*, 2008, Stuart *et al.*, 2008). Stuart *et al.* (2008) proposed that inoculative realeses, manipulation of habitat to improve control and conservation of EPNs could extend their usefulness in Florida orange groves for *D. abbreviatus* control. Other EPN species, rates and percentage mortality were summarized by Shapiro-Ilan *et al.* (2002). The use of irrigation systems for application of EPNs has been effective in delivering IJs into the zone below trees where larvae enter the soil.

Indigenous entomopathogenic fungi infect adults and larvae of *D. abbreviatus* and other weevils in the soil (McCoy *et al.*, 2007). Research conducted by Quintela and McCoy (1998) demonstrated that a commercial oil formulation of *B. bassiana* (Mycotrol) or a combination of the fungus and a sublethal concentration of imidacloprid (a chloronicotinyl insecticide) provided effective control of neonate larvae and teneral adults when applied as a soil barrier. McCoy *et al.* (2007) pointed out that the efficacy of *B. bassiana* as a weevil control agent was limited by its poor persistence in soil. Weathersbee *et al.* (2002) demonstrated larvicidal and sublethal activity of elevated concentrations *B. thuringiensis* subsp. *tenebrionis* for *D. abbreviatus* in artificial diet and potted citrus tests.

Citrus Root Weevils, Pachnaeus Spp. (Coleoptera: Curculionidae)

Pachnaeus litus (Germar) and *P. opalus* (Oliver) are native to Florida and normally are considered minor pests, although they can damage young citrus plants (Tarrant and McCoy, 1989). The adults feed on tender foliage of citrus and eggs are laid on mature leaves, often on the same trees on which the adults have been feeding. Neonate larvae drop to the ground and quickly burrow into the soil where they feed on roots (Bullock *et al.*, 1999). Field applications of 5×10^6 *H. bacteriophora* IJs/tree resulted in significant reduction of *P. opalus* adults (76%) as compared to controls (Downing *et al.*, 1991). In another experiment, two applications of *S. riobrave* or *S. carpocapsae* at 2×10^6 IJs/tree provided an overall 64% and 53% reduction of *P. litus*, respectively (Bullock *et al.*, 1999).

Fuller Rose Beetle (FRB), Asynonychus Godmani Crotch (=Pantomorus Cervinus (Boheman) (Coleoptera: Curculionidae)

Adults feed on citrus foliage and cause leaf notching. Larvae develop for 6-10 months in the soil where they feed on the roots. The beetle does not usually cause economic damage but the presence of eggs on exported fruit requires fumigation. Morse and Lindegren (1996) reported the results of field application of rather high rates of *S. carpocapsae* IJs against late-instar larvae under Valencia orange trees. A single application of the Kapow or All strain of *S. carpocapsae*, each applied at 50, 150, and 500 IJs/cm^2, reduced the number of emerging adult FRB a combined 55% and 38%, respectively, the year following treatment and 79% and 82%, respectively, the 2nd year. Based on EPN recovery six months after application and continued reduction of FRB emergence in the second year, the authors concluded that the EPNs persisted and recycled in the environment.

Citrus Fruit Borer, Ecdytolopha Auratiana (Lima) (Lepidoptera: Tortricidae)

In the late 1980's, *E. auratiana* became a key pest of citrus in São Paulo, Brazil. Females lay eggs on the fruit surface where neonate larvae enter the fruit and feed for approximately 20 days, rendering them worthless for consumption and processing. When fully grown, larvae leave the fruits and pupate in the soil. Laboratory tests with EPNs against sixth-instar larvae in pots containing sandy soil showed that *H. indica* applied at 1.6 IJ/cm^2 resulted in 92% mortality (Leite *et al.,* 2005).

False Codling Moth (FCM), Cryptophlebia Leucotreta Merge (Lepidoptera: Tortricidae)

This is an important pest of citrus in Africa and outlying islands (Moore, 2002; Smith and Peña, 2002). A granulovirus of FCM has shown promise in field trials. Fritsch (1988) conducted a small-scale field trial of the virus formulated with skimmed milk powder and a wetting agent and applied at 10^8 and 10^9 granules/ml on FCM-infested citrus in Cape Verde, resulting in a 77% reduction of FCM population. Moore (2002) provided a synopsis of research conducted on this virus in South Africa and concluded that it plays a natural regulatory role in FCM populations. Application of the granulovirus at rates of 10^{14} to 10^{15} OBs/ha provided up to 60% reduction of FCM infestations in navel oranges. Subsequently, 17 field trials have been conducted on citrus in three different provinces in South Africa. Where spray coverage was thorough and FCM pressure was moderate, infestation was reduced by 70% following a virus treatment three weeks before harvest (Moore *et al.,* 2005). The virus has been registered for use and is now commercially produced in South Africa.

Old World Bollworm, Helicoverpa Armigera (Hübner) (Lepidoptera: Noctuidae)

This insect attacks a wide variety of important crops in portions of Africa, Asia, Australia, and Europe. In South Africa and parts of Asia, it is a serious pest of citrus. *B. thuringiensis* subsp. *kurstaki* (Btk) has been applied for control of *H. armigera* in citrus, but growers have reported dissatisfaction with its efficacy (Moore *et al.,* 2004). Field trials of the nucleopolyhedrovirus of *H. armigera* conducted by Moore *et al.* (2004) demonstrated that it was superior to Btk for suppression of the bollworm. Application of 7.26 x 10^5 and 1.15 x 10^6 OBs/ml resulted in 100% reduction in bollworms within 14 days of application. Damage to fruit was reduced by up to 75-84% and rejection for export was reduced by 62-96%.

Other Lepidopteran Pests

The citrus leafminer, *Phyllocnistis citrella* Stainton (Phyllocnistidae) is a significant pest of citrus worldwide. Field application of Bt against the leafminer in the Azores resulted in significant larval mortality 48h after treatment (Dias *et al.,* 2005). Shapiro *et al.* (1998) observed that leaf damage and number of *P. citrella* larvae were significantly reduced after 21 days by treatments of Bt plus a wetting agent. Beattie *et al.* (1995) demonstrated only limited potential for control of *P. citrella* with *S. carpocapsae*. Narayanamma and Savithri (2003) and Gopalakrishnan and Gangavisalakshy (2005) reported successful control of the citrus butterfly, *Papilio demoleus* L. (Papillionidae), on sweet orange following applications of *Bt* at two locations in India. The citrus leafroller, *Cacoecia occidentalis* Walsingham (Tortricidae) is a minor pest of South African citrus. Smith *et al.* (1990) described a granulovirus with potential for its control.

Fruit Flies (Diptera: Tephritidae)

Some species, such as the Mediterranean fruit fly, *Ceratitis capitata* Weidemann, are key or major pests of citrus in tropical production areas. Several authors have reported the results of research for control of *C. capitata* and other tephritid species using EPNs and fungi in laboratory studies and field trials in other agroecosystems (see mango and guava section).

18.3.2. BANANA

Several varieties of banana and plantain (*Musa* spp.) are grown throughout the tropics and into the sub-tropics. In addition to providing indigenous populations with rich sources of carbohydrate and other nutrients, banana and plantain are valuable export crops. A wide variety of insects and mites attack banana and include species that bore into the trunk, pseudostem, rhizomes, corm and roots, and species that attack flowers, fruits and foliage (Gold *et al.* 2002).

Banana Weevil (BW), Cosmopolites Sordidus (Germar) (Coleoptera: Curculionidae)

This weevil is reported as the most important insect pest of banana and plantain (Gold *et al.* 2001). Oviposition takes place at the base of the plant and neonate larvae bore into the corm. Heavy infestations can result in crop failure in newly established stands and reduced yield and shortened life span of plants in established stands (Gold *et al.* 2001; 2002).

Strains of *B. bassiana* have shown good potential to control adult BW (reviewed by Gold *et al.,* 2002; 2003) (Figure 1.b). Mortality in adult weevils of up to 60% was reported. Godonou *et al.* (2000) evaluated two formulations of *B. bassiana* (oil palm kernel cake-based formulation of conidia [OPKC] and conidial powder) applied to the planting holes and suckers of banana. Both formulations resulted in 75% mortality in artificially released weevils. Under natural infestation conditions, the OPKC performed better (42% mortality) than the conidial powder (6%). Alves (1998b) reported up to 61% reduction of BW adults after treatment with *B. bassiana*.

A major constraint to the use of *B. bassiana* is the lack of an economic means of effectively applying the fungus. In Brazil, baits were made of sections of banana pseudostem treated in a suspension of *B. bassiana* conidia, mycelia and medium or with a paste that resulted in application of 5 x 10^9 conidia/bait. Fifty baits/ha were recommended being baits replaced 15 days, until less than 5 BW are captured/bait. Tinzaara *et al.* (2004) improved targeted delivery of the fungus in and near traps baited with an aggregation pheromone that is attractive to both sexes. *B. bassiana* is transmitted horizontally among BW individuals (Schoeman and Schoeman, 1999; Godonou *et al,.* 2000; Tinzaara *et al.,* 2004; 2007), and this greatly improves the dissemination of the fungus within populations. Also, as most mycosed adult cadavers are found in the leaf sheath at the base of plants, the likelihood that ovipositing females and mating pairs to come into contact with recently produced conidia is increased (Tinzaara *et al.,* 2004). Gold *et al.* (2003) reviewed the research of several authors on the potential of endophytic fungi (*Fusarium* spp., *Acremonium* spp., *Geotrichum* spp.) for BW suppression. The most effective species in this role was *Fusarium oxysporum* Schltdl.

Laboratory, greenhouse and field assays demonstrated the activity of EPNs against BW. Rosales and Suarez (1998) evaluated exotic and native EPNs in Venezuela and found some native isolates of *Heterorhabditis* with good potential for BW control. Figueroa (1990) evaluated *S. carpocapsae*, *S. glaseri* (Steiner) and *S. feltiae* (Filipjev) against BW in greenhouse tests in Puerto Rico. The nematodes significantly reduced the number of tunnels made by larvae in plantain corms at 400, 4,000 and 40,000 IJs/four-month-old plant. At the two higher rates, 100% larval mortality was achieved.

Treverrow and Bedding (1993) assayed 32 strains and species of *Steinernema* and *Heterorhabditis* against larvae and adults of BW and reported the greatest activity of the BW strain of *S. carpocapsae* against adults. They also described a method for introducing *S. carpocapsae* IJs into banana corms that involves removing cones (50 mm diam. by 150 mm long) from residual corms with a desuckering gouge and adding 2.5 x 10^5 IJs/cavity. The cone is then reinserted to produce a protected cavity that is attractive to adult weevils. Treverrow *et al.* (1991) also reported significant mortality (43- 68%) of BW larvae in banana rhizomes after applying *S. carpocapsae* with a water thickener into cuts or holes made in residual rhizomes. Mortality of adult BW attracted to the application sites on treated rhizomes was also observed. In contrast, bi-monthly treatments with *Heterorhabditis zealandica* Poinar and *S. carpocapsae* applied in a thickened aqueous solution into 200 mm deep incisions in the residual rhizomes of harvested plants from November to May failed to produce adequate control (Smith, 1995). The author speculated that the treatments were not effective possibly because of early nematode mortality caused by free water in the spike holes and/or because of the need for more frequent applications. Kermarrec and Mauleon (1989) demonstrated synergy between the insecticide chlordecone and *S. carpocapsae* for control of BW. Other studies on EPNs for BW control were summarized by Gold *et al.* (2002).

West Indian Sugarcane Borer, Metamasius Hemipterus Sericeus (Olivier) (Coleoptera: Curculionidae)

This pest is also known as the silky cane weevil and rotten stalk borer of sugar cane. It can be an important pest of banana in certain areas of the Americas (Giblin-Davis *et al.* 1994, Gold *et al.* 2002). As with BW, fungi and nematodes have potential to control this pest. Peña *et al.* (1995) reported infection of low density populations of the insect by *B. bassiana* in a three-year old banana field in Florida. They observed up to 70% infection when more than ten weevils were captured/trap. Unlike BW, *M. hemipterus sericeusis* are strong fliers and could disperse fungi from sources of inoculum (*i.e.* through attractant traps) into neighboring populations. Giblin-Davis *et al.* (1996) evaluated *S. carpocapsae* in palm for control of this pest (see coconut section).

Banana Moth, Opogona Sacchari (Bojer) (Lepidoptera: Tineidae)

This insect is only a minor pest of banana, but a serious pest of certain types of palms. Research on EPNs for its control are reported by Peña *et al.* (1990) in the coconut section.

Opsiphanes Tamarindi Felder (Lepidoptera: Brassolidae)

This is a major defoliator of plantains during the dry season in the region south of Lake Maracaibo, Venezuela. Broad-spectrum insecticides have been ineffective for its control. Briceno (1997) described an IPM system that combined cultural practices, application of Bt against early larval stages, and relying on natural enemies (parasitoids and predators) to control late larval and pupal stages. The seasonal application of Bt helped to eliminate first instars without affecting natural enemies.

Bagworm, Oiketicus Kirbyi Guilding (Lepidoptera: Psychidae)

This is a defoliating pest of banana in Costa Rica and Colombia (Gold *et al.* 2002). Stephens (1962) noted the occurrence of *B. bassiana* and a *Nosema* species in bagworm larvae in Costa Rican bananas.

18.3.3. Coconut and Related Palm Fruit

Coconut and palm oil are significant and sometimes predominant sources of income for several tropical countries. Although a multitude of insects and mites exert varying degrees of economic impact on it, microbial control agents have been used on relatively few of them.

Palm Rhinoceros Beetles, Oryctes Spp. (Coleoptera: Scarabaeidae: Dynastinae)

These beetles are serious pests of coconut throughout the old world tropics. *Oryctes rhinoceros* (L.) is one of the most important pests of coconut in Southeast Asia and several South Pacific islands (Bedford, 1980). Adults attack the heart of plants and feeding can reduce yield and kill trees (Bedford, 1980; Zelazny, 1983). Larvae develop mainly in rotting palm trunks.

One of the most successful uses of an entomopathogen for classical biological control is reported for *O. rhinoceros*. In 1966, Huger described a non-occluded virus of *O. rhinoceros* from Malaysia that demonstrated potential for long-term control (Huger, 1966, 2005). Adults become chronically infected via oral contact with the virus and subsequently serve as reservoirs and disseminators (Huger, 1973; Zelazny, 1973; Bedford, 1981). The midgut epithelial cells of adults become heavily infected (Huger, 1973); individual beetles may produce and excrete up to 0.3 mg of virus per day (Monsarrat and Veyrunnes, 1976). Viral transmission among adults occurs during mating, when they feed in palms contaminated with feces containing the virus, or in larval breeding sites (Zelazny, 1976; Young and Longworth, 1981; Zelazny and Alfiler, 1991). There are no external symptoms of the disease in adults and it is not immediately fatal (Zelazny, 1973). However, it shortens lifespan and reduces fecundity of infected adults. Infected beetles stop feeding, fly less frequently, and the males mate less often (Zelazny, 1977). Transmission to larvae occurs when virus-infected adults defecate in breeding sites (Zelazny, 1972; 1976). Viral infection in larvae is always lethal. Studies with other *Oryctes* species and *Strategus aloeus* (L.) revealed that the virus is cross-infective to certain other Dynastinae (Lomer, 1987). Control of *O. monoceros* (Olivier), a serious coconut pest in Africa, with the *Oryctes* virus was reported by Lomer (1986) and Purrini (1989).

The preferred method for disseminating the virus in coconut plantations has been the infection and release of *O. rhinoceros* adults (Figure 1.C), which resulted in establishment of the virus within larval and adult habitats in several locations in Asia, Africa, and the South Pacific (Bedford, 1981; Zelazny, 1978; Young and Longworth 1981; Jones *et al.,* 1998; Huger 2005) where it was previously absent. As few as ten infected beetles can successfully establish the virus on an island (Jones *et al.,* 1998). Introduction of virus into artificial and natural larval habitats has also been used successfully to inoculate beetle populations (Bedford, 1980). Since 1967, the introduction of virus into coconut plantations in several South Pacific islands and other locations has resulted in significant control of *O. rhinoceros.* Integrated measures that include removal or covering of old palm logs that serve as breeding sites along with inoculative releases of the virus have reduced the density of *Oryctes* populations to below economic thresholds in many locations (Bedford, 1980; Zelazny *et al.,* 1992; Alfiler, 1992).

Despite the successes of introducing the virus into previously virus-free islands, *O. rhinoceros* remains a serious threat in the coconut and oil palm plantations of Southeast Asia and the Pacific. The prohibition of burning palm logs in the 1990's as a method for maintaining plantation hygiene has dramatically aggravated the problem (Jackson *et al.,* 2005).

In such situations, *O. rhinoceros* outbreaks are not caused by the absence of the virus disease, but by ecological disturbances in the transmission cycle of the virus, like the availability of large number of coconut palms for insect breeding. Under these conditions there is little contact between virus-infected and healthy individuals, whereas a low density of dead-standing palms creates good conditions for spread of the virus (Zelazny and Alfiler, 1986; Alfiler 1992; Zelazny *et al.,* 1992).

Ecological methods for promoting the spread of the virus are to: 1) hide trunks of felled palms from the beetles by promoting the growth of cover crops over them, rather than attempting to burn the trunks, and 2) leave about five dead-standing coconut palms/ha (Zelazny and Moezir, 1989).

DNA analysis in Malaysia revealed several distinct viral genotypes with different virulences. The most virulent for larvae and adults (type B) was produced *in vivo* and released into healthy populations. Examination of beetles from the release site and vicinity demonstrated the spread and persistence of type B with concomitant reduction in palm damage.

Decreased control has been reported from other earlier release sites. Jackson *et al.* (2005) reported considerable genetic variation in the virus that suggests its rapid evolution. They recommended a renewed coordinated effort for the selection and distribution of virulent viral strains. Earlier work by Zelazny *et al.* (1990) showed some distinct differences in virulence among strains of the *Oryctes* virus. Marschall and Ioane (1982) demonstrated that re-release of the virus could result in an increase of the infection rate with a reduction of palm damage.

The combined use of the fungus *M. anisopliae* in larval breeding sites and release of virus has also been proposed (Young, 1974; Marschall and Ione, 1982), but the fungus does not spread well between breeding sites (Zelazny and Alfiler, 1986). The *Oryctes* virus has remained the most important biological control agent, however *O. rhinoceros* populations in Java and the southern parts of Sulawesi, Indonesia, are suspected of having developed resistance to the virus (Zelazny *et al.,* 1989).

Root Weevils (Coleoptera: Curculionidae)

The rotten sugar cane borer and the Diaprepes root weevil are reported as pests of several palm species (Weissling and Giblin-Davis, 1998). Research on the use of EPNs and fungi for control of these pests has been conducted in banana and citrus, respectively. Experiments with *S. carpocapsae* applied at 8 x 10^6 IJs/palm conducted by Giblin-Davis *et al.* (1996) on *M. hemipterus sericeus*-infested Canary Island date palms (*Phoenix canariensis* Hort. ex Chabaud) resulted in 51% mortality in weevil larvae.

Because of high weevil production/palm, Giblin-Davis *et al.* (1996) recommended that EPNs should be applied frequently and over a long period of time for effective management. The precise interval of application and duration of treatments will depend on environmental conditions and population density of the weevil.

Nettle and Slug Caterpillars (Lepidoptera: Limacodidae)

Several species of limacodids are defoliators of oil palm in Africa, Asia, and South America. Although they do not result in death of palms, they can suppress yield. Entwistle (1987) listed a variety of viruses isolated from 40 species of limacodids. Over half of these are densoviruses (parvoviruses), picornovirus, and *Nudaurelia* β with the remainder being baculoviruses and cypoviruses. Although trials conducted on Densovirus and Picornavirus showed promise for persistent management of these caterpillars in oil palm (Fédière *et al.,* 1990), their use has not been adopted. Summaries of research conducted on densoviruses and picornovirus for control of limacodids in Africa, Asia, and South America are presented by Kunjeku *et al.* (1998), Jones *et al.* 1998, Oliveira (1998), and Sosa-Gómez *et al.* (2008).

Banana Moth, Opogona Sacchari (Bojer) (Lepidoptera: Tineidae)

This insect can be a serious pest of certain palms. Damage to plants is due to feeding by larvae on roots and stems. Research on the use of steinernematid and heterorhabditid nematodes for its control are reported in other crops by Peña *et al.* (1990). Application of high rates of *S. carpocapsae* and *H. bacteriophora* to infested potted bamboo plants (5 x 10^7 IJs/plant in 500 mL drench) resulted in successful establishment of the nematodes and 58-100% reduction of *O. sacchari* larvae. Figure 1.D shows various life stages of *S. carpocapsae* in a banana moth larva, indicating the recycling potential of the nematode in this host.

18.3.4. MANGO

Over 260 species of insects and mites have been recorded as pests of mango worldwide (Peña *et al.,* 1998; Waite, 2002). Key pests that require regular control measures include fruit flies, seed weevils, tree borers and various Hemiptera.

Fruit Flies (Diptera: Tephritidae)

Fruit flies are among the most serious pests of tropical fruit and are regarded as the principal pests of mango. *Anastrepha ludens* (Loew) develops in a variety of fruit crops, but is especially damaging in mango and citrus. It is widely distributed in Mexico, most of Central America and southern United States. The Caribbean fruit fly, *Anastrepha suspensa* Loew, is a pest of mango and several other tropical fruits and is distributed within the Greater Antilles, Bahamas and Florida. *Anastrepha obliqua* (Maquart) is a significant pest of mango in Brazil and most of the new world tropics, but it has not yet been tested for susceptibility to entomopathogens. *C. capitata* and other *Ceratitis* spp. have been reported from mango and many other fruit worldwide. The oriental fruit fly, *Bactrocera dorsalis* (Hendel), is a pest of a wide range of fruit in Asia and is a key pest of mango in India. Fruit fly females oviposit in ripening fruit, and larvae burrow into the pulp. Fully grown larvae exit the fruit, usually after it has fallen to the ground, and pupate in the soil.

EPNs and fungi have been evaluated as alternatives to conventional insecticides for control of some important fruit fly pests of mango. Although *A. ludens* is susceptible to a variety of EPN species under laboratory conditions (Lezama-Gutiérrez *et al.*, 1996; Toledo *et al.*, 2001, 2005), extremely high rates are required for control in the field (2.5 x 10^2 IJs of *H. bacteriophora*/cm^2) (Toledo *et al.*, 2006). Similarly, laboratory and field research conducted on the effectiveness of EPNs against *C. capitata*, revealed susceptibility of larvae to several nematode species (Lindegren and Vail, 1986; Gazit *et al.*, 2000; Laborda *et al.*, 2003), but high application rates are required for control in the field (5 to 50 x 10^2 IJs of *S. carpocapsae*/cm^2 (Lindegren *et al.*, 1990). Research on the susceptibility of other fruit flies of mango has been limited to the laboratory. Lindegren and Vail (1986) reported on the susceptibility of *B. dorsalis* to *S. carpocapsae* and Beavers and Calkins (1984) reported on the evaluation of *A. suspensa* susceptibility to several steinernematids and heterorhabditids.

Testing of fungi on fruit flies of mango has been predominantly on *M. anisopliae* and *B. bassiana*. Laboratory and field research by Ekesi *et al.* (2002, 2003, 2005) and Dimbi *et al.* (2003a, 2003b) on *M. anisopliae* against *Ceratitis* spp. elucidated the effect of various factors on the activity of the fungus that included temperature, moisture, gender, life stage and fly species. Mochi *et al.* (2006) investigated the effect of fungicides, acaricides, insecticides and herbicides on *M. anisopliae* activity against *C. capitata* in laboratory exposures in field-collected soil. Significant pupal and adult mortality occurred in soil treated with the fungus with and without pesticides. No larval mortality was observed. Pesticides affected fungal activity slightly with the most significant effect due to the fungicides chlorothalonyl and tebuconazole. Ekesi and Maniania (2007) discuss several strategic options for using fungi for fruit fly control including aerial application, soil inoculation, and autodissemination using attractant traps.

Laboratory and field studies on *M. anisopliae* activity against *A. ludens* were reported by Lezama-Gutiérrez *et al.* (2000). When *M. anisopliae* was applied in field cages at 2 x 10^5 conidia/cm^2, adult emergence was reduced by up to 43% in loam soil. Castillo *et al.* (2000) studied the activity of strains of *M. anisopliae* and *I. fumosorosea* against *C. capitata* adults and reported LD$_{50}$ values of 5.1 x 10^3 and 6.1 x 10^3 conidia/fly, respectively, for the two most active strains. They also noted a sublethal effect of the fungi on fecundity. Konstantopoulou and Mazomenos (2005) reported on the laboratory evaluation of *B. bassiana* and *B. brongniartii* against adults of *C. capitata* and Rosa *et al.* (2002) studied the effects of *B.*

bassiana on *A. ludens*. In both studies, adult flies were very susceptible to infection by conidia, but Rosa *et al.* (2002) reported negligible effects on larvae and pupae.

Mango Seed Weevil, Sternochetus Mangiferae (F.) (Coleoptera: Curculionidae)

This is a widespread pest of mango. Eggs are usually laid on green fruit and larvae tunnel to the seed where they develop. Joubert and Labuschagne (1995) reported laboratory and field tests with two strains of *B. bassiana*, but neither strain had an effect on *S. mangiferae*. Shukla *et al.* (1984) described the isolation of a virus infecting larvae of *S. mangiferae* in India. It caused reduction in feeding, sluggishness, browning of the integument and milkiness of the haemolymph. The authors discussed the similarity of the virus to that reported in *O. rhinoceros*.

Rhytidodera Bowringii White (Coleoptera: Cerambicidae)

Zhou *et al.* (1998) reported on the evaluation of *B. bassiana* against *Rhytidodera bowringii* White. The fungus was isolated from dead *R. bowringii* adults and subsequently used to produce inoculum for field experiments in China. Treatment of two older mango orchards with *B. bassiana* resulted in 84% mortality of the beetle.

Mango Mealy Bug, Drosicha Mangiferae Green (Hemiptera: Margarodidae)

Srivastava and Fasih (1988) found *B. bassiana* infecting nymphs of *D. mangiferae* in mango orchards in five localities in Lucknow, India. In field trials on infested mango panicles, spray application of a suspension of 4.8×10^6 conidia/ml reduced populations of *D. mangiferae* by 33-100% in ten days. Masarrat and Srivastava (1998) demonstrated the dose-mortality relationship of *B. bassiana* against *D. mangiferae* first-instar nymphs in laboratory assays. Mohan *et al.* (2004) showed insecticidal activity for *Photorhabdus luminescens* (Akhurst) (the symbiotic bacterium isolated from *H. indica*). Application of a formulation of 1.4×10^6 cells/ml of *P. luminescens* on *D. mangiferae*-infested mango twigs resulted in 92.5% mortality of second-instar nymphs after 48h.

Mangohopper, Amritodus Atkinsoni Lethierry (Hemiptera: Cicadellidae)

Vyas *et al.* (1993) reported that a 75 min exposure to 10^9 *M. anisopliae* conidia/g of inert dust caused 100% mortality of *A. atkinsoni* after 96h. Concentrated aqueous suspensions of the fungus (10^9 conidia/ml) were considerably less effective. Srivastava and Tandon (1986) reported natural infection by the fungi *Lecanicillium lecanii* (Zimm.) and *B. bassiana* in populations of the leafhopper *Idioscopus clypealis* (Lethierry) on mangoes in Uttar Pradesh, India.

18.3.5. GUAVA

Psidium guajava L. is native to the American tropics but is currently grown in more than 50 subtropical and tropical countries. Brazil is the principal red guava producer followed by Mexico whereas India is the major producer of white guava (Gould and Raga, 2002). Different pests attack fruits, leaves and trunk, causing more or less damage depending on the region or country. Main pests of fruits are the guava weevil (*Conotrachelus psidii* Marshall) and fruit flies (*C. capitata* and *Anastrepha* spp.). On the leaves, the main pest is a psyllid (*Triozoida limbata* (Enderlein)) that causes damage mainly after pruning when new leaves start growing (Souza *et al.,* 2003).

Fruit Flies (Diptera: Tephritidae)

Fruit flies are very important pests in guava because the adults lay eggs in the fruit, and resulting damage by larvae lowers its quality. In Brazil, the main species in guava are *A. fraterculus* (Wied.), *A. obliqua*, *A. sororcula* (Zucchi), *A. zenildae* (Zucchi) and *C. capitata* (Souza *et al.,* 2003).

A study has been conducted in a guava orchard and it aimed to test the potential of *Heterorhabditis baujardi* Phan, Subbotin, Nguyen and Moens LPP7 for controlling *C. capitata* 3rd instar larvae. Two treatments were established (with and without nematodes) with seven replicates each. Fourteen cages were constructed over well grown guava trees using bamboo and nylon screen (2 x 2 x 2 m). Under each tree, one hundred-3rd instar *C. capitata* larvae and a suspension of half liter of water with 100,000 IJs were distributed evenly. The treatment control didn't have nematode addition. In each cage, a McPhail trap was placed and remained there for 15 days. The control efficiency was evaluated comparing the captured *C. capitata* adults in the traps in both treatments, and deducting this number from the initial population. The number of adults captured in the traps in treated trees was significantly different in relation to control (30.4% and 87.4%, respectively). When the experiment was repeated the same tendency was found (7.7% and 58.6%, respectively), showing the ability of *H. baujardi* LPP7 to be used as biological control agent against *C. capitata* larvae in the field (Minas *et al.,* 2009).

Guava Weevil, Conotrachelus Psidii Marshall (Coleopera: Curculionidae)

This is a major pest of guava in certain areas in Brazil. Females lay eggs in immature fruit (3-4 cm diameter) and larvae progress through four instars as the fruit develops. Infestation leads to acceleration in fruit maturation and fruit drop when ripe. At this moment, larvae crawl into the soil where they develop into prepupae. Individuals may remain in this stage for up to six months before pupation and development into the adult (Boscán de Martinez and Cásares, 1982; Bailez *et al.,* 2003). Control methods involve weekly applications of insecticides to suppress adults, but most of those currently in use for guava weevil control will be discontinued soon (Souza *et al.,* 2003; Agência Nacional de Vigilância Sanitária, 2004). Without chemical control, the percentage of damaged fruit in heavily infested orchards can reach 100%. The amount of fruit attacked has been increasing over the past three years possibly due to the development of insecticide resistance (C. Dolinski, personal comuunication). Poorly timed chemical applications and the tendency for adult weevils to hide in the litter around trees and avoid contact with the chemicals could also be involved (Denholm and Rolland, 1992).

The virulence of four species/strains of EPNs to fourth-instar larvae was assessed in the laboratory. In petri dish assays with sterile sand at 100 IJs/larva, larval mortality ranged from 33.5% to 84.5%, with the heterorhabditids being the most virulent. In sand column assays with *H. baujardi* LPP7, *H. indica* Hom1, and *S. riobrave* 355 at 100, 200 and 500 IJs/larva, significant mortality was observed only for *H. baujardi* (62.7%) and *H. indica* (68.3%) at the highest dose. For *H. baujardi* LPP7, the LT_{50} and LT_{90} for 100 IJs were 6.3 and 9.9 days, whereas the LC_{50} and LC_{90} over seven days were 52 and 122.2 IJs (Dolinski *et al.,* 2006). In a greenhouse study with guava trees in 20-L pots (ten weevil larvae/pot), and doses of 500, 1000 or 2000 IJs/pot, *H. baujardi* LPP7 caused 30% and 58% mortality at the two highest doses (Dolinski *et al.,* 2006).

Recent studies assessed the susceptibility of the 3rd instar larvae of the guava weevil using *H. baujardi* LPP7 infective juveniles (IJs) applied in cadavers of 7th instar *Galleria mellonella* L. (Lepidoptera: Pyralidae) larvae in the greenhouse and under field conditions. Insect cadaver concentrations of 2, 4, and 6 applied in pots in the greenhouse experiment caused significant mortality compared to the control.

Significance differences were observed in the field between control and treatments only when 6 cadavers per 0.25 m^2 were applied. It was also observed that IJs from the cadavers persisted 6 weeks after application in the field, but decreased greatly thereafter (Del Valle *et al.,* 2008a). Also, Del Valle *et al.* (2008b) observed that the highest average number of recovered IJs was found at 90 cm from the cadaver application, at 10 cm depth and on the fifth week after application, when applying 1 or 15 insect-cadavers under guava trees. There was no significant effect on the number of recovered IJs when they were applied as 1 or 15 cadavers, although the most uniform dispersal was found when 15 host cadavers were applied.

In another study different protective covering treatments including a commercial calcareous powder, a commercial talc powder, and gelatin capsules were tested. The number of emerging infective juveniles (IJs) from insect cadavers formulated with talc powder (9,722 ± 1,382) and gelatin capsules (7,892 ± 1,072) was similar to the control (6,346 ± 1,311), and indicated that these coverings do not interfere with IJ emergence.

However, the powdered calcareous covering significantly reduced IJ emergence. High infectivity was observed for IJs that emerged from cadavers in all treatments. Also, it was observed that *Ectatomma* spp. ants removed all insect cadavers from the nest entrance to a distance of 20 cm, with the exception of insect cadavers formulated in gelatin capsules, which were not removed.

In search of alternatives to chemical pesticides, a study was performed to select fungal isolates of *B. bassiana* and *M. anisopliae* as potential candidates for the control of adult *C. psidii*. Tests were carried out using three products applied with the entomopathogenic fungi: Tween 80, sunflower oil and a sub-lethal concentration of Imidacloprid (IMI) (100 ppm). The results demonstrated that *B. bassiana* LPP 19 and LPP 114 were the most effective isolates when used in combination with any of the products tested. The least virulent isolate, *M. anisopliae* ESALQ 818, when applied in Tween 80 caused only 26.6% mortality, however, this isolate showed significantly improved efficiency when applied together with either sunflower oil or IMI, causing 57.3 and 88.6% mortality, respectively. The efficiency of all the isolates tested here improved when applied together with IMI, with LT50 values of 5.3-10.3 days when compared to LT50 values in Tween alone of 9.5-17 days. The isolate that produced the highest number of conidia on the cadavers of adult *C. psidii* was *B. bassiana* LPP138, independent of the product used; however, conidial production was slightly reduced when fungi were applied together with IMI.

In small farms in Cachoeiras de Macacu, RJ, Brazil, a combination of different control methods against the guava weevil is being implemented. *H. baujardi* LPP7 is applied in orchards as infected cadavers against the larvae in the soil. Initial results indicate a 40% to 70% decrease emerging adult weevils by applying 20 cadavers/tree.

In addition, removal of all damaged fruit from their orchards helps to reduce pest population in the following year. Another alternative is the weekly application of neem oil against adults and neem cake applied to the soil for control of larvae. By eliminating pesticides, these strategies have effectively reduced production costs by 40% (Dolinski, 2006).

Leaf-Footed Bug, Leptoglossus Zonatus Dallas (Hemiptera: Coreidae)

This is usually a secondary pest on fruits and flowers. Few attempts to control this pest with microbials were published. Three *B. bassiana* isolates and one isolate of *M. anisopliae* were assessed in the laboratory against adults (Grim and Guharay, 1998), being *M. anisopliae* NB the most efficient.

In a field trial, mineral oil-based ultra low volume controlled droplet applications of *M. anisopliae* NB at 10^{10} conidia/tree caused 94% adult mortality. When *B. bassiana* was applied, there was a 28% increase in fruit yield.

18.3.6. PAPAYA

Carica papaya L. originated in southern Mexico, Central America and northern South America and is cultivated in most tropical countries (Morton, 1987a). A total of 134 species of arthropods are reported to attack papaya, some of which are important vectors of major pathogens of papaya (Pantoja *et al.,* 2002).

Since most fruit production is for exportation and the presence of pesticide residues is not tolerated, the use of alternatives to chemicals, including cultural methods and microbial control, are being increasingly employed.

Fruit Flies (Diptera: Tephritidae)

Fruit flies comprise the most important pests of papaya in most producing regions. Research on the use of microbial control agents (EPNs and fungi) against several species that attack papaya (*Anastrepha* spp., *Ceratitis* spp., *B. dorsalis*) has been conducted in other tropical fruit crops, most notably in mango (see mango section).

Twospotted Spider Mite, Tetranychus Urticae Koch (Acari: Tetranychidae)

This is the most important pest in papaya in Brazil, and is responsible for a major portion of the production costs (Alves *et al.,* 2002). Because of the high temperatures and humidity in areas where papaya is grown, fungi have potential as microbial control agents of mites. Alves *et al.* (2002) reported use of *B. bassiana* on over 1000 ha of commercial papaya production in Brazil for control of *T. urticae*.

Most of the research on *B. bassiana* and *T. urticae* has been conducted in other crops. Laboratory tests with *B. bassiana* against eggs, deutonymphs, protonynphs, larvae and adult stages of *T. urticae* on green bean showed positive results (Saenz-de-Cabezon *et al.,* 2003). The LC_{50} for juvenile stages and adults was 3184 and 1949 conidia/ml, respectively. No significant differences in mortality were observed among egg age classes (24-96h-old eggs) at the tested concentrations (1400-22,800 viable conidia/ml). Natural epizootics of fungi in the genus *Neozygites* have been responsible for spectacular declines in *T. urticae* populations in other cropping systems (reviewed in Steinkraus, 2007).

Broad Mite, *P. Latus*

This is also a significant pest of papaya in some areas (Pantoja *et al.,* 2002). Peña *et al.* (1996) investigated the use of *B. bassiana* and other fungi against this pest (see citrus section).

Aphids (Hemiptera: Aphididae)

Although aphids do not colonize papaya, they are considered important potential vectors of papaya diseases (Pantoja *et al.,* 2002). Many fungi (several Entomophthorales and *L. lecanii*) have resulted in massive natural epizootics in other cropping systems in some of the same aphids that attack papaya (*e.g., Aphis gossypii* Glover, *Myzus persicae* (Sulz.) (Latgé and Papierok, 1988; McCoy *et al.,* 1988; Steinkraus, 2007). Their potential to control aphids and other hemipteran papaya pests warrants further attention.

Other Hemipteran Pests

Papaya scale, *Philephedra tuberculosa* Nakahara and Gill (Hemiptera: Coccidae) attacks papaya and annona fruits. It is naturally infected by the fungus *L. lecanii*, which can lead to 90% mortality during summer (Peña *et al.,* 1987; Peña and Johnson, 2006). A variety of mealybugs, whiteflies, and leafhoppers attack papaya (Pantoja *et al.,* 2002). Although literature on microbial control of these pests on papaya is scant, several fungi are reported to attack these insects in other agroecosystems, including tropical fruit (McCoy *et al.,* 1988, Fransen, 1990; Lacey *et al.,* 1996, 2008b; Goettel *et al.,* 2005). Research on the use of fungi for control of these species in papaya is warranted.

18.3.7. Pineapple

Ananas comosus (L.) Merril is native to Brazil, Bolivia, Peru, and Paraguay, and is currently cultivated in most tropical countries (Morton, 1987b). It is the third largest fruit crop (after bananas and mango) harvested in the tropics (Petty *et al.,* 2002). These authors provided a summary of arthropod pests of pineapple worldwide and considered pink pineapple mealybug, Dysmicoccus brevipes (Cockerell) as the key pest worldwide.

Mealy Bugs, Dysmicoccus Brevipes and D. Neobrevipes (Hemiptera: Pseudococcidae)

Mealy bugs cause wilting due to the toxic effect of their feeding. There is potential for using EPNs against this pest based on work done with a closely related species, *D. vaccinii* Miller and Polavarapu (Stuart *et al.,* 1997).

Large Moth, Thecla Basalides (Lepidoptera: Lycaenidae)

This insect is an important pest of pineapple in Brazil, mainly in the cultivar 'Pérola'. Females lay eggs from the beginning of flowering until fruit formation. Larvae penetrate the flowers and complete development in 13 to 16 days (Fazolin, 2001). In Northern Brazil, the dose of 600 g/ha of Bt is recommended for control (Sanches, 2005). In Southern Brazil,

Lorenzato *et al.* (1997) reported on the natural enemies of *T. basalides* and effectiveness of insecticides. Application of Bt resulted in effective control.

White Grubs (Coleoptera: Scarabaeidae)

The larval stages of 23 species of scarabs in three sub-families were reported to attack the subterranean organs of pineapple plants in several locations worldwide, with the most serious pests reported from Australia and South Africa (Petty *et al.*, 2002). Although no specific studies on the use of microbial control agents have been reported for white grubs in pineapple, pathogens have been successfully applied for their control in other crops (Jackson and Glare, 1992; Klein *et al.*, 2007). Candidate control agents include fungi [*M. anisopliae*, *Beauveria brongniartii* (Sacc.) Petch], bacteria (Bt, *Paenibacillus* spp., *Serratia entomophila* Bizio), and EPNs (*Heterorhabditis* spp., *Steinernema* spp.).

18.3.8. APPLE

Codling moth, *Cydia pomonella* (L.) (Lepidoptera: Totricidae). Apple and other pome fruit are not regarded as typical tropical tree fruit. However, apple is increasingly being produced in several countries with subtropical and tropical climates including Argentina, Brazil, Chile, Mexico, South Africa and others (O'Rourke, 1996; Ramirez, 2001). A myriad of insect and mite pests attack apple, but the most serious on a worldwide basis is *C. pomonella* (Barnes, 1991). The most widely used microbial control agent for abatement of this pest is the codling moth granulovirus (CpGV).

It is one of the most virulent granuloviruses killing quickly and requiring very little virus to produce infection. A variety of research including pathology, pathogenesis and histopathology, determination of virulence, development of production methods, field use, factors that influence efficacy, commercial development, formulation, and *C. pomonella* resistance to the virus has been conducted (Lacey *et al.*, 2008a). Commercial products of CpGV are now produced in Europe and North America and used by orchardists worldwide. The number of applications of virus depends on codling moth population pressure and the number of generations per year. Research that elucidated optimal application conditions for CpGV was by Arthurs *et al.* (2005) and Lacey *et al.* (2007). With extremely high codling moth pressure, spray intervals of 7-10 days and application rates of 0.219 liters/ha (3 x 10^{13} occlusion bodies/liter) provided nearly 100% mortality in codling moth larvae. The addition of CpGV into IPM organic fruit production has provided a selective and safe method of control that enables survival of arthropod natural enemies of a variety of orchard pests. The major limitation of the virus is the requirement for frequent reapplication due to inactivation by sunlight. A potential problem is the development of resistance in codling moth after prolonged use of the virus. Resistance to CpGV has been reported in Germany, France, Italy, Switzerland and The Netherlands in organic orchards treated with multiple applications of the virus over an extended period (Fritsch *et al.*, 2005; Eberle and Jehle, 2006; Asser-Kaiser *et al.*, 2007).

The efficacy of EPNs for control of codling moth has been demonstrated under a variety of orchard and post-harvest conditions (Kaya *et al.*, 1984; Lacey *et al.*, 2005, 2006). The main caviats for its use are the requirement for sufficiently warm temperatures (15-25°C) and provision of adequate moisture to enable IJ entry into diapausing larvae before they dry.

Leafrollers (Lepidoptera: Tortricidae) and Other Lepidoptera

After *C. pomonella*, leafrollers and other lepidopteran species are the most important pests of apple production that are susceptible to microbial control. These are predominantly defoliators, but they may also feed on the surface of fruit. *B. thuringiensis* has been used routinely for control of leafrollers, budmoths, and fruitworms (Blommers, 1994; Blommers *et al.*, 1987; Lacey and Shapiro-Ilan, 2008). A number of factors can affect the performance of *B. thuringiensis* against tortricid pests of apple including temperature, other environmental factors, differences in species and instar susceptibility, spray coverage and application rate (Lacey *et al.*, 2007).

18.4. THE ROLE OF MICROBIAL CONTROL IN INTEGRATED PEST MANAGEMENT (IPM) IN TROPICAL FRUITS

Integrated Pest Management plays a significant role in crop protection being an important aspect of sustainable agriculture that attempts to minimize the negative environmental impacts and other deleterious effects due to the use of chemicals (Huffaker, 1985; Dent, 2000). Individual components of IPM are often evaluated as stand-alone tactics without consideration of their interactions with other components of the agroecosystem.

An integrated approach that is based on pest densities and their relation to economic injury thresholds will ultimately be required for each cropping system and location before agriculture will be truly sustainable. When selective insecticides are used, the negative impact on beneficial insects is reduced. Biopesticides provide an alternative means of control that further minimizes impacts on beneficials and other non-target organisms (NTOs). This is due to the specific nature of many microbial control agents. Safety testing data for entomopathogens indicate that they are generally safe for most NTOs, especially vertebrates (Laird *et al.*, 1990; Lacey and Siegel, 2000; Akhurst and Smith, 2002; Hokkanen and Hajek, 2003). However, it will be necessary to determine their effects on the beneficial organisms under the specific conditions in each agroecosystem.

The way in which entomopathogens are utilized, *i.e.* augmentation, inoculative introduction (classical biological control) or conservation, will depend on the characteristics of the pest and the fruit crop in which it causes damage or yield loss. Fruit crops are stable agroecosystems where any of the above strategies for pathogen use could be considered. In addition to the use of commercially available biopesticides, it may be useful to consider employing native entomopathogens. Surveys should be undertaken in different agro-ecological zones to identify prevailing environmental conditions and the presence of native pathogens and natural enemies that may be better suited for the targeted location than an exotic species or strain (Dolinski and Moino Jr, 2006). On the other hand, an exotic pest may require importation of natural enemies from its native range. In classical biological control, natural enemies, including entomopathogens, are sought in the region of origin of the invasive pest, imported and established in an area where they do not naturally occur. Typically this is a geographic area where the pest has invaded without its natural enemies and there are no effective native natural enemies in the invaded region.

When microbial control agents are formulated as biopesticides, they are predominantly used for inundative applications and often treated much like chemicals, with the expectations that they will perform at the same standards. In general, this has not always been possible. On the other hand, there are biological control agents capable of doing what chemicals are not able to do, *i.e.* EPNs that have a capacity to find their pest host, kill it and reproduce in it. In fact, many entomopathogens have the capacity to reproduce in the host and hence produce secondary inoculum able to attack and kill other individuals in the pest population. This numerical increase response, of which chemicals are incapable, needs to be better exploited in tropical conditions. Several other advantages of entomopathogens over chemicals are presented by Alves (1998c), Lacey *et al.* (2001) and Kaya and Lacey (2007).

The cost of producing natural enemies must be judged in terms of the value of the crop protected by using the agent and in comparison to the cost of competing control options such as chemicals (Driesche and Bellows ,1996). In Europe the costs of biological control agents used in protected crops and horticulture have proven to be economic and comparable to chemicals. In Florida, the use of EPNs is an integral part of IPM in citrus indicating that the benefit/cost relationship is positive. There are crops that have few or no registered pesticides and consumers who prefer to buy pesticide-free produce. In those cases, microbial control is strongly supported.

18.5. INTERACTION OF ENTOMOPATHOGENS AND OTHER BIOLOGICAL CONTROL AGENTS

Parasitoids and predators can interact synergistically/additively (*e.g.*, enhanced transmission and dispersal of insect pathogens) or antagonistically (*e.g.*, parasitism/infection, predation and competition) with entomopathogens. In most studies examining the interaction between entomopathogens and other natural enemies, the pathogen almost always dictates the population dynamics of other guild members (Brooks, 1993; Begon *et al.*,1999). Most studies indicate the positive nature of these interactions with respect to the control of insect populations (Brooks, 1993; Begon *et al.,* 1999; Roy and Pell, 2000). Various studies have shown the capacity of parasitoids to identify and avoid oviposition in hosts infected by the different entomopathogen groups (Brooks, 1993). This rejection is usually due to visual changes that occur in hosts and/or chemical cues associated with biochemical changes in hosts late during the disease development. As for predators, this does not appear to be true (Wraight, 2003; Koppenhöfer and Grewal, 2005). The combination of EPNs with other nematode species, fungi and viruses often results in additive effects on pest mortality, whereas nematode-bacteria interactions range from antagonistic to synergistic (Koppenhöfer and Grewal, 2005).

18.6. INTERACTION OF ENTOMOPATHOGENS AND CHEMICAL PESTICIDES

Unlike parasitoid, entomopathogens are generally compatible with chemical pesticides (Croft, 1990). Exceptions include use of several fungicides and most nematicides with entomopathogenic fungi and EPNs, respectively. Certain combinations of entomopathogens with chemical pesticides can be synergistic, such as reported by Koppenhöfer *et al.* (2000) for sublethal concentrations of imidacloprid and fungi or EPNs, respectively. The economic feasibility of such combinations will depend on how much of each component can be reduced compared with their recommended application rates when they are applied individually. The compatibility of chemical pesticides with arthropod natural enemies will be another consideration when integrating pesticides and entomopathogens into a pest management program.

18.7. Ecological Engineering and Manipulation of Environment to Enhance Activity and Persistence of Entomopathogens

Ecological engineering in the context of biological control and IPM is the manipulation of agricultural habitats to be less favorable for arthropod pests and more attractive to beneficial organisms (Gurr *et al.,* 2004; Pell, 2007). Under optimal environmental conditions many entomopathogens have the natural ability to cause disease at epizootic levels due to their persistence in the environment and efficient transmission. When an insect pathogen is capable of becoming established in an environment it also has the potential to confer long-term regulation of a pest population.

In order for a disease to become epizootic in an arthropod population three factors are required: presence of the host; presence of the pathogen; and proper environmental conditions (Ignoffo, 1985; Pell, 2007). Habitat manipulation techniques have been employed to optimize environmental conditions to increase entomopathogen (endemic, inoculative release, or inundatively applied) activity and facilitate their persistence. These include augmenting moisture in the habitat and creating habitat refuges such as mulches, hedgerows or grass banks, which maintain soil humidity and temperature favorable to microbial activity, propagation, and survival. Manipulated habitats have also been used to provide host plants for alternate arthropod hosts and nectar sources for parasitoids and other natural enemies.

Microbial agents are susceptible to ultra-violet light, heat, and desiccation, but the effects will vary with the microbial species or strain and habitat. Some are capable of remaining viable for just hours (*e.g.*, non-occluded viruses) or days (*e.g.* fungal conidia, occluded viruses). Others may persist for months (*e.g.,* EPNs, fungal resting spores) or years (*e.g.,* *Paenibacillus* spp., fungal resting spores). In addition to habitat manipulation, formulation of pathogens with humectants, nutrient sources, and UV protectants has been used to enhance persistence under field conditions (Burges, 1998).

18.8. Availability of Entomopathogens as a Limiting Factor

Factors that favor accelerated growth and use of biopesticides are their improved performance and cost competitiveness in the face of increasing insect resistance to chemical insecticides, environmental hazards and lack of selective chemical pesticides. Recently, major agrochemical companies have taken a greater interest in microbial pesticides. In all, 281 biopesticides were available on the market in 1993, with active ingredients of bacteria, EPNs, fungi, and viruses (Lisansky and Coombs, 1994). Although the market for microbial insecticides is growing, it represents less than 1-1.5% of the total crop protection market (Lacey *et al.,* 2001).

Many microbial control agents can be produced on artificial media using fairly simple methods and there are potentially ample markets for them. However, the bottleneck in the use of microbial control in many countries is their local production and availability. Although there are successful examples of important pests being controlled with biopesticides, many microbial control agents are not universally available to growers.

CONCLUSIONS

Sustainable agriculture will rely increasingly on alternatives to conventional chemical insecticides for pest management that are environmentally friendly and reduce the amount of human contact with hazardous pesticides. Microbial control of arthropod pests of tropical fruits, in conjunction with other IPM components, can provide effective control. The challenge will be to find successful combinations of entomopathogens, predators, and parasitoids along with other interventions. Aspects that warrant further study and attention are improved formulation, storage, marketing, and transfer of technology to growers.

REFERENCES

Agência Nacional de Vigilância Sanitária – ANVISA. 2004. www.anvisa.gov.br. Acessed in June 2009.

Akhurst, R. and K. Smith. 2002. Regulation and safety, p.311-332. In R. Gaugler (Ed.), *Entomopathogenic nematology.* New York, CABI Publishing, 388p.

Alfiler, A.R.R. 1992. Current status of the use of a baculovirus in *Oryctes rhinoceros* control in the Philippines, p. 261-268. In T.A. Jackson and T.R. Glare (Eds.). *Use of pathogens in scarab pest management.* Andover, Intercept, 298p.

Alves, S.B and R.B. Lopes (Eds.). 2008. Controle Microbiano de Pragas na América Latina: Avanços e Desafios. Piracicaba, Brazil: Biblioteca de Ciências Agrárias Luiz de Queiroz. 414 pp.

Alves, S.B, M.A. Tamai, L.S. Rossi and E. Castiglioni. 2005. *Beauveria bassiana* pathogenicity to the citrus rust mite *Phyllocoptruta oleivora. Exp. Appl. Acarol.* 37: 117-122.

Alves, S.B. (Ed.) 1998a. Controle microbiano de insetos. 2nd ed. Piracicaba, Brasil, Fundação de Estudos Agrários Luiz de Queiroz, 1163p.

Alves, S.B. 1998b. Fungos entomopatogênicos, p.289-381. In S.B. Alves (Ed.), *Controle microbiano de insetos.* Piracicaba, Brasil, Fundação de Estudos Agrários Luiz de Queiroz, 1163p.

Alves, S.B. 1998c. Patologia e controle microbiano: Vantagens e desvantagens, p.21-37. In S.B. Alves (Ed.), *Controle microbiano de insetos.* Piracicaba, Brasil, Fundação de Estudos Agrários Luiz de Queiroz, 1163p.

Alves, S.B., R.B. Lopes, S.A. Viera and M.A. Tamai. 2008. Fungos entomopatogênicos usados no controle de pragas na América Latina. In: Alves S.B. and Lopes R.B. (Eds.).

Controle Microbiano de Pragas na América Latina: Avanços e Desafios. Piracicaba, Brazil: Biblioteca de Ciências Agrárias Luiz de Queiroz. pp. 69-110.

Alves, S.B., R.M. Pereira, R.B. Lopes and M.A. Tamai. 2002. Use of entomopathogenic fungi in Latin America, p.193-211. In R.K. Upadhyay (Ed.), *Advances in microbial control of insect pests*. Dordrecht, The Netherlands, Kluwer Academic Publishers, 340p.

Arthurs, S., L.A. Lacey and R. Fritts, Jr. 2005. Optimizing the use of the codling moth granulovirus: effects of application rate and spraying frequency on control of codling moth larvae in Pacific Northwest apple orchards. *J. Econ. Entomol.* 98: 1459-1468.

Asser-Kaiser S., E. Fritsch, K. Undorf-Spahn, J. Kienzle, K.E. Eberle, N.A. Gund, A. Reineke, C.P.W. Zebitz, D.G. Heckel, J. Huber, J.A. Jehle. 2007. Rapid emergence of baculovirus resistance in codling moth due to dominant, sex linked inheritance. *Science* 317:1916-1918.

Bailez, O.E., A.M. Viana-Bailez, J.O.G. Lima and D.D.O Moreira. 2003. Life-history of the guava weevil, *Conotrachelus psidii* Marshall (Coleoptera: Curculionidae), under laboratory conditions. *Neotrop. Entomol.* 32: 203-207.

Barnes, M.M. 1991. Tortricids in pome and stone fruits, codling moth occurrence, host race formation and damage, in *Tortricid Pests, Their Biology, Natural Enemies and Control*, eds. L.P.S. van der Geest, and H.H. Evenhuis, Amsterdam, The Netherlands: Elsevier Science, pp. 313-27.

Beattie, G.A., V. Somsook, D.M. Watson, A.D. Clift and L. Jiang. 1995. Field evaluation of *Steinernema carpocapsae* (Weiser) (Rhabditida: Steinernematidae) and selected pesticides and enhancers for control of *Phyllocnistis citrella* Stainton (Lepidoptera: Gracillariidae). *J. Austr. Entomol. Soc.* 34: 335-342.

Beavers, J.B. and C.O. Calkins. 1984. Susceptibility of *Anastrepha suspensa* (Diptera: Tephritidae) to steinernematid and heterorhabditid nematodes in laboratory studies. *Environ. Entomol.* 13: 137-139.

Bedford, G.O. 1980. Biology, ecology, and control of palm rhinoceros beetles. *Annu. Rev. Entomol.* 25: 309-339.

Bedford, G.O. 1981. Control of rhinoceros beetle by Baculovirus, p. 409-426. In H.D. Burges (Ed.), *Microbial control of pests and plant diseases* 1970-1980. London, UK, Academic Press, 949p.

Begon, M., S.M. Sait and D.J. Thompson. 1999. Host-pathogen-parasitoid systems, p. 327-348. In B.A. Hawkins and H.V. Cornell (Eds.), *Theoretical approaches to biological control.* Cambridge, Cambridge University Press, 424p.

Blommers, L., F. Vaal, J. Freriks, H. Helsen. 1987. Three years of specific control of summer fruit tortrix and codling moth on apple in the Netherlands. *J. Appl. Entomol.* 104:353-71

Blommers, LHM. 1994. Integrated pest management in European apple orchards. *Annu. Rev. Entomol.* 39:213-41.

Boscán de Martinez, N. and R. Cásares. 1982. Distribuicion en el tiempo de las fases del gorgojo de la guayaba *Conotrachelus psidii* Marshall (Coleoptera: Curculionidae) en el campo. *Agro Tropic* 31: 123-130.

Boucias, D.G., J.M. Meyer, S. Popoonsakand, S.E. Breaux. 2007. The Genus *Hirsutella*: A Polyphyletic Group of Fungal Pathogens Infesting Mites and Insects, in *Use of Entomopathogenic Fungi in Biological Pest Management*, eds. S. Ekesi and N.K. Maniania, Kerala, India: Research Signpost, pp. 57-90.

Briceno, A.J. 1997. Perspectivas de un manejo integrado del gusano verde del platano, *Opsiphanes tamarindi* Felder (Lepidoptera: Brassolidae). Rev. Fac. Agron. Universidad del Zulia 14: 487-495.

Brooks, W.M. 1993. Host-parasitoid-pathogen interactions, p.231-272. In N.E. Beckage, S.N. Thompson and B.A. Federici (Eds.), *Parasites and pathogens of insects*. Vol. 2, Pathogens. San Diego, Academic Press, 294p.

Bullock, R.C., R.R. Pelosi and E.E. Killer. 1999. Management of citrus root weevils (Coleoptera: Curculionidae) on Florida citrus with soil-applied entomopathogenic nematodes (Nematoda: Rhabditida). *Fla. Entomol.* 82: 1-7.

Burges, H.D. 1998. Formulation of Microbial Biopesticides: Beneficial Microorganisms, Nematodes and Seed Treatments. Dordrecht, Kluwer Academic Publishers, 412p.

Cabrera, R.I., D. Domínguez and J.J. Blanco. 1987. Informe sobre *Hirsutella nodulosa*, enemigo natural del acaro blanco *Polyphagotarsonemus latus*. *Comun. Cien. Tech. Agr. Citr. Frutales* 10: 135-138.

Castillo, M.A., P. Moya, E. Hernandez and E. Primo-Yufera. 2000. Susceptibility of *Ceratitis capitata* Wiedemann (Diptera: Tephritidae) to entomopathogenic fungi and their extracts. *Biol. Control* 19: 274-282.

Childers, C.C., J.V. French, and J.C. Rodrigues. 2003. *Brevipalus californicus*, *B. obovatus*, *B. phoenicis* and *B. lewisi* (Acari: Tenuipalpidae): a review of their biology, feeding injury and economic importance. *Exp. Appl. Acarol.* 30: 5-28.

Cory, J.S. and H.F. Evans. 2007. Viruses, p. 149-174. In L.A. Lacey and H.K. Kaya (Eds.), *Field manual of techniques in invertebrate pathology: Application and evaluation of pathogens for control of insects and other invertebrate pests,* 2nd edition. Dordrecht, Springer Scientific Publishers.

Croft, B.A. 1990. Arthropod Biological Control Agents and Pesticides. New York ,Wiley and Sons., 723p.

Dekle, G.W. and T.R. Fasulo. 2001. Green Scale, *Coccus viridis* (Green) (Insecta: Hemiptera: Coccidae). University of Florida, Institute of Food and Agricultural Sciences document EENY-253 (IN436). 3 p.

Del Valle, E.E., C. Dolinski and R.M. Souza. 2008b. Dispersal of *Heterorhabditis baujardi* LPP7 (Nematoda: Rhabditida) applied to the soil as infected host cadavers. *Inter. J. Pest Management* 54: 115–122.

Del Valle, E.E., C. Dolinski, E.L.S. Barreto, R.M. Souza and R.I. Samuels. 2008a. Efficacy of *Heterorhabditis baujardi* LPP7 (Nematoda: Rhabditida) applied in *Galleria mellonella* (Lepidoptera: Pyralidae) insect cadavers to *Conotrachelus psidii*, (Coleoptera: Curculionidae) larvae. *Biocontrol Sci. Tech.*, 18: 33-41.

Denholm, I. and M.W. Rolland. 1992. Tactics for managing pesticide resistance in arthropods: Theory and practice. *Annu. Rev. Entomol.* 37: 92-112.

Dent, D. 2000. Insect pest management. 2nd edition, Ascot, CABI Publishing, 399 p.

Dias, C., P. Garcia, N. Simões and L. Oliveira. 2005. Efficacy of *Bacillus thuringiensis* against *Phyllocnistis citrella* (Lepidoptera: Phyllocnistidae). *J. Econ. Entomol.* 98: 1880-1883.

Dimbi, S., N.K. Maniania, S.A. Lux and J.M. Mueke. 2003b. Host species, age and sex as factors affecting the susceptibility of the African tephritid fruit fly species, *Ceratitis capitata, C. cosyra* and *C. fasciventris* to infection by *Metarhizium anisopliae*. *Anzeiger Schadlingskunde* 76: 113-117.

Dimbi, S., N.K. Maniania, S.A. Lux, S. Ekesi and J.K. Mueke. 2003a. Pathogenicity of *Metarhizium anisopliae* (Metsch.) Sorokin and *Beauveria bassiana* (Balsamo) Vuillemin, to three adult fruit fly species: *Ceratitis capitata* (Weidemann), *C. rosa* var. *fasciventris* Karsch and *C. cosyra* (Walker) (Diptera: Tephritidae). *Mycopathologia* 156: 375-382.

Dolinski, C. and A. Moino Jr. 2006. Utilização de nematóides entomopatogênicos nativos ou exóticos: O perigo das introduções. *Nematol. Bras.* 30: 139-149.

Dolinski, C. 2006. Developing a research and extension program for control of the guava weevil in Brazil using entomopathogenic nematodes. *J. Nematol.* 38: 270.

Dolinski, C.M., E.E. del Valle and R. Stuart. 2006. Virulence of entomopathogenic nematodes to larvae of the guava weevil *Conotrachelus psidii* (Coleoptera: Curculionidae) in laboratory and greenhouse experiments. *Biol. Control* 38: 422-427.

Downing, A.S., C.G. Erickson and M.J. Kraus. 1991. Field evaluation of entomopathogenic nematodes against citrus root weevils (Coleoptera: Curculionidae) in Florida citrus. *Fla. Entomol.* 74: 584-586.

Driesche, V.R.G. van and T.S. Bellows Jr. 1996. Biological control. New York, Chapman and Hall, 539p.

Eberle, K.E., J.A. Jehle. 2006. Field resistance of codling moth against *Cydia pomonella* granulovirus (CpGV) is autosomal and incompletely dominant inherited. *Journal of Invertebrate Pathology* 93:201-206.

Eilenberg, J., and J.K. Pell. 2007. 'Ecology' in *Arthropod-Pathogenic Entomophthorales: Biology, Ecology, Identification*, ed. S. Keller, COST office, Luxemburg, ISBN 978-92-898-0037-2, pp. 7-26.

Ekesi, S. and N.K. Maniania (Eds.). 2007. Use of Entomopathogenic Fungi in Biological Pest Management. Research Signpost. Kerala, India.

Ekesi, S., N.K. Maniania and S.A. Lux. 2002. Mortality in three African tephritid fruit fly puparia and adults caused by the entomopathogenic fungi, *Metarhizium anisopliae* and *Beauveria bassiana. Biocontrol Sci. Technol.* 12: 7-17.

Ekesi, S., N.K. Maniania and S.A. Lux. 2003. Effect of soil temperature and moisture on survival and infectivity of *Metarhizium anisopliae* to four tephritid fruit fly puparia. *J. Invertebr. Pathol.* 83: 157-167.

Ekesi, S., N.K. Maniania, S.A. Mohamed and S.A. Lux . 2005. Effect of soil application of different formulations of *Metarhizium anisopliae* on African tephritid fruit flies and their associated endoparasitoids. *Biol. Control* 35: 83-91.

El-Choubassi, W., M.A. Iparraguirre-Cruz, M.L. Sisne-Luis and H. Grillo-Ravelo. 2001. Incidencia del genero *Aschersonia* sobre la población de *Parlatoria ziziphi* (Lucas) (Homoptera: Diaspididae) en naranjo Valencia (*Citrus sinensis*) de la provincia de Ciego de Avila. Centro Agricola 28: 42-45.

Elizondo, J.M. and J.R. Quezada. 1990. Identificación y evaluación de los enemigos naturales de la mosca prieta de los citricos *Aleurocanthus woglumi* Ashby (Homoptera: Aleyrodidae) en cuatro zonas citricolas de Costa Rica. Turrialba 40: 190-197.

Entwistle, P.F. 1987. Virus diseases of Limacodidae, pp. 213-221. In M.J.W. Cock, C.H. Godfray, and J.D. Holloway (Eds.), *Slug and Nettle Caterpillars in South East Asia.* CAB International, Wallingford.

Faria, M. and S.P. Wraight. 2001. Biological control of *Bemisia tabaci* with fungi. *Crop Prot.* 20: 767-778.

Fawcett, H.S. 1944. Fungus and bacterial diseases of insects as factors in biological control. *Botan. Rev.* 10: 327-348.

Fazolin, M. 2001. Reconhecimento e manejo integrado das principais pragas da cultura do abacaxi no Estado do Acre. Embrapa Acre. Documentos n° 62, 26p.

Fédière, G., R. Phillippe, J.C. Veyrunes and P. Monsarrat. 1990. Biological control of the oil palm pest *Latoia viridissima* (Lepidoptera, Limacodidae) in Cote D'Ivoire, by a new Picornavirus. *Entomophaga* 35: 347-354.

Figueroa, W. 1990. Biocontrol of the banana root borer weevil, *Cosmopolites sordidus* (Germar), with steinernematid nematodes. *J. Agric. Univ. Puerto Rico* 74: 15-19.

Fransen, J.J. 1990. Natural enemies of whiteflies: Fungi, p.187-210. In D. Gerling, (Ed.), *Whiteflies: Their bionomics, pest status and management.* Andover, UK, Intercept, 348p.

Fredrick, J.M. 1943. Some preliminary investigations of the green scale, *Coccus viridis* (Green), in south Florida. *Florida Ent.* 26:12-15, 25-29.

Fritsch, E. 1988. Biologische bekampfung des falschen apfelwicklers, *Cryptophlebia leucotreta* (Meyrick) (Lep., Tortricidae), mit granuloseviren. *Mitt. Dtsch. Ges. Allg. Ang. Entomol.* 6: 280-283.

Fritsch, E., K. Undorf-Spahn, J. Kienzle, C.P.W. Zebitz, J. Huber. 2005. Apfelwickler Granulovirus: Erste Hinweise auf Unterschiede in der Empfindlichkeit lokaler Apfelwickler Populationen. Nachrichtenblat des Deutschen Pflanzenschutzdienstes 57:29-34.

Gao, R.X., Z. Ouyang, Z.X. Gao and J.X. Zheng. 1985. A preliminary report on the application of *Aschersonia aleyrodis* for the control of citrus whitefly. *Chinese J. Biol. Control* 1: 45-46. (in Chinese).

Gazit, Y., Y. Rössler, and I. Glazer. 2000. Evaluation of entomopathogenic nematodes for the control of Mediterranean fruit fly (Diptera: Tephritidae). *Biocontrol Sci. Technol.* 10: 157-164.

Georgis, R., A.M. Koppenhöfer, L.A. Lacey, G. Bélair, L.W. Duncan, P.S. Grewal, M. Samish, L. Tan, P. Torr and R.W.H.M. van Tol. 2006. Successes and failures in the use of parasitic nematodes for pest control. *Biol. Control* 38: 103-123.

Giblin-Davis, R.M., J.E. Peña and R .E. Duncan. 1994. Lethal pitfall trap for evaluation of semiochemical-mediated attraction of *Metamasius hemipterus sericeus* (Coleoptera: Curculionidae). *Fla. Entomol.* 77: 247-255.

Giblin-Davis, R.M., J.E. Peña and R .E. Duncan. 1996. Evaluation of an entomopathogenic nematode and chemical insecticides for control of *Metamasius hemipterus sericeus* (Coleoptera: Curculionidae). *J. Entomol. Sci.* 31: 240-251.

Glare, T.R. and M.O'Callaghan. 2000. *Bacillus thuringiensis*: Biology, Ecology and Safety. John Wiley and Sons, Chichester, West Sussex, UK. 350 pp.

Godonou, I., K.R. Green, K.A. Oduro, C.J. Lomer and K. Afreh-Nuamah. 2000. Field evaluation of selected formulations of *Beauveria bassiana* for the management of the banana weevil (*Cosmopolites sordidus*) on plantain (*Musa* spp., AAB Group). *Biocontrol Sci. Technol.* 10: 779-788.

Goettel, M.S., J. Eilenberg and T.R. Glare. 2005. Entomopathogenic fungi and their role in regulation of insect populations, p.361-406. In L.I. Gilbert, K. Iatrou and S. Gill (Eds.), *Comprehensive molecular insect science*, v. 6. Oxford, Elsevier Pergamon, 470p.

Gold, C.S., B. Pinese and J.E. Peña. 2002. Pests of banana, p. 13-56. In J.E. Peña, J.L. Sharp and M. Wysoki (Eds.), *Tropical fruit pests and pollinators: biology, economic importance, natural enemies, and control*. Wallingford, UK, CABI Publishing, 448p.

Gold, C.S., C. Nankinga, B. Niere and I. Godonou. 2003. IPM of banana weevil in Africa with emphasis on microbial control, p. 243-257. In P. Neuenschwander, C. Borgemeister and J. Langewald (Eds.), *Biological control in IPM systems in Africa*. Wallingford, UK, CABI Publishing, 400p.

Gold, C.S., J.E. Peña and E.B. Karamura. 2001. Biology and integrated pest management for the banana weevil *Cosmopolites sordidus* (Germar) (Coleoptera: Curculionidae). *Int. Pest Manage. Rev.* 6: 79-155.

Gopalakrishnan, C. and P.N. Gangavisalakshy. 2005. Field efficacy of commercial formulations of *Bacillus thuringiensis* var. *kurstaki* against *Papilio demoleus* L. on citrus. Entomon 30: 93-95.

Gould, W.P. and A. Raga. 2002. Pests of guava, p.295-313. In J.E. Peña, J.L. Sharp and M. Wysoki (Eds.), *Tropical fruit pests and pollinators*. Wallingford, UK, CABI Publishing, 400p.

Graham, J.H., D.B. Bright and C.W. McCoy. 2003. *Phytophthora-Diaprepes* weevil complex: *Phytophthora* spp. relationship to citrus rootstocks. *Plant Dis.* 87: 85-90.

Gravena, S., R.R. Leão-Neto, F.C. Moretti and G. Tozatti. 1988. Eficiência de inseticidas sobre *Selenaspidus articulatus* (Morgan) (Homoptera, Diaspididae) e efeito sobre inimigos naturais em pomar citrico. *Científica* 16: 209-217.

Grim, C and F. Guharay. 1998. Control of leaf-footed bug *Leptoglossus zonatus* and shield-backed bug *Pachycoris klugii* with entomopathogenic fungi. *Biocontrol Sci. Tech.* 8: 356-376.

Gurr, G.M., S.L. Scarrat, S.D. Wratten, L. Berndt and N. Irvin. 2004. Ecological engineering, habitat manipulation and pest management, p.1-12. In G.M. Gurr, S.D. Wratten, and M.A. Altieri and D. Pimentel (Eds.), *Ecological engineering for pest management: Advances in habitat manipulation for arthropods*. Collingwood, Australia, CSIRO Publishing, 232p.

Hall R.A. (1981), The fungus *Verticillium lecanii* as a microbial insecticide against aphids and scales, in *Microbial Control of Pests and Plant Diseases 1970-1980*, ed. H.D. Burges, London, UK: Academic Press. pp. 483-498.

Hirt, R.P., J.M. Logsdon Jr., B. Healy, M.W. Dorey, W.F. Doolittle and T.M. Embley. 1999. Microsporidia are related to fungi: Evidence from the largest subunit of RNA polymerase II and other proteins. *Proc. Natl. Acad. Sci. USA* 96: 580-585.

Hokkanen, H.M.T. and A.E. Hajek (Eds.). 2003. Environmental impacts of microbial insecticides: Need and methods for risk assessment. Dordrecht, Kluwer Academic Publishers, 269p.

Huffaker, C.B. 1985. Biological control in integrated pest management: An entomological perspective, p.13-23. In M.A. Hoy and D.C. Herzog (Eds.), *Biological control in agricultural IPM systems*. San Diego, CA, Academic Press, 589p.

Huger, A.M. 1966. A virus disease of the Indian rhinoceros beetle *Oryctes rhinoceros* caused by a new type of insect virus *Rhabdinovirus oryctes* gen. n., sp. n. *J. Invertebr. Pathol.* 8: 38-51.

Huger, A.M. 1973. Grundlagen zur biologischen Bekämpfung des Indischen Nashornkäfers, *Oryctes rhinoceros* (L.), mit *Rhabdionvirus oryctes*: *Histopathologie der Virose bei Kafern. Zeitschr. Angew. Ento*mol. 72: 309-319.

Huger, A.M. 2005. The Oryctes virus: Its detection, identification, and implementation in biological control of the coconut palm rhinoceros beetle, *Oryctes rhinoceros* (Coleoptera: Scarabaeidae). *J. Invertebr. Pathol.* 89: 78-84.

Humber, R. 1997. Fungi: Identification, p.153-185. In L.A. Lacey (Ed.), *Manual of techniques in insect pathology.* London, Academic Press, 409p.

Hunter-Fujita, F.R., P.F. Entwistle, H.F. Evans and N.E. Crook. 1998. Insect viruses and pest management. Chichester, *John Wiley and Sons*, 632p.

Ignoffo, C.M. 1985. Manipulating enzootic-epizootic diseases of arthropods, p.243-262. In M.A. Hoy and D.C. Herzog (Eds.), *Biological control in agricultural IPM systems.* San Diego, Academic Press, 589p.

Inglis, G.D., M.S. Goettel, T.M. Butt and H. Strasser. 2001. Use of hyphomycetous fungi for managing insect pests, p.27-69. In T.M. Butt, C. Jackson and N. Magan (Eds.), *Fungi as biocontrol agents: Progress, problems and potential.* Wallingford, UK, CABI Publishing, 350.

Jackson, T.A. and T.R. Glare (Eds.). 1992. Use of pathogens in scarab pest management, Andover, Intercept, 298p.

Jackson, T.A., A.M. Crawford and T.R. Glare. 2005. Oryctes virus - time for a new look at a useful biocontrol agent. *J. Invertebr. Pathol.* 89: 91-94.

Jenkins, D.A., D.I. Shapiro-Ilan and R. Goenaga. 2008. Efficacy of entomopathogenic nematodes versus *Diaprepes abbreviatus* (Coleoptera: Curculionidae) larvae in a high clay content Oxisol soil: Greenhouse trials with potted *Litchi chinensis*. Florida Entomologist 91, 75-78.

Jones, K.A., B. Zelazny, U. Ketunuti, A. Cherry and D. Grzywacz. 1998. South-east Asia and the western Pacific, p.244-257. In F.R. Hunter-Fujita, P.F. Entwistle, H.F. Evans and N.E. Crook (Eds.), *Insect viruses and pest management.* Chichester, *J. Wiley and Sons*, 620p.

Joubert, P.H. and T.I. Labuschagne. 1995. Alternative measures for controlling mango seed weevil, *Sternochetus mangiferae* (F.). Yearbook S. Afr. Mango Grow. Assoc. 15: 94-96.

Kaya, H.K. and L.A. Lacey. 2007. Introduction to microbial control, p.3-7. In L.A. Lacey and H.K. Kaya (Eds.), *Field manual of techniques in invertebrate pathology: application and evaluation of pathogens for control of insects and other invertebrate pests*, 2nd edition. Dordrecht, Springer Scientific Publishers.

Kaya, H.K. and R.Gaugler. 1993. Entomopathogenic nematodes. *Annu. Rev. Entomol.* 38: 181-206.

Kaya, H.K. and S.P. Stock. 1997. Techniques in insect nematology, p.281-324. In L. A. Lacey, (Ed.), *Manual of techniques in insect pathology*, Academic Press, London. 409 p.

Kaya, H.K., J.L. Joos, L.A. Falcon and A. Berlowitz. 1984. Suppression of the codling moth (Lepidoptera: Olethreutidae) with the entomogenous nematode, *Steinernema feltiae* (Rhabditida: Steinernematidae). *J. Econ. Entomol.* 77:1240–44

Kermarrec, A. and H. Mauleon. 1989. Synergie entre le chlordecone et *Neoaplectana carpocapsae* Weiser (Nematoda; Steinernematidae) pour le controle de *Cosmopolites sordidus* (Coleoptera: Curculionidae). *Rev. Nematol.* 12: 324-324.

Klein, M.G., P.S. Grewal, T.A. Jackson and A.M. Koppenhöfer. 2007. Lawn, turf and grassland pests, p.655-675. In L.A. Lacey and H.K. Kaya (Eds.), *Field manual of techniques in invertebrate pathology: Application and evaluation of pathogens for control of insects and other invertebrate pests,* 2nd edition. Dordrecht, Springer Scientific Publishers.

Konstantopoulou, M.A. and B.E. Mazomenos. 2005. Evaluation of *Beauveria bassiana* and *B. brongniartii* strains and four wild-type fungal species against adults of *Bactrocera oleae* and *Ceratitis capitata. BioControl* 50: 293-305.

Koppenhöfer, A.M. and P.S. Grewal. 2005. Compatibility and interaction with agrochemicals and other biocontrol agents, p.363-381. In P.S. Grewal, R.-U. Ehlers and D.I. Shapiro-Ilan (Eds.), *Nematodes as biocontrol agents.* Ascot, CABI Publishing, 505p.

Koppenhöfer, A.M., I.M. Brown, R. Gaugler, P.S. Grewal, H.K. Kaya and M.G. Klein. 2000. Synergism of entomopathogenic nematodes and imidacloprid against white grubs: Greenhouse and field evaluation. *Biol. Control* 19: 245-251.

Kunjeku, E., K.A. Jones and G.M. Moawad. 1998. Africa, the Near and Middle East, p. 280-302. In F.R. Hunter-Fujita, P.F. Entwistle, H.F. Evans, and N.E. Crook (Eds.), *Insect viruses and pest management.* Chichester, *J. Wiley and Sons,* 620p.

Laborda, R., L. Bargues, C. Navarro, O. Barajas, M. Arroyo, E.M., Garcia, E. Montoro, E. Llopis, A. Martinez and J.M. Sayagues. 2003. Susceptibility of the Mediterranean fruit fly (*Ceratitis capitata*) to entomopathogenic nematode *Steinernema* spp. ("Biorend C"). Bull. OILB/SROP 26: 95-97.

Lacey, L.A. and D.I. Shapiro-Ilan. 2008. Microbial control of insect pests in temperate orchard systems: potential for incorporation into IPM. *Annu. Rev. Entomol.* 53: 121-144.

Lacey, L.A. and H.K. Kaya. 2007. Field manual of techniques in invertebrate pathology: application and evaluation of pathogens for control of insects and other invertebrate pests, 2nd edition. Dordrecht, Springer Scientific Publishers. 868 pp.

Lacey, L.A. and J.P. Siegel. 2000. Safety and ecotoxicology of entomopathogenic bacteria, p.253-273. In J.F. Charles, A. Delécluse and C. Nielsen-LeRoux (Eds.), *Entomopathogenic bacteria: From laboratory to field application.* Dordrecht, Kluwer Academic Publishers, 524p.

Lacey, L.A., D. Thomson, C. Vincent and S.P. Arthurs. 2008a. Codling Moth Granulovirus: a comprehensive review. *Biocontrol Sci. Technol.* 1-25.

Lacey, L.A., J.J. Fransen and R. Carruthers. 1996. Global distribution of naturally occurring fungi of *Bemisia,* their biologies and use as biological control agents, p.401-433. In D. Gerling and R. Mayer (Eds.), *Bemisia* 1995: Taxonomy, biology, damage, control and management. Andover, Intercept, 702p.

Lacey, L.A., L.G. Neven, H.L. Headrick and R. Fritts, Jr. 2005. Factors affecting entomopathogenic nematodes (Steinernematidae) for the control of overwintering codling moth (Lepidoptera: Tortricidae) in fruit bins. *J. Econ. Entomol.* 98: 1863-1869.

Lacey, L.A., R. Frutos, H.K. Kaya and P. Vail. 2001. Insect pathogens as biological control agents: do they have a future? *Biol. Control* 21: 230-248.

Lacey, L.A., S.P. Arthurs, A. Knight and J. Huber. 2007. Microbial control of Lepidopteran pests of apple orchards. In L.A. Lacey and H.K. Kaya (Eds.) *Field Manual of Techniques in Invertebrate Pathology: Application and Evaluation of Pathogens for Control of Insects and Other Invertebrate Pests,* Dordrecht: Springer, 2nd edition, pp. 527-46.

Lacey, L.A., S.P. Arthurs, T.R. Unruh, H. Headrick and R. Fritts, Jr. 2006. Entomopathogenic nematodes for control of codling moth (Lepidoptera: Tortricidae) in apple and pear orchards: effect of nematode species and seasonal temperatures, adjuvants, application equipment and post-application irrigation. *Biol. Control* 37: 214–223.

Lacey, L.A., S.P. Wraight and A.A. Kirk. 2008b. Entomopathogenic fungi for control of *Bemisia* spp.: foreign exploration, research and implementation. In J.K. **Gould, K. Hoelmer and J. Goolsby (eds).**Classical Biological Control of *Bemisia tabaci* in the USA: A Review of Interagency Research and Implementation, Vol. 4 "Progress in Biological Control" (H. Hokkanen, series editor), Springer, Dordrecht, pp. 33-69.

Laird, M., L.A. Lacey and E.W. Davidson. 1990. Safety of microbial insecticides. Boca Raton, CRC Press, 259p.

Latgé, J.P. and B. Papierok. 1988. Aphid pathogens, p.323-335. In A.K. Minks and P. Harrewijn, (Eds.), *Aphids their biology, natural enemies and control*, v. B. Amsterdam, *Elsevier Science and Technology*, 322p.

Leite, L.G., F.M. Tavares, R.M. Goulart, A. Batista Filho and J.R.P. Parra. 2005. Patogenicidade de nematóides entomopatogênicos a larvas de 6° instar do bicho-furão, *Ecdytolopha aurantiana* (Lepidoptera: Tortricidae), e avaliação de dosagens de *Heterorhabditis indica* na mortalidade do inseto. *Rev. Agric.* 80: 316-330.

Lezama-Gutiérrez, R., A. Trujillo de la Luz, J. Molina-Ochoa, O. Rebolledo-Dominguez, A. R. Pescador, M. Lopez-Edwards and M. Aluja. 2000. Virulence of *Metarhizium anisopliae* (Deuteromycotina: Hyphomycetes) on *Anastrepha ludens* (Diptera: Tephritidae): laboratory and field trials. *J. Econ. Entomol.* 93: 1080-1084.

Lezama-Gutiérrez, R., O.J. Molina, O.L. Contreras-Ochoa, M. Gonzáles-Ramírez, A. Trujillo-de la Luz and O. Rebolledo-Domínguez. 1996. Susceptibilidad de larvas de *Anastrepha ludens* (Diptera: Tephritidae) a diversos nemátodos entomopatógenos (Steinernematidae y Heterorhabditidae). Vedalia 3: 31-33.

Lindegren, J.E. and P.V. Vail. 1986. Susceptibility of Mediterranean fruit fly, melon fly, and oriental fruit fly (Diptera: Tephritidae) to the entomogenous nematode *Steinernema feltiae* in laboratory tests. *Environ. Entomol.* 15: 465-468.

Lindegren, J.E., T.T. Wong and D.O. McInnis. 1990. Response of Mediterranean fruit fly (Diptera: Tephritidae) to the entomogenous nematode *Steinernema feltiae* in field tests in Hawaii. *Environ. Entomol.* 19: 383-386.

Lisansky, S.G. and J. Coombs. 1994. Developments in the market for biopesticides. Brighton Crop Protection Conference-Pests and Diseases, v. 3, 1049-1054.

Lomer, C.J. 1986. Release of Baculovirus Oryctes into *Oryctes monoceros* populations in the Seychelles. *J. Invertebr. Pathol.* 47: 237-24.

Lomer, C.J. 1987. Infection of *Strategus aloeus* (L.) (Coleoptera: Scarabaeidae) and other Dynastinae with Baculovirus Oryctes. *Bull. Entomol. Res.* 77: 45-51.

Lorenzato, D., E.C. Chouene, J. Medeiros, A.E.C. Rodrigues and R.C.D. Pederzolli. 1997. Ocorrência e controle da broca-do-fruto-do-abacaxi *Thecla basalides* (Geyer, 1837). Pesq. Agropec. Gaúcha 3: 15-19.

Marschall, K.J. and I. Ioane. 1982. The effect of re-release of *Oryctes rhinoceros* baculovirus in the biological control of rhinoceros beetles in Western Samoa. *J. Invertebr. Pathol.* 39: 267-276.

Masarrat, H. and R.P. Srivastava. 1998. Dose-mortality relationship of entomogenous fungus, *Beauveria bassiana* (Bals.) Vuill. against mango mealy bug, *Drosicha mangiferae* (Green). *Insect Environ.* 4: 74-75.

McCoy, C.W. and T.L. Couch. 1982. Microbial control of the citrus rust mite with the mycoacaricide, Mycar. *Fla. Entomol.* 65: 116-126.

McCoy, C.W. 1996. Pathogens of eriophyoid mites, p.481-490. In E.E. Linquist, M.W. Sabelis and J. Bruim (Eds.), *Eriophyoid mites. Their biology, natural enemies and control.* Amsterdam, Elsevier Press, 820p.

McCoy, C.W., A.G. Selhime, R.F. Kanavel and A.J. Hill. 1971. Suppression of citrus rust mite populations with application of fragmented mycelia of *Hirsutella thompsonii. J. Invertebr. Pathol.* 17: 270-276.

McCoy, C.W., R.A. Samson. and D.G. Boucias. 1988. Entomogenous fungi, p.151-236. In C.M. Ignoffo and N.B. Mandava (Eds.), *Handbook of natural pesticides*, v. V: Microbial insecticides, Part A: Entomogenous protozoa and fungi. Boca Raton, Fl, CRC Press, 260p.

McCoy, C.W., R.J. Stuart, D.I. Shapiro-Ilan and L.W. Duncan. 2007. Application and evaluation of entomopathogens for citrus pest control, p.567-581. In L.A. Lacey and H.K. Kaya (Eds.), *Field manual of techniques in invertebrate pathology: Application and evaluation of pathogens for control of insects and other invertebrate pests*, 2nd Edition. Dordrecht, Springer Scientific Publishers. *In press.*

McCoy, C.W., R.J. Stuart, L.W Duncan and K. Nguyen. 2002. Field efficacy of two commercial preparations of entomopathogenic nematodes against larvae of *Diaprepes abbreviatus* (Coleoptera: Curculionidae) in alfisol type soil. *Fla. Entomol.* 85: 537-544.

Meyer J.M, M.A. Hoy and D.G. Boucias. 2008. Isolation and characterization of an *Isaria fumosorosea* isolate infecting the Asian citrus psyllid in Florida. *Journal of Invertebrate Pathology* 99: 96-102.

Meyer J.M., M.A. Hoy and D.G. Boucias DG. 2007. Morphological and molecular characterization of a *Hirsutella* species infecting the Asian citrus psyllid, *Diaphorina citri* Kuwayama (Hemiptera: Psyllidae), in Florida. *Journal of Invertebrate Pathology.* 95: 101-109.

Meyerdirk, D.E., J.B. Kreasky and W.G. Hart. 1980. Whiteflies (Aleyrodidae) attacking citrus in southern Texas with notes on natural enemies. Can. Entomol. 112: 1253-1258.

Miller, L.K. 1997. The Baculoviruses. Plenum Press, New York. 447 pp.

Minas, R.S., C. Dolinski, R.M. Souza, R. Carvalho, R. Burla. 2009. Avaliação de *Heterorhabditis baujardi* LPP7 (Rhabditida: Heterorhbaditidae) contra a Mosca-do-Mediterrâneo, *Ceratitis capitata* (Wied.) (Diptera: Tephritidae) em um Goiabal. Bioassay (in press).

Mochi, D.A., A.C. Monteiro, S.A. De Bortoli, H.O.S. Dória and J.C. Barbosa. 2006. Pathogenicity of *Metarhizium anisopliae* for *Ceratitis capitata* (Wied.) (Diptera: Tephritidae) in soil with different pesticides. *Neotrop. Entomol.* 35: 382-389.

Mohan, S., A. Sirohi and H.S. Gaur. 2004. Successful management of mango mealy bug, *Drosicha mangiferae* by *Photorhabdus luminescens*, a symbiotic bacterium from entomopathogenic nematode *Heterorhabditis indica. Int. J. Nematol.* 14: 195-198.

Monsarrat, P. and J.C. Veyrunnes. 1976. Evidence of *Oryctes* virus in adult feces and new data for virus characterization. *J. Invertebr. Pathol.* 27: 387-389.

Moore, S., W. Kirkman and P. Stephen. 2005. CRYPTOGRAN: a virus for the biological control of false codling moth. *SA Fruit J.* 3: 35-39.

Moore, S.D. 2002. Entomopathogens and microbial control of citrus pests in South Africa: A review. *SA Fruit J.* 1: 30-32.

Moore, S.D., T. Pittaway, G. Bouwer and J.G. Fourie. 2004. Evaluation of *Helicoverpa armigera* nucleopolyhedrovirus (HearNPV) for control of *Helicoverpa armigera* (Lepidoptera: Noctuidae) on citrus in South Africa. *Biocontrol Sci. Technol.* 14: 239-250.

Morse, J.G. and J.E. Lindegren. 1996. Suppression of Fuller rose beetle on citrus with *Steinernema carpocapsae*. *Fla. Entomol.* 79: 373-384.

Morton, J. 1987b. Pineapple, p.18-28. In J.F. Morton (Ed.), *Fruits of warm climates*. Miami, Florida Flair Books, 505p

Morton, J.F. 1987a. Papaya, p.336-346. In J.F. Morton (Ed.), *Fruits of warm climates*. Miami, Florida Flair Books, 505p.

Moscardi, F. 1999. Assessment of the application of baculoviruses for the control of Lepidoptera. *Annu. Rev. Entomol.* 44: 257-289.

Narayanamma, V.L. and P. Savithri. 2003. Evaluation of biopesticides against citrus butterfly, *Papilio demoleus* L. on sweet orange. *Indian J. Plant Prot.* 31: 105-106.

Oliveira, M.R.V. 1998. South America, p. 339-355. In F.R. Hunter-Fujita, P.F. Entwistle, H.F. Evans and N.E. Crook, (Eds.), *Insect Viruses and Pest Management*. Chichester, *J. Wiley and Sons*, 620p.

O'Rourke, A.D. 1996. Trends in world apple production and marketing. *Good Fruit Grower.* 47(8), pp. 46-47, 49-50.

Pantoja, A., P.A. Follett and J.A. Villanueva-Jiménez. 2002. Pests of papaya, p.131-156. In J.E. Peña, J.L. Sharp and M. Wysoki (Eds.), *Tropical fruit pests and pollinators: Biology, economic importance, natural enemies, and control*. Wallingford, UK, CABI Publishing, 440p.

Pell, J.K. 2007 Ecological approaches to pest management using entomopathogenic fungi; concepts, theory, practice and opportunities. In: Ekesi, S. and N. K. Maniania (Eds.). 2007. Use of Entomopathogenic Fungi in Biological Pest Management. Research Signpost. Kerala, India. pp. 145-177.

Peña, J.E. and F.A. Johnson. 2006. Insect management in papaya. Univ. Florida Coop. Ext. Serv. Pub. ENY-414, 5 p. http://edis.ifas.ufl.edu/BODY_IG074

Peña, J.E., A.I. Mohyuddin and M. Wysoki. 1998. A review of the pest management situation in mango agroecosystems. *Phytoparasitica* 26: 1-20.

Peña, J.E., D.G. Hall and C.W. McCoy. 2000. Natural enemies of the weevil *Diaprepes abbreviatus* (Coleoptera: Curculionidae), a serious pest of citrus in Florida. Proc. Intl. Soc. *Citriculture* 2: 785-788.

Peña, J.E., L.S. Osborne and R.E. Duncan. 1996. Potential of fungi as biocontrol agents of *Polyphagotarsonemus latus* (Acari: Tarsonemidae). *Entomophaga* 41: 27-36.

Peña, J.E., R.M. Baranowski and R.E. Litz. 1987. Life history, behavior and natural enemies of *Philephedra tuberculosa* (Homoptera: Coccidae). *Fla. Entomol.* 70: 423-427.

Peña, J.E., R.M. Gilbin-Davis and R. Duncan. 1995. Impact of indigenous *Beauveria bassiana* (Balsamo) Vuillemin on banana weevil and rotten sugarcane weevil (Coleoptera: Curculionidae) populations in banana in Florida. *J. Agric. Entomol.* 12: 163-167.

Peña, J.E., W.J. Schroeder and L.S. Osborne. 1990. Use of entomopathogenic nematodes in the families Heterorhabditidae and Steinernematodae to control banana moth (*Opogona sachari*). *Nematropica* 20: 51-55.

Petty, G.J., G.R. Stirling and D.P. Bartholomew. 2002. Pests of pineapple, p.157-195 In J. Peña, J. Sharp and M. Wysoki (Eds.), *Tropical fruit pests and pollinators: Biology, economic importance, natural enemies and control.* Wallingford, UK, CABI Publishing, 440p.

Ponomarenko, N.G., H.A. Prilepskaya, M.Y. Murvanidze and L.A. Stolyarova. 1975. *Aschersonia* against whiteflies. *Zashchita Rastenii* 6: 44-45.

Poprawski, T.J., P.E. Parker and J.H. Tsai. 1999. Laboratory and field evaluation of hyphomycete insect pathogenic fungi for control of brown citrus aphid. *Environ. Entomol.* 28: 315-321.

Purrini, K. 1989. *Baculovirus oryctes* release into *Oryctes monoceros* population in Tanzania, with special reference to the interaction of virus isolates used in our laboratory infection experiments. *J. Invertebr. Pathol.* 53: 285-300.

Puterka, G.J. 1999. Fungal pathogens for arthropod pest control in orchard systems: mycoinsecticidal approach for pear psylla control. *BioControl* 44:183-210.

Puterka, G.J., R.A. Humber and T.J. Poprawski. 1994. Virulence of fungal pathogens (imperfect fungi: Hyphomycetes) to pear psylla (Homoptera: Psyllidae). *Environmental Entomology.* 23, 514-520.

Quezada, J.R. 1974. Biological control of *Aleurcanthus woglumi* (Homoptera: Aleyrodidae) in El Salvador. *Entomophaga* 19: 243-254.

Quintela, E.D. and C.W. McCoy. 1998. Synergistic effect of imidacloprid and two entomopathogenic fungi on the behavior and survival of larvae of *Diaprepes abbreviatus* in soil. *J. Econ. Entomol.* 91: 110-122.

Ramirez, H. 2001. Apple growing in Northeastern Mexico. Acta Hort. 565, 139-140. Ramle, M., M.B. Wahid, K. Norman, T.R.Glare and T.A. Jackson. 2005. The incidence and use of Oryctes virus for control of rhinoceros beetle in oil palm plantations in Malaysia. *J. Invertebr. Pathol.* 89: 85-90.

Reed, R.K. 1981. Control of mites by non-occluded viruses, p.427-432. In H.D. Burges (Ed.), *Microbial control of pests and plant diseases* 1970-1980. New York, Academic Press, 949p.

Rosa, W. de la, F.L. Lopez and P. Liedo. 2002. *Beauveria bassiana* as a pathogen of the Mexican fruit fly (Diptera: Tephritidae) under laboratory conditions. *J. Econ. Entomol.* 95: 36-43.

Rosales, L.C. and H.Z. Suarez. 1998. Nematodos entomopatogenos como posibles agentes de control del gorgojo negro del platano *Cosmopolites sordidus* (Germar 1824) (Coleoptera: Curculionidae). Bol. Entomol. Venezolana Ser. Monograf. 13: 123-140.

Rossi-Zalaf, L.S. and S.B. Alves. 2006. Susceptibility of *Brevipalpus phoenicis* to entomopathogenic fungi. *Exp. Appl. Acarol.* 40: 37-47.

Roy, H.E. and J.K. Pell. 2000. Interactions between entomopathogenic fungi and other natural enemies: Implications for biological control. *Biocontrol Sci. Technol.* 10: 737-752.

Saenz-de-Cabezon, F.J., V. Marco-Mancebón and I. Pérez-Moreno. 2003. The entomopathogenic fungus *Beauveria bassiana* and its compatibility with triflumuron: Effects on the twospotted spider mite *Tetranychus urticae*. *Biol. Control* 26: 168-173.

Samson, R.A., C. McCoy and K. O'Donnell. 1980. Taxonomy of the acarine parasite, *Hirsutella thompsonii*. *Mycologia* 72: 359-377.

Sanches, N.F. 2005. Manejo integrado da broca-do-fruto do abacaxi. Embrapa Mandioca e Fruticultura Tropical Bol. Tec. n° 36, 37p.

Schoeman, P.S. and M.H. Schoeman. 1999. Transmission of *Beuaveria bassiana* from infected adults of the banana weevil *Cosmopolites sordidus* (Coleoptera: Curculionidae). *Afr. Plant Protect.* 5: 53-54.

Shapiro, J.P., W.J. Schroeder and P.A. Stansly. 1998. Bioassay and efficacy of *Bacillus thuringiensis* and an organosilicone surfactant against the citrus leafminer (Lepidoptera: Phyllocnistidae). *Fla. Entomol.* 81: 201-210.

Shapiro-Ilan, D.I., D.H. Gouge and A.M. Koppenhöfer. 2002. Factors affecting commercial success: Case studies in cotton, turf and citrus, p.333-355. In R. Gaugler (Ed.), *Entomopathogenic nematology*. Wallingford, CABI Publishing, 388p.

Shapiro-Ilan, D.I., L.W. Duncan, L.A. Lacey and R. Han. 2005. Orchard crops, p.215-229. In P.S. Grewal, R.-U. Ehlers and D.I. Shapiro-Ilan (Eds.), *Nematodes as biological control agents*. Wallingford, CABI Publishing, 505p.

Shaw, J.G., D.L. Chambers and H. Tashiro. 1968. Introducing and establishing the non-inclusion virus of the citrus red mite in citrus groves. *J. Econ. Entomol.* 61: 1352-1355.

Shi, W.B. and M.G. Feng. 2006. Field efficacy of *Beauveria bassiana* formulation and low rate pyridaben for sustainable control of citrus red mite *Panonychus citri* (Acari: Tetranychidae) in orchards. *Biol. Control* 39: 210-217.

Shukla, R.P., P.L. Tandon and S.J. Singh. 1984. Baculovirus-a new pathogen of mango nut weevil, *Sternochetus mangiferae* (Fabricius) (Coleoptera: Curculionidae). *Current Sci.* 53: 593-594.

Simpson, S.E., H.N. Nigg, N.C. Coile and R.A. Adair. 1996. *Diaprepes abbreviatus* (Coleoptera: Curculionidae): Host plant association. *Environ. Entomol.* 25: 333-349.

Smith, D. and J.E. Peña. 2002. Tropical citrus pests, p.57-101. In J.E. Peña, J.L. Sharp and M. Wysoki (Eds.), *Tropical fruit pests and pollinators: biology, economic importance, natural enemies and control*. Wallingford, CABI Publishing, 440p.

Smith, D. 1995. Banana weevil borer control in south-eastern Queensland. Aust. *J. Exp. Agric.* 35: 1165-1172.

Smith, D.H., J.V. da Graça and V.H. Whitlock. 1990. Granulosis virus from *Cacoecia occidentalis*: Isolation and morphological description of a granulosis virus of the citrus leafroller, *Cacoecia occidentalis*. *J. Invertebr. Pathol.* 55: 319-324.

Solter, L.E. and J.J. Becnel. 2007. Entomopathogenic microsporidia, p.199-221. In L.A. Lacey and H.K. Kaya (Eds.), *Field manual of techniques in invertebrate pathology: Application and evaluation of pathogens for control of insects and other invertebrate pests*, 2nd edition. Dordrecht, Springer Scientific Publishers.

Sosa-Gómez, D.R., F. Moscardi, B. Santos, L.F. Angeli Alves and S.B. Alves. 2008. Produção e uso de vírus para o controle microbiano de pragas na América Latina. P. 49-68. In S.B. Alves (Ed.), *Controle microbiano de pragas na América Latina: Avanços e Desafios*. Piracicaba, Brazil: Biblioteca de Ciências Agrárias Luiz de Queiroz.

Souza, J.C., A. Haga and M.A. Souza. 2003. Pragas da goiabeira. Boletim Técnico 71. EPAMIG, Minas Gerais, Brazil, 60p.

Srivastava, R.P. and M. Fasih. 1988. Natural occurrence of *Beauveria bassiana*, an entomogenous fungus on mango mealy bug, *Drosicha mangiferae* Green. *Indian J. Plant Pathol.* 6: 8-10.

Srivastava, R.P. and P.L. Tandon. 1986. Natural occurrence of two entomogenous fungi pathogenic to mango hopper, *Idioscopus clypealis* Leth. *Indian J. Plant Pathol.* 4: 121-123.

Steinkraus, D.C. 2007. Documentation of naturally-occurring pathogens and their impact in agroecosystems, p.267-281. In L.A. Lacey and H.K. Kaya (Eds.), *Field manual of techniques in invertebrate pathology: application and evaluation of pathogens for control of insects and other invertebrate pests*, 2nd edition. Dordrecht, Springer Scientific Publishers.

Stephens, C.S. 1962. *Oiketicus kirbyi* (Lepidoptera: Psychidae) a pest of bananas in Costa Rica. *J. Econ. Entomol.* 55: 381-386.

Stuart, R.J., El-Borai, F.E., Duncan, L.W. 2008. From augmentation to conservation of entomopathogenic nematodes: Trophic cascades, habitat manipulation and enhanced biological control of *Diaprepes abbreviatus* root weevils in Florida citrus groves. *J. Nematol* 40, 73-84.

Stuart, R.J., S. Polavarapu, E.E. Lewis and R. Gaugler. 1997. Differential susceptibility of *Dysmicoccus vaccinii* (Homoptera: Pseudococcidae) to entomopathogenic nematodes (Rhabditida: Heterorhabditidae and Steinermatidae). *J. Econ. Entomol.* 90: 925-932.

Syvertsen, J.P. and C.W. McCoy. 1985. Leaf feeding injury to citrus by root weevil adults: Leaf area, photosynthesis, and water use efficiency. *Fla. Entomol.* 68: 386-393.

Tarrant, C.A. and C.W. McCoy. 1989. Effect of temperature and relative humidity on the egg and larval stages of some citrus weevils. *Fla. Entomol.* 72: 117-123.

Tinzaara, W., C.S. Gold, C. Nankinga, M. Dicke, A. van Huis, P.E. Ragama and G.H. Kagezi. 2004. Integration of pheromones and the entomopathogenic fungus for the management of the banana weevil. *Uganda J. Agric. Sci.* 9: 621-629.

Tinzaara, W., Gold, C.S., Dicke, M., Huis, A. van, Nankinga, C.M., Kagezi, G.H. and Ragama, P.E. 2007. The use of aggregation pheromone to enhance dissemination of *Beauveria bassiana* for the control of the banana weevil in Uganda. *Biocontrol Sci. Technol.* 17: 111-124

Toledo, J., J.E. Ibarra, P. Liedo, A. Gomez, M.A. Rasgado and T. Williams. 2005. Infection of *Anastrepha ludens* (Diptera: Tephritidae) larvae by *Heterorhabditis bacteriophora* (Rhabditida: Heterorhabditidae) under laboratory and field conditions. *Biocontrol Sci. Technol.* 15: 627-634.

Toledo, J., J.L. Gurgúa, P. Liedo, J.E. Ibarra and A. Oropeza. 2001. Parasitismo de larvas y pupas de la mosca mexicana de la fruta, *Anastrepha ludens* (Loew) (Diptera: Tephritidae) por el nemátodo *Steinernema feltiae* (Filipjev) (Rhabditida: Steinernematotidae). *Vedalia* 8: 27-36.

Toledo, J., M.A. Rasgado, J.E. Ibarra, A. Gomez, P. Liedo and T. Williams. 2006. Infection of *Anastrepha ludens* following soil applications of *Heterorhabditis bacteriophora* in a mango orchard. *Entomol. Exp. Applic.* 119: 155-162.

Treverrow, N. and R.A. Bedding. 1993. Development of a system for the control of the banana weevil borer, *Cosmopolites sordidus*, with entomopathogenic nematodes, p.41-47. In R.A. Bedding, R. Akhurst and H.K. Kaya (Eds.), *Nematodes and the biological control of insect pests*. Melbourne, CSIRO Publishing, 178p.

Treverrow, N., R.A. Bedding, E.B. Dettmann and C. Maddox. 1991. Evaluation of entomopathogenic nematodes for control of *Cosmopolites sordidus* Germar (Coleoptera: Curculionidae), a pest of bananas in Australia. *Ann. Appl. Biol.* 119: 139-145.

Uchida, M. 1970. Studies on the use of the parasitic fungus *Aschersonia* sp. for controlling citrus whitefly, *Dialeurodes citri*. *Bull. Kanagawa Hort. Exp. Sta.* 18: 66-74. (in Japanese).

Vyas, R.V., J.J. Patel, P.H. Godhani and D.N. Yadav. 1993. Evaluation of green muscardine fungus (*Metarhizium anisopliae* var. *anisopliae*) for control of mangohopper (*Amritodus atkinsoni*). *Indian J. Agric. Sci.* 63: 602-603.

Waite, G.K. 2002. Pests and pollinators of mango, p.103-129. In J.E. Peña, J.L. Sharp and M. Wysoki (Eds.). *Tropical fruit pests and pollinators: biology, economic importance, natural enemies and control.* Wallingford, UK, CABI Publishing, 440p.

Weathersbee, A.A. III, Y.Q. Tang, H. Doostdar and R.T. Mayer. 2002. Susceptibility of *Diaprepes abbreviatus* (Coleoptera: Curculionidae) to a commercial preparation of *Bacillus thuringiensis* subsp. *tenebrionis*. *Fla. Entomol.* 85: 330-335.

Weissling, T.J. and R.M. Giblin-Davis. 1998. Silky cane weevil, *Metamasius hemipterus sericeus* (Oliver) (Insecta: Coleoptera: Curculionidae: Dryophthorinae). http://edis.ifas.ufl.edu/IN210

Woodruff, R.E. 1964. A Puerto Rican weevil new to the United States (Coleoptera: Curculionidae). Florida Department of Agriculture. *Plant Industry Entomology Circular* 77: 1-4.

Wraight, S.P. 2003. Synergism between insect pathogens and entomophagous insects, and its potential to enhance biological control efficacy, p.139-163. In O. Koul and G.S. Dhaliwal (Eds.), *Advances in biopesticide research, v. 3, predators and parasitoids.* London, Taylor and Francis Group, 191p.

Yen, D.F. and Y.T. Tsai. 1969. Entomogenous fungi of citrus Homoptera in Taiwan. *Plant Prot. Bull.* 11: 1-10.

Young, E.C. and J.F. Longworth. 1981. The epizootiology of the baculovirus of the coconut palm rhinoceros beetle (*Oryctes rhinoceros*) in Tonga. *J. Invertebr. Pathol.* 38: 362-369.

Young, E.C. 1974. The epizootiology of two pathogens of the coconut palm rhinoceros beetle. *J. Invertebr. Pathol.* 24: 82-92.

Zelazny, B. and A.R.R. Alfiler. 1986. *Oryctes rhinoceros* (Coleoptera: Scarabaeidae) larva abundance and mortality factors in the Philippines. *Environ. Entomol.* 15: 84-87.

Zelazny, B. and A.R.R. Alfiler. 1991. Ecology of baculovirus-infected and healthy adults of *Oryctes rhinoceros* (Coleoptera: Scarabaeidea) on coconut palms in the Philippines. *Ecol. Entomol.* 16: 253-259.

Zelazny, B. and M. Moezir. 1989. Pengendalian hama kumbang rhinoceros pada tanman kelapa. Berita Perlindungan Tanaman Perkebunan 1: 1-6.

Zelazny, B. 1972. Studies on Rhabdionvirus rhinoceros. I. Effects on larvae of *Oryctes rhinoceros* and inactivation of virus. *J. Invertebr. Pathol.* 20: 235-241.

Zelazny, B. 1973. Studies on Rhabdionvirus rhinoceros. II. Effects on adults of *Oryctes rhinoceros*. *J. Invertebr. Pathol.* 22: 122-126.

Zelazny, B. 1976. Transmission of a Baculovirus in populations of *Oryctes rhinoceros*. *J. Invertebr. Pathol.* 27: 221-227.

Zelazny, B. 1977. *Oryctes rhinoceros* populations and behavior influenced by a baculovirus. *J. Invertebr. Pathol.* 29: 210-215.

Zelazny, B. 1978. Methods for inoculating and diagnosing the baculovirus disease of *Oryctes rhinoceros*. *FAO Plant Prot. Bull.* 26: 163-168.

Zelazny, B. 1983. *Oryctes rhinoceros* damage on coconut palm in the Maldives. *FAO Plant Prot. Bull.* 31: 119-120.

Zelazny, B., A. Lolong and A.M. Crawford. 1990. Introduction and field comparison of baculovirus strains against *Oryctes rhinoceros* (Coleoptera: Scarabaeidae) in the Maldives. *Environ. Entomol.* 19: 1115-1121.

Zelazny, B., A. Lolong and B. Pattang. 1992. *Oryctes rhinoceros* (Coleoptera: Scarabaeidae) populations suppressed by a baculovirus. *J. Invertebr. Pathol.* 59: 61-68.

Zelazny, B., A.R.R. Alfiler and A. Lolong. 1989. Possibility of resistance to a baculovirus in populations of the coconut rhinoceros beetle (*Oryctes rhinoceros*). *FAO Plant Prot. Bull.* 37: 77-82.

Zhou, Y.S., F.R. Shen and H.P. Zhao. 1998. Applying *Beauveria bassiana* to control *Rhytidodera bowringii*. *J. Beijing Forest. Univ.* 20: 65-69.

INDEX

B

D

E

F

J

K

L

M

N

S

U

V

W

Y

Z